Praise for *Signature in the Cell*:

"Meyer demolishes the materialist superstition at the core of evolutionary biology by exposing its Achilles' heel: its utter blindness to the origins of information. With the recognition that cells function as fast as supercomputers and as fruitfully as so many factories, the case for a mindless cosmos collapses. His refutation of Richard Dawkins will have all the dogs barking and the angels singing."
— George Gilder, author of *Wealth and Poverty* and *Telecosm*

"A 'must read' for all serious students of the origin-of-life debate. Not only is it a comprehensive defense of the theory of intelligent design, it is a lucid and rigorous exposition of the various dimensions of the scientific method."
— Alastair Noble, Ph.D., chemistry, former BBC education officer and Her Majesty's Inspector of Schools for Science, Scotland

"The origin of life remains one of the great unsolved mysteries of modern science. Looking beyond the biochemistry of the problem and focusing instead on the origin and information content of the 'code of life,' Meyer has written an eminently readable and engaging account of the quest to solve this mystery."
— Edward Peltzer, Ph.D., ocean chemistry, from Scripps Institution of Oceanography

"How does an intelligent person become a proponent of intelligent design? Anyone who stereotypes IDers as antiscientific ideologues or fundamentalists should read Dr. Meyer's compelling intellectual memoir."
— Dr. Marvin Olasky, provost, The King's College, New York City, and editor-in-chief, *World*

"In this engaging narrative, Meyer demonstrates what I as a chemist have long suspected: undirected chemical processes cannot produce the exquisite complexity of the living cell. He also shows compelling positive evidence for intelligent design in the digital code stored in the cell's DNA. A decisive case based upon breathtaking and cutting-edge science."
— Dr. Philip S. Skell, National Academy of Sciences and Evan Pugh Professor, emeritus, at Pennsylvania State University

"Meyer gives us a fascinating exploration of the case for intelligent design theory, woven skillfully around a compelling account of his own journey.

Along the way, he effectively dispels the most pernicious caricatures: that intelligent design is simply warmed-over creationism, the province of deluded fools and morons, or a dangerous political conspiracy. Whether you believe intelligent design is true or false, *Signature in the Cell* is a must-read book."
> —Dr. Scott Turner, professor of environmental and forest biology, State University of New York, and author of *The Tinkerer's Accomplice*: *How Design Emerges from Life Itself*

"Meyer has provided no less than a blueprint for twenty-first-century biological science—one that decisively shifts the discipline's center of gravity from nineteenth-century Darwinian preoccupations with fossils and field studies to the computerized, lab-based molecular genetics that underwrites the increasingly technological turn in the life sciences. After this book, readers will wonder whether anything more than sentimentality lies behind the continued association of Darwin's name with 'modern biology.'"
> —Dr. Steve Fuller, professor of sociology of science, University of Warwick, and author of *Dissent from Descent*

"The astonishing complexities of DNA have raised questions which the ruling scientific orthodoxy cannot begin to answer. . . . Steve Meyer guides us lucidly through that labyrinth of questions opened by discoveries in molecular biology on the frontier of scientific knowledge."
> —Christopher Booker, *The Sunday Telegraph*

"A delightful read . . . Meyer has marshaled a formidable array of evidence from fields as diverse as biochemistry, philosophy, and information theory. He deals fairly and thoroughly with even the most controversial aspects and has made a compelling case for his conclusion."
> —Dr. John C. Walton, professor of organic chemistry, Unversity of St. Andrews, Scotland

"This book is a landmark in the intelligent design debate and one which accurately draws together all relevant scientific research and information. It is elegantly written in a style that is accessible and laced with interesting historical and personal anecdotes. *Signature in the Cell* will pay rich dividends to everyne who turns its pages."
> —Dr. Norman C. Nevin, professor emeritus in medical genetics, Queen's University, Belfast, Fellow of the Royal College of Physicians

Signature
in the
Cell

Signature in the Cell

DNA and the Evidence for Intelligent Design

Stephen C. Meyer

HarperOne

An Imprint of HarperCollins*Publishers*

HarperOne

HarperCollins books may be purchased for educational, business, or sales promotional use. For information please write: Special Markets Department, HarperCollins Publishers, 10 East 53rd Street, New York, NY 10022.

HarperCollins Web site: http://www.harpercollins.com

HarperCollins®,📖 ®, and HarperOne™ are trademarks of HarperCollins Publishers

Illustrations © 2009 Ray Braun

Library of Congress Cataloging-in-Publication Data
Meyer, Stephen C.
 Signature in the cell : DNA and the evidence for intelligent design / by Stephen Meyer.
    p. cm.
 Includes bibliographical references and index.
ISBN 978–0–06–147279–4
1. Intelligent design (Teleology). 2. Evolution (Biology)—Religious aspects.
3. Religion and science. I. Title.
 BL263.M475 2009
 113'.8—dc22   2008051773

    13 14 RRD (H) 10 9 8 7

FOR ELAINE, ONCE AND FUTURE COMPANION

Contents

Prologue

"Dad, that's you!" my fourteen-year-old son exclaimed as he looked at the newspaper while we stood waiting to check out at the tiny general store. His shock at seeing my face in the front section of the *Seattle Post-Intelligencer*, when he just went to look for baseball scores, was no doubt compounded by his awareness of our location.[1] The general store on Shaw Island, one of the most remote in the San Juan chain north of Puget Sound, was the only commercial establishment on the island. This irony was not lost on my wife, whose raised eyebrow said it all. "I thought we were coming here to get away from all of this." We were. But then how was I to know that the local Seattle paper would rerun the previous day's front-page story from the *New York Times* about the program of scientists I directed and the controversy surrounding our work?[2]

The controversy about the origin of life and whether it arose from an undirected material process or from some kind of designing intelligence is not new. It goes back in Western civilization at least as far as the ancient Greeks, who produced philosophers representing both schools of thought. But the controversy over the contemporary theory of intelligent design (ID) and its implied challenge to orthodox evolutionary theory became big *news* beginning in 2004 and 2005. And, for better or worse, I found myself right in the middle of it.

Three events sparked intense media interest in the subject. First, in August 2004, a technical journal housed at the Smithsonian Institution in Washington, D.C., called the *Proceedings of the Biological Society of Washington* published the first peer-reviewed article explicitly advancing the theory of intelligent design in a mainstream scientific periodical. After the publication of the article, the Smithsonian's Museum of Natural History erupted in internal controversy, as scientists angry with the editor—an evolutionary biologist with two earned Ph.D.'s—questioned his editorial judgment and demanded his censure. Soon the controversy spilled over into the scientific press as news stories about the article and editor's decision appeared in *Science, Nature, The Scientist,* and the *Chronicle of Higher Education.*[3]

The media exposure fueled further embarrassment at the Smithsonian, resulting in a second wave of recriminations. The editor, Richard Sternberg, lost his office and his access to scientific samples and was later transferred to a hostile supervisor. After Sternberg's case was investigated by the U.S. Office of Special Counsel, a government watchdog organization, and by the U.S. House Committee on Government Reform, a congressional committee, other questionable actions came to light.[4] Both investigations found that senior administrators at the museum had interrogated Sternberg's colleagues about Sternberg's religious and political beliefs and fomented a misinformation campaign designed to damage his scientific reputation and encourage his resignation.[5] Sternberg did not resign his research appointment, but he was eventually demoted.

As word of his mistreatment spread, the popular press began to run stories about his case. Ordinarily, my reaction to such reports might have been to shake my head in dismay and move on to the next story in the news cycle. But in this case, I couldn't. As it happened, I was the author of the offending article. And some of the reporters interested in Sternberg's mistreatment were coming to me with questions. They wanted to know more about the theory of intelligent design and why it had provoked such alarm among establishment scientists.

Then in December 2004, two other events generated worldwide interest in the theory of intelligent design. First, a renowned British philosopher, Antony Flew, announced that he had repudiated a lifelong commitment to atheism, citing, among other factors, evidence of intelligent design in the DNA molecule.[6] Flew noted in his announcement that his views about the origin of life bore a striking resemblance to those of "American design theorists." Again, intelligent design was in the news. But what was it? This time I found myself on the BBC debating a prominent evolutionary biologist about the theory.

Later in the month, the American Civil Liberties Union (ACLU) announced a suit against a school board in the western Pennsylvania town of Dover. The school board had just announced its intention to let high school students learn about the theory of intelligent design. To do this, it proposed to inform students about the existence of a book in the school library—one that made the case for intelligent design in opposition to the standard evolutionary theories presented in the existing biology textbooks. When the ACLU announced its own intentions to sue, the national media descended upon the town en masse.

The press corps covering the story no doubt already knew about the 1925 Scopes "monkey trial" from the fictionalized Spencer Tracy movie *Inherit the Wind*, if from no other source. In Dover they sensed they had the makings of a sequel. During 2005, all the major American network and cable news programs ran segments about the theory of intelligent design, the Dover controversy, or both. Stories not only appeared in major U.S. newspapers, but in papers around the world, from the *Times* of London, *Sekai Nippo* (Tokyo), the *Times* of India, and *Der Spiegel* to the *Jerusalem Post*.

Then in August 2005, just as an end to the media buzz seemed near, a number of political and religious leaders—including figures as diverse as the Dalai Lama, President George W. Bush, and the pope—made public statements supportive of either intelligent design or allowing students to learn about the controversy surrounding it. When *Time* magazine followed suit with a cover story about the controversy, our phones started ringing all over again.

As summer was drawing to an end, my wife and I decided it was time for our family to get away after friends offered us the use of their island cabin. But in the two-week period corresponding to our vacation, the *New York Times* ran its two front-page stories about our program at the Discovery Institute, the *Washington Post* broke a story about the latest developments in the Sternberg case, and the *New York Times* editorial page offered criticism of Sternberg in its main staff-written editorial.[7] After Sternberg decided to appear on *The O'Reilly Factor* to tell his side of the story, we knew it was time to head back to Seattle.[8]

My temporary notoriety provided something my colleagues and I sorely needed—a platform for correcting much of the misinformation circulating about the theory of intelligent design. Many news articles and reports confused intelligent design with biblical creationism and its literal reading of the book of Genesis. Other articles echoed the talking points of our critics and portrayed our work as either "giving up on science" or a sneaky attempt to circumvent the legal prohibitions against teaching creationism in the public schools that the Supreme Court had enacted in 1987.

Yet I knew that the modern theory of intelligent design was not developed as a legal strategy, still less as one to abet creationism. Instead, it was first considered in the late 1970s and early 1980s by a group of scientists—Charles Thaxton, Walter Bradley, and Roger Olsen—as a possible explanation for an enduring mystery of modern biology: the origin of the digital information encoded along the spine of the DNA molecule.[9]

As I explained repeatedly to reporters and cable-news hosts, the theory of intelligent design is not based on a religious text or document, even if it does have implications that support theistic belief (a point to which I will return in Chapter 20). Instead, intelligent design is an evidence-based scientific theory about life's origins that challenges strictly materialistic views of evolution.

Indeed, the theory of intelligent design challenges a specific tenet of contemporary evolutionary theory. According to modern neo-Darwinists such as Oxford's Richard Dawkins, living systems "give the appearance of having been designed for a purpose." But, to Dawkins and other contemporary Darwinists, that appearance of design is entirely illusory, because wholly undirected processes such as natural selection and random mutations can produce the intricate design–like structures in living systems. In their view, natural selection can mimic the powers of a designing intelligence without being guided or directed in any way.

In contrast, the theory of intelligent design holds that there are telltale features of living systems and the universe that are best explained by an intelligent cause—that is, by the conscious choice of a rational agent—rather than by an undirected process. Either life arose as the result of purely undirected processes, or a guiding intelligence played a role. Advocates of intelligent design argue for the latter option based on evidence from the natural world. The theory does not challenge the idea of evolution defined as change over time or even common ancestry, but it does dispute the Darwinian idea that the cause of all biological change is wholly blind and undirected. Even so, the theory is not based on biblical doctrine. Intelligent design is an inference from scientific evidence, not a deduction from religious authority.

Despite the opportunity I had been given in the media to clarify our position, my experiences left me with a sense of unfinished business.

By 2005, I had devoted nearly twenty years of my life to developing a case for intelligent design based upon the discovery of the information-bearing properties—the digital code—stored in the DNA molecule. I had written a series of scientific and philosophical articles developing this idea,[10] but these articles were neither particularly accessible nor gathered into one volume. Now I repeatedly found myself in the position of having to defend an argument in sound bites that my audience did not know well enough to evaluate. How could they? Perhaps the central argument for intelligent design, the one that first induced me to consider the hypothesis, had not been explained adequately to a general, scientifically literate audience.

Of course, by 2005 many excellent books and articles—including several important peer-reviewed books—had already been published on different aspects of the theory of intelligent design. In 1996, Lehigh University biochemist Michael Behe made a detailed case for intelligent design based upon the discovery of nanotechnology in cells—such as the now famous bacterial flagellar motor with its thirty-part rotary engine. Behe's *Darwin's Black Box* sold over a quarter of a million copies and almost single-handedly put the idea of intelligent design on the cultural and scientific map. In 1998, William Dembski, a mathematician and philosopher with two Ph.D.'s (including one from the University of Chicago), followed suit by publishing a groundbreaking work on methods of design detection. Dembski's work, *The Design Inference,* published by Cambridge University Press, established a scientific method for distinguishing the effects of intelligence from the effects of undirected natural processes. His work established rigorous indicators of intelligent design, but did not make any specific argument for intelligent design based on the presence of these indicators in living organisms.

These were seminal works, but I had become convinced of intelligent design by another route. Over the years, I began to develop a related, but largely independent, case for intelligent design. Unfortunately I had a penchant for writing long, dense essays in obscure journals and anthologies. Even my article in the *Proceedings of the Biological Society of Washington* attracted more attention because of the controversy at the Smithsonian than because of controversy over the argument itself, though there had been more than a bit of that in some scientific circles.[11]

In any case, when the national media came calling, I simply could not get them to report why I thought DNA pointed to intelligent design. Reporters refused to cover the argument in their articles or backgrounders; debate partners scrupulously avoided responding to it, but instead continued to recite their talking points about the dangers of "intelligent design creationism." Even the judge in the Dover case decided the scientific validity of intelligent design without considering the DNA evidence.

Though I wasn't too keen on having federal judges decide the merit of any scientific argument, let alone one that I favored, the Dover trial and its associated media coverage made me aware that I needed to make my argument in a more prominent way. Many evolutionary biologists had acknowledged that they could not explain the origin of the first life. Leading theories failed in large measure because they could not explain where the mysterious information present in the cell came from. So it seemed there were no good counterarguments to the case I wanted to make. Yet various avoidance strategies continued to work because the argument did not have sufficient public prominence to force a response. Too few people in the public, the scientific community, and the media even knew about it. And yet it provided—arguably—one of the most important and fundamental reasons for considering intelligent design.

None of this was actually too surprising. Since World War II, scientists have stressed the importance of publishing their work in specialized peer-reviewed journals, but throughout the history of science "paradigm-shifting" ideas and theories have typically been presented in books, including many that we might now call "trade press" (rather than academic) books.

There are a couple of reasons for this. First, books allow scientists to make sustained and comprehensive arguments for synthetic new ideas. As the Italian philosopher of science Marcello Pera has shown, scientists often *argue* about competing interpretations of the evidence.[12] Although this is sometimes done successfully in short articles—as Einstein did in making his case for special and general relativity and Watson and Crick did in their nine-hundred-word article proposing a double helix structure for DNA—books have often been the go-to genre for presenting and evaluating new arguments for synthetic interpretations of a relevant body of evidence.

Perhaps the best-known example of this form of scientific discourse was provided by Charles Darwin himself, who famously described his work in *On the Origin of Species by Means of Natural Selection* as "one long argument."[13] There, Darwin proposed a comprehensive interpretation of many diverse lines of evidence. He also argued for the superior explanatory power of his theory and its two key propositions: (1) the creative power of natural selection and (2) the descent of all life from a common ancestor. As part of his case, he also argued against the explanatory adequacy of rival interpretations of the evidence and refuted arguments for them. Other scientists such as Newton, Copernicus, Galileo, and Lyell as well as a host of lesser figures have used books to advance scientific arguments in favor of novel and comprehensive interpretations of the scientific evidence in their disciplines.

There are other reasons that books are used to advance paradigm-shifting ideas. New scientific theories often synthesize a broad range of evidence from many related disciplines or subdisciplines of science. As such, they are often inherently interdisciplinary in scope. *On the Origin of Species* incorporated data from several disciplines, including embryology, paleontology, comparative anatomy, and biogeography. Modern scientific journals, typically focused as they are on topics within a narrowly defined subdiscipline, rarely permit the kind of comprehensive review and assessment of evidence that the advancement of a new interpretive framework requires.

Additionally, by creating a larger audience for a new idea, a book, and particularly a popular trade book, can go over the heads of an entrenched establishment to force the reevaluation of an established theory by creating wider interest in its standing. Darwin did this by publishing *On the Origin of Species by Means of Natural Selection* with John Murray, a prominent trade press in Victorian England. Michael Behe has done this as well. By making a case for intelligent design based upon various examples of nanotechnology in the cell, Behe's book focused international attention on the problem that complex systems have posed for neo-Darwinism. It also gave the theory of intelligent design public and, arguably, scientific standing.

This book makes a case for that same idea. It does so, however, on the basis of a different class of evidence: the information—the digital code—stored in DNA and the other large biological molecules.

The case I make for intelligent design is less well known than Professor Behe's and, therefore, to many completely new. Even so, it is not based upon a new discovery. It is, instead, based upon one of the most famous breakthroughs of modern biology: the discovery in 1953 of the information-bearing capacities of the DNA molecule, what I call the "signature in the cell."

In 2005, when I was repeatedly placed in the position of defending the theory of intelligent design in the media, the argument that I most wanted to make in its favor had little public standing. I have written this book to remedy that deficiency. This book attempts to make a comprehensive, interdisciplinary argument for a new view of the origin of life. It makes "one long argument" for the theory of intelligent design.

Before coming to work full-time at the Discovery Institute, I worked for twelve years as a college professor. In teaching I've found that it is often easier to understand a scientific theory if one can follow the historical progression of thought that led to its formulation. Following a story of discovery is not only more engaging, it can also illuminate the process of reasoning by which investigators came to their conclusions. For this reason, I've chosen to present my case for intelligent design in the context of a larger historical and personal narrative.

Thus, *Signature in the Cell* does not just make an argument; it also tells a story, a mystery story and the story of my engagement with it. It tells about the mystery that has surrounded the discovery of the digital code in DNA and how that discovery has confounded repeated attempts to explain the origin of the first life on earth. Throughout the book I will call this mystery "the DNA enigma."

A brief word about the organization of the book: in Chapters 1 and 2 I define the scientific and philosophical issues at stake in the DNA enigma and give some historical background about the larger origin-of-life debate. In Chapters 3, 4, and 5 I describe the mystery surrounding DNA in more detail in order to establish what it is that any theory of the origin of life must explain. After a short interlude in Chapters 6 and 7 in which I examine what scientists in the past have thought about biological origins and how scientists currently investigate these questions, I examine (in Chapters 8 through 14) the competing explanations for the origin of biological information. Then, in Chapters 15 and 16, I present a positive case for intelligent design as the best explanation for the origin of the information necessary to produce the first life. Finally, in Chapters 17 through 20, I defend the theory of intelligent design against various popular objections to it. In the Epilogue, I show that intelligent design offers a fruitful approach to future scientific research. Not only does it illuminate some very recent and surprising discoveries in genomics, but it also suggests productive new lines of scientific investigation for many subdisciplines of biology.

My interest in the DNA enigma stretches back nearly twenty-five years. And though there were times (particularly in 2005) when I was frustrated with myself for not having already produced this work, my protracted production schedule has had at least two unintended advantages. First, it has given me the opportunity to engage in both private conversation and public debate with some of the leading scientific figures involved in this controversy. That has made it possible for me to present what I hope is an unusually thorough analysis of the competing explanations for the origin of the information in living cells.

Second, because of the timing of its release, this book may contribute to the ongoing assessment of Darwin's legacy just when many scientists, scholars, reporters, and others will be doing so. This year marks the 200th anniversary of Darwin's birth and the 150th anniversary of the publication of *On the Origin of Species*. In the *Origin*, Darwin accomplished many things. He introduced a new framework for understanding the history of life. He identified a new mechanism of biological change. And, according to many scholars and scientists, he also refuted the scientific argument for design. He did this by explaining away any presumed vestiges of an actual designing intelligence, showing instead that these "appearances of design" had been produced by a purely undirected process—indeed, one that could mimic the powers of a designing mind. As evolutionary biologist Francisco Ayala has recently explained, Darwin explained the appearance of design without recourse to an actual designer. He gave us "design without a designer."[14] But is this really true? Even if we grant Darwin's argument in the *Origin*, does it really follow that he refuted the design hypothesis? This book will present a fresh perspective on this question by examining one of the most enduring mysteries of modern biology.

1

DNA, Darwin, and the Appearance of Design

When James Watson and Francis Crick elucidated the structure of DNA in 1953, they solved one mystery, but created another.

For almost a hundred years after the publication of *On the Origin of Species* by Charles Darwin in 1859, the science of biology rested secure in the knowledge that it had explained one of humankind's most enduring enigmas. From ancient times, observers of living organisms had noted that living things display organized structures that give the appearance of having been deliberately arranged or designed for a purpose, for example, the elegant form and protective covering of the coiled nautilus, the interdependent parts of the eye, the interlocking bones, muscles, and feathers of a bird wing. For the most part, observers took these appearances of design as genuine. Observations of such structures led thinkers as diverse as Plato and Aristotle, Cicero and Maimonides, Boyle and Newton to conclude that behind the exquisite structures of the living world was a designing intelligence. As Newton wrote in his masterpiece *The Opticks:* "How came the Bodies of Animals to be contrived with so much Art, and for what ends were their several parts? Was the Eye contrived without Skill in Opticks, and the Ear without Knowledge of Sounds? . . . And these things being rightly dispatch'd, does it not appear from Phænomena that there is a Being incorporeal, living, intelligent . . . ?"[1]

But with the advent of Darwin, modern science seemed able to explain this appearance of design as the product of a purely undirected process.

In the *Origin*, Darwin argued that the striking appearance of design in living organisms—in particular, the way they are so well adapted to their environments—could be explained by natural selection working on random variations, a purely undirected process that nevertheless mimicked the powers of a designing intelligence. Since then the appearance of design in living things has been understood by most biologists to be an illusion—a powerfully suggestive illusion, but an illusion nonetheless. As Crick himself put it thirty-five years after he and Watson discerned the structure of DNA, biologists must "constantly keep in mind that what they see was not designed, but rather evolved."[2]

But due in large measure to Watson and Crick's own discovery of the information-bearing properties of DNA, scientists have become increasingly and, in some quarters, acutely aware that there is at least one appearance of design in biology that may not yet have been adequately explained by natural selection or any other purely natural mechanism. Indeed, when Watson and Crick discovered the structure of DNA, they also discovered that DNA stores information using a four-character chemical alphabet. Strings of precisely sequenced chemicals called nucleotide bases store and transmit the assembly instructions—the information—for building the crucial protein molecules and machines the cell needs to survive.

Crick later developed this idea in his famous "sequence hypothesis," according to which the chemical parts of DNA (the nucleotide bases) function like letters in a written language or symbols in a computer code. Just as letters in an English sentence or digital characters in a computer program may convey information depending on their arrangement, so too do certain sequences of chemical bases along the spine of the DNA molecule convey precise instructions for building proteins. Like the precisely arranged zeros and ones in a computer program, the chemical bases in DNA convey information in virtue of their "specificity." As Richard Dawkins notes, "The machine code of the genes is uncannily computer-like."[3] Software developer Bill Gates goes further: "DNA is like a computer program but far, far more advanced than any software ever created."[4]

But if this is true, how did the information in DNA arise? Is this striking appearance of design the product of actual design or of a natural process that can mimic the powers of a designing intelligence? As it turns out, this question is related to a long-standing mystery in biology—the question of the origin of the first life. Indeed, since Watson and Crick's discovery, scientists have increasingly come to understand the centrality of information to even the simplest living systems. DNA stores the assembly instructions for building the many crucial proteins and protein machines that service and maintain even the most primitive one-celled organisms. It follows that building a living cell in the first place requires assembly instructions stored in DNA or some equivalent molecule. As origin-of-life researcher Bernd-Olaf Küppers explains, "The problem of the origin of life is clearly basically equivalent to the problem of the origin of biological information."[5]

Much has been discovered in molecular and cell biology since Watson and Crick's revolutionary discovery more than fifty years ago, but these

Figure 1.1. James Watson and Francis Crick at the Cavendish Laboratory in Cambridge. *Courtesy of Barrington Brown/Photo Researchers, Inc.*

discoveries have deepened rather than mitigated the enigma of DNA. Indeed, the problem of the origin of life (and the origin of the information needed to produce it) remains so vexing that Harvard University recently announced a $100 million research program to address it.[6] When Watson and Crick discovered the structure and information-bearing properties of DNA, they did indeed solve one mystery, namely, the secret of how the cell stores and transmits hereditary information. But they uncovered another mystery that remains with us to this day. This is the DNA enigma—the mystery of the origin of the information needed to build the first living organism.

In one respect, of course, the growing awareness of the reality of information within living things makes life seem more comprehensible. We live in a technological culture familiar with the utility of information. We buy information; we sell it; and we send it down wires. We devise machines to store and retrieve it. We pay programmers and writers to create it. And we enact laws to protect the "intellectual property" of those who do. Our actions show that we not only value information, but that we regard it as a real entity, on par with matter and energy.

That living systems also contain information and depend on it for their existence makes it possible for us to understand the function of biological organisms by reference to our own familiar technology. Biologists have also come to understand the utility of information, in particular, for the operation of living systems. After the early 1960s advances in the field of molecular biology made clear that the digital information in DNA was only part of a complex information-processing system, an advanced form of nanotechnology that mirrors and exceeds our own in its complexity, storage density, and logic of design. Over the last fifty years, biology has advanced as scientists have come to understand more about how information in the cell is stored, transferred, edited, and used to construct sophisticated machines and circuits made of proteins.

The importance of information to the study of life is perhaps nowhere more obvious than in the emerging fields of genomics and bioinformatics. Over the last decade, scientists involved in these disciplines have begun to map—character by character—the complete sequence of the genetic instructions stored on the human genome and those of many other species. With the completion of the Human

Genome Project in 2000, the emerging field of bioinformatics entered a new era of public interest. News organizations around the world carried President Clinton's announcement of the project's completion on the White House lawn as Francis Collins, scientific director of the project, described the genome as a "book," a repository of "instructions," and the "book of life."[7] The Human Genome Project, perhaps more than any discovery since the elucidation of the structure of DNA in 1953, has heightened public awareness of the importance of *information* to living things. If Watson and Crick's discovery showed that DNA stores a genetic text, Francis Collins and his team took a huge step toward deciphering its message. Biology has irrevocably entered an information age.

In another way, however, the reality of information within living things makes life seem more mysterious. For one thing, it is difficult to understand exactly what information *is*. When a personal assistant in New York types a dictation and then prints and sends the result via fax to Los Angeles, some *thing* will arrive in L.A. But that thing—the paper coming out of the fax machine—did not originate in New York. Only the information on the paper came from New York. No single physical substance—not the air that carried the boss's words to the dictaphone, or the recording tape in the tiny machine, or the paper that entered the fax in New York, or the ink on the paper coming out of the fax in Los Angeles—traveled all the way from sender to receiver. Yet something did.

The elusive character of information—whether biological or otherwise—has made it difficult to define by reference to standard scientific categories. As evolutionary biologist George Williams notes, "You can speak of galaxies and particles of dust in the same terms because they both have mass and charge and length and width. [But] you can't do that with information and matter."[8] A blank magnetic tape, for example, *weighs* just as much as one "loaded" with new software—or with the entire sequence of the human genome. Though these tapes differ in information content (and value), they do not do so because of differences in their material composition or mass. As Williams concludes, "Information doesn't have mass or charge or length in millimeters. Likewise matter doesn't have bytes. . . . This dearth of shared descriptors makes matter and information two separate domains."[9]

When scientists during the late 1940s began to define information, they did not make reference to physical parameters such as mass, charge, or watts. Instead, they defined information by reference to a psychological state—the reduction of uncertainty—which they proposed to measure using the mathematical concept of probability. The more improbable a sequence of characters or signals, the more uncertainty it reduces, and thus the more information it conveys.[10]

Not surprisingly, some writers have come close to equating information with thought itself. The information technology guru George Gilder, for example, notes that developments in fiber optics have allowed more and more information to travel down smaller and smaller (and lighter and lighter) wires. Thus, he notes that as technology advances, we convey ever more thought across ever less matter—where the numerator in that ratio, namely, thought, corresponds precisely to information.[11]

So should we think of information as thought—as a kind of mental chimera etched in stone or burned onto compact discs? Or can we define information less abstractly as, perhaps, just an improbable arrangement of matter?

Whatever information is—whether thought or an elaborate arrangement of matter—one thing seems clear. What humans recognize as information certainly *originates* from thought—from conscious or intelligent activity. A message received via fax by one person first arose as an idea in the mind of another. The software stored and sold on a compact disc resulted from the design of a software engineer. The great works of literature began first as ideas in the minds of writers—Tolstoy, Austen, or Donne. Our experience of the world shows that what we recognize as information invariably reflects the prior activity of conscious and intelligent persons.

What, then, should we make of the presence of information in living organisms? The Human Genome Project, among many other developments in modern biology, has pressed this question to the forefront of public awareness. We now know that we do not just create information in our own technology; we also find it in our biology—and, indeed, in the cells of every living organism on earth. But how did this information arise? And what does the presence of information in even the simplest living cell imply about life and its origin? Who or what "wrote" the book of life?

The information age in biology officially began in the mid-1950s with the elucidation of the chemical structure and information-bearing properties of DNA (deoxyribonucleic acid)—the molecule of heredity. Beginning in 1953 with their now famous communication to the British scientific journal *Nature,* James Watson and Francis Crick identified DNA as the molecular repository of genetic information.[12] Subsequent developments in the field of molecular biology confirmed this idea and showed that the precisely sequenced bases attached to the helical backbone of DNA store the information for building proteins—the sophisticated enzymes and machines that service the cells in all living things.

Though the discovery of the information-bearing properties of DNA dates back over a half century, the recognition of the full significance of this discovery has been slow in coming. Many scientists have found it difficult to relinquish an exclusive reliance upon the scientific categories of matter and energy alone. As George Williams (himself an evolutionary biologist) notes, "Evolutionary biologists have failed to realize that they work with two more or less incommensurable domains: that of information and that of matter. . . . The gene is a package of information, not an object. The pattern of base pairs in a DNA molecule specifies the gene. But the DNA molecule is the medium, it's not the message."[13]

Yet this recognition begs deeper questions. What does it mean when we find information in natural objects—living cells—that we did not ourselves design or create? As the information theorist Hubert Yockey observes, the "genetic code is constructed to confront and solve the problems of communication and recording by the same principles found . . . in modern communication and computer codes." Yockey notes that "the technology of information theory and coding theory has been in place in biology for at least 3.85 billion years," or from the time that life first originated on earth.[14] What should we make of this fact? How did the information in life first arise?

Our commonsense reasoning might lead us to conclude that the information necessary to the first life, like the information in human technology or literature, arose from a designing intelligence. But modern evolutionary biology rejects this idea. Many evolutionary biologists admit, of course, that living organisms "appear to have been carefully and artfully designed," as Richard Lewontin puts it.[15] As Richard Dawkins states, "Biology is the study of complex things that appear to have been designed for a purpose."[16] Nevertheless, Lewontin and Dawkins, like evolutionary biologists generally, insist that the appearance of design in life is illusory. Life, they say, looks designed, but was not designed by an actual intelligent or purposive agent.

Darwin's Designer Substitute

Why do evolutionary biologists so confidently assert that the appearance of design in living organisms is illusory? Of course, the answer to this question is well known. Evolutionary biologists have a theory that can apparently explain, or explain away, the appearance of design without invoking an actual designer. According to classical Darwinism, and now modern neo-Darwinism, the mechanism of natural selection acting on random variations (or mutations) can mimic the effects of intelligence, even though the mechanism is, of course, entirely blind, impersonal, and undirected.[17]

Figure 1.2. English naturalist Charles Robert Darwin (1809–82), age seventy-two. *Courtesy of SPL/Photo Researchers, Inc.*

Darwin developed his principle of natural selection by drawing on an analogy with artificial selection: the process of selective breeding to change the characteristics (whether anatomical, physiological, or behavioral) of a group of organisms. For example, a farmer might observe that some of his young stallions are faster than others. If he allows only the fastest of these to breed with the fastest mares, then, after several generations of selective breeding, he will own a small group of speedy "thoroughbreds" suitable for racing on the Downs.

Darwin realized that nature could imitate this process of selective breeding. The presence of unusually fast predatory wild cats would imperil all but the fastest horses in a wild herd. After several generations of such predatory challenge, the speed of the remaining herd might exhibit a discernible increase. Thus, environmental forces (predators, changes in weather, competition for food, etc.) could accomplish the work of a human breeder. By causing a population to adapt to its environment, blind forces of nature could come to mimic, over time, the action of a selecting or designing intelligence.

Yet if natural selection, as Darwin called this process, could improve the speed of a horse or an antelope, why couldn't it also produce those animals in the first place? "Reason," wrote Darwin "ought to conquer . . . imagination"[18]—namely, our incredulity about the possibility of such happenings and our impression that living things appear to have been designed. According to Darwin, if given enough time, nature's selective power might act on any variation perfecting any structure or function far beyond what any human could accomplish. Thus, the complex systems in life that we reflexively attribute to intelligence have wholly natural causes. As Darwin explained, "There seems to be no more design in the variability of organic beings, and in the action of natural selection, than in the course which the wind blows."[19] Or as evolutionary biologist Francisco Ayala explains, "The functional design of organisms and their features would . . . seem to argue for the existence of a designer. It was Darwin's greatest accomplishment [however] to show that the directive organization of living beings can be explained as the result of a natural process, natural selection, without any need to resort to a Creator or other external agent."[20] Thus, Ayala and other Darwinian biologists not only affirm that natural selection can produce "design without a designer," they also assert that it is "creative without being conscious."[21]

The Appearance of Design

To many outside evolutionary biology, the claim that design arises without a designer may seem inherently contradictory. Yet, in theory at least, the possibility that life is not what it seems represents nothing particularly unusual. Science often shows that our perceptions of nature do not match reality. A straight pencil appears bent when inserted in a glass of water; the sun appears to circle the earth; and the continents appear immobile. Perhaps, living organisms only appear to be designed.

Even so, there is something curious about the scientific denial of our ordinary intuition about living things. For almost a hundred and fifty years, since its putative explanation by Darwinian theory, this impression of design persists as incorrigibly as ever. Public opinion polls suggest that nearly 90 percent of the American public does not accept the full-fledged neo-Darwinian account of evolution with its denial of any role for a purposeful creator.[22] Though many of these people accept some form of evolutionary change and have a high view of science generally, they apparently cannot bring themselves to repudiate their deepest intuitions and convictions about the design of the living world. In every generation since the 1860s, scientific critics of Darwinism and neo-Darwinism have arisen marshaling serious evidential objections to the theory. Since the 1980s a growing number of scientists and scholars have expressed deep reservations about both biological and chemical evolutionary theory, each with their implicit denial of design. And even orthodox evolutionary biologists admit the overwhelming *impression* of design in modern organisms. To quote Francis Crick again, "Biologists *must constantly keep in mind* that what they see was not designed, but rather evolved."[23]

Perhaps more curiously, modern biologists can scarcely describe living organisms without resorting to language that seems to imply the very thing they explicitly deny: intentional and purposive design. As philosopher of science Michael Ruse notes, biologists ask about "the *purpose* of the fins on the back of the stegosaurus" or "the *function* of the bird's feathers" and discuss whether "the Irish elk's antlers did or did not exist *in order* to intimidate rivals." "It is true," Ruse continues, "that during the nineteenth century [some physicists] suggested that the moon exists in order to light the way home of lonely travelers, but no physicist would use such language today. In biology, however, especially evolutionary biology, this kind of talk is commonplace." He concludes, "The world of the evolutionist is drenched in the anthropomorphism of intention." And yet "paradoxically, even the severest critics" of such intentional language slip into it "for the sake of convenience."[24]

In theory, at least, the use of such metaphor in science derives from ignorance. Physicists talk about gravitational "attraction," because they don't really know what causes action at a distance. Metaphors reign where mystery resides. Yet, on these grounds, we might have expected that as biology advanced, as new discoveries explicated the molecular basis of biological functions, biology's reliance upon the language of purpose, upon teleological metaphor, might have diminished. Yet the very opposite has taken place. The advent of the most reductionistic subdiscipline of modern biology—molecular biology—has only deepened our dependence on teleological language.

In fact, molecular biologists have introduced a new "high-tech" teleology, taking expressions, often self-consciously, from communication theory, electrical engineering, and computer science. The vocabulary of modern molecular and cell biology includes apparently accurate descriptive terms that nevertheless seem laden with a "metaphysics of intention": "genetic code," "genetic information," "transcription," "translation," "editing enzymes," "signal-transduction circuitry," "feedback loop," and "information-processing system." As Richard Dawkins notes, "Apart from differences in jargon, the pages of a molecular-biology journal might be interchanged with those of a computer-engineering journal."[25] As if to underscore the point, University of Chicago cell biologist James Shapiro describes the integrated system of proteins that constitutes the mammalian blood-clotting system "as a powerful real-time distributed computing system." In the same context he notes that many biochemical systems within the cell resemble "the wiring diagram for an electronic circuit."[26] As the historian of biology Timothy Lenoir observes, "Teleological thinking has been steadfastly resisted by modern biology. And yet in nearly every area of research, biologists are hard pressed to find language that does not impute purposiveness to living forms."[27]

Thus, it seems that an acquaintance with biological organisms, to say nothing of the molecular biology of the cell, leads even those who repudiate design to use language that seems incompatible with their own reductionistic and Darwinian perspective—with their official denial of actual design. Although this may ultimately signify nothing, it does at least raise a question. Does the persistence of our perception of design, and the use of incorrigibly teleological language, indicate anything about the origin of life or the adequacy of scientific theories that deny (actual) design in the origin of living systems?

As always, in science the answer to such questions depends entirely on the justification that scientists can provide for their theories. Intuitions and perceptions can be right or wrong. It might well be, as many in biology assure us, that public and even scientific doubts about evolutionary theory derive solely from ignorance or religious prejudice, and that teleological language reflects nothing more than a metaphor of convenience, like saying the sun has set behind the horizon. Yet the persistence of dissenting scientific opinion and the inability of biologists to avoid the language of purpose raise a pardonable curiosity. Have evolutionary biologists discovered the true cause of the appearance of design in living systems, or should we look for another? Should we trust our intuitions about living organisms or accept the standard evolutionary account of biological origins?

The Origin of Biological Information

Consider the following sequence of letters:

```
AGTCTGGGACGCGCCGCCGCCATGATCATCCCTGTACGCTGCTTCACTTGT-
GGCAAGATCGTCGGCAACAAGTGGGAGGCTTACCTGGGGCTGCTGCAGG
CCGAGTACACCGAGGGGTGAGGCGCGGGCCGGGGCTAGGGGCTGAGTCC-
GCCGTGGGGCGCGGGCCGGGGCTGGGGGCTGAGTCCGCCCTGGGGTGCGCG
CCGGGGCGGGAGGCGCAGCGCTGCCTGAGGCCAGCGCCCCATGAGCAGCTTCAG-
GCCCGGCTTCTCCAGCCCCGCTCTGTGATCTGCTTTCGGGAGAACC
```

This string of alphabetic characters looks as if it could be a block of encoded information, perhaps a section of text or machine code. That impression is entirely correct, for this string of characters is not just a random assortment of the four letters A, T, G, and C, but a representation of part of the sequence of genetic assembly instructions for building a protein machine—an RNA polymerase[28]—critical to gene expression (or information processing) in a living cell.

Now consider the following string of characters:

```
010101110110100001100101011011100010000001101001011
011100010000001110100011010000110010100100000100
0011011011110111010101011001001110011011001010010000
00011011110110011000100000011010000111010101101101
011000010110111000010000000110010101110110011001001
011011100111010001110011001000000011010010101110100
```

This sequence also appears to be an information-rich sequence, albeit written in binary code. As it happens, this sequence is also not just a random array of characters, but the first words of the Declaration of Independence ("When in the course of human events . . .")[29] written in the *binary conversion* of the American Standard Code for Information Interchange (ASCII). In the ASCII code, short specified sequences of zeros and ones correspond to specific alphabetic letters, numerals, or punctuation marks.

Though these two blocks of encoded information employ different conventions (one uses the ASCII code, the other the genetic code), both are complex, nonrepeating sequences that are highly specified relative to the functional or communication requirements that they perform. This similarity explains, in part, Dawkins's observation that, "The machine code of the genes is uncannily computer-like." Fair enough. But what should we make of this similarity between informational software—the undisputed product of conscious intelligence—and the informational sequences found in DNA and other important biomolecules?

Introduction to an Enigma

I first encountered the DNA enigma as a young scientist in Dallas, Texas, in 1985. At the time, I was working for one of the big multinational oil companies. I had been hired as an exploration geophysicist several years earlier just as the price of oil had spiked and just as I was graduating from college with degrees in physics and geology. My job, as the Texas oilmen put it, was to "look for *awl* out in the *guff.*"

Though I had been a physics and geology student, I had enough exposure to biology to know what DNA did. I knew that it stored the instruction set, the information, for building proteins in the cell and that it transmitted hereditary traits in living things using its four-character chemical alphabet. Even so, like many scientists I had never really thought about where DNA—or the information it contained—came from in the first place. If asked, I would have said it had something to do with evolution, but I couldn't have explained the process in any detail.

On February 10, 1985, I learned that I wasn't the only one. On that day I found myself sitting in front of several world-class scientists who were discussing a vexing scientific and philosophical question: How did the first life on earth arise? As recently as the evening before, I had known nothing about the conference where this discussion was now taking place. I had been attending another event in town, a lecture at the Southern Methodist University by a Harvard astronomer discussing the big-bang theory. There I learned of a conference taking place the following day that would tackle three big scientific questions—the origin of the universe, the origin of life, and the nature of human consciousness. The conference would bring together scientists from competing philosophical perspectives to grapple with each of these issues. The next morning I walked into the downtown Hilton where the conference was being held and heard an arresting discussion of what scientists knew they didn't know.

I was surprised to learn—contrary to what I had read in many textbooks—that the leading scientific experts on the origin of life had no satisfactory explanation for how life had first arisen. These experts, many of whom were present that weekend in Dallas, openly acknowledged that they did not have an adequate theory of what they called "chemical evolution," that is, a theory of how the first living cell arose from simpler chemicals in the primordial ocean. And from their discussions it was clear that DNA—with its mysterious arrangements of chemical characters—was a key reason for this impasse.

The discussion changed the course of my professional life. By the end of that year, I was preparing to move to the University of Cambridge in England, in part to investigate questions I first encountered on that day in February.

On its face, my change of course looked like a radical departure from my previous interests, and that's certainly how my friends and family took it. Oil-company geophysics was a highly practical, commercially relevant form of applied science. A successful study of the subsurface of the earth could net the company millions of dollars of revenue from the resulting discovery of oil and gas. The origin of life, however, was a seemingly intractable—even arcane—theoretical question, with little or no direct commercial or practical import.

Nevertheless, at the time, the transition seemed entirely natural to me. Perhaps it was because I had long been interested in scientific questions and discoveries that raised larger philosophical issues. In college, I had taken many philosophy courses while pursuing my scientific training. But perhaps it was what I was doing at the oil company itself. By the 1980s looking for oil required the use of sophisticated computer-assisted seismic-imaging techniques, at the time a cutting-edge form of information technology. After sending artificial seismic waves down into the earth, geophysicists would time the resulting echoes as they traveled back to the surface and then use the information from these signals to reconstruct a picture of the subsurface of the earth. Of course, at every stage along the way we depended heavily on computers and computer programs to help us process and analyze the information we received. Perhaps what I was learning about how digital information could be stored and processed in machines and about how digital code could direct machines to accomplish specific tasks made life itself—and the digital code stored in its DNA—seem less mysterious. Perhaps this made the problem of the origin of life seem more scientifically tractable and interesting. In any case, when I learned of the enigma confronting origin-of-life researchers and why DNA was central to it, I was hooked.

A controversy that erupted at the conference added to my sense of intrigue. During a session on the origin of life, the scientists were discussing where the information in DNA had come from. How do chemicals arrange themselves to produce code? What introduced drama into what might have otherwise been a dry academic discussion was the reaction of some of the scientists to a new idea. Three of the scientists on the panel had just published a controversial book called *The Mystery of Life's Origin* with a prominent New York publisher of scientific monographs. Their book provided a comprehensive critique of the attempts that had been made to explain how the first life had arisen from the primordial ocean, the so-called prebiotic soup. These scientists, Charles Thaxton, Walter Bradley, and Roger Olsen, had come to the conclusion that all such theories had failed to explain the origin of the first life. Surprisingly, the other scientists on the panel—all experts in the field—did not dispute this critique.

What the other scientists did dispute was a controversial new hypothesis that Thaxton and his colleagues had floated in the epilogue of their book in an attempt to explain the DNA enigma. They had suggested that the information in DNA might have originated from an intelligent source or, as they put it, an "intelligent cause." Since, in our experience, information arises from an intelligent source, and since the information in DNA was, in their words, "mathematically identical" to the information in a written language or computer code, they suggested that the presence of information in DNA pointed to an intelligent cause. The code, in other words, pointed to a programmer.

That was where the fireworks started. Other scientists on the panel became uncharacteristically defensive and hostile. Dr. Russell Doolittle, of the University of California at San Diego, suggested that if the three authors were not satisfied with the progress of origin-of-life experiments, then they should "do them." Never mind that another scientist on the panel who had favored Thaxton's hypothesis, Professor Dean Kenyon, of San Francisco State University, was a leading origin-of-life researcher who had himself performed many such experiments. It was clear that Doolittle regarded the three scientists, despite their strong credentials, as upstarts who had violated some unspoken convention. Yet it was also clear, to me at least, that the authors of the new book had seized the intellectual initiative. They had offered a bold new idea that seemed at least intuitively plausible, while those defending the status quo offered no plausible alternative to this new explanation. Instead, the defenders of the status quo were forced to accept the validity of the new critique. All they could do was accuse the upstarts of giving up too soon and plead for more time.

I left deeply intrigued. If my sense of the scientific status of the problem was accurate—if there was no accepted or satisfactory theory of the origin of the first life—then a mystery was at hand. And if it was the case that evolutionary theory could not explain the origin of the first life *because it could not explain the origin of the genetic information in DNA,* then something that we take for granted was quite possibly an important clue in a mystery story. DNA with its characteristic double-helix shape is a cultural icon. We see the helix in everything from music videos and modern art to science documentaries and news stories about criminal proceedings. We know that DNA testing can establish guilt, innocence, paternity, and distant genealogical connections. We know that DNA research holds the key to understanding many diseases and that manipulating DNA can alter the features of plants and animals and boost food production. Most of us know roughly what DNA is and

Figure 1.3. Charles Thaxton.
Printed by permission from
Charles Thaxton.

what it does. But could it be that we do not know anything about where it came from or how it was first formed?

The controversy at the conference served to awaken me to the strange combination of familiarity and mystique that surrounds the double helix and the digital code it contains. In the wake of the conference, I learned that one of the scientists who participated in the origin-of-life discussion was living in Dallas. It was none other than Charles Thaxton, the chemist who with his coauthors had proposed the controversial idea about an intelligence playing a role in the origin of biological information. I called him, and he offered to meet with me. We began to meet regularly and talk, often long after work hours. As I learned more about his critique of "origin-of-life studies" and his ideas about DNA, my interest in the DNA enigma grew.

These were heady and exciting days for me as I first encountered and grappled with these new ideas. If Thaxton was right, then the classical design argument that had been dismissed first by Enlightenment philosophers such as David Hume in the eighteenth century and then later by evolutionary biologists in the wake of the Darwinian revolution might have legitimacy after all. On a visit back home to Seattle, I described what I had been learning to one of my earlier college mentors whose critical faculties I greatly respected, a philosophy professor named Norman Krebbs. He surprised me when he told me that the scientific idea I was describing was potentially one of the most significant *philosophical* developments in three hundred years of Western thought. Could the design argument be resuscitated based upon discoveries in modern science? And was DNA the key?

As intriguing as this new line of thinking was for me, I had a growing list of questions. I wondered, what exactly is information in a biological context? When biologists referred to the sequences of chemicals in the DNA molecule as "information," were they using the term as a metaphor? Or did these sequences of chemicals really function in the same way as "code" or "text" that humans use? If biologists were using the term merely as a metaphor, then I wondered whether the genetic information designated anything real and, if not, whether the "information" in DNA could be said to point to anything, much less an "intelligent cause."

But even if the information in DNA was in some important sense similar to the information that human agents devise, it didn't necessarily follow that a prior intelligent cause was the only explanation of such information. Were there causes for information that had not yet been considered at the conference that day? Maybe some other cause of information would be discovered that could provide a better explanation for the information necessary for the origin of life. In short, I wondered, is there really evidence for the intelligent design of life, and if so, just how strong is that evidence? Was it, perhaps, scientifically premature or inappropriate to consider such a radical possibility, as Thaxton's critics had suggested?

My concerns about this were heightened because of some of the things that Thaxton and his colleagues had written to justify their conclusion. *The Mystery of Life's Origin* had made the radical claim that an intelligent cause could be considered a legitimate *scientific* hypothesis for the origin of life. To justify this claim Thaxton and colleagues argued that a mode of scientific inquiry they called *origins science* allowed for the postulation of singular acts of intelligence to explain certain phenomena. Thaxton and his colleagues distinguished what they called "origins sciences" from "operation sciences." Operation sciences, in their view, focus on the ongoing operation of the universe. These sciences describe recurring phenomena like the motions of the planets and chemical reactions that can be described by general laws of physics and chemistry. Origins sciences, on the other hand, deal with unique historical events and the causes of those events— events such as the origin of the universe, the formation of the Grand Canyon, and the invention of ancient tools and agriculture. Thaxton and his colleagues argued that inferring an intelligent cause was legitimate in *origins science,* because such sciences deal with singular events, and the actions of intelligent agents are usually unique occurrences. On the other hand, they argued that it was not legitimate to invoke intelligent causes in operations sciences, because such sciences only deal with regular and repeating phenomena. Intelligent agents don't act in rigidly regular or lawlike ways and, therefore, cannot be described mathematically by laws of nature.

Though their terminology was admittedly cumbersome, it did seem to capture an intuitively obvious distinction. But still I had questions. Thaxton had argued that theories in the operation sciences are readily testable against the repeating phenomena they describe. Regularity enables prediction. If a theory describing a repeating phenomenon was correct, then it should be able to predict future occurrences of that phenomenon at a specific time or under controlled laboratory conditions. Origins theories, however, do not make such predictions, because they deal with unique events. For this reason, Thaxton thought that such theories could not be tested. Theories about the past can produce plausible, but never decisive conclusions. As a geophysicist, I knew that earth scientists often formed hypotheses about past events, but I wasn't sure that such hypotheses were never testable or decisive. We have very good scientific reasons for thinking that dinosaurs existed before humans and that agriculture arose after the last ice age. But if Thaxton was right, then such conclusions about the past were merely plausible—no more than possibly true—and completely untestable.

Yet I wondered if a hypothesis about the past couldn't be tested—if there is no way to judge its strength or compare it against that of competing hypotheses—then why regard the claims of historical or "origins" theories as significant? It is provocative to claim that the evidence from DNA and our best scientific reasoning points strongly to an intelligent cause of life. It is not very interesting to claim that it is possibly true ("plausible") that DNA owes its origin to such cause. Many statements are merely plausible or possibly true. But that doesn't mean we have any reason to think them likely to be true. Rigorous scientific testing usually provides evidence-based reasons for making such claims or for preferring one hypothesis over another. Absent such testability, I wasn't sure how significant, or *scientific,* Thaxton's argument really was.

Even so, I was deeply fascinated with the whole issue. In September 1985, I learned that I was to be laid off from my oil-company job, as the price of oil had dropped from $32 to $8 per barrel. I was strangely relieved. I used the rather generous severance the company provided to begin supporting myself as a freelance science writer. But soon after I started, I also learned that I had received a Rotary scholarship to study in England. The following spring a thin airmail letter arrived informing me that I had been accepted to study the history and philosophy of science at the University of Cambridge. This course of study would enable me to explore many of the questions that had long fascinated me at the intersection of science and philosophy. It would also allow me to investigate the questions that had arisen in my discussions with Charles Thaxton.

What methods do scientists use to study biological origins? Is there a distinctive method of historical scientific inquiry? And what does the scientific evidence tell us about the origin of biological information and how life began? Is it possible to make a rigorous scientific argument for the intelligent design of life? I eventually completed a Ph.D. dissertation on the topic of origin-of-life biology. In it, I was able to investigate not only the history of scientific ideas about the origin of life, but also questions about the definition of science and about how scientists study and reason about the past.

The Current Controversy

I couldn't have known as I was leaving for England, but the two main questions I had about Dr. Thaxton's idea—"Is it scientific?" and "How strong is the evidence for it?"—would resurface with a vengeance twenty years later at the center of an international controversy, indeed, one that would engage the attention of the mainstream media, the courts, the scientific establishment, and the publishing and movie industries. In 2005, a federal judge would rule that public-school science students in Dover, Pennsylvania, could not learn about the idea that life pointed to an intelligent cause, because the idea was neither scientific nor testable. Mainstream scientific organizations—such as the National Academy of Sciences and American Association for the Advancement of Science—would issue similar pronouncements.

In 2006 and 2007, a spate of books with titles like *The God Delusion* and *God Is Not Great* would argue there is no evidence for design in biology and, therefore, no good evidence for the existence of God. According to Oxford evolutionary biologist Richard Dawkins and other New Atheists, the lack of evidence for design has made the idea of God tantamount to a "delusion." In 2008, the controversy surrounding what is now known as the "theory of intelligent design" moved into movie theaters, video stores, and candidate press conferences. And this year, with the celebration of the 200th anniversary of Darwin's birth and the 150th anniversary of the publication of *On the Origin of Species,* the main question that Darwin himself addressed—"Was life designed or does it merely *appear* designed?"—has reemerged as scientists, scholars, teachers, and media commentators evaluate his legacy.

Yet in all of this discussion—from Dover to Dawkins to Darwin's big anniversary—there has been very little discussion of DNA. And yet for me and many other scientists and scholars, the question of whether science has refuted the design argument or resuscitated it depends critically upon the central mystery of the origin of biological information. This book examines the many successive attempts that have been made to resolve this enigma—*the DNA enigma*—and will itself propose a solution.

2

The Evolution of a Mystery and Why It Matters

Few schoolchildren commit to memory the name of nineteenth-century chemist Friedrich Wöhler, nor is the waste product associated with his most famous experiment easily romanticized. Yet in 1828 the German scientist performed an experiment that revolutionized our understanding of life.

As a professor at the Polytechnic School in Berlin, he had begun investigating substances that released cyanide when heated. One day he heated some ammonium cyanate, figuring it would release cyanide. It didn't. The heat transformed the ammonium cyanate crystals, altering both their melting point and appearance. Indeed, the resulting material, a white crystalline substance, possessed none of the properties typical of cyanates. What had happened? The new material seemed familiar somehow. Where had he encountered it before? At first he thought it might be an alkaloid, but he had to discard this idea after the mysterious substance failed to respond to tests in ways typical of alkaloids. Wöhler cast about in his memory, rifling through his extensive learning in both chemistry and medicine. Then he had it. Urea![1] Wöhler dashed off a letter to fellow chemist Jöns Jakob Berzelius: "I can no longer, as it were, hold back my chemical water; and I have to let out that I can make urea without needing a kidney, or even of an animal, whether of man or dog: the ammonium salt of cyanic acid (*cyansäures Ammoniak*) is urea."[2]

The experiment, eventually replicated in laboratories around the world, showed that the chemical compounds in living organisms could be artificially synthesized.[3] Though chemists before Wöhler had synthesized naturally occurring mineral substances, many assumed it was impossible to synthesize compounds found in organisms, since it was thought that organic matter contained mysterious and immaterial "vital forces."[4] As Sir Frederick Gowland Hopkins later suggested, Wöhler's discovery marked the beginning of a challenge to the "primitive faith in a boundary between the organic and the inorganic which could never be crossed."[5] For this reason, Wöhler's work would also exert a profound influence on scientific ideas about the origin of life for over a century and would serve as a starting point for my own investigation of the topic.

Beginning at the Beginning

By the time I arrived in England I was fascinated with the origin of life and wanted to learn everything I could about the history of scientific thinking on the subject. I also wanted to investigate, following my discussions with Charles Thaxton, whether scientists who studied origin events in the remote past used a distinctive method of scientific investigation, and if so, what that method of investigation entailed.

Unfortunately, being an American untutored in the intricacies of the university system in the United Kingdom, I found it difficult to find information about the British academic programs that best fit my interests. The Rotary scholarship I had received allowed me to attend any one of five overseas universities, provided I could gain admittance. Several of them offered programs in the history or philosophy of science, but in a pre-Internet era it was difficult to extract detailed information from them about the specializations of their faculties. In the end, I set my hopes on Cambridge, since it had more of a reputation for science than the other universities on my list.

When my wife, Elaine, and I arrived in the fall of 1986, parking our rental car underneath the imposing Gothic architecture on Trumpington Street near the center of Cambridge, I was more than a little intimidated. Yet within a few weeks I began to settle into my life as a graduate student. I soon discovered that I had made a far better choice of programs than I could have known while making my decision. Not only were many of the critical discoveries about DNA and molecular biology made in Cambridge, but it also had an excellent program in the history and philosophy of science that included a kindly Dutch scholar named Harmke Kamminga, who happened to be an expert on the history of scientific theories about the origin of life.

During my first year of study—in between tutorials and lectures on everything from the history of molecular biology to the philosophy of physics and the sociology of science—I began to meet regularly with Harmke to discuss the origin of life. Under her supervision, I began to investigate some current theories about the origin of life, but also the early theories that gave rise to them. Thus, I began at the beginning—with a study of how origin-of-life studies first emerged as a scientific enterprise in the nineteenth century at the time of Darwin and his scientific contemporaries.

I was soon confronted with an interesting historical puzzle. With the acceptance of Darwin's theory of evolution, most biologists agreed that natural selection could explain the appearance of design in biology. For this reason, most philosophers and scientists have long thought that Darwin's theory of evolution by natural selection destroyed the design argument. Yet I also discovered that Darwin himself admitted that his theory did not explain the origin of life itself. In fact, one day Peter Gautry, an archivist in the manuscripts room of the university library, allowed me in to read a letter by Charles Darwin on the subject written in 1871, twelve years after the publication of *On the Origin of Species*. The letter, handwritten on brittle paper, made clear that Darwin had little more than vague speculations to offer as to how the first life on earth had begun.[6]

This was consistent with what I knew. In the *Origin,* Darwin did not try to explain the origin of the first life. Instead, he sought to explain the origin of new forms of life from simpler preexisting forms, forms that already possessed the ability to reproduce. His theory assumed rather than explained the origin of the first living thing. Since this limitation of Darwin's theory was widely recognized, it raised a question: Why were nineteenth- and twentieth-century biologists and philosophers so sure that Darwin had undermined the design argument from biology? If scientists at the time had no detailed explanation for how life had first arisen, how did they know that design—that is, actual intelligent design—played no role in this critically important event?

This chapter tells the story of what I learned as I sought to answer these questions. In the process, it describes some of the earliest scientific theories about the origin of life. This background will later prove helpful, since many contemporary theories have been formulated on the foundation of these earlier approaches. This chapter highlights something else I learned in my investigations as well. From the beginning, scientific theories about the origin of life have inevitably raised deeper philosophical issues not only about life, but also about the nature of ultimate reality. As I discuss at the close of the book, these philosophical issues remain with us today and are an integral part of the DNA enigma.

Of course, during the late nineteenth century, scientists were not trying to explain the origin of biological information, let alone the information stored in DNA. They did not know about DNA, at least not by that name, nor were they thinking about biological information even as a concept. But they did seek to explain how life began and were keenly aware of the philosophical implications of the theories they proposed. And despite their lack of knowledge about the inner workings of the cell, they were often oddly confident about the adequacy of these theories. That confidence had much to do with Friedrich Wöhler's "Eureka!"—or "Urea!"—moment and how scientists at the time viewed the nature of life.

Setting the Philosophical Stage

Since the time of the ancient Greeks, there have been two basic pictures of ultimate reality among Western intellectuals, what the Germans call a *Weltanschauung,* or worldview. According to one worldview, mind is the primary or ultimate reality. On this view, material reality either issues from a preexisting mind, or it is shaped by a preexistent intelligence, or both. Mind, not matter, is, therefore, the prime or ultimate reality— the entity from which everything else comes, or at least the entity with the capacity to shape the material world. Plato, Aristotle, the Roman Stoics, Jewish philosophers such as Moses Maimonides, and Christian philosophers such as St. Thomas Aquinas each held some version of this perspective.[7] Most of the founders of modern science during the period historians of science call the scientific revolution (1300–1700) also held this mind-first view of reality. Many of these early modern scientists thought that their studies of nature confirmed this view by providing evidence, in Sir Isaac Newton's words, of "an intelligent and powerful Being" behind it all.[8] This view of reality is often called *idealism* to indicate that ideas come first and matter comes later. *Theism* is the version of idealism that holds that God is the source of the ideas that gave rise to and shaped the material world.

The opposite view holds that the physical universe or nature is the ultimate reality. In this view, either matter or energy (or both) are the things from which everything else comes. They are self-existent and do not need to be created or shaped by a mind. Natural interactions between simple material entities governed by natural laws eventually produce chemical elements from elementary particles, then complex molecules from simple chemical elements, then simple life from complex molecules, then more complex life from simpler life, and finally conscious living beings such as ourselves. In this view matter comes first, and conscious mind arrives on the scene much later and only then as a by-product of material processes and undirected evolutionary change. The Greek philosophers who were called atomists, such as Leucippus and Democritus, were perhaps the first Western thinkers to articulate something like this view in writing.[9] The Enlightenment philosophers Thomas Hobbes and David Hume also later espoused this matter-first philosophy.[10] Following the widespread acceptance of Darwin's theory of evolution in the late nineteenth century, many modern scientists adopted this view. This worldview is called either *naturalism* or *materialism,* or sometimes scientific materialism or scientific naturalism, in the latter case because many of the scientists and philosophers who hold this perspective think that scientific evidence supports it.

The age-old conflict between the mind-first and matter-first worldviews cuts right through the heart of the mystery of life's origin. Can the origin of life be explained purely by reference to material processes such as undirected chemical reactions or random collisions of molecules? Can it be explained without recourse to the activity of a designing intelligence? If so, then such an explanation would seem to make a materialistic worldview—with its claim that all of reality can be explained solely by undirected material processes—all the more credible. Who needs to invoke an unobservable designing intelligence to explain the origin of life, if observable material processes can produce life on their own? On the other hand, if there is something about life that points to the activity of a designing intelligence, then that raises other philosophical possibilities. Does a matter-first or a mind-first explanation best explain the origin of life? Either way, the origin of life was not only an intrinsically interesting scientific topic, but one that raised incorrigibly philosophical issues as well. For me, that was part of what made it interesting.

The Mystery of the Missing Mystery

By the close of the nineteenth century, many scientists had accepted the matter-first view. Whereas many of the founders of early modern science—such as Johannes Kepler, Robert Boyle, and Isaac Newton—had been men of deep religious conviction who believed that scientific evidence pointed to a rational mind behind the order and design they perceived in nature, many late-nineteenth-century scientists came to see the cosmos as an autonomous, self-existent, and self-creating system, one that required no transcendent cause, no external direction or design.

Several nineteenth-century scientific theories provided support for this perspective. In astronomy, for example, the French mathematician Pierre Laplace offered an ingenious theory known as the "nebular hypothesis" to account for the origin of the solar system as the outcome of purely natural gravitational forces.[11] In geology, Charles Lyell explained the origin of the earth's most dramatic topographical features—mountain ranges and canyons—as the result of slow, gradual, and completely naturalistic processes of change such as erosion or sedimentation.[12] In physics and cosmology, a belief in the infinity of space and time obviated any need to consider the question of the ultimate origin of matter. And, in biology, Darwin's theory of evolution by natural selection suggested that an undirected process could account for the origin of new forms of life without divine intervention, guidance, or design. Collectively, these theories made it possible to explain all the salient events in natural history from before the origin of the solar system to the emergence of modern forms of life solely by reference to natural processes—unaided and unguided by any designing mind or intelligence. Matter, in this view, had always existed and could—in effect—arrange itself without the help of any preexisting intelligence.

But the origin of the first life remained a small hole in this elaborate tapestry of naturalistic explanation. Although Laplace's nebular hypothesis provided additional support for a materialistic conception of the cosmos, it also complicated attempts to explain life on earth in purely material terms. Laplace's theory suggested that earth had once been too hot to sustain life, since the environmental conditions needed to support life existed only after the planet had cooled below the boiling point of water. For this reason, the nebular hypothesis implied that life had not existed eternally, but instead appeared at a definite time in earth's history.[13] To scientific materialists, life might be regarded as an eternal given, a self-existent reality, like matter itself. But this was no longer a credible explanation for life on earth. There was a time when there was no life on earth. And then life appeared. To many scientists of a materialistic turn of mind, this implied that life must have evolved from some nonliving materials present on a cooling prebiotic earth. Yet no one had a detailed explanation for how this might have happened. As Darwin himself noted in 1866, "Though I expect that at some future time the [origin] of life will be rendered intelligible, at present it seems to me beyond the confines of science."[14]

The problem of the origin of life was, at this time, rendered more acute by the failure of "spontaneous generation," the idea that life originates continually from the remains of once living matter. This theory suffered a series of setbacks during the 1860s because of the work of Louis Pasteur. In 1860 and 1861, Pasteur demonstrated that microorganisms or germs exist in the air and can multiply under favorable conditions.[15] He showed that if air enters sterile vessels, contamination of the vessels with microorganisms occurs. Pasteur argued that the observed "spontaneous generation" of mold or bacterial colonies on rotting food or dead meat, for example, could be explained by the failure of experimenters to prevent contamination with preexisting organisms from the atmosphere.[16] Pasteur's work seemed to refute the only naturalistic theory of life's origin then under experimental scrutiny.[17]

Despite the impasse, late-Victorian-era biologists expressed little, if any, concern about the absence of detailed explanations for how life had first arisen. The obvious question for me was, Why? From my vantage point in 1986, having just learned about the current impasse in contemporary origin-of-life research, the nonchalance of the Victorians seemed itself a bit mysterious.

As I began investigating these questions during my first year at Cambridge, I discovered that these scientists actually had several reasons for holding this point of view. Even though many scientists knew that Darwin had not solved the origin-of-life problem, they were confident that the problem could be solved because they were deeply impressed by the results of Friedrich Wöhler's experiment. Before the nineteenth century many biologists had taken it as almost axiomatic that the matter out of which life was made was qualitatively different than the matter in nonliving chemicals. These biologists thought living things possessed an immaterial essence or force, an *élan vital,* that conferred a distinct and qualitatively different kind of existence upon organisms.[18] Scientists who held this view were called "vitalists," a group that included many pioneering biologists.

Since this mysterious *élan vital* was responsible for the distinctive properties of organic matter, vitalists also thought that it was impossible to change ordinary inorganic matter into organic matter. After all, the inorganic matter simply lacked the special ingredient—the immaterial right "stuff." That's why Wöhler's experiment was so revolutionary. He showed that two different types of inorganic matter could be combined to produce organic matter, albeit of a somewhat inglorious kind. Though some scientists continued to support vitalism well into the twentieth century, they had to do so on other grounds.

Thus, Wöhler's experiment had a direct influence on thinking about the origin of life. If organic matter could be formed in the laboratory by combining two inorganic chemical compounds, then perhaps organic matter could have formed the same way in nature in the distant past. If organic chemicals could arise from inorganic chemicals, then why couldn't life itself arise in the same way? After all, if vitalism was as wrong as it now appeared, then what is life but a combination of chemical compounds?

Developments in other scientific disciplines reinforced this trend in thought. In the 1850s, a German physicist named Hermann von

Helmholtz, a pioneer in the study of heat and energy (thermodynamics), showed that the principle of conservation of energy applied equally to both living and nonliving systems. The conservation of energy is the idea that energy is neither created nor destroyed during physical processes such as burning or combustion, but merely converted to other forms.

The chemical energy in gasoline, for example, is used by an engine to propel a car. The engine burns the gasoline and uses it up. But the energy contained in the gasoline is not destroyed; it is converted into heat (or thermal) energy, which in the cylinders is turned into mechanical or kinetic energy to propel the car. Helmholtz demonstrated that this same principle of energy conservation applied to living systems by measuring the amount of heat that muscle tissues generated during exercise.[19] His experiment showed that although muscles consume chemical energy, they also expend energy in the work they perform and the heat they generate. That these processes were in balance supported what became known as the "first law of thermodynamics"—energy is neither created nor destroyed.

Even before this first law of thermodynamics had been refined, Helmholtz used a version of it to argue against vitalism. If living organisms are not subject to energy conservation, if an immaterial and immeasurable vital force can provide energy to organisms "for free," then perpetual motion would be possible.[20] But, argued Helmholtz, we know from observation that is impossible. Other developments supported this critique of vitalism. During the 1860s and 1870s scientists identified the cell as the energy converter of living organisms. Experiments on animal respiration established the utility of chemical analysis for understanding respiration and other energetic processes in the cell.[21] Since these new chemical analyses could account for all the energy the cell used in metabolism, biologists increasingly thought it unnecessary to refer to vital forces.[22]

As new scientific discoveries undermined long-standing vitalist doctrines, they also bolstered the confidence of scientific materialists. German materialists, such as the biologist Ernst Haeckel, denied any qualitative distinction between life and nonliving matter: "We can no longer draw a fundamental distinction between organisms and anorgana [i.e., the nonliving]."[23] In 1858, in an essay entitled "The Mechanistic

Interpretation of Life," another German biologist, Rudolf Virchow, challenged vitalists to "point out the difference between chemical and organic activity."[24] With vitalism in decline, Virchow boldly asserted his version of the materialist credo: "Everywhere there is mechanistic process *only*, with unbreakable necessity of cause and effect."[25] Life processes could now be explained by various physical or chemical mechanisms. Since, in our experience, mechanisms—like cogged wheels turning axles—involve material parts in motion and nothing more, this meant that the current function of organisms could be explained by reference to matter and energy alone.

This outlook encouraged scientific materialists to assume they could easily devise explanations for the origin of life as well. Haeckel himself would be one of the first scientists to try. If life was composed solely of matter and energy, then what else besides matter in motion—material processes—could possibly be necessary to explain life's origin? For materialists such as Haeckel, it was inevitable that scientists would succeed in explaining how life had arisen from simpler chemical precursors and that they would do so only by reference to materialistic processes. For Haeckel, finding a materialistic explanation for the origin of life was not just a scientific possibility; it was a philosophical imperative.[26]

Evolution on a Roll

If the imperative for many scientists during this time was matter first, the central image was increasingly that of evolution, of nature unfolding in an undirected way, with the nebular and Darwinian hypotheses suggesting the possibility of an unbroken evolutionary chain up to the present. Yes, the origin of life was a missing link in that chain, but surely, it was thought, the gap would soon be bridged. Darwin's theory, in particular, inspired many evolutionary biologists to begin formulating theories to solve the origin-of-life problem. My supervisor, Dr. Kamminga, had a memorable way of describing this phenomenon. She noted that the success of Darwin's theory inspired attempts at "extending evolution backward" in order to explain the origin of the first life.

Darwin's theory inspired confidence in such efforts for several reasons. First, Darwin had established an important precedent. He had shown that there was a plausible means by which organisms could gradually produce new structures and greater complexity by a purely undirected material process. Why couldn't a similar process explain the origin of life from preexisting chemicals?

Darwin's theory also implied that living species did not possess an essential and immutable nature. Since Aristotle, most biologists had believed that each species or type of organism possessed an unchanging nature or form; many believed that these forms reflected a prior idea in the mind of a designer. But Darwin argued that species can change—or "morph"—over time. Thus, his theory challenged this ancient view of life. Classification distinctions among species, genera, and classes did not reflect unchanging natures. They reflected differences in features that organisms might possess only for a time. They were temporary and conventional, not set in stone.[27] If Darwin was right, then it was futile to maintain rigid distinctions in biology based on ideas about unchanging forms or natures. This reinforced the conviction that there was no impassable or unbridgeable divide between inanimate and animate matter. Chemicals could "morph" into cells, just as one species could "morph" into another.[28]

Darwin's theory also emphasized the importance of environmental conditions on the development of new forms of life. If conditions arose that favored one organism or form of life over another, those conditions would affect the development of a population through the mechanism of natural selection.[29] This aspect of Darwin's theory suggested that environmental conditions may have played a crucial role in making it possible for life to arise from inanimate chemistry. It was in this context that Darwin himself first speculated about the origin of life. In the 1871 letter to botanist Joseph Hooker, which I had seen in the Cambridge library archive, Darwin sketched out a purely naturalistic scenario for the origin of life. He emphasized the role of special environmental conditions and just the right mixture of chemical ingredients as crucial factors in making the origin of life possible: "It is often said that all the conditions for the first production of a living organism are present. . . . But if (and Oh! what a big if!) we could conceive in some warm little pond, with all sorts of ammonia and phosphoric salts, light, heat, electricity, etc., that a protein compound was chemically formed ready to undergo still more complex changes, at the present day such matter would be instantly devoured or absorbed, which would not have been the case before living creatures were formed."[30] Although Darwin conceded that his speculations ran well ahead of available evidence, the basic approach he outlined would seem increasingly plausible as a new theory about the nature of life came to prominence in the 1860s and 1870s.

The Protoplasmic Theory of Life

In my first year of research, I came across a statement by Russian scientist Aleksandr Oparin. Oparin was the twentieth century's undisputed pioneer of origin-of-life studies, and his comment helped me to identify another key reason for the Victorian lack of concern about the origin-of-life problem. "The problem of the nature of life and the problem of its origin have become inseparable," he said.[31]

To explain how life originated, scientists first have to understand what life is. That understanding, in turn, defines what their theories of the origin of life must explain. The Victorians weren't especially concerned with the origin-of-life problem because they thought simple life was, well, simple. They really didn't think there was much *to* explain. Biologists during this period assumed that the origin of life could eventually be explained as the by-product of a few simple chemical reactions.

Then, as now, scientists appreciated that many intricate structures in plants and animals appeared designed, an appearance that Darwin explained as the result of natural selection and random variation. But for Victorian scientists, single-celled life didn't look particularly designed, most obviously because scientists at the time couldn't see individual cells in any detail. Cells were viewed as "homogeneous and structure-less globules of protoplasm,"[32] amorphous sacs of chemical jelly, not intricate structures manifesting the appearance of design.

In the 1860s, a new theory of life encouraged this view. It was called the "protoplasmic theory," and it equated vital function with a single, identifiable chemical substance called protoplasm.[33] According to this theory, the attributes of living things derived from a single substance located inside the walls of cells. This idea was proposed as a result of several scientific developments in the 1840s and 1850s.[34] In 1846, a

German botanist named Hugo von Mohl demonstrated that plant cells contained a nitrogen-rich material, which he called protoplasm.[35] He also showed that plant cells need this material for viability. Mohl and Swiss botanist Karl Nägeli later suggested that protoplasm was responsible for the vital function and attributes of plant cells and that the cell wall merely constituted an "investment lying upon the surface of the [cell] contents, secreted by the contents themselves."[36]

This turned out to be fantastically inaccurate. The cell wall is a separate and fascinatingly intricate structure containing a system of gates and portals that control traffic in and out of the cell. Nevertheless, Mohl and Nägeli's emphasis on the importance of the cell contents received support in 1850 when a biologist named Ferdinand Cohn showed that descriptions of protoplasm in plants matched earlier descriptions of the "sarcode" found in the cavities of unicellular animals.[37] By identifying sarcode as animal-cell protoplasm, Cohn connected his ideas to Mohl's. Since both plants and animals need this substance to stay alive, Cohn established that protoplasm was essential to all living organisms. When, beginning in 1857, a series of papers by scientists Franz Leybig, Heinrich Anton de Bary, and Max Shultze suggested that cells could exist without cellular membranes (though, in fact, we now know they cannot), scientists felt increasingly justified in identifying protoplasm as life's essential ingredient.[38] Thus, in 1868 when the famous British scientist Thomas Henry Huxley declared in a much publicized address in Edinburgh that protoplasm constituted "*the* physical basis or matter of life" (emphasis in original), his assertion expressed a gathering consensus.[39]

With the protoplasmic theory defining the chemical basis of life, it seemed plausible that the right chemicals, in the right environment, might combine to make the simple protoplasmic substance. If so, then perhaps the origin of life could be explained by analogy to simple processes of chemical combination, such as when hydrogen and oxygen join to form water. If water could emerge from the combination of two ingredients as different from water as hydrogen and oxygen, then perhaps life could emerge from the combination of simple chemical ingredients that by themselves bore no obvious similarity to living protoplasm.

Early Theories of Life's Origin: The Chemical Two-Step

I discovered another reason that scientists had maintained their confidence in a completely materialistic account of life and the cosmos. Beginning in the late 1860s, scientists began to offer materialistic theories of the origin of life. And for the better part of the next eighty-five years or so (with the exception of one gap after the turn of the century), these theories kept pace with new scientific discoveries about the complexity of life. That is, for the most part, these new theories about how life came to be were able to explain what scientists were learning about what life is.

Two scientists, Thomas Henry Huxley and Ernst Haeckel, were first to offer theories of how life had arisen from nonliving chemicals. Though Huxley was British and Haeckel German, the two men had much in common intellectually. Both men rejected vitalism. Both men were staunch defenders of Darwin's evolutionary approach to the origin of species. Both were ardent scientific materialists. And both had articulated or defended the protoplasmic theory of life. In these respects, Huxley and Haeckel embodied the various reasons for Victorian insouciance about the origin-of-life problem. Each man would formulate a theory of abiogenesis (life arising from nonliving matter) that reflected this intellectual posture.

Huxley imagined that the origin of life had occurred by a simple two-step chemical process in which simple elements such as carbon, hydrogen, nitrogen, and oxygen first reacted to form common compounds such as water, carbonic acid, and ammonia.[40] He believed that these compounds then combined, under some unspecified conditions, to form protoplasm, the chemical essence of life.

Meanwhile, in Germany, Haeckel[41] offered a bit more detail, though not much. He identified "constructive inner forces" or "formative tendencies" inherent in matter—like those we find in inorganic crystal formation—as the cause of the self-development of life. He asserted that the causes that produce form are the same in both inorganic crystals and living organisms.[42] Thus, for Haeckel, the origin of life could be explained by the spontaneous crystallization of "formless lumps of protein" from simpler carbon compounds.[43] Haeckel believed that, once formed, the first one-celled organisms, which he called Monera, would have gradually attained the relatively simple structure he assumed them to possess as they assimilated new material from the environment. Then, due to their semifluid constitution, these primitive cells would continue to rearrange themselves internally over time.[44] Even so, he clearly regarded the essential step in the process of abiogenesis as complete after the spontaneous crystallization of the "homogeneous and structure-less globules of protoplasm."[45]

Huxley also viewed the nature of life as scarcely distinguishable from inorganic crystals. Many other biologists adopted similar views. Eduard Pflüger, Karl Wilhelm von Nägeli, August Weismann, and Oscar Loew each attributed the essential properties of life to a single chemical entity rather than to complex processes involving many interrelated parts.[46] Pflüger, for example, thought the presence of carbon and nitrogen (in the form of the cyanogen radical, $-CN$) distinguished "living" proteins from "dead" ones.[47] By equating the essence of life with a single chemical unit such as "living proteins" (Pflüger), "active proteins" (Loew), "biophors" (Weismann), "probionts" (Nägeli), or "homogeneous protoplasm" (Haeckel and Huxley), scientists during the 1870s and 1880s made it easy to explain the origin of life. Yet only as long as their simplistic conceptions of the nature of life held sway did their equally simplistic models of the origin of life seem credible.

Over the next sixty years biologists and biochemists gradually revised their view of the nature of life. During the 1890s, scientists began to learn about enzymes and other types of proteins. Before 1894, scientists had only observed enzymes catalyzing reactions outside the cell.[48] With the advance of laboratory techniques allowing scientists to gather evidence of the activity of enzymes within cells, and with the discovery of enzymes responsible for such metabolic reactions as oxidation, fermentation, and fat and protein synthesis, a new theory called the "enzymatic theory" displaced the protoplasmic theory of life.[49] By the turn of the century, most biologists came to see the cell as a highly complex system of integrated chemical reactions, not at all the sort of thing that could be adequately explained by vague references to processes of crystallization. For a time, the growing awareness of this chemical complexity impeded attempts to explain the origin of life. But by the 1920s and 1930s, a pioneering Russian scientist formulated a new theory to keep pace with this growing scientific awareness of the cell's complexity.

Oparin to the Rescue

A new theory of *evolutionary* abiogenesis that envisioned a multibillion-year process of transformation from simple chemicals to a complex metabolic system[50] was proposed to the Russian Botanical Society in May 1922 by the young Soviet biochemist Aleksandr I. Oparin (1894–1980).[51] Oparin first published his theory in Russian in 1924 and then refined and developed it, publishing it again in English in 1938. Both books were simply called *The Origin of Life*.

Oparin's interest in the origin of life was first awakened after hearing lectures on Darwinism from the plant physiologist Kliment Arkadievich Timiriazev, who was himself a staunch Darwinian. "According to Oparin," writes historian of science Loren Graham, "Timiriazev described Darwinian evolution and revolutionary political thought as being so intimately connected that they amounted to the same thing. In this view Darwinism was materialistic, it called for change in all spheres, it was atheistic, it was politically radical, and it was causing a transformation of thought and politics."[52]

Figure 2.1. Aleksandr Oparin (1894–1980), pioneering chemical evolutionary theorist. *Courtesy of Novosti/Photo Researchers, Inc.*

Oparin was a fascinating figure from a fascinating time. He published his first theory of the origin of life just five years after the Bolshevik Revolution while living in Moscow, where Marxist slogans and thinking were popular, especially in intellectual circles.[53] It seemed a little odd to me at first that anyone could think about something as seemingly remote as the origin of the first life while such cataclysmic changes were taking place in society, but I discovered many of the early Marxists were quite interested in the subject of biological origins. Marx himself had corresponded with Darwin, and he thought that Darwin's theory of evolution put his own theory about how societies evolved on a firm materialistic and scientific footing.[54]

Friedrich Engels, Marx's intellectual collaborator, actually wrote an essay on the origin of the first life.[55] Like Marx, he was convinced that major societal changes took place in sudden spurts in response to changes in the material conditions of life and society. He wanted to show that a similar "revolution" had taken place to produce life, so that he could demonstrate the plausibility of Marxist doctrine. A key Marxist idea was that a small quantitative increase in the intensity of some condition or situation could suddenly produce a qualitative or revolutionary change. Dissatisfaction and alienation with the capitalist system among workers, for example, might gradually increase over time, but it would eventually grow to a point where a revolutionary change would suddenly occur, ushering in a completely new way of ordering society. Engels thought he could illustrate this key Marxist concept if he showed that a quantitative increase in the complexity of a system of chemicals could suddenly produce a qualitative (i.e., revolutionary) change in that system, resulting in the first life.[56]

Was Oparin influenced or motivated by such specifically Marxist ideas? In addition to Timiriazev, whose politics Oparin described as "very progressive" and Leninist, Oparin was also closely associated with an older Marxist biochemist and former revolutionary, A. N. Bakh, after 1920.[57] Even so, it's not clear how much Marxism per se influenced Oparin's thinking about the origin of life. It is clear, however, that Oparin rejected all forms of idealism. Instead, he embraced a materialistic view of reality. Accordingly, he saw that the origin-of-life problem needed to be solved within a materialistic framework of thought.[58]

At the same time, Oparin thought there were a number of scientific reasons for supposing that the origin of life could be explained by reference to purely chemical processes. First of all, there was Wöhler's famous synthesis of urea, which showed that both living and nonliving matter share a common chemical basis. It was clear from Oparin's writing that, a hundred years after Wöhler's experiment, it continued to have a profound influence on thinking about the nature and origin of life. To Oparin, Wöhler's experiment established that "there is nothing peculiar or mysterious" about the processes at work in a living cell "that cannot be explained in terms of the general laws of physics and chemistry." Oparin also noted that several nonliving materials, not just urea, manifest attributes like those once thought to characterize only living organisms.[59] For example, carbon, the element common to all living protoplasm and organisms, also occurs naturally in inanimate minerals such as graphite, diamond, marble, and potash. Besides, argued Oparin, like living organisms, many inorganic materials display chemical organization and structure. Inanimate materials like crystals and magnets have a very definite and orderly organization. Crystals even reproduce themselves, though not in the same way cells do. Although Oparin admitted inanimate materials like crystals do not have the kind of "complicated order" observed in living cells, the similarities he did identify between life and nonlife made him optimistic that scientists could explain the origin of life by reference to ordinary chemical processes.[60]

Even so, given the complexity of the chemical reactions going on inside the cell, Oparin thought any return to spontaneous generation was untenable. As he stated, "The idea that such a complicated structure with a completely determinate fine organization could arise spontaneously in the course of a few hours . . . is as wild as the idea that frogs could be formed from the May dew or mice from corn."[61] Instead, in his view, biological organization must have evolved gradually from simpler chemistry over a long period of time.[62]

Oparin Sets the Stage

Oparin's theory envisioned many discrete events along the way to the development of life. Nevertheless, his theory described processes that can be divided into two basic stages. The first part of his theory described how the chemical building blocks of life arose from much

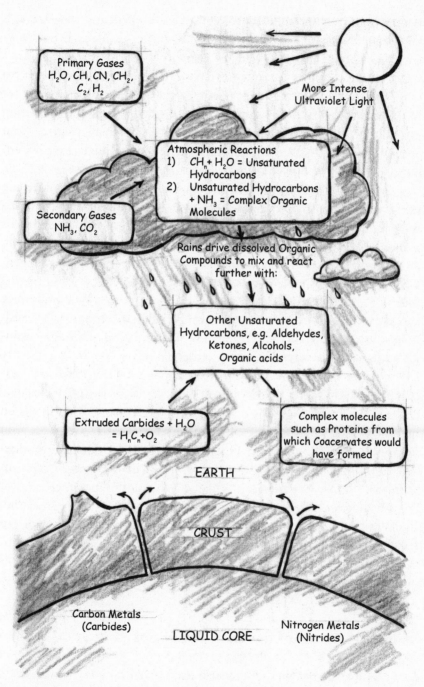

Figure 2.2. Oparin's 1936 early earth origin-of-life scenario.

simpler chemicals in the earth's atmosphere and oceans. The second stage tells how the first organism arose from these molecular building blocks. Let's look at the first part of Oparin's scenario first.

Oparin thought that the early earth had a core made of heavy metals.[63] As the early earth cooled after its initial formation, he postulated, its core would have contracted, exposing cracks and fissures in the surface of the earth. Heavy metals from the core would have then combined with carbon forming compounds called iron carbides. These compounds were squeezed to the surface of the earth like toothpaste through a tube (see Fig. 2.2).

After arriving at the surface these carbide compounds would have begun to react with the atmosphere. By 1936, Oparin had come to think that the early earth's atmosphere was devoid of free oxygen. Instead, he envisioned an early atmosphere containing a noxious mixture of energy-rich gases such as ammonia (NH_3), dicarbon (C_2), cyanogen (CN), steam, and simple hydrocarbons like methene (CH) and methylene (CH_2). He then envisioned these simple hydrogen-rich molecules in the atmosphere reacting with the iron carbides arriving at the surface of the earth. This would have resulted in the formation of heavy energy-rich hydrocarbons, the first organic molecules.[64]

The compounds[65] produced in this way would have then reacted with ammonia (NH_3) in the atmosphere to form various nitrogen-rich compounds.[66] This was a significant step, because Oparin knew that the amino acids out of which protein molecules are made are themselves rich in nitrogen. Oparin also thought that energy-rich hydrocarbon derivatives in water could participate in every type of chemical change occurring in the cell, including polymerization. This was important because polymerization is the kind of reaction by which amino acids link up to form proteins. Thus, Oparin suggested that these hydrocarbon derivatives reacted with one another and with other chemicals in the oceans to produce amino acids, which then linked together to form proteins.

Oparin's Account of the First Organisms

The second stage of Oparin's scenario used specifically Darwinian evolutionary concepts to explain the transformation of organic molecules into living things. In particular, he suggested that competition for survival arose between little enclosures of protein molecules. This competition eventually produced primitive cells with all kinds of complex chemical reactions going on inside them. But before he could describe how competition between protocells produced life, he had to find a chemical structure that could function as a primitive cell, or at least as a primitive cell membrane. He needed an inanimate structure that could enclose proteins and separate them from the environment.

He found what he was looking for in the work of an obscure Dutch chemist named H. G. Bungenberg de Jong. In 1932, Bungenberg de Jong described a structure called a "coacervate" (from the Latin *coacervare*, meaning "to cluster"). A coacervate is a little cluster of fat molecules that clump together into a spherical structure because of the way they repel water. (See Fig. 2.3.) Because these fat molecules, or lipids, have a water-repelling side and a water-attracting side, they will form a structure that both repels water on the outside and encloses water on the inside. Thus, these coacervates define a distinct boundary with the surrounding environment. They even allow organic molecules to pass in and out of the coacervate cluster, thus simulating the function of a cell membrane.

Oparin suggested that biologically significant molecules like carbohydrates and proteins could have been enclosed in such structures in the prebiotic ocean. As these molecules began to react with one another inside the coacervate clumps, they developed a kind of primitive metabolism. For this reason, Oparin regarded them as intermediate structures between animate and inanimate chemistry: "With certain reservations

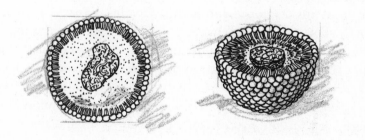

Figure 2.3. Two-dimensional cross section of a coacervate (*left*) and three-dimensional cutaway of half of a coacervate (*right*).

we can even consider the first piece of organic slime which came into being on the earth as being the first organism."[67]

Oparin proposed that the attributes of these coacervate clumps would enable the complex biochemical organization that now characterizes living cells to arise gradually through a process of natural selection. As some coacervates grew, they would develop increasingly efficient means for assimilating new substances from the environment, causing their growth rates to increase. Those that failed to develop efficient means of assimilating essential nutrients would languish. The good "eaters" thrived, while the poor "eaters" did not. As the relative abundance of nutrients in the environment changed, conditions arose that favored more highly organized organic bodies. Less efficient protocells would soon exhaust their stored potential energy and decompose as their supplies of nutrients diminished. But those primitive organisms that had (by chance) developed crude forms of metabolism would continue to develop. A Darwinian-style competition developed, eventually resulting in the first living cell.[68] (See Fig. 2.4.)

So Aleksandr Oparin explained the origin of life using Darwinian principles. He showed how complex structures could arise gradually from simpler ones after environmental changes had occurred that favored the complex structures in their competition for survival.

The Miller-Urey Experiment

Oparin's theory stimulated considerable scientific activity following the English publication of his book; several scientists during the 1940s and early 1950s developed and refined Oparin's scenario in pursuit of a more detailed theory of chemical evolution. Perhaps the most significant attempts to advance Oparin's research program came in the form of laboratory experiments, including several attempts to simulate an important step in his historical narrative, the production of biological building blocks from simpler atmospheric gases. The most famous, immortalized in high-school biology textbooks around the world, is the Miller-Urey experiment.

In December 1952, while doing graduate work under Harold Urey at the University of Chicago, Stanley Miller conducted the first experimen-

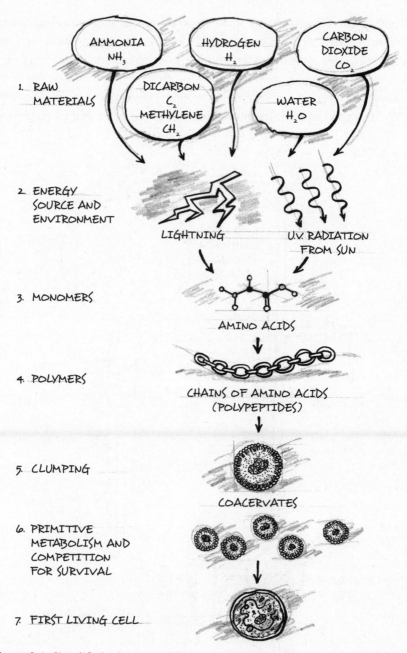

Figure 2.4. Simplified schematic of Oparin's 1936 chemical evolutionary scenario showing the main steps from simple chemicals to a more complex living cell. *Adapted by permission from John Wiester.*

tal test of the Oparin chemical evolutionary model. Using boiling water, Miller circulated a gaseous mixture of methane, ammonia, water, and hydrogen through a glass vessel containing an electrical discharge chamber.[69] Miller sent a high-voltage charge into the chamber via tungsten filaments to simulate the effects of lighting on prebiotic atmospheric gases. (See Fig. 2.5.) After two days, Miller found amino acids in the U-shaped water trap

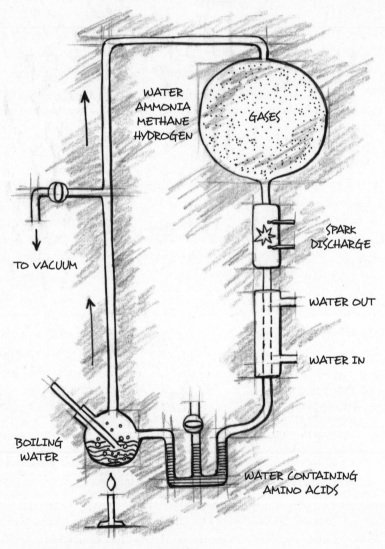

Figure 2.5. The Miller-Urey experiment.

he used to collect reaction products at the bottom of the vessel. Using a technique for analyzing mixtures called paper chromatography, he identified the amino acids glycine, alpha-alanine, and beta-alanine.

Miller's success in producing these protein building blocks was heralded as a breakthrough and as powerful experimental support for Oparin's theoretical work. After the publication of his findings in 1953, others replicated Miller's results, and soon hopes were running high that a comprehensive theory of life's origin was within reach. Miller's experiment received widespread coverage in popular publications such as *Time* magazine and gave chemical evolutionary theory the status of textbook orthodoxy almost overnight.[70] As science writer William Day reflected, "It was an experiment that broke the logjam. The simplicity of the experiment, the high yields of the products and the specific biological compounds . . . produced by the reaction were enough to show the first step in the origin of life was not a chance event, but was inevitable."[71]

By the Darwinian centennial in 1959, spirits were running high. The final holdout in the grand materialistic story of life, earth, and the cosmos seemed at last to be falling into line. With the modern "neo-Darwinian" version of Darwin's theory firmly established and the emergence of an experimentally validated theory of chemical evolution, evolutionary science had now provided a comprehensive and fully naturalistic explanation for every appearance of design in the living world from the humble single-celled bacterium to the most intricate neural structure of the human brain. A seamless and fully naturalistic account of the origin and development of life-forms appeared, if not complete, then at least sketched in enough detail to preclude anachronistic speculations about a designing hand. The problem of the origin of life had at last been solved. Or at least so it seemed, until scientists began to reflect more deeply on the other great discovery of 1953.

3

The Double Helix

The information revolution in biology officially began in 1953 with the elucidation of the structure of the DNA molecule. Yet the scientists who eventually ignited this revolution were a seemingly unimpressive pair. Two unknowns in the developing field of biochemistry, James Watson and Francis Crick possessed no firsthand experimental data and a very limited knowledge of the relevant chemistry. Crick hadn't even finished his Ph.D., and the degree he did have was in physics.[1]

There were three teams in the running to unlock the mystery of the structure of DNA, which by now most biologists assumed would help explain how hereditary traits are passed from one generation to another. The obvious front-runner was Nobel laureate Linus Pauling. Next was an imposing team of Cambridge scientists headed by Lawrence Bragg and Max Perutz. The third was a team with the most sophisticated imaging equipment in the world, headed by Maurice Wilkins at King's College, University of London. As for Watson and Crick, no scientist at the time would have even put them in the race. By most appearances, they were mere hangers-on at the Cavendish lab in Cambridge, a couple of young men lurking around pilfering other people's data, out of their depth and out of the race.

While the London team leader, Maurice Wilkins, and its X-ray specialist, Rosalind Franklin, did the painstaking work of collecting the hard data on the mystery molecule, it looked as if Watson and Crick did little more than play with toy models. One they presented was greeted with laughter by Rosalind Franklin. Drawing on her extensive knowledge of the X-ray images that she had generated of crystals made of

DNA material, she quickly convinced Watson, Crick, and everyone else in the room that their toy model was far wide of the mark.

Watson, with his wild hair and perfect willingness to throw off work for a Hedy Lamarr film, and Crick, a dapper and no longer especially young fellow who couldn't seem to close the deal on his dissertation—who were these guys? They even got their lab space stripped away at one point. Eventually they got it back, but a peek into the Cavendish lab months later would have done little to inspire confidence. Crick hadn't arrived yet, and there was wild-haired Watson at the table tinkering around with cardboard cutouts—a far cry from the sophisticated technology on display at the King's lab.

But it was in the end Watson and Crick who sparked a revolution. The molecular biological revolution, as it came to be called, would redefine our understanding of the nature of life by highlighting the importance of information to the inner workings of living things. This revolution would also redefine the questions that scientists investigating the origin of life would, from that time forward, have to answer.

Of Natures and Origins

During my Ph.D. studies, I learned that scientists investigating the past often reason much like detectives in a whodunit. Detectives consider a number of suspects in order to determine the culprit as they try to reconstruct the scene of the crime. In a similar way, historical scientists—such as geologists, archaeologists, paleontologists, cosmologists, and evolutionary biologists—weigh the merits of competing explanations as they try to figure out what caused a particular event in the past to occur or what caused a given structure or piece of evidence to arise. In doing so, historical scientists use a scientific method called the "method of multiple working hypotheses."[2] But before scientists can evaluate competing ideas about the cause of a given event or structure, they must have a clear understanding of what it is that needs to be explained.

For scientists trying to explain the origin of life, one of the most important clues we have is life itself—its structure, function, and composition. That's why Aleksandr Oparin, the first scientist to propose a comprehensive scientific theory of the origin of life, said, "The problem

of the nature of life and the problem of its origin have become inseparable."[3] Harmke Kamminga puts it this way: "At the heart of the problem
of the origin of life lies a fundamental question: What is it exactly that
we are trying to explain the origin of?"[4]

Watson and Crick's discovery, and those that soon followed in its
wake, revolutionized our understanding of the nature of life. These discoveries also defined the features of life that scientists are now "trying
to explain the origin of." This chapter tells the story of the discovery
that inaugurated this revolution in biological understanding—the story
of the double helix. This historical background will prove indispensable
in later chapters. In order to evaluate competing ideas about the origin
of life and biological information, it is important to know what DNA
is, what it does, and how its shape and structure allow it to store digital
information. As I show in subsequent chapters, some recent theories
of the origin of life have failed precisely because they have failed to account for what scientists have discovered over the last century about the
chemical structure of DNA and the nature of biological information.

The Mystery of Heredity

From ancient times, humans have known a few basic facts about living
things. The first is that all life comes from life. *Omne vivum ex vivo*. The
second is that when living things reproduce themselves, the resulting
offspring resemble their parents. Like produces like. But what inside a
living thing ensures that its offspring will resemble itself? Where does
the capacity to reproduce reside?

This was one of the long-standing mysteries of biology, and many
explanations have been proposed over the centuries. One theory proposed that animals contained miniature replicas of themselves stored in
the reproductive organs of males. Another theory, called pangenesis,
held that every tissue or organ of the body sent parts of itself—called
gemmules—to the reproductive organs to influence what was passed on
to the next generation. But by the mid-nineteenth century the target
began to narrow as scientists increasingly focused on the small spherical
enclosures called cells, only recently within reach of the best microscopes of the day. In 1839, Matthias Schleiden and Theodor Schwann

proposed the "cell theory," which asserted that cells are the smallest and most fundamental unit of life. In the wake of their proposal, biologists increasingly focused their search for the secret of heredity on these seemingly magical little entities and their critical contents. But through much of the rest of the nineteenth century, the structure of cells was a complete mystery, which is why prominent scientists like Ernst Haeckel could describe the cell as "homogeneous and structure-less globules of protoplasm."[5]

Meanwhile, however, scientists began to notice that the transmission of hereditary traits—wherever the capacity for producing these traits might be stored—seemed to occur in accord with some predictable patterns. The work of Gregor Mendel in the 1860s was particularly important in this regard. Mendel studied the humble garden pea. He knew that some pea plants have green seeds, while some have yellow. When he crossed green peas with yellow peas, the second-generation plants always had yellow peas. If Mendel had stopped there, he might have assumed that the capacity for making green seeds in the next generation had been lost. But Mendel didn't stop there. He crossed the crosses. Each of these parent plants had yellow seeds, but their offspring had 75 percent yellow seeds and 25 percent green. Apparently the first generation of crossed seeds, the all-yellow batch, nevertheless had something for making "green" seeds tucked away inside of them, waiting to emerge in a subsequent generation, given the right circumstances.[6]

Mendel called the yellow trait "dominant" and the green trait "recessive." The latter might disappear in a given generation, but it hadn't dropped out of existence. It was stored inside the seed in the form of some sort of signal, memory, or latent capacity, waiting to express itself in a future generation. Mendel showed that the entity or factor responsible for producing a trait (which was later called a "gene") has some kind of existence of its own independent of whether the trait is seen in an individual plant.

Mendel's discovery raised an obvious question: Where and how was this hereditary memory or signal being stored? Beginning with experiments done in the years after the Civil War, biologists began to focus on the cell nucleus. In 1869, Friedrich Miescher, the son of a Swiss physician, discovered what would later be called DNA. Miescher was interested in the chemistry of white blood cells. To find such cells, he collected pus from postoperative bandages. He then added hydrochloric acid to the pus, dissolving all the material in the cell except the nuclei. After that he added alkali and then acid to the nuclei. Miescher called the gray organic material that formed from this procedure "nuclein," since it was derived from the nucleus of the cell. Other scientists using staining techniques soon isolated banded structures from the nucleus. These came to be called "chromatin" (the material we now know as *chromosomes*) because of the bright color they displayed once stained. When it was later shown that chromatin bands and Miescher's nuclein reacted to acid and alkali in the same way, scientists concluded that nuclein and chromatin were the same material. When biologists observed that an equal number of chromatin strands combine when an egg and sperm fuse into a single nucleus, many concluded that chromatin was responsible for heredity.[7]

To make further progress toward a solution to the mystery of heredity, geneticists needed to study these chromatin bands more closely. In 1902 and 1903, Walter Sutton published two papers suggesting a connection between the laws of Mendelian genetics and chromosomes.[8] Sutton suggested that Mendel's laws could be explained by observing chromosomes during reproduction. Since offspring receive an equal number of chromosomes from each parent, it was possible that they were receiving the capacity for different characteristics—Mendel's traits—from separate maternal and paternal chromosomes. Since traits often occurred in pairs, and chromosomes occurred in pairs, perhaps the capacity for producing these traits was carried on chromosomes.

Some scientists thought that this idea could be tested by altering the composition of the chromatin bands to see what effect various changes would have on the creatures that possessed them. What was needed was a creature that reproduced quickly, possessed a relatively simple set of features, and could be bathed in change-producing or "mutation-inducing" radiation without raising ethical concerns. Fruit flies were the perfect choice. They had a fourteen-day life cycle and only four pairs of chromosomes, and nobody was likely to start picketing on their behalf.

Beginning in 1909 at Columbia University, Thomas Hunt Morgan undertook experiments with large populations of fruit flies, subjecting them to a variety of mutagens (i.e., substances that cause mutations), increasing their mutation rate manyfold. Then he bred them, steadily assembling a mountain of data about the resulting mutations and how often they were passed from one generation to the next. He encountered all of the fruit-fly mutations found in natural populations, such as "white eye" and "vestigial wing" (see Fig. 3.1). After studying many generations, Morgan found that some of these traits were more likely to occur in association. Specifically, he noticed four linkage groups, suggesting that information-bearing entities responsible for passing along these mutations were located physically next to each other on the chromosome. Morgan devised a number of experiments to show that genes have a definite, linear order on the chromosome.[9]

By 1909, scientists had been able to separate an acidic material from other proteinaceous material in the chromatin bands. Chemists soon determined the chemical composition of this acidic material. They called it a "nucleic acid," because it had come from the nucleus. They called it a "deoxyribose nucleic acid," because they were able to identify a deoxygenated sugar molecule called ribose in the molecule (see Fig. 3.2, comparing the structure of deoxyribose sugar and ribose sugar). Scientists also determined that the molecule was made of phosphates and four bases, called adenine, cytosine, guanine, and thymine, the chemical formulas and structures of which had been known for a while. (Figure 3.3 shows

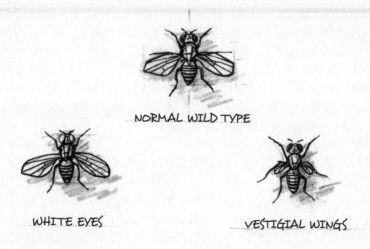

NORMAL WILD TYPE

WHITE EYES

VESTIGIAL WINGS

Figure 3.1. A normal fruit fly and two mutant fruit flies of the kind studied by Thomas Morgan.

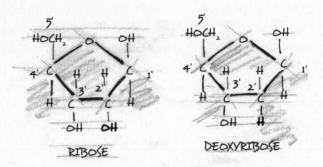

Figure 3.2. Structure of ribose sugar (*left*) and deoxyribose sugar (*right*).

the formulas and structures of each of the chemical parts of deoxyribonucleic acid, or DNA.)

Science historians often describe the process that leads to a great
discovery as "putting pieces of a puzzle together." In the case of DNA,
that metaphor is unusually apt. By 1909, the composition and structure of the chemical parts of DNA were mostly known. But the structure of the whole molecule was not. Further progress in the search
for the secret of hereditary information required scientists to piece together the constituent parts of the molecule in various different ways
in search of a solution to this puzzle. When the pieces locked in place
properly—in accord with all that was known about the dimensions,
shapes, and bonding proclivities of the constituent parts—a solution
to the puzzle would be obvious. Everything would fit. But in 1909,
scientists were far from understanding how all the pieces of the DNA
molecule fit together. In fact, for years, many showed little interest in
determining the structure of DNA, because they did not think that
DNA had anything to do with heredity.

Many scientists overlooked DNA because they were convinced that
proteins played the crucial role in the transmission of hereditary traits.
They favored proteins over DNA primarily due to a misunderstanding
about the chemical structure of DNA. By the early part of the twentieth
century, scientists knew that in addition to containing sugars and phosphates, nucleic acid was composed of the four bases adenine, thymine,
cytosine, and guanine, but in 1909 chemist P. A. Levene incorrectly
reported that these four nucleotide bases always occurred in equal quan-

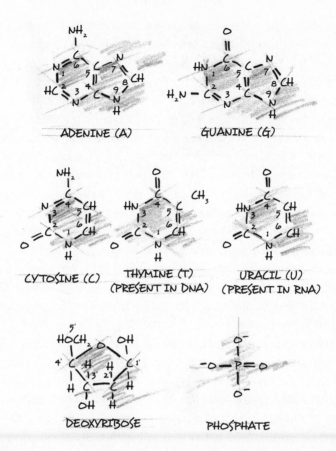

Figure 3.3. The structural formulas of each of the chemical parts of DNA (as well as the nucleotide base uracil, which is present in RNA). RNA consists of phosphates; the bases uracil, cytosine, guanine, and adenine; and ribose sugar (see Fig. 3.2).

tities within the DNA molecule.[10] To account for this putative fact, he formulated what he called the "tetranucleotide hypothesis." According to this hypothesis the four nucleotide bases in DNA linked together in the same repeating order to form a sequence such as ATCGATC-GATCGATCGATCGATCG.

Levene's model threw many scientists off the scent, but for understandable reasons. For DNA to be the material responsible for producing hereditary traits, it had to have some feature that could account for, or produce, the great variety of physiological traits found in living organisms. Even the humble fruit flies that Morgan used in his mutation studies had many different features—different kinds of eyes, legs, wings, bristles, and body proportions. If the capacity for building these structures and traits was something like a signal, then a molecule that simply repeated the same signal (e.g., ATCG) over and over again could not get the job done. At best, such a molecule could produce only one trait. Instead, scientists knew that they needed to discover some source of variable or irregular specificity, a source of *information,* within the heritable material (or germ line) of organisms to account for the many different features present in living things. Since the sequence of bases in DNA was, according to Levene, rigidly repetitive and invariant, DNA's potential seemed inherently limited in this regard.

That view began to change in the mid-1940s for several reasons. First, a scientist named Oswald Avery successfully identified DNA as the key factor in accounting for heritable differences between different bacterial strains.[11]

When Avery was working at the Rockefeller Institute in New York he became intrigued by an experiment on *Pneumococci* bacteria performed by Frederick Griffith. The experiment moved from the unsurprising to the surprising. If a deadly strain of the bacteria was first heated to death, the strain was harmless when injected into mice. No surprise there. The mice were also unharmed when injected with a living but nonvirulent strain of the bacteria. No surprise there either. But then Griffith injected mice with both the lethal strain of bacteria that had been heated to death and the living but harmless strain of bacteria. The mice died. This was surprising. One would expect the mice to be unaffected, since both forms of bacteria had proven totally harmless before. Injected with either of the two strains separately, the mice lived. But when the strains were injected together, the mice died as if the dead bacteria had become suddenly lethal again.[12] (See Fig. 3.4.)

It was almost too strange to believe. It was like those old zombie movies, where the walking dead attack and convert nice ordinary people into killing machines. Avery wanted to get to the bottom of this strange phenomenon. His laboratory started by taking the mice out of the equation. The scientists prepared a rich medium for the bacteria, then placed the two strains of bacteria—the living harmless *Pneumococci* bacteria along with the once lethal but now dead strain—into direct contact with each other in the rich medium. After several life cycles, Avery was able to begin detecting living versions of the lethal, but previously dead strain.

There were two possibilities. Either the dead strain was coming back to life—but this was absurd—or something in the dead strain of bacteria was

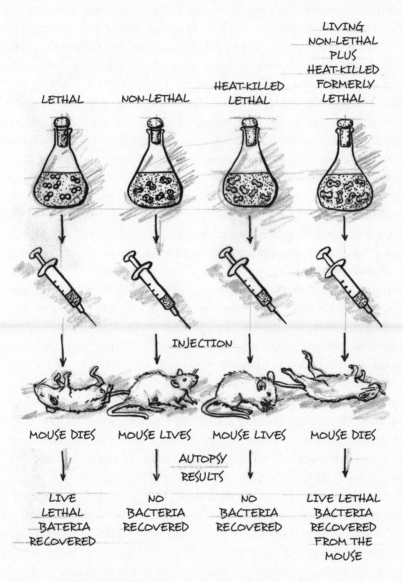

Figure 3.4. Frederick Griffith's injected mice experiment.

being transferred to the living strain, making it suddenly lethal. With the complicating variable of the mice out of the picture, the way was clear for Avery to locate the culprit; he set out to isolate the material responsible for this surprising transformation. In 1944, Avery and two of his colleagues, Colin MacLeod and Maclyn McCarty, published their findings in the *Journal of Experimental Medicine*. What was the transforming agent? To everyone's surprise, including Avery's, it appeared to be Levene's seemingly uninteresting nucleic acid, DNA. DNA from the dead strain was being transferred to the living strain, rendering the once harmless living strain suddenly lethal.[13]

When Erwin Chargaff, of Columbia University, read Avery's paper, he immediately sensed its importance. He saw "in dark contours the beginning of a grammar of biology," he recounted. "Avery gave us the first text of a new language or rather he showed us where to look for it. I resolved to search for this text."[14]

Chargaff's experimental work eventually provided additional evidence that DNA could be the source of biological variability. It also provided an important clue about the structure of the DNA molecule. Chargaff purified samples of DNA and then separated its chemical constituents, the sugars, the phosphates, and the four bases. Using techniques of quantitative chemical analysis, he determined the relative proportions of each of these constituents and, in the process, discovered a puzzling regularity. The quantity of adenine always equaled the quantity of thymine, while the quantity of the guanine always equaled the quantity of cytosine.[15]

This discovery was puzzling, in part because Chargaff also discovered a surprising irregularity—one that contradicted the earlier "tetranucleotide hypothesis" of Levene. Chargaff discovered that individual nucleotide frequencies actually differ between species, even if they often hold constant within the same species or within the same organs or tissues of a single organism.[16] More important, Chargaff recognized that even for nucleic acids with the same proportion of the four bases (A, T, C, and G), "enormous" numbers of variations in sequence were possible. As he put it, different DNA molecules or parts of DNA molecules might "differ from each other . . . in the sequence, [though] not the proportion, of their constituents."[17]

In other words, a strand of DNA might be like a strand of binary computer code. The information-rich string of zeros and ones will have a completely irregular and nonrepetitive sequence of the two characters, but given a fairly long string of the binary code, one can expect to find very close to the same number of zeros as ones. Thus Chargaff argued that, contrary to the tetranucleotide hypothesis, base sequencing in DNA might well display the high degree of variability and irregularity required by any potential carrier of heredity.[18] And so, by the late 1940s many had begun to suspect that DNA was a good candidate for transmitting hereditary information. But still, no one knew how it did this.

Unlikely Heroes

James Watson crossed an ocean to find out. As a twenty-three-year-old who had once spent much of his spare time bird watching, he did not appear, on casual inspection, to possess either the experience or the fire to solve the mystery. Yet there was more to the seemingly easygoing young man than met the eye. Watson was a former Chicago *Quiz Kid* game-show prodigy who had entered the University of Chicago at the tender age of fifteen. By nineteen, he had finished his bachelor's degree in biology. By twenty-two, he had earned a Ph.D. from the University of Indiana under Salvador Luria, an expert on the genetics of viruses. Watson's doctoral studies focused on viral genetics, but along the way he learned a fair amount of biochemistry and radiation genetics. At one point, he took a course from Hermann J. Muller, of fruit fly fame. Though he "aced" the course, Watson concluded that the best days for gene-radiation studies were past. New methods were needed to get DNA to divulge its secrets.[19]

After he graduated, Watson was almost constantly ruminating about biology and keeping his ears open for any word about new ways to study DNA. He traveled to Copenhagen for postdoctoral research. While there, he performed experiments alongside Danish scientist Ole Maaløe that strengthened his growing sense that DNA, not protein, was the carrier of genetic information. Then in the spring of 1951, at a conference in Naples on X-ray crystallography, he met Maurice Wilkins, the head of the lab at King's College, London. From their conversations, Watson was seized with the idea of moving to Cambridge, England, where various experts were gathering to discover the secret of heredity.[20]

He landed a position with the Cavendish Laboratory in 1951 under the Austrian Max Perutz and Englishman William Lawrence Bragg, both distinguished experts in the use of X rays to study large biological molecules. Watson quickly formed a partnership with Francis Crick, a theoretical physicist who knew precious little chemistry, but who had used advanced mathematics to develop theoretical insights about how to study the structure of proteins by using X rays.[21] In Crick, Watson found a partner who shared his thinking about DNA. Both men were interested in genetics, but both thought that a deeper understanding of heredity would emerge only after scientists understood "what the genes were and what they did."[22] For Watson, at least, that meant understanding the structure of DNA. And he soon convinced Crick that cracking this problem would make it possible to understand the transmission of genetic information. Crick's expertise in understanding protein structure and X-ray imaging techniques would come in handy, as would his knack for taking insights from disparate fields and finding significant patterns in them that other, more specialized scientists missed.

Watson and Crick also possessed an important quality rarely appreciated in scientists, but vital to those attempting to make discoveries or challenge an outmoded framework of thought. The two men were perfectly willing to ask questions exposing their own ignorance, to embarrass themselves, if necessary, in pursuit of answers.[23] They had no world-class reputations to risk and little to lose as they pursued answers unfettered by concerns about their own respectability. Nor could they be shamed into pouring their energies into gathering original data when what was needed was new thinking. They would leave data gathering to others while they focused on the big picture, constantly reassembling the pieces of a growing puzzle in search of an elegant and illuminative synthesis.[24]

In 1951, after only a few months working on the problem, Watson and Crick presented their first model in a seminar at the Cavendish Laboratory in Cambridge. Maurice Wilkins, Rosalind Franklin, and two other scientists from King's College, London, attended, as did Professor Lawrence Bragg, their supervisor at the Cavendish. The meeting went poorly. Watson and Crick represented DNA as a triple-stranded helix. Franklin objected. Though the sugar-phosphate backbone of the molecule might form a helix, she insisted, there was not yet "a shred of evidence" for this idea from X-ray studies.[25] Such ideas were merely in the air as the result of recent discoveries about the presence of helical structures in proteins.

Other aspects of their model were more obviously mistaken. Watson had miscalculated the density of water in the DNA molecule. (The amount of water absorbed by DNA determines its dimensions and whether it will adopt one of two structures, the "A-form" or "B-form.") When the correct water density was used to calculate the spatial dimensions of DNA, the justification for their model dissolved. Watson and Crick also had placed the sugar phosphate backbone in the center of the molecule with the bases sticking outward, thus producing a structure that looked something like a gnarled tree with stubby branches. Franklin correctly pointed out that DNA could take up water as easily as it did only if the phosphate groups were on the outside, not the inside, of the structure. The phosphate groups had to be on the outside of the model where they could attract and hold water easily.[26]

Embarrassed for them, if not for himself, Professor Bragg asked Watson and Crick to stop work on DNA. Crick was to finish his Ph.D. dissertation; Watson was assigned to study viruses.

Collecting the Clues

Undaunted by their failure, the two gradually wormed their way back into the action. Together they mused and snooped, tinkered with toy models, and picked the brains of the various specialists in England and abroad. By 1952 a growing number of scientists had set aside the distraction of protein molecules and were focusing squarely on deoxyribonucleic acid, including two-time Nobel laureate Linus Pauling, the Caltech chemist who had earlier determined the form of an important structure within proteins, the alpha-helix. Watson and Crick sensed that time was short. At any moment somebody might beat them to the discovery of the structure of the mystery molecule. But that fear raised an exhilarating possibility: perhaps collectively the scientific community knew enough already. Perhaps it wasn't fresh evidence that was needed, but rather a flash of insight about how all the pieces of evidence fit together.

While others approached the problem methodically, steadily gathering data in their labs, Watson and Crick behaved more like gumshoe detectives, heading from one place to another looking for clues to help them weigh the merits of competing hypotheses. Then on January 28, 1953, Watson procured a draft copy of a scientific manuscript written by Linus Pauling from Pauling's son, Peter, who was working in Cambridge as a Ph.D. student under John Kendrew.[27] The senior Pauling was proposing a triple-helix structure for DNA that was very similar to the model Watson and Crick had proposed the year before. He sent the draft to Peter, who showed it to Watson and Crick.

Like Watson and Crick's original model, Pauling's envisioned a triple-sugar-phosphate backbone running up the middle (or inside) of the molecule with the nucleotide bases attached on the outside and pointing outward. Watson exhaled in relief. He knew all too well that Pauling's model couldn't be right. But what was? Two days later, with Pauling's manuscript in hand, he traveled to the lab at the University of London to see what he could turn up.[28] The King's College lab was the place to go to see the best current X-ray images of DNA.

Although the lab employed a cutting-edge technology, conceptually the essence of its technique was a simple old trick from physics: throw something at an object and see what bounces back or passes through. Then collect the resulting signal and see what you can tell about the object under study by analyzing it. Bats navigate the airways using this technique. Their echo-location system sends sound waves at objects and then times the returning signals so the bats can "see," or locate the objects around them. As noted in Chapter 1, geophysicists use a similar technique. They send sound waves deep underground and then collect the resulting echoes in order to image the subsurface of the earth.

The key technology in the search for the structure of DNA employed a variation of this strategy as well. Instead of projecting sound waves at DNA, the scientists at King's College directed X rays at fibers of DNA. And instead of analyzing what bounced back, they collected the rays that passed through the molecules. By seeing how the direction of the X rays changed—how they were diffracted by their target—these scientists were eventually able to learn about the structure of DNA.

Rosalind Franklin was the acknowledged expert on this technique for studying DNA. That's why Watson was there. Franklin had earlier discovered that DNA has two distinct forms with different dimensions depending upon whether water is present. This in itself was a huge breakthrough, since previously the two forms were jumbled together, muddying the results anytime someone tried to use X-ray diffraction to discern the form of DNA. Armed with this fresh insight, Franklin set about developing a method for separating the two forms. The technique she arrived at was highly demanding, but also highly effective. Now she could and did obtain revealing diffraction patterns of the "B-form DNA." When Watson showed up at King's lab, he had a tense conversation with Franklin. He lectured her about helical theory and why DNA

Figure 3.5. Portrait of Rosalind Franklin (1920–58). *Courtesy of Science Source/ Photo Researchers.*

must be a helix. Franklin insisted that there was no proof of that yet. Franklin rose in anger from behind her lab bench, visibly annoyed at Watson's presumption and condescension. Watson made a hasty retreat from Franklin's lab, later saying he feared that she might strike him.[29]

But before leaving the lab, Watson stopped by to see Maurice Wilkins. After a little prodding, Wilkins gave Watson a look at Franklin's best X ray of DNA in the B-form. The picture showed, quite distinctly, a pattern known as a Maltese cross (see Fig. 3.6). Watson was elated. What Crick had taught him told him he was looking at evidence of a helix. On the train ride home, Watson sketched the image from memory.[30] Upon seeing Watson's sketch and after questioning Watson, Crick agreed it must be a helix. But what kind, and how did the chemical constituents of DNA fit into this structure?

Clues gleaned from another quarter would ultimately help Watson and Crick answer these questions. The year before, they had shared a meal with the gruff, brilliant chemist Erwin Chargaff while he was visiting in Cambridge. During the meal they asked a number of questions that exposed

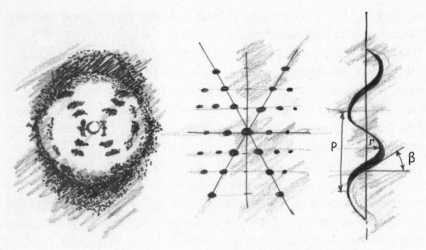

Figure 3.6. Maltese cross X-ray image of DNA crystal. Helix on the right represents the structure that scientists think produces the Maltese cross X-ray image.

their own ignorance of some of the relevant chemistry. In particular, it became apparent during the meal that they did not know about Chargaff's famous correspondences, or "rules," establishing that the amount of guanine in DNA equaled the amount of cytosine (G = C) and the amount of adenine equaled the amount of thymine (A = T).[31] The eminent biochemist, who at the time knew virtually everything there was to know about the chemistry of DNA, was aghast that this breezy, ambitious pair didn't know these basics. Chargaff even got Crick to admit that he did not know the differences in the chemical structure of the four nucleotide bases in the DNA molecule, which by then was common knowledge to everyone else in the race.[32]

Chargaff would later muse about the irony of these seemingly uninitiated scientists making the discovery of the ages: "I seem to have missed the shiver of recognition of a historical moment; a change in the rhythm of the heartbeats of biology. As far I could make out, they wanted, unencumbered by any knowledge of the chemistry involved, to fit DNA into a helix. I do not remember whether I was actually shown their scale model of a polynucleotide chain, but I do not believe so, since they still were unfamiliar with the chemical structures of the nucleotides."[33] Despite his contempt for this ambitious pair of know-nothings, or perhaps because of it, Chargaff passed on to them the cor-respondences he had discovered.[34]

So by the time Watson returned from his visit to Franklin and Wilkins at the end of January 1953, he and Crick knew several key facts about nucleic acid. In addition to Chargaff's rules, they knew from Franklin's X-ray images that DNA almost certainly did form a helix with a back-bone made of sugars linked to phosphates. From X-ray studies of the molecule, they also knew the key dimensions of the B-form of the DNA molecule—20 angstroms across and 34 angstroms long for one full turn of the helix (an angstrom is the length of one hydrogen atom, about one ten-billionth of a meter). And they knew that Franklin was convinced that the sugar-phosphate backbone would have to have the phosphates on the outside.[35]

They also knew they had competition. Watson's trip to London no doubt had been partly motivated by his serendipitous acquisition of Linus Pauling's manuscript. Though Watson was enormously relieved that Pauling had blundered, he suspected that it would not take long for Pauling to discover his mistakes. Franklin also knew, immediately upon reading Pauling's proposal, that it was wrong. Pauling had proposed a structure in which the sugar-phosphate strands of the helix ran through the center of the molecule. Thus, his model required the bases to stick out horizontally from the rising helices. This meant that the molecule would not define a smooth or definite edge, but instead a jagged series of knobby protrusions. Yet Franklin knew that the X-ray data showed that the molecule had a definite 20 angstrom diameter. Such a precise measurement could never have been established if the molecule had the kind of uneven edge Pauling envisioned.[36]

Pauling's model also failed to account for Chargaff's correspon-dences. And it had density problems. With three helices running right through the middle of the molecule, the density of the atoms in the center was too high to be consistent with available data. X-ray and chemical studies had revealed the number of bases (about 10) present in the molecule per turn of a single helix. A triple helix should, there-fore, have a proportionately higher density of bases per unit length than a double helix. When Watson discovered that density measurements (bases per unit length of a fiber of DNA) agreed more closely with the values calculated for a double helix than a triple helix, he and Crick not only rejected Pauling's triple-helix model, but all triple helices. DNA, they were convinced, was much more likely to be some sort of double helix.[37] They were much closer to a solution, but so too were the others in the race. Watson and Crick sensed that time was short, and they had spacing problems of their own.

Models and Molecules

To solve the puzzle, Watson began in haste to build a series of models. He first tried to make a double helix with the helices again running up the center of the molecule, ignoring Franklin's earlier insistence about the need for keeping the phosphates exposed on the outside of the molecule. Crick recalls that they persisted in this approach because they thought the demands of biology required it. If DNA was, indeed, the molecule of hereditary information and the arrangement of bases conveyed genetic instructions, then the structure of DNA must allow some way of copying these instructions. Putting the bases on the outside made it easier to envision how the bases were copied during cell division. Additionally, placing them on the outside of the molecule eliminated the need to figure out how the bases fit together within the strands. "As long as the bases were on the outside, we didn't have to worry about how to pack them in," said Crick.[38]

Unfortunately, an outside placement made it more difficult to explain other facts. First, a jagged bases-on-the-outside model would not produce consistent 20-angstrom-diameter X-ray measurements—a point Franklin had made earlier. Also, positioning the bases on the outside of the sugar-phosphate backbone made it difficult to explain how DNA could form crystals at all. Crystals form when repeating chemical structures pack together in what is called a lattice. DNA crystals form when long strands of (A-form) DNA molecules align themselves alongside each other. To get a visual image of what a crystal of DNA would look like, imagine a series of parallel lanes at a drag racing track with each lane of equal width. Now instead of those lanes being perfectly straight, imagine them curving back and forth in parallel to each other all the while maintaining that even spacing.

The jagged edged molecule Watson envisioned in his bases-on-the-outside model was not conducive to the formation of regularly spaced, parallel strands of DNA. The jagged edges might stick together, but only in an irregular tongue-in-groove sort of way. Yet the ability to study DNA structure with X rays is possible only because of the regular spacing DNA molecules exhibit when aligned in parallel.

Placing the bases on the outside and the backbone on the inside also made it more difficult to produce a structure that matched known measurements of the length between the bases. X-ray measurements indicated that there are about 3.4 angstroms of space between each base on the backbone. But running the backbone up the middle of the molecule, as opposed to along the outside edge, generates a helix with a steeper pitch (angle of ascent). Watson soon found that the pitch of his internal backbone model created too much space between the sugar groups (and thus the bases) to match the 3.4-angstrom measurement derived from X-ray studies. As Watson worked with the model, he also realized that the chemical properties of the backbone limited how much it could be compressed or stretched to accommodate definite spacing requirements. Further, as historian of science Robert Olby recounts, the structure as a whole looked "awful."[39]

A change of direction was clearly needed. Watson began to try models with the backbones on the outside. Initially, things went no better. But then two breakthroughs came.

"Aha" Times Two

The first breakthrough came as a flash of insight to Francis Crick. As Watson began trying to fit the bases into the interior of the helices, leaving the backbone on the outside of the model molecule, he still encountered difficulties. Watson didn't know, however, that how he oriented the two sugar-phosphate backbones in relation to each other would matter. He knew that there is a structural asymmetry in the construction of the backbone of the DNA molecule. The phosphate groups attach to a different carbon molecule on one side of the ribose sugar than they do on the other. On one side of the ring structure of the sugar, the phosphate attaches to what is called its 5' carbon (five prime carbon); on the other, the phosphate attaches to what is called its 3' carbon (see Fig. 3.7). Chemists know these differences well, and so did Watson.

But as Watson was constructing his models, he had assumed that the two backbones should run *in the same direction*, parallel to each other with each backbone starting from the 5' end and running to the 3' end

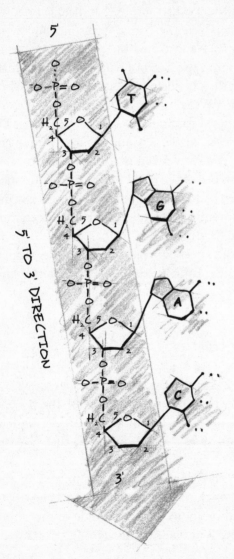

Figure 3.7. The sugar phosphate backbone of the DNA showing how phosphates are bonded to the 5' carbon of one sugar and the 3' carbon of another. The genetic text (sequence of bases) is transcribed in the 5' to 3' direction of the DNA molecule.

of the chain. So try as he might, he could not get the bases to lock into place—to connect with each other—in a stable structure *across* the molecule. Then a crucial insight came.

Crick, who was studying the structure of proteins in his Ph.D. dissertation, was familiar with helical structures in proteins. He knew that sometimes sections of the alpha helices of a protein will run "antiparallel" to each other. That is, one section of a helix in a protein will run up, then turn back the other direction and run down right alongside the first section of the helix, making an upside-down "U-shape." One day while reading a highly technical description of the X-ray diffraction pattern of the B-form of DNA in a Medical Research Council report summarizing Rosalind Franklin's data,[40] Crick realized that the pattern she was describing—called a "monoclinic C2 symmetry" pattern—was indicative of an antiparallel double helix structure. Instead of the two strands running from 5' to 3' in the same direction, Crick realized that Franklin's data indicated that one helix was running *up* in the 5' to 3' direction and the other strand was running *down* in the 5' to 3' direction *the other way.*

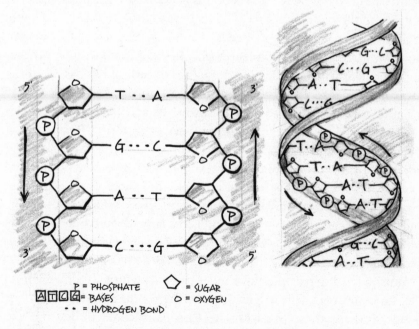

P = PHOSPHATE   ⬠ = SUGAR
🄰🄃🄲🄶 = BASES   O = OXYGEN
• • = HYDROGEN BOND

Figure 3.8. Antiparallel strands of DNA entwined around each other. Represented in two dimensions (*left*) and three dimensions (*right*).

He also realized by studying the dimensions indicated by the X-ray data that these antiparallel backbones were not separate molecules or separate stretches of the same molecule bent back on itself as occurred in proteins. Instead, these were two separate strands entwined around each other. They were simply too close together to be anything else (see Fig. 3.8).

Crick's insight was possible only because of his specialized expertise in crystallography. At the time, it was nearly beyond Watson to understand Crick's argument, and he apparently took quite a long time to get it. But once he did and once he started using antiparallel sugar-phosphate backbones, his model building would proceed in an entirely more satisfying and fruitful direction—though he would first need to have a flash of insight of his own.

Eureka!

Watson resumed his model building in late February 1953 using cardboard cutouts of his own construction as he waited for the metal pieces to come from the machine shop.[41] The challenge he faced was to find a way to fit these four irregularly shaped bases into the available space between the two twisting sugar-phosphate backbones and to do so in a way that was consistent with the chemical properties of each of the bases and the sugar molecules to which the bases must attach. He had studied every scientific paper he could about the dimensions and chemical proclivities of the bases. Watson learned that the bases often form a weak kind of chemical bond with each other called a hydrogen bond. Perhaps he could bridge the gap between the two backbones with two bases connected by hydrogen bonds. This seemed plausible. But there was a complicating factor. The bases weren't the same size. Two of the bases, adenine and guanine, called the purines, exhibited a large double-ring structure, while the two others, cytosine and thymine, called pyrimidines, exhibited a smaller single-ring structure (see Fig. 3.9).

Watson tried to pair the two similar bases with each other. He found that he could get them to join together by forming double hydrogen bonds. Perhaps these like-with-like base pairs could bridge the gap between the two backbones. Indeed, he could envision them forming rungs on the twisting ladder.

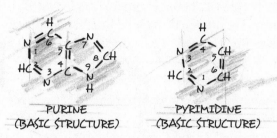

PURINE
(BASIC STRUCTURE)

PYRIMIDINE
(BASIC STRUCTURE)

Figure 3.9. The basic chemical structures of purines and pyrimidines. Notice that pyrimidines have a single-ring structure and purines have a double-ring structure. The bases adenine and guanine are purines. Thymine and cytosine are pyrimidines. Since complementary base pairing always involves one pyrimidine and one purine, the rungs on the DNA ladder formed by these pairs are closely equivalent in size and shape, whether they are adenine-thymine pairs or guanine-cytosine pairs.

But there were problems. Watson recognized that if he used the like-with-like pairs, he would have a hard time producing a helix with a consistent diameter. The pairs of large purine bases would cause the helix to bulge in some places, while the pairs of the smaller pyrimidines would cause it to narrow in others. And there was another difficulty. Lab associate Jerry Donohue, who knew a lot about the chemical structure and properties of the nitrogenous bases, insisted that Watson was mistaken.[42] He was. Most of the nitrogenous bases come in either one of two variant forms, a common "keto" form and a rare "enol" form. These two forms have slightly different structural formulas, reflecting slight differences in the configurations of their constituent atoms. For three of the four bases, Watson had unknowingly used the rare alternative form of the bases, instead of the common "keto" form that is actually present in DNA. Deferring to Donohue's expertise, Watson tried again using the correct (keto) forms of the bases. But this made hydrogen bonding much more difficult for like-with-like pairings.[43]

That's when Watson's insight came. He remembers that it occurred the morning of February 28, 1953. Still waiting for the tin models to arrive from the Cavendish shop, he cleared the papers off the desktop and, constructing a makeshift set of cardboard models, he began trying to fit the pieces together this way and that. Initially, he reverted to his "like-with-like prejudices," but very quickly he realized this was a dead end. He began to try other pairing schemes: "Suddenly I became aware that an adenine-thymine pair held together by two hydrogen bonds was identical in shape to a guanine-cytosine pair held together by at least two hydrogen bonds. All the hydrogen bonds seemed to form naturally; no fudging was required to make the two types of base pairs identical in shape."[44]

The AT pair formed a rung across the helix that was the exact length and shape of the GC pair. He placed these pairs inside the two backbones. Now the twisting ladder maintained a consistent diameter. No bulging. No pinching. Everything fit.

The Double Helix

Watson now saw the double helix rising before him and marveled at how "two irregular sequences of bases could be regularly packed in the center of a helix."[45] This new structure also explained Chargaff's rules. If A bonded only to T and if C bonded only to G, then of course the amount of A would always equal T and the amount of C would always equal G. It was also so easy to see the two strands parting and new As, Ts, Gs, and Cs tumbling into place, seeking out hydrogen bonds with their natural opposites on the single strands. Soon each single strand would be its own double strand and, presto, the winding staircase structure would replicate itself. Crick arrived at the lab and quickly discovered that Watson's revised structure explained several other significant facts of DNA chemistry as well.[46]

The new model was beautiful and fit the data exceptionally well. Crick later observed that he and Watson "were looking for gold" and by "blundering about" had found the gold they sought.[47] At lunch, Crick was telling everyone at the Eagle pub, around the corner from the Cavendish on Free School Lane in Cambridge, that they had found the secret of life.[48] On April 25, 1953, a seemingly modest paper appeared in the journal *Nature*. The article was only nine hundred words long, was signed by a pair of unknowns, J. D. Watson and F. H. C. Crick, and featured the anodyne title "Molecular Structure of Nucleic Acids: A Structure for Deoxyribose Nucleic Acid." It revolutionized biology.

Figure 3.10. Watson and Crick presenting their model of the DNA double helix. *Courtesy of A. Barrington Brown/Photo Researchers, Inc.*

The Watson-Crick model made it clear that DNA had an impressive chemical and structural complexity. It was a very long molecule composed on the outside of a regular arrangement of sugar and phosphate groups. But on the inside it could contain many potentially different arrangements of the four bases. Thus, it had an impressive potential for variability and complexity of sequence as required by any potential carrier of hereditary information. As Watson and Crick later explained: "The phosphate-sugar backbone of our model is completely regular, but any sequence of the pairs of bases can fit into the structure. It follows that in a long molecule, many different permutations are possible, and it therefore seems likely that the precise sequence of the bases is the code which carries the genetic information."[49] Thus, their paper not only described the structure of DNA; it also anticipated what later discoveries would confirm: DNA was a repository of information.

Watson and Crick's discovery would forever change our understanding of the nature of life. At the close of the nineteenth century, most biologists thought life consisted solely of matter and energy. But after Watson and Crick, biologists came to recognize the importance of a third fundamental entity in living things: information. And this discovery would redefine, from that point forward, what theories about the origin of life would need to explain.

4

Signature in the Cell

In Darwin's time few, if any, biologists talked about biological or genetic information, but today they routinely refer to DNA, RNA, and proteins as carriers or repositories of information. Biologists tell us that DNA stores and transmits "genetic information," that it expresses a "genetic message," that it stores "assembly instructions," a "genetic blueprint," or "digital code."[1] Biology has entered its own information age, and scientists seeking to explain the origin of life have taken note. Life does not consist of just matter and energy, but also information. Since matter and energy were around long before life, this third aspect of living systems has now taken center stage. At some point in the history of the universe, biological information came into existence. But how? Theories that claim to explain the origin of the first life must answer this question.

But what exactly is information? What is *biological* information? Beginning in the late 1940s, mathematicians and computer scientists began to define, study, measure, and quantify information. But they made distinctions between several distinct types or conceptions of information. What kind of information does DNA have? What kind of information must origin-of-life researchers "explain the origin of"?

As we will see, it is important to answer these questions because DNA contains a particular kind of information, one that only deepens the mystery surrounding its origin.

Defining Information: Two Distinctions

Most of us use the term "information" to describe some piece of knowledge. When we say so-and-so passed on some interesting information, we mean that so-and-so told us something that we didn't know before, but that we now know, thanks to what we were told. In other words, information equals knowledge. The first definition of information in Webster's dictionary reflects this idea: information is "the communication or reception of knowledge or intelligence." Because many of my students had this idea of information firmly in mind, they were often confused at first when I talked about information stored in a molecule. There is a sense in which it could be said that DNA stores the "know-how" for building molecules in the cell. Yet since neither DNA nor the cellular machinery that receives its instruction set is a conscious agent, equating biological information with knowledge in this way didn't seem to quite fit.

But our English dictionaries point to another common meaning of the term that does apply to DNA. Webster's, for instance, has a second definition that defines information as "the attribute inherent in and communicated by alternative sequences or arrangements of something that produce specific effects." Information, according to this definition, equals an arrangement or string of characters, specifically one that accomplishes a particular outcome or performs a communication function. Thus, in common usage, we refer not only to a sequence of English letters in a sentence, but also to a block of binary code in a software program as information. Information, in this sense, does not require a conscious recipient of a message; it merely refers to a sequence of characters that produces some specific effect. This definition suggests a definite sense in which DNA contains information. DNA contains "alternative sequences" of nucleotide bases and can produce a specific effect. Of course, neither DNA nor the cellular machinery that uses its information is conscious. But neither is a paragraph in a book or a section of software (or the hardware in the computer that "reads" it). Yet clearly software contains a kind of information.

This seemed reasonably clear to me in 1985 after I first began to think about the DNA enigma. But at the time something else puzzled me. I had been working in a technical field in which the processing of

Figure 4.1. Portrait of the U.S. mathematician and information theorist Claude Shannon (1916–2001).
Courtesy of the Estate of Francis Bello/Photo Researchers, Inc.

information on computers was part of our stock and trade. I was familiar with the science of information storage, processing, and transmission called "information theory." Information theory was developed in the 1940s by a young MIT engineer and mathematician named Claude Shannon. Shannon was studying an obscure branch of algebra and, not surprisingly, few people were paying any attention. He would later gain notoriety as the inventor of the rocket-powered Frisbee and for juggling four balls while cruising down the halls of Bell Laboratories on a unicycle, but in 1937 he was putting the finishing touches on a master's thesis that, to some, may have seemed hopelessly dull. Shannon had taken nineteenth-century mathematician George Boole's system of putting logical expressions in mathematical form and applied its categories of "true" and "false" to switches found in electronic circuits.

"I've always pursued my interests without much regard to financial value or value to the world," Shannon confessed in 1983. "I've spent lots of time on totally useless things."[2] His master's thesis, however, wasn't one of them. Called "possibly the most important, and also the most famous, master's thesis of the century,"[3] it eventually became the foundation for digital-circuit and digital-computer theory. Nor was he finished laying these foundations. He continued to develop his ideas and eventually published "A Mathematical Theory of Communication." *Scientific American* later called it "the Magna Carta of the information age."[4]

Shannon's theory of information provided a set of mathematical rules for analyzing how symbols and characters are transmitted across communication channels. After the conference in 1985 awakened my interest in the origin of life, I began to read more about Shannon's theory of information. I learned that his mathematical theory of information applied to DNA, but there was a catch.

Shannon's Information Theory

Shannon's theory of information was based upon a fundamental intuition: information and uncertainty are inversely related. The more informative a statement is, the more uncertainty it reduces.

Consider this illustration. I live in Seattle, where it rains a lot, especially in November. If someone were to tell me that it will rain in November, that would not be a very informative statement. I am virtually certain that it will rain then. Saying so does not reduce uncertainty. On the other hand, I know very little about what the weather will be like in Seattle on May 18. I'm uncertain about it. If a time-traveling weather forecaster with a flawless track record were to tell me that on May 18 next year Seattle will have an unseasonably cold day resulting in a light dusting of snow, that would be an informative statement. It would tell me something I couldn't have predicted based upon what I know already—it would reduce my uncertainty about Seattle weather on that day.

Claude Shannon wanted to develop a theory that could quantify the amount of information stored in or conveyed across a communication channel. He did this first by linking the concepts of information and uncertainty and then by linking these concepts to measures of probability.

According to Shannon, the amount of information conveyed (and the amount of uncertainty reduced) in a series of symbols or characters is inversely proportional to the probability of a particular event, symbol, or character occurring. Imagine rolling a six-sided die. Now think about flipping a coin. The die comes up on the number 6. The coin lands on tails. Before rolling the die, there were six possible outcomes. Before flipping the coin, there were two possible outcomes. The cast of the die thus eliminated more uncertainty and, in Shannon's theory, conveyed more information than the coin toss. Notice here that the more improbable event (the die coming up 6) conveys more information. By equating information with the reduction of uncertainty, Shannon's theory implied a mathematical relationship between information and probability. Specifically, it showed that the amount of information conveyed by an event is *inversely* proportional to the probability of its occurrence. The greater the number of possibilities, the greater the *im*probability of any one being actualized, and thus the more information that is transmitted when a particular possibility occurs.

Shannon's theory also implied that information increases as a sequence of characters grows, just as we might expect. The probability of getting heads in a single flip of a fair coin is 1 in 2. The probability of getting four heads in a row is $1/2 \times 1/2 \times 1/2 \times 1/2$, that is, $(1/2)^4$ or $1/16$. Thus, the probability of attaining a specific sequence of heads and tails decreases as the number of trials increases. And the quantity of information increases correspondingly.[5] This makes sense. A paragraph contains more information than the individual sentences of which it is composed; a sentence contains more information than the individual words in the sentence. All other things being equal, short sequences have less information than long sequences. Shannon's theory explained why in mathematical terms: improbabilities multiply as the number of characters (and combination of possibilities) grows.

For Shannon, the important thing was that his theory provided a way of measuring the amount of information in a system of symbols or characters. And his equations for calculating the amount of information present in a communication system could be readily applied to any sequence of symbols or coding system that used elements that functioned like alphabetic characters. Within any given alphabet of x possible characters (where each character has an equal chance of occurring), the probability of any one of the characters occurring is 1 chance in x. For instance, if a monkey could bang randomly on a simplified typewriter possessing only keys for the 26 English letters, and assuming he was a perfectly random little monkey, there would be 1 chance in 26 that he would hit any particular letter at any particular moment.

The greater the number of characters in an alphabet (the greater the value of x), the greater the amount of information conveyed by the occurrence of a specific character in a sequence. In systems where the value of x is known, as in a code or language, mathematicians can generate precise measures of information using Shannon's equations. The greater the number of possible characters at each site and the longer the sequence of characters, the greater the Shannon information associated with the sequence.

What Shannon's Theory Can't Say

But, as I said, there was a catch. Though Shannon's theory and his equations provided a powerful way to measure the amount of information stored in a system or transmitted across a communication channel, it had important limits. In particular, Shannon's theory did not, and could not, distinguish merely improbable sequences of symbols from those that conveyed a message or "produced a specific effect"— as Webster's second definition puts it. As one of Shannon's collaborators, Warren Weaver, explained in 1949, "The word *information* in this theory is used in a special mathematical sense that must not be confused with its ordinary usage. In particular, information must not be confused with meaning."[6]

Consider two sequences of characters:

"Four score and seven years ago"
"nenen ytawoi jll sn mekhdx nnx"

Both of these sequences have an equal number of characters. Since both are composed of the same 26-letter English alphabet, the amount of uncertainty eliminated by each letter (or space) is identical. The probability of producing each of those two sequences at random is identical. Therefore, both sequences have an equal amount of information as measured by Shannon's theory. But one of these sequences communicates something, while the other does not. Why?

Clearly, the difference has something to do with the way the letters are arranged. In the first case, the letters are arranged in a precise way to take advantage of a preexistent convention or code—that of English vocabulary—in order to communicate something. When Abraham Lincoln wrote the first words of the Gettysburg address, he arranged the letters—f-o-u-r—s-c-o-r-e—a-n-d—s-e-v-e-n—y-e-a-r-s—a-g-o—in a specific sequence. When he did so he invoked concepts—the concept of "four," the concept of "years," and so on—that had long been associated with specified arrangements of sounds and characters among English speakers and writers. The specific arrangement he chose allowed those characters to perform a communication function. In the second sequence, however, the letters are not arranged in accord with any established convention or code and, for that reason, are meaningless. Since both sequences are composed of the same number of equally improbable characters, both sequences have a quantifiable amount of information as calculated by Shannon's theory. Nevertheless, the first of the two sequences has something—a specificity of arrangement—that enables it "to produce a specific effect" or to perform a function, whereas the second does not.

And, *that's* the catch. Shannon's theory cannot distinguish functional or message-bearing sequences from random or useless ones. It can only measure the improbability of the sequence as a whole. It can quantify the amount of functional or meaningful information that *might be present* in a given sequence of symbols or characters, but it cannot determine whether the sequence in question "produces a specific effect." For this reason, information scientists often say that Shannon's theory measures the "information-carrying capacity," as opposed to the functionally specified information or "information content," of a sequence of characters or symbols. This generates an interesting paradox. Long meaningless sequences of letters can have more information than shorter meaningful sequences, as measured by Shannon's information theory.

All this suggested to me that there are important distinctions to be made when talking about information in DNA. In the first place, it's important to distinguish information defined as "a piece of knowledge known by a person" from information defined as "a sequence of characters or arrangements of something that produce a specific effect." Whereas the first of these two definitions of information doesn't apply to DNA, the second does. But it is also necessary to distinguish Shannon information from information that performs a function or conveys a meaning. We must distinguish sequences of characters that are (a) merely improbable from sequences that are (b) improbable and also *specifically* arranged so as to perform a function. That is, we must distinguish information-carrying capacity from functional information.

So what kind of information does DNA possess, Shannon information or some other? To answer this question we will need to look at what molecular biologists have discovered since 1953 about the role of DNA within the miniature world of the cell.

Genetic Information: What Does It Do?

In the wake of Watson and Crick's seminal 1953 paper, scientists soon realized that DNA could store an immense amount of information. The chemistry of the molecule allowed any one of the bases to attach to any one of the sugar molecules in the backbone, allowing the kind of variable sequencing that any carrier of genetic information must have. Further, the weak hydrogen bonds that held the two antiparallel strands together suggested a way the molecule might unzip—like two opposing strips of Velcro—to allow the exposed sequence of bases to be copied. It seemed that DNA was ideal for storing information-rich sequences of chemical characters. But how DNA expressed this information and how it might be using that information remained uncertain. Answers would soon come, however, thanks to developments in the field of protein chemistry.

The Puzzle of Proteins

Scientists today know that protein molecules perform most of the critical functions in the cell. Proteins build cellular machines and structures, they carry and deliver cellular materials, and they catalyze chemical reactions that the cell needs to stay alive. Proteins also process genetic information. To accomplish this critical work, a typical cell uses thousands of different kinds of proteins. And each protein has a distinctive shape related to its function, just as the different tools in a carpenter's toolbox have different shapes related to their functions.

By the 1890s, biochemists had begun to recognize the centrality of proteins to the maintenance of life. They knew proteins were heavy ("high molecular weight") molecules that were involved in many of the chemical reactions going on inside cells. During the first half of the twentieth century, chemists were also able to determine that proteins were made of smaller molecules called amino acids. During this time, many scientists thought that proteins were so important that they, rather than DNA molecules, were the repositories of genetic information. Even so, until the 1950s, scientists repeatedly underestimated the complexity of proteins.

One reason for this mistaken view was a discovery made by a distinguished British scientist. William Astbury was an outstanding scientist with impeccable credentials. Formerly chosen "head boy" at a distinguished English boarding school, Astbury had studied physics at Cambridge during and after World War I. He then worked with William Bragg, the pioneering X-ray crystallographer whose son Lawrence later supervised Watson and Crick. Astbury had an infectious enthusiasm for laboratory work and a gift for persuasion. But he also had a reputation for sometimes letting his enthusiasm get the better of him and for jumping to premature conclusions.

Astbury was convinced that proteins held the key to understanding life. And partly for that reason, he thought they should exhibit a simple, regular structure that could be described by a mathematical equation or some general law. Perhaps his occupation biased him toward this view. As a crystallographer, he was used to studying highly regular and orderly structures. For example, salt crystals, the first structures determined using X-ray techniques, have a highly repetitive or regular structure of sodium and chlorine atoms arranged in a three-dimensional grid, a pattern in which one type of atom always has six of the other types surrounding it (see Fig. 4.2). Astbury was convinced that proteins—the secret of life—should exhibit a similar regularity. During the 1930s he made a discovery that seemed to confirm his expectation.

Astbury used X rays to determine the molecular structure of a fibrous protein called keratin, the key structural protein in hair and skin.[7] Astbury discovered that keratin exhibits a simple, repetitive molecular structure, with the same pattern of amino acids repeating over and over again—just like the repeating chemical elements in a crystal (or the

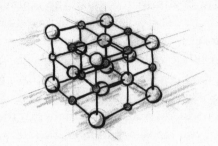

Figure 4.2. Three-dimensional structure of a salt crystal. Sodium atoms are represented by the dark-colored balls; the chlorine atoms by the light-colored balls.

repeating pattern of bases that P. A. Levene had mistakenly proposed as a structure for DNA). Astbury concluded that *all* proteins, including the mysterious globular proteins so important to life, would exhibit the same basic pattern he discovered in keratin. Many of Astbury's contemporaries shared the same view. In 1937, for example, two leading American biochemists, Max Bergmann and Carl Niemann, of the Rockefeller Institute, argued that the amino acids in all proteins occurred in regular, mathematically expressible proportions.[8]

During the early 1950s, about the same time that Watson and Crick were on the hunt to solve the structure of DNA, another Cambridge scientist made a discovery that would challenge the dominant view of proteins. While working just a couple of miles from Watson and Crick at the now famous Laboratory for Molecular Biology (or LMB), biochemist Fred Sanger determined the structure of the protein molecule insulin. Sanger's discovery would later earn him the first of two Nobel prizes in chemistry. Sanger showed that insulin consisted of irregular sequences of various amino acids, rather like a string of differently colored beads arranged with no discernible or repeating pattern.[9] Subsequent work on other proteins would show the same thing: the sequencing of amino acids is usually highly irregular and defies description by any general rule.[10]

But old prejudices die hard. Many biologists at the time still expected proteins, considered by many the fundamental unit of life, to exhibit regularity, if not in the arrangement of their amino acids, then at least in their overall three-dimensional shapes or structures. Most thought the three-dimensional structures of proteins would end up exhibiting some sort of geometric regularity. Some imagined that insulin and hemoglobin proteins, for example, would look like "bundles of parallel rods."[11] As Johns Hopkins biophysicist George Rose recounts, "Protein structure was a scientific *terra incognita*. With scant evidence to go on, biologists pictured proteins vaguely as featureless ellipsoids: spheres, cigars, Kaiser rolls."[12]

Then came the publication of John Kendrew's paper. Kendrew was the son of academics, an Oxford scientist and an art historian. He studied chemistry at Cambridge and graduated in 1939 just as World War II was beginning. After doing research on radar technology during the war, he took up the study of molecular biology at the Medical Research Council Laboratory in Cambridge in 1946. There Kendrew began to work closely with Max Perutz, the Austrian crystallographer who, along with Lawrence Bragg, had officially supervised Watson and Crick. In 1958, Kendrew made his own contribution to the molecular biological revolution when he published a paper on the three-dimensional structure of the protein myoglobin.

"The report in *Nature*," historian Horace Judson recalls, "was marked by a comical tone of surprise, of eyebrows raised at eccentricity."[13] Far from the simple structure that biologists had imagined, Kendrew's work revealed an extraordinarily complex and irregular three-dimensional shape, a twisting, turning, tangled chain of amino acids. Whereas protein scientists had anticipated that proteins would manifest the kind of regular order present in crystals (see Fig. 4.2), they found instead the complex three-dimensional structure shown in Figure 4.3. As Kendrew put it, with characteristic British understatement, "The big surprise was

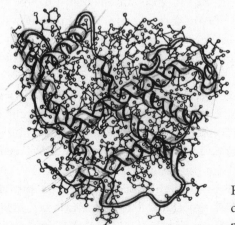

Figure 4.3. The complex three-dimensional structure of the protein myoglobin.

that it was so irregular."[14] In the *Nature* paper, he wrote, "Perhaps the most remarkable features of the molecule are its complexity and its lack of symmetry. The arrangement seems to be almost totally lacking in the kind of regularities which one instinctively anticipates, and it is more complicated than has been predicted by any theory of protein structure."[15]

But biochemists soon recognized that proteins exhibited another remarkable property. In addition to their complex shapes and irregular arrangements of amino acids, proteins also exhibit *specificity*. By specificity, biologists mean that a molecule has some features that have to be what they are, within fine tolerances, for the molecule to perform an important function in the cell.

Proteins are specified in two ways. First, proteins display a specificity of *shape*. The strangely irregular shapes of proteins that Kendrew and others discovered turned out to be essential to the function of the proteins. In particular, the three-dimensional shape of a protein gives it a hand-in-glove fit with other equally specified and complex molecules or with simpler substrates, enabling it to catalyze specific chemical reactions or to build specific structures within the cell. Because of its three-dimensional specificity, one protein cannot usually substitute for another. A topoisomerase can no more perform the job of a polymerase than a hatchet can perform the function of soldering iron or a hammer the job of a wrench.

Figure 4.4 illustrates how three-dimensional specificity of fit determines the function of a protein, in this case, an enzyme. Enzymes are proteins that catalyze specific chemical reactions. The figure shows an enzyme called a beta-galactosidase and a two-part sugar molecule (a disaccharide) called lactose. Because the enzyme's shape and dimensions exactly conform to the shape and dimensions of the disaccharide molecule, the sugar can nestle into the pockets of the enzyme. Once it does, a chemically active part of the enzyme, called an active site, causes a chemical reaction to occur. The reaction breaks the chemical bonds holding the two parts of the sugar together and liberates two individual molecules of glucose, each of which the cell can use easily.

Consider another example of how the specific shape of proteins allows them to perform specific functions. The eukaryotic cell has an uncanny way of storing the information in DNA in a highly compact

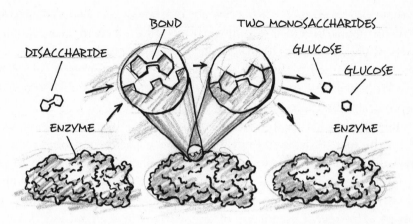

Figure 4.4. An enzyme (ß-D-galactosidase) breaking apart a two-part sugar molecule (a disaccharide). Notice the tight three-dimensional specificity of fit between the enzyme and the disaccharide at the active site where the reaction takes place. Notice also that the active site is small in relation to the size of the enzyme as a whole.

way. (*Eukaryotes* are cells that contain a nucleus and other membrane-bound organelles; *prokaryotic* cells lack these features.) Strands of DNA are wrapped around spool-like structures called nucleosomes. These nucleosomes are made of proteins called histones. And, again, it is the specific shape of the histone proteins that enables them to do their job. Histones 3 and 4, for example, fold into well-defined three-dimensional shapes with a precise distribution of positive electrical charges around their exteriors. This precise shape and charge distribution enables DNA strands to coil efficiently around the nucleosome spools and store an immense amount of information in a very small space.[16] Thanks in part to nucleosome spooling, the information storage density of DNA is many times that of our most advanced silicon chips.[17]

To visualize how this works, imagine a large wooden spool with grooves on the surface. Next, picture a helical cord made of two strands. Then visualize wrapping the cord around the spool so that it lies exactly in the hollowed-out grooves of the spool. Finally, imagine the grooves hollowed in such a way that they exactly fit the shape of the coiled cord—thicker parts nestling into deeper grooves, thinner parts into shallower ones. The irregularities in the shape of the cord exactly match irregularities in the hollow grooves. In the case of nucleosomes, instead of grooves in the spool, there is an uncanny distribution of positively charged regions on the surface of the histone proteins that exactly matches the negatively charged regions of the double-stranded DNA that coils around it.[18] (See Fig. 4.5.)

But proteins have a second type of specificity—one that helps to explain the first. Proteins do not just display a specificity of shape; they also display a specificity of *arrangement*. Whereas proteins are built from rather simple amino-acid "building blocks," their various functions depend crucially on the specific arrangement of those building blocks. The specific sequence of amino acids in a chain and the resulting chemical interactions between amino acids largely determine the specific three-dimensional structure that the chain as a whole will adopt. Those structures or shapes determine what function, if any, the

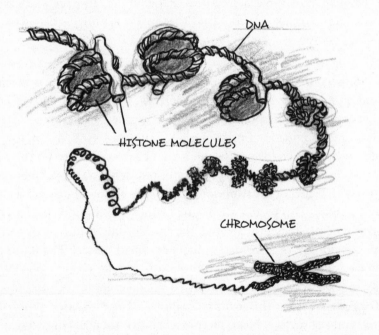

Figure 4.5. A highly specific pattern of matching positive and negative charges allows information-rich DNA to wrap tightly around histone proteins in a process known as nucleosome spooling. Chromosomes are made of DNA spooled on histones.

amino-acid chain can perform in the cell—whether as an enzyme, a structural component, or a machine for processing information.[19]

When I was teaching college students, I used to illustrate this relationship between amino-acid sequencing and protein structure and function with a children's toy called Snap-Lock Beads. Snap-Lock Beads are made of plastic and come in many colors and shapes. Each bead has a hole on one end and a plug on the other that allows the beads to be snapped together to form a long chain. In my illustration, the individual beads with their distinctive shapes represent the twenty different protein-forming amino acids; the chain of beads represents a *potential* protein; and the connection point between the beads represents a chemical bond. As I snapped the different-shaped beads together, I would explain to my students that proteins are formed from amino acids linked together by a specific kind of chemical bond called a peptide bond. As the amino acids "snap" together, the resulting chains are called *polypeptides*. If one of these polypeptide chains folds into a specific shape that enables it to perform tasks within the cell, it is called a *protein*.

But that, it turns out, is a big "if." Only if amino acids are arranged in very specific ways will the chain fold into useful shapes or *conformations*. Individual amino acids have distinguishing features (called side chains) that exert forces on each other. The resulting constellation of forces will cause the chain as a whole to twist and turn and, sometimes, fold into a stable structure. Most amino-acid arrangements, however, produce chains that do not fold into stable structures at all. Other arrangements generate stable structures, but do not perform a function. Relatively few sequences will produce functional structures. In each case, it is the arrangement of the amino acids that determines the difference. *Specificity of sequence* distinguishes proteins from ordinary, useless polypeptides.

I would explain the importance of this specificity to my students by arranging and rearranging my Snap-Lock Beads. I would twist the chain to show how one sequence might cause the chain to fold into a specific three-dimensional shape. I would then rearrange the beads to show how a different sequence and constellation of forces might cause the chain to collapse into a limp or amorphous jumble. I emphasized that proteins have a type of "specificity"—called "sequence specificity"—that needed to be explained. A system or sequence of characters manifests "sequence specificity" if the function of the system as a whole depends upon the specific arrangement of the parts. Language has this property. Software has this property. And so too do proteins.

The revolution in molecular biology produced some exciting and surprising discoveries about proteins and raised an intriguing question. If the structure of proteins depends upon the specific arrangement of its amino acids, what determines the arrangement of the amino acids?

The Sequence Hypothesis

Discoveries in the 1950s showed that both DNA and proteins are long, linear molecules composed of many irregularly arranged subunits of smaller molecules. Could these two otherwise very different kinds of molecules be connected in some important, but as yet undiscovered way?

Following Sanger's discovery, most molecular biologists assumed that proteins could not assemble themselves from free-floating amino acids in the cell. Too many processes in the cell depend upon particular proteins having just the right shape and sequence of amino acids to leave the assembly of these molecules to chance. Nor did it seem likely that some general chemical law or uniform force of attraction could explain the assembly of these irregular and oddly shaped structures. Instead, as the French molecular biologist Jacques Monod would later recall, molecular biologists began to look for some source of information or "specificity" within the cell that could direct the construction of these highly specific and complex molecules.[20]

Watson and Crick's seminal discovery suggested a source for that information or "specificity." Perhaps it lay along the spine of DNA's sugar-phosphate backbone.[21] But what did the information in DNA have to do with the specific arrangements of amino acids in proteins? In 1958, Francis Crick proposed an answer in what came to be called the "sequence hypothesis."[22]

The sequence hypothesis suggested that the nucleotide bases in DNA functioned just like alphabetic letters in an English text or binary digits in software or a machine code. According to Crick's hypothesis, it is *the precise arrangement* or *sequence* of these bases that determines the arrangement of amino acids—which, in turn, determines protein folding and structure. In other words, the sequence specificity of amino acids in proteins derives from a prior specificity of arrangement in the nucleotide bases on the DNA molecule.[23]

Crick's sequence hypothesis was not the kind of hypothesis that could be confirmed by a single experiment. Nor was it based upon a single discovery. Rather, it capped a series of other discoveries and insights and would require many more experiments and discoveries to confirm. It was also a natural extension of what was already known, or at least suspected.

Since the early 1940s, following the publication of a research paper by American geneticists George Beadle and Edward Tatum, many geneticists had begun to suspect a link between genes and proteins even before it was clear that genes were made of DNA. Beadle and Tatum had been studying fruit-fly eyes and bread molds. They used X rays to induce mutations in the chromosomal material of these organisms. They discovered that these mutations often had a discernible effect on their features. In the case of a bread mold called *Neurospora,* Beadle and Tatum discovered that the mutant forms of mold were missing an essential chemical compound—a compound that was produced with the help of a particular enzyme. Beadle and Tatum proposed that the mutated gene was normally responsible for synthesizing the enzyme in question, but that the mutations in the gene had destroyed its ability to do so. They further suggested that it is the job of genes generally to produce enzymes; in fact, they suggested that, as a rule, one gene produces one enzyme.[24]

Crick's sequence hypothesis was built on this presumed linkage between genes and proteins and on other recent insights, including Avery's identification of DNA as the likely carrier of genetic information, Sanger's discovery of the irregularity and linearity of amino-acid sequences, and Watson and Crick's own understanding of the chemical structure of DNA. If a sequence of bases constitutes a gene, if genes direct protein synthesis, and if specific sequences of amino acids constitute proteins, then perhaps the specific sequencing of bases determines the specific sequencing of amino acids. By 1958, this seemed to Crick to be far more than just a reasonable supposition. The sequence hypothesis explained many of the central facts then in currency among the elite circle of British, American, and French molecular biologists trying to decipher exactly what DNA does.

Even so, confirming Crick's sequence hypothesis would eventually involve at least two additional kinds of experimental results. In a narrow sense, confirmation of the sequence hypothesis meant establishing that changes in genes made a difference in the amino-acid sequence of proteins.[25] By the early 1960s, scientists had developed many techniques for studying the effects of changes in DNA sequences on proteins. These techniques enabled scientists to establish a definitive link between base sequences in DNA and the sequences of amino acids. Experiments using such techniques eventually revealed a set of correspondences between specific groups of bases and individual amino acids. These correspondences came to be called the "genetic code" (see Fig. 4.6).

	U	C	A	G
U	UUU PHENYLALANINE UUC PHENYLALANINE UUA LEUCINE UUG LEUCINE	UCU SERINE UCC SERINE UCA SERINE UCG SERINE	UAU TYROSINE UAC TYROSINE UAA STOP UAG STOP	UGU CYSTEINE UGC CYSTEINE UGA STOP UGG TRYPTOPHAN
C	CUU LEUCINE CUC LEUCINE CUA LEUCINE CUG LEUCINE	CCU PROLINE CCC PROLINE CCA PROLINE CCG PROLINE	CAU HISTIDINE CAC HISTIDINE CAA GLUTAMINE CAG GLUTAMINE	CGU ARGININE CGC ARGININE CGA ARGININE CGG ARGININE
A	AUU ISOLEUCINE AUC ISOLEUCINE AUA ISOLEUCINE AUG METHIONINE (START)	ACU THREONINE ACC THREONINE ACA THREONINE ACG THREONINE	AAU ASPARAGINE AAC ASPARAGINE AAA LYSINE AAG LYSINE	AGU SERINE AGC SERINE AGA ARGININE AGG ARGININE
G	GUU VALINE GUC VALINE GUA VALINE GUG VALINE	GCU ALANINE GCC ALANINE GCA ALANINE GCG ALANINE	GAU ASPARTIC GAC ASPARTIC GAA GLUTAMIC GAG GLUTAMIC	GGU GLYCINE GGC GLYCINE GGA GLYCINE GGG GLYCINE

Figure 4.6. The standard genetic code showing the specific amino acids that DNA base triplets specify after they are transcribed and translated during gene expression.

In a broader sense, however, the confirmation of the sequence hypothesis would await the full explication of what is now called the gene-expression system—the system by which proteins are constructed from the information in DNA. It was one thing to know *that* there is a correlation between specific base sequences in DNA and specific amino acids in proteins. It was quite another to learn *how* that linkage is produced. Nevertheless, by the early 1960s, a series of experiments established how the information in DNA produces proteins and, in the process, provided a deeper and more comprehensive confirmation of Crick's bold hypothesis.[26]

In the next chapter, I describe the intricate process of gene expression in more detail. Here I simply give a brief sketch of what scientists discovered about how the process works, since knowing that will enable me to answer the question I raised about the nature of biological information at the beginning of this chapter.

Gene expression begins as long chains of nucleotide bases are copied during a process known as "transcription." During this process, the genetic assembly instructions stored on a strand of DNA are reproduced on another molecule called "messenger RNA" (or mRNA). The resulting single-stranded copy or "transcript" contains a sequence of RNA bases precisely matching the sequence of bases on the original DNA strand.[27] (RNA also uses chemicals called bases to store genetic information, but it uses a slightly different chemical alphabet from DNA. RNA substitutes a base called uracil for the base thymine used in the DNA.)

After it is produced, the messenger-RNA molecule travels to the ribosome, a molecular machine that helps translate the mRNA assembly instructions. These instructions consist of a series of three-letter genetic "words" called "codons." Each codon consists of three bases and directs the cell to attach a specific amino acid to a growing chain of other amino acids. For example, the mRNA word UUA directs the ribosome to attach the amino acid leucine, whereas AGA specifies the amino acid arginine. Other codons direct the ribosome to start or stop building proteins. This translation process occurs with the aid of specific adapter molecules (called transfer RNAs, or tRNAs) and specific enzymes (called aminoacyl-tRNA synthetases).[28] (See Fig. 4.7. To view an animation of this process of gene expression, see this book's Web site at www.signatureinthecell.com.)

By the way, the elucidation of the gene-expression system shed further light on the experiment with the dead mice discussed in the last

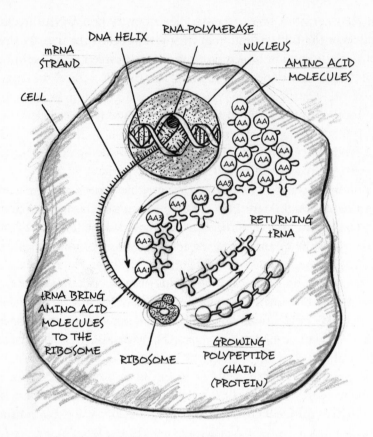

Figure 4.7. A simplified schematic of gene expression showing both transcription and translation. (The enzymes called tRNA synthetases are not pictured, although they are essential to the process.) *Courtesy of I. L. Cohen of New Research Publications.*

chapter. Oswald Avery correctly determined it was DNA, not bacteria living or dead, which caused the mice to die. But when scientists established that DNA directs protein synthesis, Avery's results were viewed in a new light. DNA caused the mice to die, all right, but it did so by directing the production of a polysaccharide capsule that was toxic to the mice (though, obviously, not to the bacteria from which the DNA was derived).

In any case, with the explication of gene expression complete, Francis Crick's sequence hypothesis was formally confirmed. But what kind of information does DNA contain? The discovery that DNA directs protein synthesis enabled molecular biologists working in the two decades after the formulation of Shannon's theory to answer this question.

Shannon Information or Shannon Plus?

A parable may help to illustrate a crucial distinction and help define some key terms we will need throughout the rest of this book. Mr. Smith has promised to pick up Mr. Jones at Reagan National Airport in Washington when he returns from Chicago, but he has asked Jones to call just after he arrives. When he steps off the plane, Jones reaches into his pocket for Smith's number in order to make the call. He finds that he has lost the slip of paper with the number on it. Nevertheless, Jones doesn't panic. He knows that all long-distance phone numbers have ten characters and are constructed from the same ten digits. He writes down the digits 0, 1, 2, 3, 4, 5, 6, 7, 8, 9 on another piece of paper. He gives a sigh of relief. Now he has the raw materials, the building blocks, to generate Smith's ten-digit phone number.

He quickly arranges the digits on his list at random into a sequence of ten characters. Now he has some information, some Shannon information. Being mathematically inclined, he quickly calculates how much information he has using Shannon's familiar formula. He's relieved and impressed. His new ten-digit phone number has a lot of information—33.2 bits to be exact. He calls the number. Unfortunately, he has a problem. The sequence of characters is not arranged "to produce a specific effect." Instead of reaching Smith, he gets a recording from the phone company: "The number you have dialed is not in service . . ."

After a while Mr. Smith gets worried. He hasn't received the call from Jones that he was expecting. But then he remembers that he has some information that can help him reach Jones. He has Jones's cell phone number. It also has ten digits. These digits are arranged in a definite sequence: 202–555–0973. He enters them in his own cell phone and presses the call button. Jones answers and tells Smith of his arrival and location. They connect, and there is a happy ending.

In our illustration, both Smith and Jones have an equally improbable sequence of ten characters. The chance of getting either sequence at random is the same: $1/10 \times 1/10 \times 1/10 \times 1/10 \times 1/10 \times 1/10 \times 1/10 \times 1/10 \times 1/10 \times 1/10$, or 1 chance in 10^{10}. Both sequences, therefore, have information-carrying capacity, or Shannon information, and both have an equal amount of it as measured by Shannon's theory. Clearly, however, there is an important difference between the two sequences. Smith's number is arranged in a particular way so as to produce a specific effect, namely, ringing Jones's cell phone, whereas Jones's number is not. Thus, Smith's number contains *specified information* or *functional information,* whereas Jones's does not; Smith's number has *information content,* whereas Jones's number has only *information-carrying capacity* (or Shannon information).

Both Smith's and Jones's sequences are also *complex.* Complex sequences exhibit an irregular, nonrepeating arrangement that defies expression by a general law or computer algorithm (an algorithm is a set of instructions for accomplishing a specific task or mathematical operation). The opposite of a complex sequence is a highly ordered sequence like ABCABCABCABC, in which the characters or constituents repeat over and over due to some underlying rule, algorithm, or general law.

Information theorists say that repetitive sequences are *compressible,* whereas complex sequences are not. To be compressible means a sequence can be expressed in a shorter form or generated by a shorter number of characters. For example, a computer programmer could write two commands that would generate a 300-character sequence of repeating ABCs simply by writing the commands "write ABC" and then "repeat 100 times." In this case, a simple rule or algorithm could generate this exact sequence without each individual character having to be specified. Complex sequences, however, cannot be compressed to, or expressed by, a shorter sequence or set of coding instructions. (Or rather, to be more precise, the complexity of a sequence reflects the extent to which it cannot be compressed.)

Information scientists typically equate "complexity" with "improbability," whereas they regard repetitive or redundant sequences as highly probable. This makes sense. If you know that there is a reason that the same thing will happen over and over again, you are not surprised when that thing happens; nor will you regard it as improbable when it does. Based upon what we know about the way the solar system works, it is very probable that the sun will rise tomorrow. If you know that a computer is executing a command to "repeat"—to generate the same sequence over and over again—then it is highly probable that the sequence will arise. Since information and probability are inversely related, high-probability repeating sequences like ABCABCABCAB-CABCABC have very little information (either carrying capacity or content). And this makes sense too. Once you have seen the first triad of ABCs, the rest are "redundant"; they convey nothing new. They aren't informative. Such sequences aren't complex either. Why? A short algorithm or set of commands could easily generate a long sequence of repeating ABCs, making the sequence compressible.

In our parable, both Smith and Jones have ten-digit numbers that are *complex;* neither sequence can be compressed (or expressed) using a simpler rule. But Smith's sequence was specifically arranged to perform a function, whereas Jones's was not. For this reason, Smith's sequence exhibits what has been called *specified* complexity, while Jones's exhibits *mere* complexity. The term *specified complexity* is, therefore, a synonym for *specified information* or *information content.* (See Fig. 4.8.)

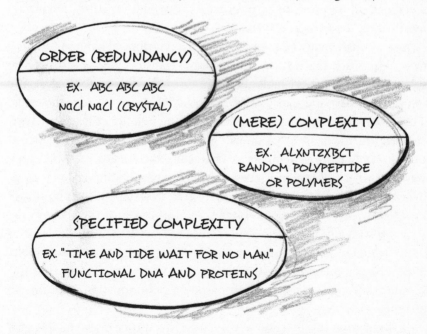

Figure 4.8. Examples of order, complexity, and specified complexity.

So what kind of information does DNA contain, Shannon informa-tion or specified information? Mere complexity or specified complexity? The answer is—*both*.

First, DNA certainly does have a quantifiable amount of information-carrying capacity as measured by Shannon's theory. Since DNA contains the assembly instructions for building proteins, the gene-expression system of the cell functions as a communication channel.[29] Further, the nucleotide bases function as alphabetic characters within that system. This enables scientists to calculate the information-carrying capacity of DNA using Shannon's equations. Since, at any given site along the DNA backbone, any one of four nucleotide bases may occur with equal ease,[30] the probability of the occurrence of a specific nucleotide at that site equals $1/4$. For the occurrence of two particular nucleotide bases, the odds are $1/4 \times 1/4$. For three, $1/4 \times 1/4 \times 1/4$, or $1/64$, or $(1/4)^3$, and so on.[31] The information-carrying capacity of a sequence of a specific length n can then be calculated using Shannon's familiar expression $(I = -\log_2 p)$ once one computes a probability value (p) for the occurrence of a particular sequence n nucleotides long where $p = (1/4)^n$. The p value thus yields a corresponding measure of information-carrying capacity or syntactic information for a sequence of n nucleotide bases.

Just as mathematicians and engineers can apply Shannon's theory to analyze a written text, a cryptographic transmission, or a section of software, mathematical biologists can apply the theory to analyze the information-carrying capacity of a DNA, RNA, or protein molecule. That is what Charles Thaxton meant when he told me in Dallas in 1985 that the treatment of DNA and English text is "mathematically identi-cal." Both systems of symbols or characters can be analyzed the same way. As I write this sentence, the placement of each additional letter elim-inates twenty-five other possible letters and a corresponding amount of uncertainty. It, therefore, increases the information of the sentence by a quantifiable amount as measured by Shannon's theory. Similarly, at each site along the DNA molecule any one of the four bases is possible. Thus, the placement or presence of any one of the bases eliminates un-certainty and conveys a quantifiable amount of information according to Shannon's theory.

Is this significant? In several ways, it is. It is certainly remarkable that DNA can carry or encode information using chemical subunits that function as alphabetic characters. It is also remarkable that DNA forms part of a communication channel that can be analyzed so readily using the mathematical tools of Shannon's information theory. Further, scientists have applied Shannon's information theory to generate quantitative measures of the information-carrying capacity (or brute complexity) of DNA sequences and their corresponding proteins. These analyses have shown that these molecules are highly complex, and quantifiably so. DNA, RNA, and proteins have a tremendous capacity, at least, to store and transmit information.

Nevertheless, the ease with which information theory applies to molecular biology has also created some confusion. The sequences of nucleotide bases in DNA and the sequences of amino acids in proteins are highly improbable and, therefore, have large information-carrying capacities. Knowing this, some scientists have mistakenly described DNA and proteins as if they contained *only* Shannon information or possessed *mere* information-carrying capacity.

Yet, like meaningful sentences, lines of computer code, or the phone number that Smith used to ring Jones's cell phone, the arrangements of bases in DNA and amino acids in proteins are also *specified* with respect to function. This judgment is a direct consequence of the experimental verification of Francis Crick's sequence hypothesis. Indeed, since the confirmation of the sequence hypothesis in the early 1960s, biologists have known that the ability of the cell to build functional proteins depends upon the precise sequential arrangement of the bases in DNA. Thus, molecular biologists beginning with Francis Crick have equated *biological information* not only with improbability (or complexity), but also with "specificity," where "specificity" or "specified" has meant "necessary to function."[32]

Thus, in addition to a quantifiable amount of Shannon information (or complexity), DNA also contains information in the sense of Webster's second definition: it contains "alternative sequences or arrangements of something *that produce a specific effect.*" Although DNA does not convey information that is received, understood, or used by a conscious mind, it does have information that is received and used by the cell's machinery to build the structures critical to the maintenance of life. DNA displays a property—functional specificity—that transcends the merely mathematical formalism of Shannon's theory.

Is this significant? In fact, it is profoundly mysterious. Apart from the molecules comprising the gene-expression system and machinery of the cell, sequences or structures exhibiting such specified complexity or specified information are not found anywhere in the natural—that is, the nonhuman—world. Sequences and structures exhibiting either redundant order or *mere* complexity are common in the chemical substrate of nature. But structures exhibiting specified complexity are completely unknown there apart from DNA, RNA, and proteins. As the origin-of-life biochemist Leslie Orgel observes: "Living organisms are distinguished by their specified complexity. Crystals . . . fail to qualify as living because they lack complexity; mixtures of random polymers fail to qualify because they lack specificity."[33] Nevertheless, human artifacts and technology—paintings, signs, written text, spoken language, ancient hieroglyphics, integrated circuits, machine codes, computer hardware and software—exhibit specified complexity; among those, software and its encoded sequences of digital characters function in a way that most closely parallels the base sequences in DNA.

Thus, oddly, at nearly the same time that computer scientists were beginning to develop machine languages, molecular biologists were discovering that living cells had been using something akin to machine code[34] or software[35] all along. To quote the information scientist Hubert Yockey again, "The genetic code is constructed to confront and solve the problems of communication and recording by the same principles found . . . in modern communication and computer codes."[36] Like software, the coding regions of DNA direct operations within a complex material system via highly variable and improbable, yet also precisely specified, sequences of chemical characters. How did these digitally encoded and specifically sequenced instructions in DNA arise? And how did they arise within a channel for transmitting information?

Explaining the "Origin of"

As we saw in Chapter 2, early theories of the origin of life did not need to address, nor did they anticipate, this problem. Since scientists did not know about the information-bearing properties of DNA, or how the cell uses that functionally specified information to build proteins, they did not worry about explaining these features of life. But after 1953 the landscape changed irrevocably. Any idea of a repetitive, crystalline order at the foundation of life gave way to a picture of life in which information-rich macromolecules direct the metabolism and machinery of living things. Origin-of-life researchers have accordingly turned their attention to this crucially specified form of information, the origin of which constitutes the central aspect of the DNA enigma. But before examining how scientists have attempted to explain this mystery, it is important to examine another feature of the informational system at work within the cell, since there is another facet of the DNA enigma that scientists investigating the origin of life must now explain.

5

The Molecular Labyrinth

A military surgeon at a primitive field hospital is treating a battalion commander crucial to the success of a coming campaign. The surgeon needs to remove shrapnel lodged in the soldier's internal organs, but to do so effectively, the surgeon needs some advanced software and imaging technology unavailable at the field hospital. He radios his superior and tells him the problem. The supervisor in turn relates the situation to a bureaucrat at a major hospital in Washington, D.C. Unfortunately, the bureaucrat is Mr. Jones from the previous chapter.

Jones quickly goes to work procuring the necessary software for running the diagnostic imaging technology. With it, he boards a military plane and flies to the remote location. He hands the surgeon a high-density storage device and raises a hand to forestall expressions of gratitude. "Don't mention it," he says. "Solving problems is what I do."

And yet the problem remains. It's not the software. The information in the file is error-free and contains precisely the program and instructions needed to conduct the diagnostic procedure. The problem is the surgeon does not have either the computer or the operating system that would allow him to use the software. These are built into the diagnostic equipment—the imaging machine—that the surgeon was missing in the first place. The surgeon has a software file that he cannot access, translate, or use. The information is useless to him. The soldier dies. The war drags on.

This story illustrates what in the realm of human technology and experience is obvious: encoded information is worthless without a system—whether mind or machine—that can use, read, translate, or otherwise process that information. What does this tell us about the DNA enigma?

As scientists began to discover more about how the cell uses the information in DNA to build proteins, they realized that DNA is only part of a complex system for expressing and processing information. They realized that without this whole system, DNA by itself could do nothing.

In the previous chapter, we saw that the cell uses the information stored in DNA to build proteins. Establishing this enabled us to answer a question about the kind of information that DNA contains, namely, functionally specified information. Nevertheless, the previous chapter did not say much about how the cell processes the information in DNA or how scientists came to learn about it. This is worth looking at more closely because it turns out that the cell's information-processing system is itself one of the key features of life that any theory of the origin of life must explain.

Unlike the unfortunate surgeon in our opening story, the cell does have a system for processing stored information. As for how it works: the key insight came when Francis Crick realized that the cell must be using a code to translate information from one form to another and that this code was mediated by an interdependent system of molecules. This insight raised an age-old question: the question of the chicken and the egg.

Deducing a Code

By the late 1950s the relationship between DNA and protein was coming into focus. By then leading molecular biologists understood that the three-dimensional specificity of proteins depended upon the one-dimensional specificity of their amino-acid sequences. They also suspected that specific arrangements of amino acids in protein chains derived in turn from specific sequences of nucleotide bases on the DNA molecule. Yet the question remained: *How* does the sequence of bases on the DNA direct the construction of protein molecules? How do specific sequences in a four-character alphabet generate specific sequences in a twenty-character alphabet?

Francis Crick again anticipated the answer: the cell is using some kind of a code. Crick first began to suspect this as he reflected on a proposal by George Gamow, a Russian-born theoretical physicist and cosmologist who had, in the postwar years, turned some of his prodigious intellectual powers to reflect on new discoveries in molecular biology.

Gamow had immigrated to the United States in the 1930s to take an appointment at George Washington University after working at the famed Theoretical Physics Institute in Copenhagen.[1] In 1953 and 1954, he proposed a model to explain how the specific sequences in DNA generate the specific sequences in proteins.[2] According to Gamow's "direct template model," as it was called, protein assembly occurred directly on the DNA strand. Gamow proposed that proteins formed as amino acids attached directly to the DNA molecule at regularly spaced intervals. Gamow thought the amino acids could nestle into diamond-shaped cavities in DNA that formed in the space between the two backbones (see Fig. 5.1). In this model, a group of four nucleotide bases from two parallel strands of DNA made a specifically shaped hollow into which one and only one amino acid could fit.[3] As each group of nucleotides acquired an amino-acid partner, the amino acids would also link with each other to form a chain. According to Gamow, the nearly identical spacing of amino acids in protein and of bases in DNA enabled the direct "matching" of amino acids to nucleotide groups on the DNA

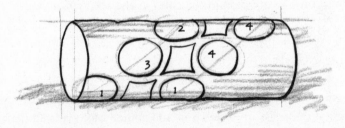

Figure 5.1. A simplified version of George Gamow's own drawing depicting his direct template model of protein synthesis. The diamonds represent places where he thought various of the twenty amino acids would bond to the DNA double helix. The circles represent one of the four nucleotide bases. Gamow thought that the amino acids would nestle into the grooves between the nucleotide bases and that the structure of the DNA molecule would somehow determine where individual amino acids would bond, thereby directly determining protein sequencing.

template. This matching occurred because of the supposed fit between the cavities produced by the bases and the shape of the side chain of the amino acids and the chemical affinity between the bases and amino acids.

Francis Crick first recognized the futility of this scheme. In a famous communication to members of the "RNA Tie Club," Crick explained that there was nothing about either the chemical properties or shapes of the bases to ensure that one and only one amino acid would fit into, or attach to, the cavities created by a group of bases.[4] In the first place, many of the amino acids were difficult to differentiate structurally, because they had similar side chains. Second, the bases themselves did not necessarily create shapes that either matched these shapes or, still less, differentiated one from another. As Crick put it, "Where are the nobby hydrophobic surfaces to distinguish valine from leucine from isoleucine? Where are the charged groups, in specific positions, to go with the acidic and basic amino acids?"[5] As Judson explains, "Crick was a protein crystallographer, and knew of no reason to think that Gamow's holes in the helix could provide the variety or precision of shapes necessary to differentiate a score or more of rather similar objects."[6]

Yet in the absence of such spatial matching and the formation of corresponding chemical attachments, there could be no reliable transmission of information. For the direct template model to explain the irregularity and specificity of the amino-acid sequences in proteins, individual bases (or groups of bases) needed to manifest discriminating spatial geometries. Figuratively speaking, DNA bases not only needed to "zig" in a way that matched the specific "zags" of each amino acid, but DNA needed to do so at irregular intervals in order to produce the irregular sequencing of amino acids that characterizes proteins. Yet as Crick realized, both the individual bases themselves and their various combinations lacked distinguishing physical features that could account for the specificity of amino-acid sequencing. Further, the geometry of the DNA molecule as a whole, at the level of its gross morphology, presents a highly repetitive series of major and minor grooves (see Fig. 5.2). Therefore, it could not function as a direct template for protein synthesis. As Crick explained, "What the DNA structure *does* show . . . is a specific pattern of *hydrogen bonds,* and very little else."[7]

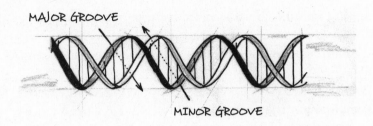

Figure 5.2. The large-scale form of the DNA helix showing its major and minor grooves.

If the chemical features and shapes of the DNA bases do not directly account for the specific sequencing of proteins, what does? Crick remained adamant that the specific arrangement of the nucleotide bases, not anything about their physical or chemical features per se, dictated amino acid sequencing.[8]

Crick's insight had profound implications. If a single protein could not copy the information in DNA directly, as the direct template model suggested, then, as Jacques Monod would later explain, "you absolutely needed a code."[9] And so Crick postulated a third factor consistent with his original sequence hypothesis. He proposed the existence of a genetic code—a means of translating information from one chemical domain into another.

To envision what Crick had in mind, imagine having a human alphabet that uses four and only four distinct shapes that combine in various specific ways to form not just words, but words that correspond to individual letters in a larger alphabet roughly the size of the English alphabet; this larger alphabet then uses its letters (each of which is one of those words composed of the four shapes) to build sentences. Of course, we actually have a symbol system that does much the same thing. The binary code that computer programs use has a translation key that enables programmers to produce English text from sequences of binary digits. Each letter in the English alphabet is represented by a unique combination of two character types, 0's and 1's. For example, in ASCII code the letter A is represented by the sequence 100 0001, the letter B by the sequence 100 0010, and so on. Each of the letters of the twenty-six-letter English alphabet has a corresponding representation in the two-digit numeric alphabet of this binary system (see Fig. 5.3).[10] Crick realized that if his sequence hypothesis was true, then there must

THE AMERICAN STANDARD CODE FOR INFORMATION INTERCHANGE
(ASCII)

A	100 0001	H	100 1000	O	100 1111	V	101 0110
B	100 0010	I	100 1001	P	101 0000	W	101 0111
C	100 0011	J	100 1010	Q	101 0001	X	101 1000
D	100 0100	K	100 1011	R	101 0010	Y	101 1001
E	100 0101	L	100 1100	S	101 0011	Z	101 1010
F	100 0110	M	100 1101	T	101 0100	a	110 0001
G	100 0111	N	100 1110	U	101 0101	b	110 0010

ALPHABETIC CHARACTERS FOLLOWED BY CORRESPONDING
BINARY REPRESENTATION

Figure 5.3. Part of the ASCII code.

be some similar translation system in the cell—one that determined how sequences written in the four-character alphabet of DNA are converted into sequences that use a twenty-letter amino-acid alphabet. The DNA detective was now in the business of code breaking as well.

Yet in a physical system, as opposed to a social or linguistic one, a code must have a physical expression. Crick postulated the existence of a third molecule, an adapter molecule functioning as a translating device that could recognize and convert the information in the sequential arrangements of the bases into specific amino-acid sequences.[11] More specifically, he proposed the existence of twenty separate adapter molecules corresponding to each of the twenty protein-forming amino acids. Each adapter would, by the familiar mechanism of complementary base pairing, bind to a section of DNA text at one end and to a specific amino acid at the other (see Fig. 5.7). Crick also proposed the existence of specific enzymes (one for each of the twenty adapter–amino acid pairs) to connect the specific amino acids and their corresponding adapters.[12] The set of correspondences between sections of genetic text, on the one hand, and a specific amino acid, on the other, constituted a genetic code (see Fig. 4.6).

Though these correspondences were mediated physically by adapter molecules and enzymes, this complex system, as conceived by Crick, would be governed as much by the functional requirements of information transfer as by rules of chemical affinity—as much by a set of chemically arbitrary conventions as by the necessary relations of physical-chemical law. Indeed, as Crick imagined this system, nothing about the physical or chemical features of the nucleotides or amino acids directly dictated any particular set of assignments between amino acids and bases in the DNA text. The code could not be deduced from the chemical properties of amino acids and nucleotide bases. It had to be cracked. Just as a specific letter of the English language can be represented by any combination of binary digits, so too could a given amino acid correspond to any combination of nucleotide bases. The assignments are, in both cases, arbitrary. For this reason, the progressive elimination of the many chemically possible types of codes would eventually prove a laborious task. Yet precisely this feature of chemical underdetermination would also later prove the key to understanding DNA's information-carrying capacity. As Judson explains, "Crick freed the nucleic acid template of its most stringent physical limitation, allowing it to be thought about formally"[13] as an information carrier rather than merely a chemical substance.

Crick's proposal was striking in its sheer theoretical audacity. Biochemistry had not a shred of direct evidence for the existence of adapter molecules or their corresponding enzymes. Crick simply deduced the need for a code by thinking about what would be needed to make the cell's communication system work. Only a code could facilitate the translation of the information from DNA's four-character base sequences into the twenty-character "language" of proteins. His adapter hypothesis followed logically, and intuitively, from an understanding of the functional requirements of information transfer and the limited informational capacity of the chemical constituents of the relevant molecules themselves.

Simple regular geometries plus rules of chemical affinity did not and could not generate the specific complexity present in functional proteins. But complex combinations of a few physically simple bases functioning as characters or symbols could specify a vast array of possible structures, just as only two symbols, when variously arranged, can specify many characters by means of the ASCII code. Yet if the sequential arrangement—and nothing else—about the bases dictated protein structure, then the cell needed a means of translating and expressing the information stored in DNA. The cell needed a code, albeit one physically instantiated as part of an integrated system for translating and expressing genetic information. And so Crick conceived the "adapter hypothesis"—described by Judson as "the postulation, from theoretical necessity, of a new biochemical entity."[14]

What Crick would postulate on the grounds of functional necessity took nearly five years of intensive research and many transatlantic communications and conferences to verify and elucidate. But during the late 1950s and early 1960s, a series of experiments enabled scientists to show how the information in DNA directs protein synthesis. First, Paul Zamecnik, a researcher at Massachusetts General Hospital, discovered that proteins were produced in cellular structures called ribosomes, located in the outer cytoplasm "far" from the nucleus where DNA resides.[15] This discovery suggested the need for mechanisms to transcribe, transport, and translate the information in DNA so that amino-acid chains could be constructed at these sites. Because Zamecnik had discovered that ribosomes contained no DNA, Francis Crick, along with two other molecular biologists, François Jacob and Sydney Brenner, suggested that another type of molecule, a copy of the original DNA gene, actually directed the synthesis of proteins at the ribosomes. They proposed that this "messenger" molecule was made of RNA, or ribonucleic acid. Soon Brenner and others found this messenger RNA operating as he, Crick, and Jacob had anticipated.[16]

About the same time, Severo Ochoa, a Spanish-born physician working at the New York University Medical School, and Marshall Nirenburg, an American at the National Institutes of Health in Maryland, performed experiments that enabled scientists to decipher the genetic code.[17] Ochoa identified an enzyme that enabled him to synthesize RNA molecules. Nirenburg used Ochoa's technique to make a synthetic form of messenger RNA (mRNA) that he then used to direct protein synthesis.[18] Together their work showed that groups of three nucleotides (called codons) on the mRNA specify the addition of one of the twenty protein-forming amino acids during the process of protein synthesis. Other scientists discovered that the cell uses a set of adapter molecules to help convert the information on mRNA into proteins just as Crick expected.[19] Indeed, Crick's adapter molecules and their corresponding enzymes functioned much as he had originally envisioned, albeit as part of a far more complex process than even he had foreseen.

"CAD-CAM" and Protein Synthesis

In April 2008, a film called *Expelled: No Intelligence Allowed,* starring the Hollywood comic actor Ben Stein, was released in theaters all over the United States. Part of the film was about the current scientific controversy over the origin of life. When the producers came to our offices to plan interviews, they told us they wanted to find a way to represent what DNA does visually, so that a general audience could follow the scientific discussion they planned to incorporate into the film. They commissioned a visually stunning three-dimensional animation of DNA and the inner workings of the cell and retained a team of molecular biologists to work closely with the animators. When the film opened in theaters, audiences were impressed by the beauty of the animation and the intricacy of the processes that it depicted. The producers have kindly made some of their animation, including some that didn't make it into the film, available to me at www.signatureinthecell.com. This animation beautifully depicts the process by which the information in DNA directs the synthesis of proteins. I recommend viewing it in tandem with the discussion that follows.

Though the animation in *Expelled* provided a visual representation of how the cell uses information to direct its manufacturing operations, the film producers thought it might also help audiences understand the process if they could see footage of an analogous manufacturing procedure used in many industries. "Computer-assisted design and manufacturing," or "CAD-CAM" for short, uses digital information to manufacture various machines and products, from airplanes to automobiles to garage doors.[20]

At the Boeing plant in Seattle, engineers use CAD-CAM to direct the production of many key parts of airplanes.[21] A CAD program is used to design some part of an airplane, such as a wing. Engineers provide specifications that enable the CAD program to produce a three-dimensional drawing, a visual display of the specific part. The engineers can examine this visual image to check that they have achieved their design objectives. The CAD program also stores the information for producing this visual display in binary code. In the CAD-CAM process, the digital information stored in the CAD program is then transferred to another computer program, called an NC ("numerical code") interface. This program then translates the instructions from the CAD program into a machine code. This machine code then directs the manufacturing machinery—a robotic rivet arm, for example—to make the parts of the airplane.

In the film, the producers showed a manufacturing plant operating with CAD-CAM technology to illustrate how the cell uses digital code to manufacture its proteins and protein machines. In fact, the similarities between the two processes are striking. Like a production facility at Boeing or Ford, the cell uses digitally encoded information to direct the manufacture of the parts of its machines. In addition, the process of gene expression involves the conversion of information from one digital format to another before the information is used in manufacturing. In CAD-CAM, the original digitally encoded information in the CAD program is translated into another machine code by the NC interface and then used to direct the manufacture of airplane parts. In the gene-expression system, the original digital information in DNA is converted into an RNA format, which then directs the construction of proteins.

The system is a marvel of sophistication and merits a closer look. What follows is a short tour of the molecular labyrinth, the cell's sophisticated information-processing system. This will help to clarify what origin-of-life researchers have to explain. For those without a background in biology, don't worry if some of the details get past you. The details of the process are fascinating, but the big picture that emerges from this overview is the key. The cell's information-processing system is strikingly similar to CAD-CAM technology, though it differs from CAD-CAM in at least one important respect: the cell's information-processing system not only produces machines, it also reproduces itself.

Gene Expression

Molecular biologists describe the process of protein synthesis, or "gene expression," as a two-stage process of information transfer involving many smaller discrete steps and many molecular machines. This process proceeds as long chains of nucleotide triplets (the genetic message) are first copied during a process known as "transcription" and then transported (by the molecular messenger mRNA) to a complex organelle called a ribosome. At the ribosome site, the genetic message is then "translated" with the aid of a suite of adapter molecules called transfer RNAs to produce growing amino-acid chains—chains that fold into the functional proteins the cell needs to survive. Let's look at each of the two stages of gene expression in turn.

Transcription

The first stage in the process of protein synthesis is called transcription. During transcription, a copy, or transcript, of the DNA text is made by a large protein complex, known as RNA polymerase, that moves down the DNA chain and "reads" the original DNA text. As RNA polymerase proceeds, it makes an identical copy of the DNA transcript in an RNA format. (Like DNA, RNA contains four chemical bases, called nucleotide bases. These bases are the same as those in DNA with one exception: RNA uses a base called uracil instead of thymine.) The resulting single-stranded RNA copy, or transcript, then moves from the chromosomes to the ribosome in the outer cytoplasm to begin translation, the next step in the processing of genetic information.[22] (See Fig. 5.4.)

Transcription can be thus described in a few simple sentences. Yet any such description conceals an impressive complexity. In the first place, RNA polymerase is an extraordinarily complex protein machine of great specificity. The RNA polymerases present in the simplest bacteria (*Mycoplasma*) comprise several separate protein subunits with (collectively) thousands of specifically sequenced amino acids.[23]

RNA polymerase performs several discrete functions in the process of transcription. First, it recognizes (and binds to) specific regions of the DNA that mark the beginning of genes. Second, it unwinds (or

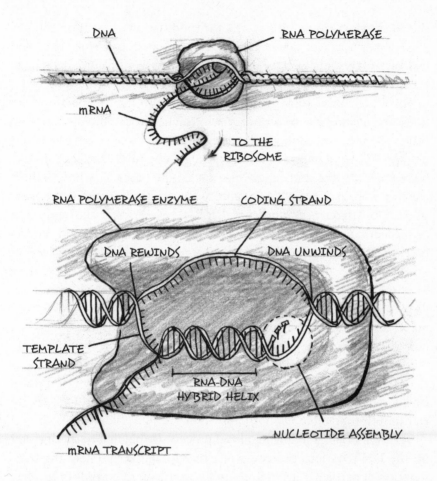

Figure 5.4. The process of transcription. The top view shows a view from outside the RNA polymerase. The bottom view is a close-up showing (in simplified form) what is happening inside the RNA polymerase.

helps unwind) the DNA text, exposing the strand that will serve as the template for making the RNA copy. Third, it sequesters and positions RNA bases (A, U, G, C) with their complementary partners on the DNA template (T, A, C, G, respectively). Fourth, it polymerizes or links together the separate RNA nucleotides, thus forming a long message-bearing ribbon of mRNA.[24] As molecular biologist Stephen Wolfe explains: "The structure of the RNA polymerases reflects the complexity of their activities in RNA transcription. The enzymes have sites that recognize promoters, react with initiation, elongation and termination factors, recognize DNA bases for correct pairing, bind and hydrolyze RNA nucleotides, form phospho-diester linkages, terminate transcription and perhaps unwind and rewind DNA."[25]

Yet for all its complexity and specificity, RNA polymerase alone does not ensure accurate transcription. The process involves several other complex and functionally integrated parts and steps. For example, for RNA polymerase to access the genetic text, the DNA double helix must unwind and expose the nucleotide bases. Further, to initiate transcription, the RNA polymerase must bind to the correct part of the DNA sequence in order to begin transcribing at the beginning of a genetic message, rather than in the middle or at the end. For its part, the DNA text provides a promoter sequence or signal upstream of the actual coding sequence to facilitate recognition of the correct location by the RNA polymerase. Yet the RNA polymerase cannot, on its own, find this site with any reliability. In prokaryotes (cells without nuclei), a protein known as sigma factor combines with the core RNA polymerase enzyme (itself a composite of enzymes) to form a larger "holoenzyme." The addition of this sigma-factor protein increases the accuracy of RNA polymerase–DNA binding by roughly a million times, making recognition of promoter sequences and thus accurate transcription possible.[26]

Transcription is a highly regulated process. By binding to specific sites on the DNA, various proteins will either inhibit or promote the transcription of particular genes in response to the varying needs of the cell. For example, when bacteria have no lactose available to metabolize, the protein "lactose repressor" binds near the gene that produces proteins for consuming or metabolizing lactose. This prevents RNA polymerase from transcribing the gene and producing an unnecessary protein. When lactose enters the cell, however, a chemical derivative of lactose binds to the lactose repressor, producing a change in the repressor protein that causes the repressor to fall off the gene, allowing the gene to be transcribed and the protein for metabolizing lactose to be synthesized.[27] By regulating transcription, repressor and activator proteins ensure that the cell maintains appropriate levels of proteins. Thus, even in prokaryotic organisms, many separate proteins are necessary to facilitate—and regulate—transcription.

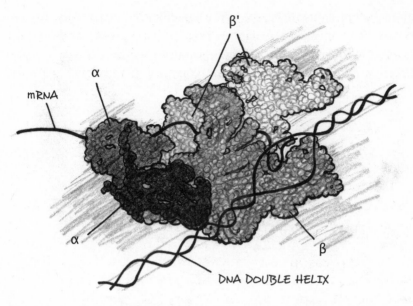

Figure 5.5. Parts of the RNA polymerase holoenzyme (sigma factor not pictured).

In eukaryotic cells (cells with nuclei), the process of transcription is considerably more complex.[28] Here recognition is promoted by a massive complex of several necessary initiation factors, enzymes that, like a jigsaw puzzle, fit together with each other, with the promoter sequence on the DNA molecule, and with the RNA polymerase (see Fig. 5.5). After transcription, the mRNA transcript must be heavily edited by still other proteins before transport to the ribosome site for translation. (In prokaryotes, this editing process takes place to a much lesser, but still necessary degree; see Fig. 5.6).[29]

As it turns out, the original DNA text in eukaryotic organisms has long sections of text called "introns" that do not (typically) encode proteins. Although these introns were once thought to be nonfunctional "junk DNA," they are now known to play many important functional roles in the cell.[30] (I discuss some of these newly discovered functions of introns and other types of nonprotein coding DNA in Chapter 18 and the Epilogue.) In any case, the initial transcript, being a copy of the DNA, also contains sections of text that do not encode proteins, but that are interspersed with coding text. To excise these noncoding regions before translation, the cell must edit the initial mRNA transcript so that only coding regions remain. This process requires the existence of other specific enzymes—exonucleases, endonucleases, and spliceosomes, for example—that can correctly identify and excise the

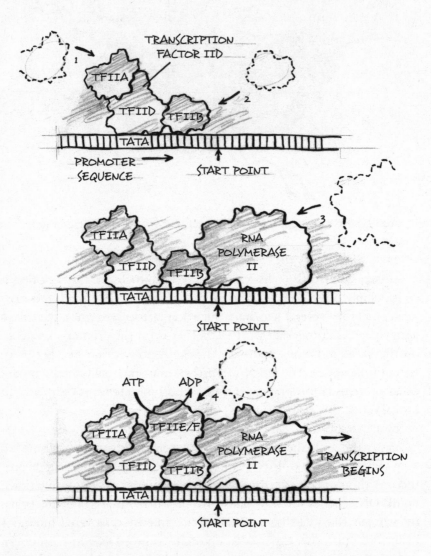

Figure 5.6. The initiation phase of transcription showing a promoter sequence (TATA), the various transcription factors that help position the RNA polymerase correctly along the DNA and the RNA polymerase enzyme.

nonprotein coding text from the initial RNA transcript and then splice together the resulting fragments of coding text in the correct order.[31]

Thus, in both prokaryotes and eukaryotes, transcription constitutes a complex, functionally integrated process involving several specialized and individually necessary proteins. Yet production of each of these separate proteins is itself dependent on the very process of transcription that they make possible. To build RNA polymerase, for example, the cell must transcribe the genetic texts with the instructions for building RNA polymerase. Yet to transcribe this information requires RNA polymerase. The information necessary to manufacture RNA polymerase and all the other associated enzymes and protein cofactors of transcription is stored on the DNA template. But expressing that information on the DNA template for building the proteins of the transcription system requires most of the proteins of the transcription system.

Translation

The next step in gene expression, called translation, exhibits even greater integrated complexity. Whereas transcription makes a single-stranded copy—a transcript—of DNA in an RNA format, translation uses that information to build a protein. Since many biologists think of protein molecules themselves as information-rich molecules constructed from a twenty-character amino-acid alphabet, they think of the process of protein synthesis as a process of translating information from the four-character alphabets of DNA and RNA into the twenty-character amino-acid alphabet; hence the name "translation."

Even in the simplest prokaryotic cells, the process of translation utilizes many dozens of separate proteins or protein machines, each one of which is produced during translation.[32] After the messenger RNA reaches the outer cytoplasm, it arrives at the site of a large chemical-processing unit known as a ribosome, the site of protein synthesis. In prokaryotes, the ribosome alone contains fifty separate proteins and three long RNA molecules combined in two distinct but associated subunits.[33]

The process of translation begins as the ribosome subunits dissociate and the messenger RNA (mRNA) binds to the smaller of the two subunits (see Fig. 5.7). Auxiliary proteins known as initiation factors catalyze this disassociation and temporarily stabilize the second subunit in its disassociated state. At the same time, a group of three RNA bases on a transfer-RNA (tRNA) molecule binds to the first triplet of RNA bases on the mRNA molecule as it docks in the ribosome. The groups of three bases on mRNA are called codons or triplets. The groups of three bases to which they bind on the tRNA are called anticodons. The sequence AUG constitutes the "initiator codon" at the head of the mRNA transcript.[34]

After the initiator codon (AUG) on the mRNA transcript binds to the anticodon triplet on the corresponding tRNA, then the second and larger subunit of the ribosome rejoins the first, forming a large complex of molecules including both ribosomal subunits, the mRNA, and a tRNA molecule carrying its corresponding amino acid. The protein chain can now begin to form. An additional amino acid–tRNA combination (known as an aminoacyl-tRNA molecule) binds to a second and adjacent active site on the ribosome, bringing its amino acid into close proximity to the first. A protein within the ribosome known as a peptidyl transferase then catalyzes a polymerization (linking) reaction involving the two (tRNA-borne) amino acids. In the process, the first amino acid detaches from its tRNA and attaches to the second amino acid, forming a short dipeptide chain. The ribosome then ejects the first and empty tRNA

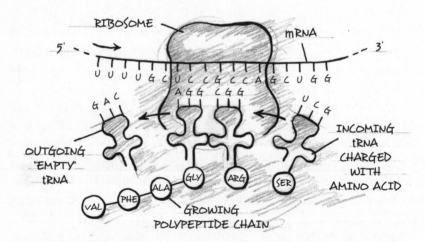

Figure 5.7. The translation of messenger RNA at the ribosome resulting in a growing amino acid chain.

molecule and moves along to "read" the next triplet of bases on the mRNA. Another tRNA–amino acid carrier pairs with the next mRNA codon, bringing a new amino acid into close proximity with the growing chain, and the process repeats itself until the signal for termination is reached on the mRNA. Then a protein termination factor, rather than an aminoacyl tRNA, binds to the second ribosome site and catalyzes breakage of the bond holding the peptide chain to the tRNA at the first ribosome site. The newly assembled protein then detaches.[35]

At each step in the translation process, specialized proteins perform crucial functions. For example, the initial coupling of specific amino acids to their specific tRNA molecules (Crick's adapters) depends upon the catalytic action of twenty specific enzymes, one for each tRNA–amino acid pair. The integrity of the genetic code depends upon the specific properties of these enzymes, known as aminoacyl-tRNA synthetases.[36]

These synthetases are necessary because, as Francis Crick anticipated, there is nothing about the chemical properties of the bases in DNA (or those in mRNA) that favors forming a chemical bond with any

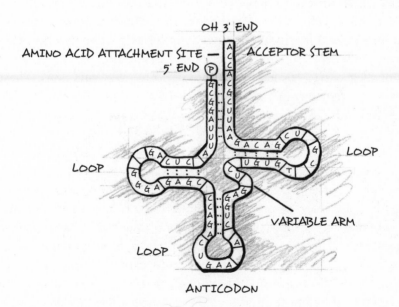

Figure 5.8. The transfer-RNA molecule showing the anticodon on one end of the molecule and the amino acid attachment site on the other.

specific amino acid over another. In fact, the cloverleaf-shaped tRNA molecule attaches to the mRNA transcript on one end and carries a specific amino acid on the other. Figure 5.8 shows that the amino acid and the codon-anticodon pairs are at opposite ends of the tRNA molecule. This distance ensures that neither the codons on mRNA nor the anticodons on tRNA interact chemically with the amino acids. As Crick anticipated, direct chemical interactions between bases (codons) and amino acids do not determine the assignments that constitute the genetic code.

Instead, these associations are mediated indirectly by the enzymatic action of the aminoacyl-tRNA synthetases. The synthetases have several active sites that enable them to: (1) recognize a specific amino acid, (2) recognize a specific corresponding tRNA (with a specific anticodon), (3) react the amino acid with ATP (adenosine triphosphate) to form an AMP (adenosine monophosphate) derivative, and then, finally, (4) link the specific tRNA molecule in question to its corresponding amino acid. Current research suggests that the synthetases recognize particular three-dimensional or chemical features (such as methylated bases) of the tRNA molecule. In virtue of the specificity of the features they must recognize, individual synthetases have highly distinctive shapes that derive from specifically arranged amino-acid sequences. In other words, the synthetases are themselves marvels of specificity.[37]

For their part, ribosomes must also perform many functions. These include: (1) enhancing the accuracy of codon-anticodon pairing between the mRNA transcript and the aminoacyl-tRNAs, (2) polymerizing (via peptidyl transferase) the growing peptide chain, (3) acting as energy transducers converting chemical energy into the mechanical energy during translocation of amino acids from tRNA carriers, (4) protecting the growing protein from attack by proteases (protein-degrading enzymes) possibly by forming a long protective tunnel, and (5) assisting in the breakage of the amino acid–tRNA bond during termination. Further, several separate protein factors and cofactors facilitate various specialized chemical transformations during the three discrete steps of translation: initiation, elongation, and termination. In eukaryotes, initiating translation alone requires a dozen separate protein cofactors. In prokaryotes, for each of the three steps of translation, three specialized protein cofactors perform specific (and in several cases necessary) functions. Thus, here, as in the transcription system, origin-of-life researchers find themselves confronted by a chicken-egg problem.

Making Copies

Besides transcribing and translating, the cell's information-processing system also replicates DNA. This happens whenever cells divide and copy themselves. As with the processes of transcription and translation, the process of DNA replication depends on many separate protein catalysts to unwind, stabilize, copy, edit, and rewind the original DNA message. In prokaryotic cells, DNA replication involves more than thirty specialized proteins to perform tasks necessary for building and accurately copying the genetic molecule. These specialized proteins include DNA polymerases, primases, helicases, topoisomerases, DNA-binding proteins, DNA ligases, and editing enzymes.[38] DNA needs these proteins to copy the genetic information contained in DNA. But the proteins that copy the genetic information in DNA are themselves built from that information. This again poses what is, at the very least, a curiosity: the production of proteins requires DNA, but the production of DNA requires proteins.

To complicate matters further, proteins must catalyze formation of the basic building blocks of cellular life such as sugars, lipids, glycolipids, nucleotides, and ATP (adenosine triphosphate, the main energy molecule of the cell). Yet each of these materials is also constructed with the help of specific enzymes. For example, each of the systems involved in the processing of genetic information requires energy at many discrete steps. In the cell, ATP (adenosine triphosphate) or a similar molecule (GTP, guanosine triphosphate) supplies this energy whenever one of its three phosphate bonds are broken. The cell manufactures ATP from glucose by a process known as glycolysis. Yet glycolysis involves ten discrete steps each catalyzed by a specific protein. These proteins (e.g., hexokinase, aldolase, enolase, pyruvate kinase) are, in turn, produced from genetic information on DNA via the processes of transcription and translation. Thus, the information-processing system of the cell requires ATP, but ATP production (via glycolysis) requires the

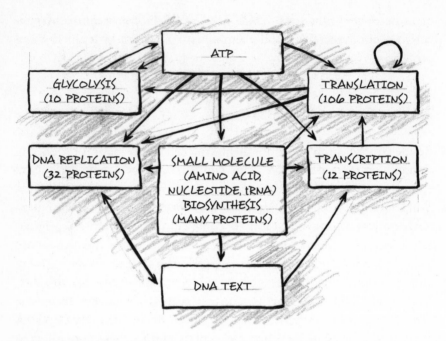

Figure 5.9. Schematic of the functional integration of the prokaryotic information-processing system listing the number of proteins involved in each process.

cell's information-processing system, again forming a "closed loop." Indeed, it even takes ATP to make ATP during glycolysis.[39] Figure 5.9 shows some of the functionally interdependent relationships among the systems of proteins that compose the cell's information-processing system.

Integrated Complexity and the Origin of Life

Following the elucidation of the structure and function of DNA during the 1950s and early 1960s, a radically new conception of life began to emerge. Not only did molecular biologists discover that DNA carried information; they soon began to suspect that living organisms must contain systems for processing genetic information. Just as the digital information stored on a disc is useless without a device for reading the disc, so too is the information on DNA useless without the cell's information-processing system. As Richard Lewontin notes, "No living molecule [i.e., biomolecule] is self-reproducing. Only whole cells may contain all the necessary machinery for self-reproduction. . . . Not only is DNA incapable of making copies of itself, aided or unaided, but it is incapable of 'making' anything else. . . . The proteins of the cell are made from other proteins, and without that protein-forming machinery nothing can be made."[40]

Crick was right: the cell contains not just molecular repositories of genetic information, but a code for translating the information in the DNA molecule (and its RNA transcript) into the construction of a protein. But this requires some physical medium of information transfer. After Crick and others realized that this transfer is not achieved via the direct attraction of amino acids to individual nucleotide bases or groups of bases—as Gamow had proposed—it became evident that the transcription and translation of genetic information is mediated by a complex information-processing system composed of many types of nucleic acids (such as mRNAs and tRNAs) and many specific enzymes.

These and other developments in molecular biology since the 1960s have shown that the information-processing system of the cell depends on a "tightly integrated" system of components—indeed, a system of systems. Both the transcription and translation systems depend upon numerous proteins, many of which are jointly necessary for protein synthesis to occur at all. Yet all of these proteins are made by this very process. Proteins involved in transcription such as RNA polymerases, for example, are built from instructions carried on an RNA transcript. Translation of the RNA transcript depends upon other specialized enzymes such as synthetases, yet the information to build these enzymes is translated during the translation process that synthetases themselves facilitate.

Biochemist David Goodsell describes the problem: "The key molecular process that makes modern life possible is protein synthesis, since proteins are used in nearly every aspect of living. The synthesis of proteins requires a tightly integrated sequence of reactions, most of which are themselves performed by proteins."[41] Or as Jacques Monod noted in 1971: "The code is meaningless unless translated. The modern cell's translating machinery consists of at least fifty macromolecular components *which are themselves coded in DNA: the code cannot be translated otherwise than by products of translation.*"[42] (Scientists now know that translation actually requires more than a hundred proteins.)[43]

The integrated complexity of the cell's information-processing system has prompted some profound reflection. As Lewontin asks, "What makes the proteins that are necessary to make the protein?"[44] As David Goodsell puts it, this "is one of the unanswered riddles of biochemistry: which came first, proteins or protein synthesis? If proteins are needed to make proteins, how did the whole thing get started?"[45] The end result of protein synthesis is required before it can begin.

The interdependence of proteins and nucleic acids raises many obvious "chicken and egg" dilemmas—dilemmas that origin-of-life theorists before the 1960s neither anticipated nor addressed. The cell needs proteins to process and express the information in DNA in order to build proteins. But the construction of DNA molecules (during the process of DNA replication) also requires proteins. So which came first, the chicken (nucleic acids) or the egg (proteins)? If proteins must have arisen first, then how did they do so, since all extant cells construct proteins from the assembly instructions in DNA. How did either arise without the other?

As the late British philosopher Sir Karl Popper mused, "What makes the origin of life and the genetic code a disturbing riddle is this: the code cannot be translated except by using certain products of its translation. This constitutes a really baffling circle: a vicious circle it seems, for any attempt to form a model, or a theory, of the genesis of the genetic code."[46] Even the simplest living things operate as complex systems of multiple interdependent parts. Yet how this feature of life could have arisen is, as Jacques Monod put it, "exceedingly difficult to imagine."[47]

The Question Redefined

The picture of the cell provided by modern molecular biology has led scientists to redefine the question of the origin of life. The discovery of life's information-processing systems, with their elaborate functional integration of proteins and nucleic acids, has made it clear that scientists investigating the origin of life must now explain the origin of at least three key features of life. First, they must explain the origin of the system for storing and encoding digital information in the cell, DNA's capacity to store digitally encoded information. Second, they must explain the origin of the large amount of specified complexity or functionally specified information in DNA. Third, they must explain the origin of the integrated complexity—the functional interdependence of parts—of the cell's information-processing system.

Later chapters in the book (8 through 16) describe the various attempts scientists have made to explain the origin of the information and information-processing system that the cell requires to maintain itself. As we will see, there have been three basic approaches for explaining the informational properties of life. The first relies heavily on random molecular interactions (chance). The second explains by reference to lawlike forces (necessity). The third approach combines chance and necessity. But before I investigated these possible explanations in detail I studied more about what scientists in the past thought about questions of origins and about how scientists in the present investigate them. I did this in part because I wanted to know if there was a fourth option—the design hypothesis—that could be legitimately considered as a possible scientific explanation for the DNA enigma.

6

The Origin of Science and the Possibility of Design

At the beginning of my graduate studies, I did not know the extent of the cell's complexity and, therefore, did not fully appreciate the challenge facing origin-of-life research. But as I was learning more about various origin-of-life scenarios, I also began to learn everything I could about molecular biology and the extraordinary discoveries that had taken place in that field during the preceding thirty-five years. Fortunately, the Cambridge system allowed graduate students to attend lectures in various disciplines, and our supervisors in the history and philosophy of science particularly encouraged us to attend lectures in scientific fields relevant to our topics. I found the challenge of retooling in molecular biology exciting and was further aided in my quest by a judicious choice of new friends, several of whom were molecular biologists working at the Medical Research Council Laboratory, including two American postdoctoral researchers who shared my interest in the origin of life and who more or less willingly indulged my persistent inquiries about the inner workings of the cell.

As I learned more about the chemistry of DNA, the sequence hypothesis, the gene-expression system, and what molecular biology generally had to say about the origin-of-life problem, I also learned more about how Watson and Crick had made their famous discovery. One day, Maurice Wilkins, the crystallographer who had received the Nobel Prize along with Watson and Crick for his work on the double helix, came to lecture in our department. In his talk, Wilkins told about his sometimes tense working relationship with Rosalind Franklin, the importance of her work, his encounters with Watson and Crick, and the key scientific insights that led to their eventual discovery. He was personally modest about the part he played in this historical achievement and clearly energized by the opportunity to relive the excitement of the discovery, although he did comment, rather plaintively, that if only he had had a bit more time, he might himself have solved the structure first.

Afterward, several scholars questioned Wilkins about the ethics of Watson and Crick's use of what they had learned from Wilkins about Franklin's X-ray images in order to determine the structure of DNA. Wilkins's response was interesting. He acknowledged that Franklin, who died before the Nobel Prize was awarded, had not received her share of credit for the important discovery, but he defended his rivals from Cambridge for their willingness and ability to synthesize evidence from diverse sources and disciplines in order to put all the pieces of a puzzle together into a coherent new picture of reality.

I had a chance to talk with Wilkins afterward and his defense of Watson and Crick left me with a key insight about how science works. Many people think that scientists spend all their time doing experiments in the laboratory in order to test hypotheses. The discovery of the structure of DNA showed that science certainly involves careful laboratory work. Rosalind Franklin's painstaking collection and analysis of X-ray data played an indispensable role in the discovery of the double helix, one for which she doubtless did not receive sufficient credit. But the discovery of the double helix showed that science also depends upon the synthesis of diverse lines of evidence in order to make sense of the big picture. In this, Watson and Crick excelled. Seen in this light, there was more method—more scientific method—to their mad dash to collect and synthesize all available data than some of their critics have allowed.

Though Watson and Crick were relatively unknown and certainly undercredentialed, they had solved one of the great scientific mysteries of the ages. Moreover, they achieved this feat not by working their way up through the establishment, which typically involves publishing a series of narrowly focused technical papers based on their own experimental research, but by explaining an array of preexisting evidence in a new and more coherent way. Could a similar approach help to crack the other mystery surrounding DNA, the mystery of how the digital information in the molecule had originated in the first place?

As I addressed these questions I found Watson and Crick's example instructive and encouraging. I was not an experimentalist, but a former applied scientist and philosopher of science. In my investigations of the DNA enigma, I began to marshal every relevant intellectual resource and insight—scientific, historical, mathematical, and philosophical— that I could. Watson and Crick showed that science involves such inter-disciplinary activity. Many scientists working on the origin of life are particularly aware of this. As Freeman Dyson, a Princeton physicist who has written extensively on the origin of life, explains, "The origin of life is one of the few scientific problems which is broad enough to make use of ideas from almost all scientific disciplines."[1]

Unfortunately, this understanding of how science works has not pen-etrated public understanding. Nor do scientists themselves always ac-knowledge it. Years later, as those of us in the intelligent design (ID) research community began to advance what we regarded as a new syn-thetic understanding of the origin and nature of life, we were repeat-edly maligned for not "doing science." In 2004–5, as the program I led at Discovery Institute suddenly found itself at the center of a frenzy of hostile media coverage, reporters kept repeating the same criticism, namely, "ID advocates aren't really scientists, because they don't do any experiments of their own." Reporters would then demand to see our laboratory, as if doing experiments was the only kind of activity that scientists pursued.

At the time, I knew there were scientists in labs around the world who supported the theory of intelligent design. Some of these scientists had done experiments testing different aspects of the theory, sometimes with our financial support. Beyond that, we actually *had* started our own laboratory, though we had chosen to keep it out of public view initially to protect some scientists and sensitive projects from prema-ture attempts to discredit them. Later when we publicly unveiled a lab, called Biologic Institute, even our critics acknowledged that our scien-tists were addressing significant scientific questions, though many still attempted to stigmatize Biologic's experimental research as "religiously motivated."[2] In any case, because of my role overseeing ID research at

Discovery Institute, I knew firsthand that this critique of the theory of intelligent design was not accurate: many ID scientists do in fact work in labs and conduct experiments.

But as Watson and Crick's discovery showed, even if advocates of ID weren't doing experiments in labs—as I personally was not—it didn't follow that we weren't "doing science." To say otherwise betrayed a blinkered view of the scientific enterprise. Watson and Crick performed many experiments during their long careers. But the work for which they are best known came as the result of building models based on data they acquired almost exclusively from other sources—from scientific journals, other scientists, and other labs.

Many of the great discoveries in science were achieved not just by experimentalists who produced new factual knowledge, but by theoreticians who taught us to think differently about what we already know. Examples of this kind of scientific work leaped to mind: Copernicus's *De revolutionibus orbium coelestium,* Newton's *Principia,* and the papers Einstein produced in his *annus mirabilis,* his miracle year of 1905. While working as a patent clerk without access to any experimental apparatus, Einstein rethought the whole framework of modern physics and, in the process, explained many previously confounding factual anomalies.[3]

Charles Darwin also did little experimental science. He did make several descriptive studies of barnacles and worms and some experimental studies about how species spread through seed dispersal and other processes. Yet his masterpiece, *On the Origin of Species by Means of Natural Selection,* contains neither a single mathematical equation nor any report of original experimental research. Yet he formulated a great scientific theory. He did this by drawing together disparate lines of observational evidence and presenting an argument for a novel interpretation of that evidence. Of course Darwin made some of the observations that supported his theory himself, but even if he had not done so, this would have hardly detracted from his theory. Darwin's method of investigation typified that of many other historical scientists who functioned more like detectives solving a mystery by collecting clues and developing a case than like stereotypical experimental scientists who test hypotheses under carefully controlled laboratory conditions.

But I had other reasons for liking the story of Watson and Crick, and particularly the story of their encounter with Erwin Chargaff, in front of whom they had grievously embarrassed themselves in their headlong attempt to acquire information about the chemical structure of the mystery molecule. During my years in Cambridge and even after, I had a number of similar experiences—although I can't say I wasn't warned.

One day I attended a spellbinding lecture in my department by a visiting scholar who was perhaps the world's leading authority on Immanuel Kant. During the ensuing discussion, I asked him if he would direct me to a good source whereby I could learn more about a particular aspect of Kant's philosophy that he had just discussed. He answered me graciously enough, but I had a vague sense from others in the room that I may have created an awkward moment. As I was leaving the lecture hall, the head of my department, Michael Hoskin, a kindly Cambridge don of the old school, gently pulled me aside. "Meyer," he intoned in his high Oxbridge accent, "I know that in the States you've learned that the only stupid question is the one you don't ask. But it's different here. Everyone here is bluffing, and if you're to succeed, you must learn to bluff too. So never admit you haven't read something. And in future, if you have a question that reveals ignorance, please, come ask me privately."

Professor Hoskin's advice was no doubt good, but I was temperamentally disinclined to follow it. As I met Cambridge scientists, philosophers, and fellow graduate students over high-table dinners or in common-room discussions, I often shared my interest in the DNA enigma. When I met someone with relevant expertise, I would probe for insights. Often, I learned things as a result, but sometimes I felt the sting of embarrassment at having asked one too many questions.

Foolishly perhaps, I also sometimes let it slip that I was interested in the question of design and, specifically, in the question of whether a rigorous scientific argument for design could be reformulated based upon the discovery of the information-bearing properties of DNA. It did not take me long, naïve American though I was, to understand that this was not always a welcome turn in the conversation. Sometimes fellow graduate students would warn me about not venturing into such a dangerous topic, though more often I received blank and impassive stares. Cambridge was, after all, Darwin's alma mater, and the design hypothesis had not been seriously debated in biology for almost a hundred and fifty years.

Oddly, I did find willingness to discuss these topics among some of the more prominent scientists whom I had the good fortune to meet. The year before coming to Cambridge, I had the opportunity to interview the Nobel Prize–winning physicist Eugene Wigner at a conference at Yale. I asked him directly about the origin-of-life problem, and he explained why he thought the odds were overwhelmingly against any process of undirected chemical evolution producing life. Wigner further expressed openness to, even sympathy for, the idea that the information present in DNA pointed to an intelligent source.

Upon arriving in Cambridge I had a few similar experiences, but again only with very prominent scientists, those who, perhaps, felt secure enough in their reputations to entertain heterodox ideas openly. Once, before attending a high-table dinner at Churchill College as the guest of Professor Hoskin, a charming Hungarian man broke with Cambridge etiquette and sat down cross-legged in front of me on the carpet in the senior common room. He introduced himself merely as Herman and began asking me questions about my research interests. As I explained them, he expressed his fascination with my ideas about DNA and shared some of his own questions about the origin-of-life problem. Later, as we resumed our conversation after a dinner in which anti-American invective was flowing as freely as the port and Madeira, Professor Hoskin circled around to make sure that we had made proper introductions.

"Oh, Bondi," he said, "I'm delighted to see that you've met young Meyer." I gulped as I suddenly realized who it was that I had been regaling with my various musings about the origin of life. "Herman" was Sir Professor Herman Bondi, the mathematician and cosmologist who had, with Sir Fred Hoyle, formulated the steady-state theory of the universe and who was also famous for his work on general relativity. Beyond that, I knew that Bondi was a well-known secular humanist. He had every philosophical inclination to dismiss my ideas out of hand. Yet there he sat, cross-legged, in front of a graduate student, openly discussing a radical proposal for solving the problem of the origin of life. I'll never forget him.

Later Bondi's collaborator, Fred Hoyle, also came to Cambridge to discuss his views on the origin of life. This was a lecture I couldn't miss. Hoyle was famous for his correct prediction of the resonance levels of the carbon atom and for his pioneering work on the fine-tuning problem in physics and cosmology. But he hadn't come to Cambridge to discuss any of this. He was here to explain why he had come to reject chemical evolutionary theory. Afterward I asked him directly about whether he thought the information stored in DNA might point to an intelligent source. His eyes brightened, and he motioned me to continue walking with him after his lecture. "That would certainly make life a lot easier to explain," he said.

Modern Cambridge, Ancient Cambridge

Apart from a few such encounters, contemporary Cambridge did not share my interest in design. Nevertheless, it didn't take me long to find that ancient Cambridge did. Cambridge was world-famous as a city where the history of science was not just studied, but made. I soon discovered that those who had established the scientific enterprise and, for several hundred years, made its history did not share the disdain for the design hypothesis that I had encountered. The founders of the scientific revolution (ca. 1300–1700) were often deeply religious men who expressed a profound appreciation for the design of life and the universe. Moreover, for these scientists, the concept of design was not just a pious sentiment. For them it was an indispensable assumption upon which the whole of the scientific enterprise rested.

As many historians of science have noted, the founders of modern science needed to assume that if they studied nature carefully, it would reveal its secrets. Their confidence in this assumption was grounded in the Greek and Judeo-Christian idea that the universe is an orderly system—a cosmos, not a chaos. As the British philosopher Alfred North Whitehead explained, "There can be no living science unless there is a widespread instinctive conviction in the existence of an *Order of Things*. And, in particular, of an *Order of Nature*." Whitehead argued that confidence in this proposition was especially inspired by the "medieval insistence upon the rationality of God."[4]

Other historians of science have amplified Whitehead's observation. They have insisted that modern science was specifically inspired by the conviction that the universe is the product of a rational mind who designed the universe to be understood and the human mind to understand it. As sociologist of science Steve Fuller notes, Western science is grounded in "the belief that the natural order is the product of a single intelligence from which our own intelligence descends."[5] This foundational assumption gave rise to the idea that nature was "intelligible," that it had been designed in accord with discernible laws that could be understood by those who subjected nature to careful scrutiny. Or as the astronomer Johannes Kepler said, scientists have the job of "thinking God's thoughts after him."[6]

Though Greek, Jewish, and Christian philosophers agreed about the rationality of nature, they did not necessarily agree about how to discover it. Many historians of science have noted that Greek ideas about nature tended to induce a sterile armchair philosophizing unconstrained by actual observations. For science to advance, it needed to develop a more empirical approach. This began to occur during the scientific revolution after medieval philosophers and the early modern scientists made a decisive break with one aspect of Greek thinking.

Although the Greek philosophers thought that nature reflected an underlying order, they nevertheless believed that this order issued not from a designing mind, but from an underlying and self-evident logical principle. For this reason, many assumed that they could deduce how nature ought to behave from first principles without actually observing nature. In astronomy, for example, the Greeks (Aristotle and Ptolemy) assumed that planets must move in circular orbits. Why? Because according to the Greek cosmology, the planets moved in the "quintessential" realm of the crystalline spheres, a heavenly realm in which only perfection was possible. Since, they deduced, the most perfect form of motion was circular, the planets must move in circular orbits. What could be more logical?

The idea of design helped liberate Western science from such fact-free reasoning. Like the Greek philosophers, the early modern scientists thought that nature exhibited an underlying order. Nevertheless, they thought this natural order had been impressed on nature by a designing mind, in particular, the mind of the Judeo-Christian God.

For this reason they thought that the order in nature was the product not of logical necessity, but of rational deliberation and choice. Because nature had been designed by the same mind that had designed the human mind, the early scientists assumed that nature was intelligible, that it could be understood by the human intellect. But because the order in nature was also contingent on the will of its designer, they realized they had to observe nature in order to get it to reveal its secrets.

Just as there are many ways to paint a picture or design a clock or organize the books in a library, there were many ways to design and organize a universe. Because it had been chosen by a rational mind, the order in nature could have been otherwise. Thus, the natural philosophers could not merely deduce the order of nature from logical first principles; they needed to observe nature carefully and systematically. As the chemist Robert Boyle explained, the job of the natural philosopher (the scientist) was not to ask what God must have done, but (as far as possible) to inquire into what God actually did. Boyle argued that God's absolute freedom as designer and creator requires of us an empirical and observational, not a deductive, approach to the natural order: *look*—observe—*to find out*.[7]

Thus, the assumption that a rational mind had designed the universe gave rise to two ideas—intelligibility and contingency—which, in turn, provided a powerful impetus to study nature and to feel confident that such study would yield understanding. As the Oxford physicist and historian of science Peter Hodgson observes: "According to Judeo-Christian beliefs the world is the free creation of God from nothing. The structure of the world cannot therefore be deduced from first principles; we have to look at it, to make observations and experiments to find out how God made it. This reinforces the Aristotelian principle that all knowledge comes through the senses, but requires that it be situated within a wider set of beliefs concerning the nature of the world that is implicit in the doctrine of creation."[8] Hodgson notes that early scientists assumed that the world was both rational—because it was created by a Mind—and contingent—because that Mind had acted freely. These assumptions led to "a fresh style of scientific thinking," one that "was made possible by the Judeo-Christian vision of the world."[9]

Everywhere I went in the city of Cambridge, I encountered evidence of this long dominant viewpoint. Each day as I walked to my department on Free School Lane, I passed by the entrance to the old Cavendish Laboratory in which thirty-odd years before Francis Crick and James Watson realized that their model of DNA was so beautiful it had to be right.[10] On the archway across the great wooden door of the Cavendish was a Latin inscription that reads, *Magna opera Domini exquisita in omnes voluntates ejus.* The inscription had been placed there at the insistence of the physicist James Clark Maxwell, the first Cavendish professor in 1871. The inscription quotes a Psalm that reads, "Great are the works of the Lord, sought out by all who take pleasure therein." The inscription summarized Maxwell's inspiration for scientific study: the thought that works of nature reflect the work of a designing mind. In this belief he had been joined by many of the leading scientists of Western civilization for over four hundred years—Copernicus, Kepler, Ray, Linnaeus, Cuvier, Agassiz, Boyle, Newton, Kelvin, Faraday, Rutherford—on and on the list could go.

As I studied the history of science, I soon discovered, however, that many of these scientists did not just assume or assert by faith that the universe had been designed; they also argued for this hypothesis based on discoveries in their disciplines. Johannes Kepler perceived intelligent design in the mathematical precision of planetary motion and the three laws he discovered that describe that motion. Other scientists perceived design in many of the structures or features of the natural world upon which the laws of nature operated. Louis Agassiz, the leading American naturalist of the nineteenth century, for whom the Agassiz Chair is named at Harvard, believed that the patterns of appearance in the fossil record pointed unmistakably to design.[11] Carl Linnaeus argued for design based upon the ease with which plants and animals fell into an orderly groups-within-groups system of classification.[12] Robert Boyle insisted that the intricate clocklike regularity of many physical mechanisms suggested the activity of "a most intelligent and designing agent."[13]

Newton, in particular, was noteworthy in this regard. As I discussed in Chapter 1, he made specific design arguments based upon discoveries in physics, biology, and astronomy. He argued for the intelligent design of the eye in his classic work, *Opticks.* He also argued for the intelligent design of the planetary system in his masterpiece, *The Mathematical Principles of Natural Philosophy* (often cited in brief by part of its Latin title, the *Principia*).[14] Writing in the General Scholium (introduction) to the *Principia*, Newton suggested that the stability of the planetary system depended not only upon the regular action of universal gravitation, but also upon the precise initial positioning of the planets and comets in relation to the sun. As he explained: "Though these bodies may, indeed, persevere in their orbits by the mere laws of gravity, yet they could by no means have, at first, derived the regular position of the orbits themselves from those laws. . . . [Thus] this most beautiful system of the sun, planets, and comets could only proceed from the counsel and dominion of an intelligent and powerful being."[15]

So central was the idea of design to Newton's scientific perspective that I was repeatedly cautioned not to miss it as I researched Newton's ideas about gravity in the *Principia*. In one tutorial, one of my supervisors, a staunch atheist, reminded me not to read modern secular biases into my interpretation of Newton's work. "If you miss Newton's theism," he exhorted, "you've missed everything."

What I learned about Newton would come in handy years later. In 1999, I was asked to testify at a hearing of the U.S. Commission on Civil Rights about the alleged viewpoint discrimination in the teaching of biological origins. My opposing expert witness at this hearing was Eugenie Scott, an anthropologist who heads the National Center for Science Education. Scott had distinguished herself as an ardent critic of the scientific legitimacy of intelligent design and has consistently argued that the very idea of intelligent design violates the rules of science. According to Scott, these rules preclude invoking any nonnaturalistic or nonmaterialistic causes (such as the action of a designing intelligence) to explain features of the natural world.

At the hearing, one of the commissioners asked me an interesting question. He wondered, given Newton's views on intelligent design in physics and Dr. Scott's widely accepted definition of science, whether Newtonian physics would qualify as "scientific" in our present educational climate. Eugenie Scott responded by insisting that the commissioner had misunderstood the context in which Newton discussed intelligent design. She explained that, for Newton, intelligent design was merely a theological belief—one that he took pains to keep separate from his science so as to avoid subjecting it to empirical test. Here's what she said according to the transcript of the hearing: "Newton made a very clean distinction—as a very religious scientist—he made a very clear distinction about how science should work. Newton's view was that we should understand the natural world solely by using natural processes. And he said this for religious reasons because he didn't want God's existence or God's transcendence, shall we say, to be tested by the base methods of science." Scott then proceeded to reassert that invoking intelligent design would violate the central rule of science, namely, as she put it, that "we explain the natural world restricting ourselves only to natural cause."[16]

As it happened, I had just finished an essay on the history of scientific design arguments and had quoted Newton's argument in the *Principia*—an indisputably scientific treatise—at length. The essay was at that moment sitting in my briefcase. As the transcript of the hearing makes clear, I immediately challenged Scott's portrayal of Newton. "The historical point on Newton, I'm afraid, is just simply incorrect. If one opens the General Scholium, the introduction to the *Principia,* arguably the greatest book of science ever written, one finds an exquisite design argument by Newton in which he makes clear that the arrangements of the planets can only be explained not, he says, by natural law, but only by the contrivance of a most wise artificer. He's very explicit about this. This is in the introduction to his magnum opus." Several of the commissioners appeared intrigued, so I pressed on, noting that Newton was not alone in his formulation of such design arguments. I continued, "You find these kinds of design arguments all throughout the scientific revolution, from Boyle, from Kepler, from others."[17]

The Abolition of Design

Based on my research (and from noticing many of the landmarks at Cambridge), I had learned that the historical record on this point was irrefutable. I began to wonder: How could the act of invoking something so foundational to the history of science as the idea of design now completely violate the rules of science itself, as I had repeatedly heard many scientists assert? If belief in intelligent design first inspired modern scientific investigation, how could mere openness to the design hypothesis now act as a "science stopper" and threaten to put an end to productive scientific research altogether, as some scientists feared?

Clearly, the idea of intelligent design had played a formative role in the foundation of modern science. Many great scientists had proposed specific design hypotheses. This seemed to suggest to me that intelligent design could function as a possible scientific hypothesis. But many contemporary scientists rejected this idea out of hand. Why?

The Nature of Historical Scientific Reasoning

I remembered that my Dallas mentor, Charles Thaxton, thought that many scientists today rejected design out of hand because they failed to recognize that there were different types of scientific inquiry, specifically, that there was a distinctive kind of scientific inquiry concerned with investigating and explaining the past. His distinction between origins and operations science suggested a reason to consider an "intelligent cause" as a possible scientific explanation for the origin of life. Thaxton thought that since the origin of life was a unique historical event, and since the singular actions of intelligent agents often produce unique events, the act of a designing agent might provide an appropriate kind of scientific explanation for this event. This made sense when I first encountered Thaxton's distinction. It seemed to me that invoking an "intelligent cause" as the explanation for a past event might be a perfectly reasonable thing to do for the simple reason that the activity of an intelligent agent might have actually caused the event in question.

It now occurred to me that the great scientists who had proposed design hypotheses during the scientific revolution had typically done so to explain the *origin* of some system or feature of the world—the eye, solar system, the universe, the "correlation of parts" in biological systems.[18] So perhaps Thaxton was right. Perhaps the design hypothesis could be formulated as a rigorous scientific explanation for events such as the origin of life. But to decide that question I would need to know how scientists investigated the past and how they formulated and justified their theories. And so the focus of my investigation shifted from an examination of the features of life that needed to be explained (see Chapters 3 through 5) to a study of how historical scientists go about explaining the origin of particular features and events in the remote past (see Chapter 7).

7

Of Clues to Causes

When I left Dallas in 1986 I was seized by questions that had emerged in my discussions with Charles Thaxton. Could the inference to design from DNA be formulated as a rigorous scientific argument? Is there a distinctive scientific method for studying the remote past? When I left for Cambridge, I was intrigued by Thaxton's ideas, but not fully convinced. For one thing, I didn't find his account of the distinction between the two types of science fully satisfying. The distinction seemed plausible, but needed further clarification and justification.

Even so, I soon discovered that other scientists had made comparable distinctions, though usually using different terminology. Stephen Jay Gould, the Harvard paleontologist and historian of science, insisted that the "historical sciences" such as geology, evolutionary biology, and paleontology used different methods than did "experimental sciences" such as chemistry and physics. Interestingly, he also argued that understanding how historical sciences differed from experimental sciences helped to legitimate evolutionary theory in the face of challenges to its scientific rigor by those who questioned its testability. Gould argued that historical scientific theories were testable, but not necessarily by experiments under controlled laboratory conditions. Instead, he emphasized that historical scientists tested their theories by evaluating their explanatory power.[1]

Could the same thing be true of the design hypothesis? Perhaps a theory of intelligent design could be formulated as a historical scientific theory about what had happened in the past. If so, perhaps such a theory could be tested by assessing its explanatory power rather than its ability to generate predictions in a controlled laboratory setting.

In my second year at Cambridge, after I had acquired a better understanding of what origin-of-life researchers needed to explain, I began to investigate the questions that first emerged in my discussions with Thaxton. Is there a distinctive mode of historical scientific investigation? If so, how do scientists reason and make inferences about the past? Are these inferences testable? If so, how? And in the back of my mind another question continued to linger: If there is a historical way of scientific reasoning, could such reasoning be used to make a rigorous scientific case for intelligent design?

As I began to study how scientists investigate the past, I examined the works of the nineteenth-century scientists and philosophers of science who first developed and refined these methods.[2] This eventually led me to the masters: to Charles Darwin and his mentor, the geologist Charles Lyell, and to a few lesser known luminaries who, along with Darwin and Lyell, pioneered a method of scientific investigation designed to illuminate the history of life, the earth, and the cosmos.

The Science of Past Causes

Visit the Royal Tyrrell Museum in Alberta, Canada, and you will find a beautiful reconstruction of the Cambrian seafloor with its stunning assemblage of ancient organisms. Or read the fourth chapter of Cambridge paleobiologist Simon Conway Morris's book on the Burgess Shale and you will be taken on a vivid guided tour of an ancient marine environment teeming with exotic life-forms.[3] How do scientists come to know about such ancient environments? What methods of reasoning do they use to investigate what happened on the earth so long ago?

I soon discovered that I wasn't the first person to ask such questions. During the 1830s and 1840s, William Whewell, a distinguished scientist and philosopher and the master of Trinity College, Cambridge, published two books about the nature of science that addressed this issue.[4] The publication of Whewell's work coincided with a surge of interest in natural history in Victorian England. By the 1830s, Charles Lyell had published his seminal *Principles of*

Geology, the first scientific descriptions of dinosaur fossils had been made, and new evolutionary ideas about the history of life were circulating in elite scientific societies.

In his volumes *History of the Inductive Sciences* (1837) and *The Philosophy of the Inductive Sciences* (1840), Whewell distinguished sciences such as physics and chemistry from what he called palaetiology—literally, the "study of past causes." And he argued that "palaetiological," or historical, sciences could be distinguished from nonhistorical sciences in three ways.[5] First, the palaetiological, or historical, sciences have a distinctive *objective:* to determine ancient conditions or past causes, as opposed to establishing universal laws by which nature generally operates.[6]

Second, such sciences *explain* present events ("manifest effects") by reference to past (causal) events, rather than by reference to general laws (though Whewell acknowledged that laws often play a supporting role in such explanations).[7] Historical scientists cite the occurrence of an event or series of events in the past as the explanation for some observable phenomenon in the present.[8] For example, a geologist might explain the origin of the Alps as the result of a series of geological events involving the collision of specific tectonic plates, the overthrusting of sedimentary layers, and then the subsequent folding and faulting of those layers. As science historian Jonathan Hodge explains, Whewell realized that historical sciences do not study "forces that are permanent causes of motion, such as gravitational attraction," but "causes that have worked their effects in temporal succession."[9]

Third, in their attempt to reconstruct "ancient conditions," Whewell argued that historical scientists also utilized a distinctive mode of reasoning.[10] Using knowledge of cause-and-effect relationships, historical scientists "calculate backwards" and "infer" past conditions and causes from "manifest effects."[11] As Gould later put it, historical scientists proceed by "inferring history from its results."[12] For example, in order to reconstruct the Cambrian environment, paleontologists such as Gould and Conway Morris made inferences about the past based on present-day fossils and other clues.[13] They inferred a past environment and set of conditions as the cause of the evidence they found. Like other historical scientists, they reasoned from clues back to causes.[14]

Abductive Reasoning

This type of reasoning is called abductive reasoning or abduction. It was first described by the American philosopher and logician Charles Sanders Peirce. He noted that, unlike inductive reasoning, in which a universal law or principle is established from repeated observations of the same phenomena, and unlike deductive reasoning, in which a particular fact is deduced by applying a general law to another particular fact or case, abductive reasoning infers unseen facts, events, or causes in the past from clues or facts in the present.[15]

As Peirce himself showed, there is a problem with abductive reasoning.[16] Consider the following syllogism:

If it rains, the streets will get wet.
The streets are wet.
Therefore, it rained.

In this syllogism, a past condition (it was raining) is inferred from a present clue (the streets are wet). Nevertheless, this syllogism commits a logical fallacy known as *affirming the consequent*. Given that the street is wet (and without additional evidence to decide the matter), one can only conclude that *perhaps* it rained. Why? Because there are other possible ways the street may have gotten wet. A street-cleaning machine, an uncapped fire hydrant, or rain might have caused the street to get wet. It can be difficult to infer the past from the present because there are often many possible causes of a given effect. When this is the case, abductive inferences yield plausible, but not certain, conclusions.

For Peirce, this raised an important question: How is it that, despite the logical problem of affirming the consequent, we nevertheless frequently make conclusive inferences about the past? He noted, for example, that no one doubts the existence of Napoleon. Yet we use abductive reasoning to infer Napoleon's existence. That is, we infer his past existence not by traveling back in time and observing him directly, but by inferring his existence from our study of present effects, namely, artifacts and records. But despite our dependence on abductive reasoning to make this inference, no sane and educated person would doubt that Napoleon Bonaparte actually lived. How could this be if the problem of affirming the consequent bedevils our attempts to reason abductively? Peirce's answer was revealing: "Though we have not seen the man [Napoleon], yet we *cannot explain* what we have seen without the hypothesis of his existence."[17] Peirce suggested that a particular abductive hypothesis can be firmly established if it can be shown that it represents the best or only explanation of the "manifest effects" in question.

As Peirce noted, the problem with abductive reasoning is that there is often more than one cause that can explain the same effect. To address this problem in geology, the late-nineteenth-century geologist Thomas Chamberlain delineated a method of reasoning he called the "method of multiple working hypotheses."[18] Geologists and other historical scientists use this method when there is more than one possible cause or hypothesis to explain the same evidence. In such cases, historical scientists carefully weigh the relevant evidence and what they know about various possible causes to determine which best explains it. Contemporary philosophers of science call this the method of "inference to the best explanation." That is, when trying to explain the origin of an event or structure from the past, historical scientists compare various hypotheses to see which would, if true, best explain it. They then provisionally affirm the hypothesis that best explains the data as the most likely to be true.

Peter Lipton and Inference to the Best Explanation

In the spring of 1990, as I was pressing to finish my Ph.D. dissertation, I had the good fortune to meet an American philosopher of science who helped me understand abductive reasoning. Peter Lipton had just been hired away from Williams College in Massachusetts by the Cambridge Department of History and Philosophy of Science. After assuming the position of lecturer (the British equivalent of assistant professor), he eventually rose to become professor and chair of the department before his premature death in 2007 at the age of fifty-three. While Lipton was visiting Cambridge in the spring of 1990, one of the Cambridge faculty members, seeing that we shared some common interests, introduced us. Lipton had just finished writing a major treatise called *Inference to the Best Explanation,* which later would become something of a minor classic in contemporary philosophy of science. Given what I had learned about how historical scientists needed to weigh competing hypotheses, I was immediately interested in Lipton's ideas. He kindly sent me a pre-publication copy of his manuscript.

Lipton had made a systematic study and defense of a way of reasoning that he called "inference to the best explanation" (IBE). He argued that this way of reasoning was used commonly in science and ordinary life. For example, he noted, that "faced with tracks in the snow of a peculiar shape, I infer that a person with snowshoes has recently passed this way. There are other possibilities, but I make this inference because it provides the best explanation of what I see."[19] From the examples he used to illustrate "inference to the best explanation," it was clear that this method of reasoning was especially useful to scientists who were trying to provide causal explanations of events or circumstantial evidence.

I was impressed by Lipton's work. He wrote clearly, using practical scientific case studies to illustrate his ideas about this scientific method. His work employed none of the jargon or penchant for abstraction that

Figure 7.1. Peter Lipton. *Printed by permission from Howard Guest.*

sometimes gave philosophers of science a bad name with working scientists. Instead, he obviously respected the practice of science and wanted to give an accurate account of how scientists actually formulate and justify their hypotheses and inferences.

Interestingly, in his Oxford doctoral dissertation, Lipton initially had argued *against* inference to the best explanation as an accurate understanding of scientific reasoning. He did this, he said, because he didn't have time as a doctoral student to mount a sufficiently comprehensive case for it. Instead, he decided to make the best case against it, so that he would fully understand the objections to it as a way of preparing himself to write a book defending it. His story revealed a person capable of weighing the merits of competing arguments before deciding a question. Lipton practiced what he preached.

According to Lipton, "beginning with the evidence available to us" we generally "infer what would, if true, best explain that evidence."[20] This echoed what Peirce had said about how an abductive inference can be strengthened by showing that it alone can explain the evidence. But Lipton made the connection between inference and explanation more explicit and showed how considerations of explanatory power often influenced assessments of competing inferences or hypotheses. Peirce and Gould had convinced me that historical scientists used a distinctive kind of inference—in which past causes are inferred from present effects. Whewell and other philosophers of science showed that historical scientists formulated distinctive types of explanations—ones in which past events or causes are invoked to explain particular facts. Lipton now showed how the two intellectual activities of inferring and explaining are connected.[21]

He showed that scientists often evaluate competing inferences or hypotheses by comparing their explanatory power. That is, assessments of an inference's explanatory power determine how much stock we place in it. In making this argument, Lipton challenged a popular conception of science, namely, that scientists test their theories only by evaluating the accuracy of the predictions that their theories make about future events.[22] Lipton demonstrated that the ability to explain known facts often mattered as much or more than predictive success in the evaluation of an inference or hypothesis.[23]

But there was a problem with this kind of reasoning, one that Lipton and many philosophers of science had noted. Many regarded "inference to the best explanation" as little more than a slogan, because no one could say exactly what made an explanation best. "Sure," many argued, "we often infer hypotheses that, if true, best explain the evidence, but what does it mean to explain something and to explain it best?"[24]

As a first step toward answering the question, "What makes an explanation best?" Lipton gave a general account of what a good explanation should look like. He noted that scientists sometimes explain general classes of phenomena (such as condensation) and sometimes a particular fact or event (such as the extinction of the dinosaurs). According to Lipton, good explanations—whether of particular one-time events or general phenomena—are typically causal.

In the case of general phenomena such as condensation, good explanations typically cite causal mechanisms. For example, a physicist would explain condensation by citing a general mechanism of heat exchange. (When warm water vapor contacts a cold solid, the colder solid draws heat out of the gas, causing the water to change from a gaseous to a liquid state.)

In the case of particular events, Lipton showed that a good explanation cites a prior cause, typically an event that is part of the "causal history" of the event in question. Further, Lipton showed that to identify the cause of an event, scientists must identify something within the causal history of the event that accounts for a crucial difference—the difference between what did occur and what we might have otherwise expected. Good explanations answer questions of the form, "Why this rather than that?" A good (or best) explanation cites an event that makes a "causal difference" in outcome.[25]

Here's an example of what Lipton had in mind. Imagine that a paleontologist wants to explain why the dinosaurs went extinct in the Cretaceous period. He realizes that many events led up to the extinction of the dinosaurs, from the big bang to the formation of continents to the origin of the dinosaurs themselves. All these events are part of the "causal history" of dinosaur extinction. But of these events, the one that best explains the extinction is the event—the cause—that accounts for the difference between what happened and what we might have otherwise expected to happen—in this case, the difference between the extinction of the dinosaurs and the dinosaurs continuing to roam the earth. The big bang is part of the causal history of extinction of the dinosaurs, but it doesn't explain the extinction. Why? It does not account for the difference in outcome. Whether the dinosaurs had continued to live or became extinct, the big bang would still have been part of either causal history.

As it happens, paleontologists have proposed a theory that cites a causal difference. As an explanation for a variety of evidence, they have inferred that a massive meteorite hit the earth at the end of the Cretaceous period, causing an environmental catastrophe. They postulate that the meteorite impact generated dust and debris that blocked sunlight and cooled the earth, eventually killing the cold-blooded dinosaurs by destroying their food supply.

This explanation illustrates Lipton's conception of a sound causal explanation. It explains one event (the extinction of the dinosaurs) by citing another prior event (the meteorite impact and subsequent environmental catastrophe) as the cause. It also accounts for the difference between what happened (the extinction of the dinosaurs) and what otherwise might have been expected to happen (the dinosaurs continuing to thrive). Indeed, had the meteorite not hit the earth, the dinosaurs would, presumably, have continued to live beyond the Cretaceous period. Thus, the meteorite impact could have made a causal difference. The impact hypothesis also could explain many other evidences—such as the iridium levels in rocks from the period and the presence of an impact crater in the Yucatan Peninsula—and so it has broad explanatory power. None of these facts alone prove the hypothesis is correct, but they provide support for it and put it in contention as a possible best explanation.

In *Inference to the Best Explanation,* Lipton went on to develop an even more comprehensive account of causal explanation in order to enhance his characterization of how the method of inference to the best explanation works. In the process, he also provided a general, if partial, answer to the question, "What makes an explanation best?" Because of his more general focus on how IBE works, however, Lipton did not address the specific questions about the nature of historical scientific reasoning that most interested me. Nevertheless, his general description of how scientists evaluate inferences by assessing their explanatory power aptly characterized how historical scientists evaluate their theories. Indeed, many of his examples involved scientists trying to explain particular events by reference to past causes.

Moreover, his method directly addressed the main problem that often arose in historical scientific investigations. Lipton noted that sometimes more than one possible explanation might satisfy his general criteria of "best" by citing a past cause that could conceivably make a causal difference. For example, paleontologists have proposed other possible causes for the extinction of the dinosaurs—such as a sharp increase in volcanic activity—that could explain why the dinosaurs went extinct rather than continuing to exist. Lipton noted that in such cases, where more than one appropriately causal explanation is available, scientists use a comparative method of evaluation and a process of elimination to evaluate competing possible causal hypotheses.[26]

At the same time, Lipton did not attempt to establish criteria for determining which *past* cause among a group of possible causes constituted the best explanation in such cases. Fortunately, however, I discovered that the founders of several historical sciences had developed their own practical criteria for determining what made an explanation of a past event "best." These criteria supplemented Lipton's work in a way that made them practically relevant to scientists investigating the past and helped to answer the question, "What makes an explanation of a past event 'best'?"

Causes Now in Operation

Historical scientists have developed two key criteria for deciding which cause, among a group of competing possible causes, provides the best explanation for some relevant body of evidence. The most important of these criteria is called "causal adequacy." This criterion requires that historical scientists, as a condition of a successful explanation, identify causes that are known to have the power to produce the kind of effect, feature, or event in need of explanation. In making these determinations, historical scientists evaluate hypotheses against their present knowledge of cause and effect. Causes that are known to produce the effect in question are judged to be better candidates than those that are not.

For instance, a volcanic eruption provides a better explanation for an ash layer in the earth than an earthquake, because eruptions have been observed to produce ash layers, whereas earthquakes have not. A receding lakeshore offers a better explanation of ripple marks in sedimentary rock than does a volcanic eruption, because evaporating lakes have been observed to produce ripple marks, whereas volcanic eruptions have not.

One of the first scientists to develop this principle was the geologist Charles Lyell, who exerted tremendous influence on the development of nineteenth-century historical science—and on Charles Darwin, in particular. Darwin read Lyell's magnum opus, *The Principles of Geology*, on the voyage of the *Beagle* and employed its principles of reasoning in *On the Origin of Species*. The subtitle of Lyell's *Principles* summarized the geologist's central methodological principle: *Being an Attempt to Explain the Former Changes of the Earth's Surface, by Reference to Causes Now in Operation*. Lyell argued that when historical scientists are seeking to explain events in the past, they should not invoke unknown or exotic causes, the effects of which we do not know; instead, they should cite causes that are known from our uniform experience to have the power to produce the effect in question.[27] Historical scientists should cite *"causes now in operation,"* or presently acting causes. This was the idea behind his uniformitarian principle and the dictum, "The present is the key to the past." According to Lyell, our *present* experience of cause and effect should guide our reasoning about the causes of *past* events. Or as Whewell explained, "Our knowledge respecting [past] causes . . . must be arrived at by ascertaining what the causes of change in such matters *can* do."[28]

Darwin himself adopted this methodological principle. His term for a presently acting cause was a *vera causa*, that is, a true, known, or actual cause.[29] Darwin thought that when explaining past events, scientists should seek to identify established causes—causes known to produce the effect in question. Darwin appealed to this principle to argue that presently observed microevolutionary processes of change could be used to explain the origin of new forms of life in the past.[30] Since the observed process of natural selection can produce a small amount of change in a short time, Darwin argued that it was capable of producing a large amount of change over a long period of time.[31] In that sense, natural selection was "causally adequate."

In the late 1950s and early 1960s, a leading philosopher of science from the University of California at Berkeley named Michael Scriven made a detailed study of how historical scientists reconstruct the past. He argued that historical scientists use a distinctive method of reasoning that he called "retrospective causal analysis."[32] In his description of this method of reasoning, he echoed the importance of the causal-adequacy criterion affirmed by Darwin and Lyell. According to Scriven, in order to establish the cause of a past event, historical scientists must first show that any candidate cause "has on other occasions clearly demonstrated its capacity to produce an effect of the sort here under study."[33] As Scriven put it, historical scientists must show that "*independent* evidence supports the claim that it [the cause] can produce this effect."[34]

And Then There Was One

Both philosophers of science and leading historical scientists have emphasized causal adequacy as the main criterion by which competing hypotheses are adjudicated. But philosophers of science also have noted that assessments of explanatory power lead to sound inferences only when it can be shown that there is one known cause for the effect or evidence in question. Michael Scriven, for example, points out that historical scientists can make inferences about the past with confidence when they encounter evidence or a condition for which there is only one known cause.[35] When historical scientists know that there is only one known cause of a given effect, they can infer the cause from the occurrence of its effect. When scientists can infer a *uniquely* plausible cause, they can avoid the fallacy of affirming the consequent—the error of ignoring other possible causes with the power to produce the same effect.[36]

In his book *Reconstructing the Past*, Elliott Sober shows that the tasks of inferring the past may be easy or difficult (or even impossible) depending upon whether there are many causes that produce the effect in question or just one. He suggests that the severity of the problem confronting historical scientists will depend on whether the processes linking the present and past are "information preserving" or "information destroying."[37] To Sober, an information-preserving process is one that maps a present state to a single past state, whereas an information-destroying process is one that maps a present state to a multiplicity of possible past states. He illustrates these two general cases with the following idealized diagram:

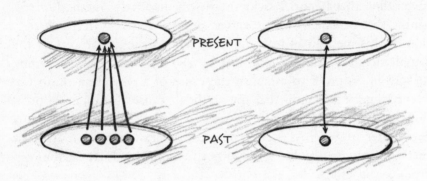

Figure 7.2. Schematic of the logical problem of retrodiction. Whether reconstructing the past is easy or difficult depends upon whether there is a single cause or condition that gives rise to a given present state or whether there are many possible past causes or conditions that give rise to a given present state. The diagram on the left portrays an information-destroying situation, in which many past causes (or conditions) correspond to a given present state. The diagram on the right portrays an information-preserving situation, in which only one past cause (or condition) corresponds to a present state. *Adapted from Sober,* Reconstructing the Past, *4.*

Sober uses a simple mechanical system to illustrate what makes reconstructing the past either easy or difficult. He notes that if someone places a metal ball on the rim of a concave metal bowl and releases it, it may traverse a complex path before coming to rest at the bottom of the bowl. Observing the ball in its final resting position (at equilibrium) will reveal little about the ball's initial position, since the ball could have been released from many places on the rim and still ended up at the bottom of the bowl. Imagine, however, that a ball is rolled down an indented track with high boundaries on either side and a large hole at the bottom, so that the ball will always come to rest in the hole just below the place where it is released. In this case, the final resting place of the ball will tell far more about the initial position of the ball. Indeed, the final state will correspond to just a single initial state, making it easy to reconstruct the ball's initial position from the observation of its final position.[38]

Sober's illustration suggests that if scientists can discover an effect for which there is only one plausible cause, they can infer the presence or action of that cause in the past with great confidence.[39] But if they can't, establishing a past cause is much more difficult. Two logical syllogisms illustrate the nub of the issue. First form:

> If X, then Y.
> Y occurred.
> Therefore X.

Arguments of this form commit the fallacy of affirming the consequent. Nevertheless, such arguments can be restated in a logically compelling form if it can be shown that Y has only one known cause (namely, X) or that X is a necessary condition of Y:

> X is necessary to the occurrence of Y.
> Y exists.
> Therefore, X existed.

Thus, this form of argument is a logically valid form of historical inference in which a past cause or event can be established from the effect alone.[40] For instance, the archaeologist who knows that scribes are the only known cause of linguistic inscriptions will, upon discovering tablets containing ancient writing, infer scribal activity. The historical geologist who knows that a volcanic eruption is the only known cause of widespread deposits of volcanic ash will infer that a past eruption caused the ash layer discovered today. As British philosopher Walter Bryce Gallie noted, where a particular past cause is known to be necessary to produce a subsequent effect, the occurrence of "the effect is taken as sufficient to establish the occurrence of the cause."[41]

Another Case

But what do historical scientists do in the more difficult case where there *is* more than one possible cause—more than one causally adequate explanation—of the evidence in question? Scientists will look for additional evidence that can help discriminate between the explanatory power of the remaining explanations. Ideally in such situations, scientists try to discover an additional fact or group of facts for which there is only one adequate causal explanation. In other words, if scientists investigating the past can't find a single fact that can be explained uniquely well (by reference to the only cause known of that type of fact), they try to find a wider group of facts for which only one adequate explanation exists.[42]

In many such cases, of course, the investigators will have to work their way back to the actual cause one painstaking step at a time. For instance, both wind shear and compressor-blade failure could explain an airline crash, but the forensic investigator will want to know which one did or if the true cause lies elsewhere. If both causes can explain the brute fact of the crash, the investigator will need to seek additional evidence about some specific aspect of the crash that only one of the candidate explanations can explain—maybe some evidence from the flight recorder or at the crash scene. Ideally, the investigator will find an additional piece of evidence or suite of evidences for which there is only one known cause, allowing him or her to distinguish between competing explanations and to eliminate every explanation but the correct one.

Here's an idealized illustration that I developed with some colleagues to show how this process works.[43] Suppose I wake up from a nap and see that the driveway at my house is glistening with water. The car in the driveway is also wet. From these two facts, or pieces of evidence, what can I conclude? There are actually several "causally adequate" explanations. It might have rained while I was napping, or the automatic sprinklers may have come on, or someone may have sprayed the car and driveway with the hose. With only the data that the driveway and car are wet, all these explanations are credible possibilities. Each cause cited is known to be capable of producing the effect in question.

How do I decide which of these "causally adequate" explanations best explains the data? Clearly, I must look for more evidence. As I do, I discover that the lawn and the street are perfectly dry and there isn't a cloud in the sky. What should I think now? Although the sprinkler theory and the rainstorm theory are still possible, these explanations are now much less likely in the light of the new evidence. Rainstorms produce wet driveways, but usually not wet driveways surrounded by dry yards, streets, and houses. And cloudless skies aren't particularly good candidates for producing rain. Thus, although the rainstorm and sprinkler hypotheses provide causally adequate explanations of the first group of facts, they fail to do so for the second.

But now, suppose I look a little harder and see a bucket with soapy water and a sponge sitting behind the car. With this additional piece of data, the best explanation for the whole group of observations becomes obvious: someone probably washed the car and sprayed the driveway with the hose. The car-wash hypothesis now provides the only "causally adequate" explanation of the whole ensemble of facts. It does this by providing a (merely) adequate explanation of some facts (the wet car and the wet driveway) and a uniquely adequate explanation of others (the bucket of soapy water).

This homespun example shows how we often use the method of inference to the best explanation to determine past causes of events when more than one cause can explain the same evidence. In my research, I came across many examples of historical scientists using precisely this strategy.

Here's a famous example from cosmology. During the late 1940s, two theories—the big bang and the steady state—could explain the main facts of cosmology. According to the big-bang theory, the universe had a definite beginning a finite time ago and has been expanding outward ever since as a result of its original explosive beginning. According to the steady-state theory, the universe is expanding, but it has existed eternally in a state of constant density. Space has been expanding eternally, and as it does, new matter is continuously created, thus maintaining the average density of matter in the universe. Thus, both theories could explain the evidence of an expanding universe, such as the "red shift" of light coming from distant receding galaxies.

Faced with this situation, cosmologists sought to make additional observations to determine which theory had superior explanatory power. When cosmic background radiation was discovered in 1965, it allowed cosmologists to decide this question. The big-bang theory postulated an early universe in which all matter was densely compacted, whereas the steady-state theory did not. If matter had existed in an early high-density state, it subsequently would have produced a ubiquitous background radiation that would have filled the universe as it expanded. Thus, the discovery of the background radiation helped to establish the superior explanatory power of the big-bang theory. In conjunction with other discriminating pieces of evidence, it also resulted in the rejection of the steady-state model. Despite an initial parity in explanatory power, the big bang ultimately provided a better, more causally adequate explanation of the full range of relevant evidence than did the steady state.

These illustrations show that the process of determining the best explanation often involves generating a list of possible hypotheses, comparing their known (or theoretically plausible) causal powers against the relevant evidence, looking for additional facts if necessary, and then, like a detective, progressively eliminating potential but inadequate explanations until, finally, one causally adequate explanation for the ensemble of relevant evidence remains. As Scriven explained, such "retrospective causal analysis" "proceeds by the elimination of possible causes."[44]

Causal Existence Criterion

There is another way of thinking of the problem of reconstructing the past when there are several possible causes of the same effect. If there is more than one causally adequate explanation of the same evidence, then scientists need to establish which of the adequate causes was *actually present* and responsible for the event in question. For that reason, some philosophers of science have argued that "best" explanations of past events must meet two criteria: causal adequacy and causal existence. To meet this second condition of a "best" explanation, a scientist must show that a proposed cause is not only capable of producing the event in question, but that the cause was present—that it existed at the right time and place—so as to have had the opportunity to produce the event in question. As Michael Scriven insisted, historical scientists not only have to show that a postulated cause could have produced the effect in need of explanation; they need "to show . . . that this cause was present."[45]

Scriven notes that, for any given effect or phenomenon, there might be "a list of possible causes," each of which is known to be capable of producing the effect or event in question.[46] Citing one of the possible causes on such a list, however, would not necessarily explain the event, even if the cause is known to have the power to produce the same kind of event on other occasions. Why? Because the event cited as the cause (as opposed to another possible cause) may not have actually occurred. Thus, he states, "We do not explain [a particular event] . . . merely by mentioning some *possible* cause."[47]

An example will help to illustrate. In order to explain the extinction of the Irish elk, a biologist must do more than just mention events that could have caused the extinction (such as an invasion of predators or the pollution of the elk's habitat). After all, these possible causal events might not have occurred. And obviously, if these events didn't happen, they cannot have caused the extinction and won't provide the best explanation of it. Instead, a biologist must also show that one of these events actually occurred, that predators actually invaded the elk's environment or that an increase in pollution ruined the elk's habitat. Scientists need to establish the past *existence* of the cause (in proximity to the effect) as well as the cause's adequacy.

So how do historical scientists establish that a given cause existed in the past? How can scientists meet this condition of causal existence? They have two ways of doing this, both of which I have already described. First, historical scientists can show that a presently acting cause must have been present in the past because the proposed candidate is the *only known cause* of the effect in question. If there is only one possible cause of a salient piece of evidence, then clearly the presence of that evidence establishes the past existence of its cause. Second, historical scientists can establish the existence of a cause by examining a wider class of facts to show that only one of the possible causes explains the whole ensemble (i.e., the main fact that needs explanation as well as other associated facts). If there is only one cause capable of producing all the relevant facts, then, again, that cause must have been present in the past in order to produce them.

This latter strategy was used in the wet-driveway illustration. Since there were many possible causes by which the driveway could have gotten wet, determining which cause was actually responsible for the wet driveway required comparing the explanatory power of the various hypotheses against a group of associated facts. This evaluation established that only one of the otherwise adequate causes of the wet driveway also provided an adequate explanation of all the other relevant facts. Thus, it also established that someone had washed the car and that the causal factors implied by that hypothesis—a spraying hose and sloshing bucket—had been present in the past. Thus, the car-wash hypothesis met the causal existence criterion (and, having already met the causal-adequacy criterion, also qualified as the best explanation).[48]

Clearly, the way historical scientists go about meeting the causal-existence criterion involves further assessments of causal adequacy, so in practice the two criteria are closely related. Meeting one criterion (causal existence) depends upon how well a hypothesis meets the other (casual adequacy)—either with respect to a single salient fact or a wider ensemble of associated facts. For this reason, many historical scientists may not make any explicit mention (or even be aware) of seeking to meet a separate causal-existence criterion. Indeed, they need not do so, because they can ensure that this criterion gets met just by identifying a *uniquely* adequate causal explanation, either with respect to a single salient fact or with respect to a wider ensemble. If a single fact does not sufficiently discriminate between the explanatory power of competing hypotheses, then historical scientists intuitively seek additional facts until they eliminate all but the best (most causally adequate) explanation. In so doing, they meet the causal-existence criterion and identify the best explanation at the same time.

The Nature of Historical Science

Even so, Michael Scriven describes the historical method of "retrospective causal analysis" as a simple three-part method of evaluating proposed causal explanations. Many historical scientists find it helpful to think of it this way. According to Scriven, in order to establish a causal claim, the historical scientist needs: (1) "evidence that his candidate [cause] was present" and (2) evidence that "it has on other occasions clearly demonstrated its capacity to produce an effect of the sort here under study." Additionally, the historical scientist needs to establish that there is (3) an "*absence* of evidence (despite a thorough search) . . . of . . . other possible causes."[49]

Many fields that seek to explain events by reference to past causes—such as forensic science, evolutionary biology, paleontology, geology, archeology, and cosmology—use this kind of "retrospective causal analysis."[50] Indeed, many scholars think that Charles Darwin structured his case in the *Origin* to show that natural selection met both the causal-adequacy and causal-existence conditions of a best explanation[51] and, in so doing, that he was able to test his theory as well.

According to Darwin, his theory of "universal common descent" could be tested not by using it to predict future outcomes under controlled experimental conditions, but by showing that it could explain already known facts in a better, a more causally adequate way than rival hypotheses. As he explained in a letter to Asa Gray: "I . . . test this hypothesis [universal common descent] by comparison with as many general and pretty well-established propositions [facts] as I can find. . . . And it seems to me that, supposing that such a hypothesis were to explain such general propositions, we ought, in accordance with the common way of following all sciences, to admit it till some *better hypothesis* be found out."[52]

Taking Stock

And so my study of how historical scientists reconstruct past causes led me to conclude that Charles Thaxton had been on to something. There does, indeed, seem to be a distinctive method of historical scientific reasoning and investigation. Historical scientists have a distinctive objective (to identify the causes of past events); they formulate distinctive types of explanation (in which they cite past events as causes); and they make inferences with a distinctive (abductive) logical form.

But I concluded that Thaxton was mistaken about something as well. The historical sciences were not quite as unique as he thought. Historical scientists use a common method of evaluating their theories in which they evaluate the relative explanatory power of an inference to determine its strength, plausibility, or likelihood. This meant that historical scientific theories, like other scientific theories, were testable, albeit by reference to distinctively historical criteria for determining when an explanation qualifies as "best." Historical scientists test their theories against the explanatory power of their competitors and against our knowledge of the "causal adequacy" or "causal powers" of various entities. They make these assessments based upon our present knowledge of established cause-and-effect relationships. The criteria of causal adequacy and to a lesser extent causal existence, developed by Lyell and used by Darwin, constitute critical experience-based *tests* of historical scientific theories, tests that can be used to discriminate between the explanatory power (and merit) of competing hypotheses.

Clearly, this method of testing scientific ideas is different from that used by experimental scientists, who test their theories by making predictions about what will happen under controlled laboratory conditions. Even so, historical scientists are not the only scientists to use it. Arguably, Watson and Crick used this method to test their ideas about the structure of DNA against competing models. And many scientists—theoretical physicists, biochemists, psychologists, astronomers, pathologists, medical diagnosticians—as well as historians, detectives, and thinking people everywhere use this method of reasoning every day to make sense of their experiences.

In any case, whether the historical sciences use an absolutely distinctive scientific method now seemed less important to me than understanding exactly how conclusions about past events and causes could be rigorously established. And my study of how historical scientists formulated their inferences and arguments gave me a clear understanding of this. Historical scientists evaluate the strength of competing abductive inferences by comparing their explanatory power. In the best of cases, they will infer a clearly best explanation—one that cites the "only known cause" of the effect or effects in question.

Intelligent Design as a Possible Scientific Explanation?

What did all this have to do with the DNA enigma? Quite simply this. I wondered if a case for an intelligent cause could be formulated and justified in the same way that historical scientists justify any other causal claim about an event in the past. What I learned about how historical scientists determine the causes of past events gave me another reason to think that intelligent design was at least a possible (historical) scientific explanation for the origin of biological information. As conceived by advocates of the theory, the activity of a conscious designing agent in the past constituted a causal event (or series of such events). Moreover, an "intelligent cause" might well have made a causal difference in outcome. The activity of a designing intelligence might account, for example, for the difference between a chaotic prebiotic environment in which simple chemical constituents interacted with each other in accord with natural entropic (randomness-producing) processes and an environment in which systems with highly specific arrangements of chemicals arose that stored information for building proteins and protein machines. Intelligent design could have made a causal difference between continuing chemical chaos and the origin of information-rich molecules.

Moreover, my study of historical scientific reasoning also suggested to me that it was at least possible to formulate a rigorous case for intelligent design as an inference to the best explanation, specifically, as the best explanation for the origin of biological information. Not only was it was possible to conceive of the purposeful act (or repeated action) of an intelligent agent as a causal event, the action of a conscious and intelligent agent clearly represented both a known (presently acting) and adequate cause for the origin of information. Uniform and repeated experience affirms that intelligent agents produce information-rich systems, whether software programs, ancient inscriptions, or Shakespearean sonnets. Minds are clearly capable of generating specified information.

Conclusion: The Causes Now in Operation

When I first noticed the subtitle of Lyell's book, "By Reference to *Causes Now in Operation*," a light came on for me. I immediately asked myself: What causes *now in operation* produce digital code or specified information? Is there a known cause—a *vera causa*—of the origin of such information? What does our uniform experience tell us? As I thought about this further, it occurred to me that by Lyell and Darwin's own rule of reasoning and by their test of a sound scientific explanation, intelligent design must qualify as, at least, a possible scientific explanation for the origin of biological information. Why? Because we have independent evidence—"uniform experience"—that intelligent agents are capable of producing specified information. Intelligent activity is known to produce the effect in question. The "creation of new information is habitually associated with conscious activity."[53]

But is intelligent design the "only known cause" of the origin of specified information? I was now more than intrigued with this possibility, especially given the implications for the status of the design argument in biology. If intelligent design turned out to be the only known or adequate cause of the origin of specified information, then the past action of a designing intelligence could be established on the basis of the strongest and most logically compelling form of historical inference—an inference from the effect in question (specified information) to a single necessary cause of that effect (intelligent activity). Moreover, if intelligent design were shown to be the only known cause, if it were shown to be a uniquely adequate cause, then intelligent design would also automatically meet the causal-existence criterion of a best explanation as well.

Is intelligent design in fact the best explanation for the origin of life? I knew that the central question facing scientists trying to explain the origin of the first life was precisely: How did the sequence-specific digital information necessary to building the first cell arise?[54] At the same time, I knew that origin-of-life researchers had proposed many explanations for the origin of biological information, including some new and potentially promising approaches. Though I had studied both Oparin's classical theory of chemical evolution and several contemporary alternatives, much of my research had focused on methodological issues—on *how* historical scientists reason about past events and causes. These studies convinced me that intelligent design was a possible—a causally adequate—explanation for the origin of biological information. But to determine whether intelligent design was the best—the *only* causally adequate explanation—I would need to know more about other scientific possibilities. I would need to follow Scriven's directive to "make a thorough search" for and evaluation of other possible causes. Over the next several years, as I assumed my duties as an assistant professor, I set out to do exactly that.

8

Chance Elimination and
Pattern Recognition

My work in England on historical scientific reasoning convinced me that an intelligent cause could function as a *possible* scientific explanation for the origin of biological information. But I wondered if it could provide a better explanation than the alternatives. Could it be inferred as the *best explanation* for the origin of the specified information in the cell? To answer this question, I would need to undertake a rigorous examination of the various alternatives.

By the time I finished my Ph.D. studies, in 1990, I was familiar with most of the main theories then current for explaining the origin of life. These theories exemplified a few basic strategies of explanation. Some relied heavily on chance—that is, on random processes or events. Some invoked lawlike processes—deterministic chemical reactions or forces of attraction. Other models combined these two approaches. In this respect, research on the origin of life followed a well-established pattern in the sciences, one explicitly recommended by the leading scientists of our time.

During the period of explosive discovery from 1953 through 1965 now known as the molecular biological revolution, one of Francis Crick's colleagues was the French biologist Jacques Monod. In 1971 he wrote an influential book called *Chance and Necessity*, extolling the powers of *chance* variation and lawlike processes of *necessity* in the history of life.[1] As an example of a chance process, Monod noted how random mutations in the genetic makeup of organisms can explain how variations in populations arise during the process of biological evolution. In Monod's shorthand, "chance" referred to events or processes that produce a range of possible outcomes, each with some probability of occurring. The term "necessity," on the other hand, referred to processes or forces that produce a specific outcome with perfect regularity, so that the outcome is "necessary" or inevitable once some prior conditions have been established. Since, for example, Newton's law of gravity stipulates that "all unsuspended bodies fall," scientists say that any unsuspended body will "of necessity" fall to the earth. If enough heat is applied under a pan of water, the water will eventually—of necessity—boil.

In *Chance and Necessity,* Monod did not try to explain the origin of biological information or the cell's system for processing information. Instead, he presupposed this system and then argued, following Darwin, that the subsequent history of life could be explained solely by reference to the twin factors of chance and necessity. For scientists reluctant to consider intelligent design as an explanation for the origin of life, Monod's work codified a normative set of guidelines for approaching this and every other scientific problem. His message was clear: scientists can, and should, explain all phenomena by reference to chance, necessity, or the combination of the two. And, indeed, most scientists have sought to explain the origin of life using one of these approaches.

Following Monod's order of presentation and the rough sequence in which ideas about the origin of biological information appeared after 1953, I began to examine scenarios that relied primarily on chance. Could the random interaction of the molecular building blocks of life explain the mysterious origin of the biological information needed to produce life? Various scientists have proposed this idea, but many origin-of-life researchers have expressed reservations about it, because they thought that it amounted to little more than saying, "We don't know what caused life to arise." And, indeed, as I scrutinized the chance-based explanations for life's origin, I often found them disappointing. Specific chance-based proposals were few and far between. And where present, they were hopelessly vague. Upon reading them, my reaction was often the same: "That's it?"

This situation posed a dilemma for me in my investigation of the DNA enigma. Though chance was assumed to have played a significant role in the origin of life, few scientists could say exactly what that meant or offer any rigorous criteria for evaluating such explanations. Is "chance" a cause that can be cited to explain something? Is chance even a "something" that can cause anything? Chance-based explanations were, as a rule, so thin on detail that it was difficult to assess them.

What Is Chance?

My concerns about chance as an explanation boiled down to a few basic issues. What does it mean to say that something happened by chance? When is it reasonable to invoke chance in an explanation? And what justifies excluding such an explanation from consideration? Considering a couple of different situations clarified my thinking about these questions.

In the Pacific Northwest, a now infamous bridge called the Tacoma Narrows (nicknamed Galloping Gertie) collapsed into the Puget Sound in 1940 after the bridge began to undulate more and more violently during a stiff, but not unusually high, wind. Now imagine that the engineers investigating the collapse had come back and, without citing any other factors, told the political authorities that the bridge had collapsed "by chance." Would the authorities have found that explanation satisfying or informative?

Consider a second scenario. A man named Slick has just won at a Las Vegas roulette table where the odds of winning are 1 in 38. Although Slick's name does not exactly inspire confidence, the croupier at the game table certifies that he did not cheat and the roulette wheel is operating properly. The victory was legitimate. Slick is paid in accord with his wager. Why did he win? What explains the ball falling into the pocket corresponding to Slick's bet? The croupier spun the wheel and the ball fell—by chance—into the pocket that Slick had chosen.

A bridge collapsed "by chance." A gambler won at the roulette table "by chance." Which of these two explanations seems more satisfactory or reasonable? Clearly the second, and not the first. But why?

In the second case, it seems legitimate to attribute the event to chance, because there is a known process, namely, the spinning roulette wheel with its 38 compartments, which can generate the event of interest—the ball landing on a red 16—with some regular or predictable frequency. To say that this event occurred by chance does not mean that chance caused something to happen, but rather that, given the construction of the roulette wheel, there is a definite probability—a 1 in 38 chance, to be precise—that the ball will fall in the red 16 pocket. Absent indicators of other influences at work, we say that the ball fell in the red 16 pocket "by chance" to indicate that we think this event occurred as one of the normal possible outcomes of a regular underlying process.

The case of the investigating engineers, however, is different. They did not specify an underlying process capable of producing bridge failure as a possible outcome. They cited chance purely to conceal what they did not know about the cause of the bridge failure. Any official receiving this answer would have correctly regarded it as an admission of ignorance, a fancy way of saying, "We don't know what happened" or "We can't explain it."

There's a second aspect to this, however. When scientists say that something happened by chance, they do not usually mean to deny that *something* caused the event in question. (Some interpretations of quantum physics would stand as an exception.) Instead, they usually mean that the event in question occurred because of a combination of factors so complex that it would have been impossible to determine the exact ones responsible for what happened or to have predicted it. Imagine that I roll a die and it turns up a five. That outcome was caused or determined by a number of factors—the exact force applied to the die, the orientation of the die as it left my hand, the angle at which it first hit the table, and so on. Saying that the five turned up "by chance" does not deny that the event was determined or caused by physical factors, but instead that the exact combination of physical factors was so complex that we cannot know them exhaustively and, therefore, could not have predicted the precise outcome.

Thus, when scientists say that something happened by chance, they do not usually mean that some entity called "chance" caused something to happen. Instead, they mean that there is a process in play that produces a range of outcomes each of which has a *chance* or probability of occurring, including the outcome in question.

Chance as Negation

But there is usually more to the notion of chance than that. When scientists attribute an event to chance, they also usually mean there is no good reason to think the event happened either by design or because of a known physical process that must, of necessity, generate only one possible outcome.

Imagine that a statistician is trying to determine whether a coin is fair by observing the distribution of heads and tails that result from flipping it. The "chance" hypothesis in this case is that the coin is fair. The alternative hypothesis is that the coin is biased. If the distribution of heads and tails comes up roughly even, the statistician will conclude that the coin is fair and elect the "chance hypothesis." In saying this, the statistician is not saying that "chance" caused the 50–50 distribution or even that she knows what caused any particular outcome. Instead, the affirmation of the chance hypothesis mainly serves to negate the alternative hypothesis of bias (in this case, a loaded coin produced by design). The essentially negative character of the chance hypothesis is suggested by its other common name: the *null* hypothesis (i.e., the hypothesis to be nullified or refuted by alternative hypotheses of design or lawlike necessity).[2]

Such negations, implicit or explicit, are part of what give substantive chance hypotheses content. To say that the roulette ball fell in the red 16 pocket by chance is also to say that the ball was *not* placed there intentionally and that there was nothing about the construction of the roulette wheel that forced the ball to land in the red 16 *of necessity*. On the other hand, the engineers who appeal to chance because they don't know anything about the cause of a bridge collapse don't make such a negation. Because they know nothing about the cause, they are unable to rule out alternatives, whether foul play (design) or some underlying structural failure induced by environmental conditions acting on a structural defect (necessity). Vacuous appeals to chance neither affirm a cause nor negate one. But substantive appeals to chance specify the operation of a relevant outcome-producing process, and they also implicitly or explicitly negate (or *null*ify) other possible types of hypotheses.

William Dembski, R. A. Fisher, and Chance Elimination

Of course, just because a chance hypothesis avoids being vacuous by making definite claims does not necessarily mean that its claims are true. Successful appeals to chance must be justified, including their implicit negations of design and necessity. So how do statisticians and scientists test substantive (rather than vacuous) chance hypotheses? How do they decide when to accept and when to reject chance as an explanation?

About the time I was asking myself these questions, I met William Dembski, a mathematician who later would become one of the best-known proponents of intelligent design. In the summer of 1992, I returned to Cambridge to work on a research project with Dembski and another colleague, Paul Nelson, who was then working toward a Ph.D. in the philosophy of biology at the University of Chicago.

During the preceding year Nelson and I had been approached by a research institute about the feasibility of developing a rigorous theory of intelligent design. The institute wanted to support a small interdisciplinary team and invited a proposal. Nelson and I thought it would be desirable to find someone with a background in probability and information theory to round out our team. A week before the proposal was due Nelson stumbled across two articles by a young mathematician with a recently minted Ph.D. from the University of Chicago. The articles were impressive. One broke important new ground in understanding pattern recognition. The other described how probability theory could be used to assess chance hypotheses. At the end of one of the articles

Figure 8.1. William
Dembski.
Photo by Laszlo Bencze.

was a biosketch describing Dembski's considerable qualifications. In one line it noted that Dembski wanted to use his expertise in probability theory to investigate the feasibility of the design hypothesis. We had found our man, or so we thought.

Unfortunately, despite concerted effort, Nelson could not locate Dembski, who was traveling in Germany. Then, the day before the deadline, I received a call from a friend in Texas who wanted to tell me about someone he thought I should know. Sure enough, he wanted to tell me about Bill Dembski. He described Dembski as unusually brilliant, yet unassuming and pleasantly eccentric. He also confirmed that Dembski definitely shared our interests. Upon hearing of our dilemma, he encouraged me to "pencil him in" on the proposal and make explanations later. We did. After receiving word that our proposal had been accepted—contingent on the willingness of all investigators to participate—Nelson and I finally located Dembski and told him what we had done. He was surprised, but delighted by our audacity.

My first meetings with Dembski in Cambridge confirmed my friend's assessment. Dembski had an extraordinary facility with logical and mathematical formalism and a desire to use these tools to address real scientific questions. He also exhibited some rather endearing eccentricities. He had prepacked a supply of health food for the summer in order to avoid the heavy fare in the Cambridge colleges. He also had an odd penchant for answering questions with illustrations involving coins and dice without explaining what it was he was using these examples to illustrate. By summer's end, I had enticed him to join me for "beans and chips" in college and to begin his answers with straightforward assertions before bringing out the dice and coins to illustrate his point. He, in turn, introduced me to a way of assessing chance as an explanation for the origin of life.

During the summer, Dembski, Nelson, and I talked frequently about the problem of assessing chance as an explanation. Dembski introduced us to the work of the pioneering statistician Ronald A. Fisher. During the 1920s, Fisher had developed a method of statistical hypothesis testing that brought greater rigor to discussions of chance.[3] On the one hand, Fisher's method acknowledged that some phenomena could be reasonably explained by, or at least attributed to, random processes or chance events. On the other hand, Fisher's work established a clear method for deciding when chance alone could be eliminated from consideration as the best explanation. I was intrigued.

Here's an example that shows how Fisher's method of chance elimination works. Apparently encouraged by his success at the casino, Slick returns the next day to try his hand at the roulette wheel again. The croupier smiles to himself. This is how casinos make their money back from the previous day's big winners—overconfidence. But then a most unusual morning unfolds. Over the next few hours, a preternaturally confident Slick bets only on the red 16. He wins not just once at the roulette wheel, but 100 times in a row (fortunately for the casino, he keeps betting only small amounts, thus not breaking the bank).

The croupier knows, based upon the construction of the wheel, that the probability of the ball falling into a given pocket is exactly 1 in 38. Thus, the probability of the ball falling into a given pocket 100 times consecutively is extremely small, 1 chance in 38^{100}, or 1 chance in 10^{158}. In this situation, the casino rightly suspects (and long before the 100th spin) that something more than chance is at work. Either the roulette wheel has a mechanical defect that causes the ball to catch and stop in the same place every time or someone is cheating in some way that has not yet been detected. Either way, given their knowledge of the odds in roulette and the construction of the wheel, the casino's managers will conclude that Lady Luck is not the culprit. But why?

On average the ball should land in the red 16 pocket once every 38 spins. (This is the expected waiting time between successes).[4] Dumb luck might allow it to land there two or three times in 100 spins, but since it has landed in the winning pocket in every one of the last 100 spins, this will excite concern. This event cannot reasonably be attributed to the normal function of the roulette wheel, because the frequency of the occurrence of the event in question has deviated too much from the frequency that the house rightly expects. Thus, "chance"—even chance understood in the more substantive way defined above—no longer seems an appropriate explanation.

According to Fisher, chance hypotheses can be eliminated precisely when a series of events occurs that deviates too greatly from an expected statistical distribution of events based on what we know about the processes that generate those events or what we know from sampling about how frequently those events typically occur. Fisher developed a statistical method for identifying these variations from expectation. That method depends crucially on what he calls a "rejection region."[5]

Fisher knew that any truly random process would produce a range of events with some predictable distribution. In the case of roulette, gamblers will win the game about once every 38 times. A few will win less frequently than that; some will win more frequently. A very few will win either quite frequently or rarely. These frequencies can be plotted on a curve that shows the frequency of winning on one axis (the x axis) and the number of players who achieve a given frequency on the other (the y axis). The resulting distribution or pattern is the familiar "bell curve" known to students of statistics everywhere (see Fig. 8.2). Events that conform to our expectations about what should happen fall near the center of the bell curve. Events that deviate from expectation fall on the sides of the curve or on the tails. Events that deviate dramatically from statistical expectation—like Slick winning at roulette 100 times in a row—will fall way out on the tails of the curve, often so far out that we have no other examples of such events occurring, at least not by chance alone.

And this was Fisher's insight. He argued that statisticians could "catch" or "identify" events that resulted from factors other than chance by prespecifying a rejection region. A rejection region is a place on a bell curve or some similar statistical plot that defines a set of outcomes that deviate dramatically from our statistical expectation—so dramatically, in fact, that statisticians think they can reasonably conclude that

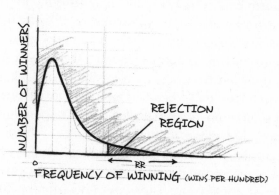

Figure 8.2. A bell-curve distribution with a prespecified rejection region shaded.

these events did not occur from the normal outworking of the statistical process in play. For example, a statistician working for a casino may specify as a rejection region any statistically significant set of outcomes in which a roulette player wins more than 50 percent of the time. When an event occurs that falls within that rejection region—like Slick winning 100 times in a row—the statistician will reject the chance hypothesis. Further, by specifying in advance where those rejection regions lie, Fisher thought statisticians could guard against reading patterns or significance into a statistical distribution that wasn't there. Prespecifying rejection regions could prevent "cherry-picking" the data.[6]

Fisher's method made intuitive sense to me, and I wondered if it could shed light on the question of whether random processes were responsible for the origin of biological information. But Dembski had other questions on his mind. He was wondering *why* Fisher's method worked in light of a puzzling feature of probabilistic reasoning. As a probability theorist, Dembski knew that intuitively we tend to reject chance as an explanation for extremely improbable events. Fisher's method of chance elimination illustrated this. Clearly, the improbability of Slick's winning every time contributes to our skepticism about the chance hypothesis. But there was something puzzling about this intuitive conviction as well—something that Dembski pointed out to me. Extremely improbable events—events that don't necessarily indicate anything other than chance—happen all the time. Indeed, many extremely improbable events don't seem to justify skepticism about chance at all.

Dembski illustrated this by asking me to imagine flipping a coin 100 times and then writing down the exact sequence of heads and tails that turned up. He pointed out that if I was bored enough to do this, I would actually participate in an incredibly unlikely event. The precise sequence that occurred would have a probability of 1 chance in 2^{100} (or approximately 1 in 10^{30}). Yet the improbability of this event did not mean that something other than chance was responsible. After all, some sequence of that improbability had to happen. Why not this one?

This same puzzle is evident in my example of Slick at the roulette wheel. When Slick won roulette 100 times in a row, he participated in an incredibly improbable event. The odds of his winning 100 times in a row were incredibly low, 1 chance in 10^{158}. (There are only 10^{80} elementary particles in the observable universe.) Now imagine that Slick has not won every time, but instead only two or three times out of 100, more in keeping with the expected frequency for winning by chance. Imagine also that each time the ball falls in a pocket, Slick records the outcome in hopes of finding a pattern to guide his future betting. After 100 spins, Slick has compiled a series of 100 outcomes—red 10, black 17, red 28, black 5, and so on.

What are the odds of this exact series of 100 outcomes occurring? They are *exactly the same* as the odds of Slick winning 100 times in a row. Since the probability of the ball falling in any one of the pockets on any given spin is the same each time (1 in 38), any specific combination of 100 outcomes will be the same, 1 chance in 38^{100}, or 1 chance in 10^{158}. Even in defeat, Slick will witness an incredibly improbable event.

The Improbability Paradox

Here we have a paradox. On the one hand, the improbability of Slick's winning 100 times in a row seemed to be an important factor in determining that something other than chance was at work. After all, had Slick won only once or twice in a row—something much less improbable—no one would have suspected anything. On the other hand, it was easy to identify another event just as improbable as his winning 100 times in a row that would not trigger the same recognition.

The essence of this paradox is that two equally improbable events nevertheless seem as though they should be explained in two different ways. This paradox led Dembski to think that assessments of probability alone could not be the whole reason for rejecting chance. It also led him to ask a series of important questions.

What, in addition to improbability, justifies excluding chance as an explanation for an improbable event? What was Fisher's method identifying, other than just an improbable event, that made it a reliable method for eliminating chance—for catching "something else going on"? Clearly, Slick's winning 100 times in a row aroused justifiable suspicion of cheating, whereas his winning only twice out of 100 did not. What was his method identifying in one case, but not in the other?

As Dembski thought about this, he realized that Fisher was introducing another factor into his analysis. Fisher's method was not just analyzing an improbable event (or sequence of outcomes); it was looking for a significant pattern within that larger event (or sequence). Fisher's method compared the frequency of occurrence of certain *kinds* of outcomes within a larger event to a statistical norm. For example, it compared the frequency of Slick's winning bets to the statistical norm for roulette players. Dembski realized that by defining certain *kinds* of outcomes, the frequency of winning, for example, as significant, Fisher had introduced an extra-probabilistic factor into his analysis. In other words, Fisher's method required statisticians to deem certain *kinds* of events as significant. But Dembski realized that "a significant kind of event" is the same thing as a pattern.

Events at the roulette wheel again prove instructive. Consider two kinds of outcomes—two patterns—that would lead statisticians to reject a chance hypothesis. First, while testing a new roulette table, the croupier at the gaming house finds that the ball repeatedly falls in the same pocket. Yet on the chance hypothesis that the wheel is working properly and there has been no outside interference, the odds of the ball repeatedly falling in exactly the same pocket—say, 100 times in a row—are very low. The croupier suspects that something beyond chance is in play, but again, not just because the probability of the observed outcome is incredibly small. Clearly, in this case, a pattern has emerged. The same physical event keeps recurring, though not one that has any effect on a game, since the roulette wheel is being tested only at the moment. What should the croupier conclude? The recurrence of the exact same physical outcome suggests to him some physical defect with the wheel. He investigates. Sure enough, there is an inordinately high lip on the backside of the red 16 pocket. Each time the ball rolls over the red 16 it collides with the lip and falls—by necessity, not chance—into the same pocket.

Now consider a second case in which a different kind of pattern triggers the rejection of the chance hypothesis. In this case, Slick repeatedly changes his bet, but nevertheless always wins. The exact sequence of bets manifests no obvious physical pattern (such as the ball repeatedly falling in the red 16). Instead, a different kind of pattern emerges. In this case, a seemingly random sequence of different physical outcomes produces a significant outcome *in the game*. A specific sequence of physical events combines to achieve a significant objective—Slick winning a lot of money. Unlike the previous case, where the pattern was evident simply by observing the event itself, the pattern in this case corresponds to an improbable series of events that has significance beyond the physical boundaries of the roulette wheel, a significance based on other things the croupier or statistician knows (i.e., the object and rules of the game of roulette).

In this case, the statistician or croupier will again reject the chance hypothesis (that Slick won fairly and the wheel is working properly) and for a good reason. Once again an outcome has occurred that would be—if nothing but chance were at work—incredibly improbable. But beyond that the statistician also recognizes in that outcome a pattern of events that advances a *goal* or performs a *function*, namely, winning the game and money. Rejection regions often implicitly identify patterns of events that advance some goal or perform some function. Thus, the rejection region that a statistician uses to eliminate chance in this case will not only correspond to an improbable sequence of events, but also to an improbable sequence of *victories* in the game. It identifies a frequency of *winning* that deviates dramatically from expectation. And since winning is a significant objective that agents routinely seek to achieve, a highly improbable string of winning bets suggests the activity of agents. It suggests cheating, not chance. Conversely, the gaming commission might suspect that the casino was cheating if the roulette wheel inflicted a disproportionately high rate of losses on customers, a rate consistently and significantly higher than the roulette wheel would generate if it were governed purely by chance.[7]

Thus, Dembski realized that when we use a statistical plot to track the frequency of some event occurring, we are implicitly looking for a kind of pattern, a pattern of events that suggests something other than chance alone. A pattern of events may present itself as interesting because some physical state of affairs occurs repeatedly. Or a pattern of events may appear interesting because it has some independent functional significance apart from the physical features of the events. Rejection regions sometimes identify repeating patterns; other times they correspond to what I call "functionally significant patterns." Either way, Dembski realized that implicit in Fisher's method of defining rejection regions around "a frequency of occurrence that deviates dramatically from expectation" was the identification of *specific kinds of events*—events that manifested significant or salient patterns.

The role of pattern recognition is even easier to see in cases where we reject the chance hypothesis without the use of a formal statistical method such as Fisher's. Consider another example that illustrates the importance of pattern recognition to the elimination of chance.

During my first year as a college professor, I was assigned to help teach a team-taught course on the history of Western philosophy. During a grading session with the other professors, my senior colleagues read samples of various student papers aloud to showcase both model papers and typical student errors. As I listened, I had a sudden recognition. "Hey," I said, "I've read that paper." One of the other profs chuckled and said, "I know, they all sound the same after a while." But I persisted. "No," I said, "I mean, I've read that *exact* paper, really." As my colleague continued to read portions of this student essay on Plato, I rifled through my stack. About halfway down the pile, I saw the line he was reading and I finished his sentence for him. He looked up at me and read some more. I finished his sentence again. An antiphonal reading ensued. Groans arose from the teaching assistants and the other professors on the team. The two papers were exactly the same. The students were eventually confronted. After some emotion and vague denials, one of the two students confessed. It was exactly what it seemed—plagiarism.

After meeting Dembski, I realized that my close encounter with a student plagiarist aptly illustrated his ideas about why we eliminate chance hypotheses—in this case, the hypothesis that the two students had just happened to write the same three typed pages of the same English characters in the same sequence *without* one having copied the other or both having copied a common source.

Why did my colleagues and I eliminate the chance hypothesis in this case? Clearly, the odds did have something to do with it. A typical three-page double-spaced college essay has about 1,000 precisely sequenced English words, about 6,000 characters and spaces. The probability of one student generating the exact same sequence of 1,000 words independently, even if writing on the same topic, was almost incalculably small. Our awareness of the incredible improbability of this event occurring by chance contributed to our judgment that plagiarism was involved. After all, if the students had just replicated two or three short words or a short phrase, we would have had lesser grounds for suspicion, no doubt because the probability of generating a few short repeats was much, much higher.

Even so, my discussions with Dembski made me aware that the odds weren't the whole of the matter in such cases. The odds of getting *any* specific sequence of 6,000 characters are roughly the same as the odds of getting a sequence of 6,000 characters that exactly matches the sequence of 6,000 characters in another text. So what was the difference? In one case an improbable event matched another, and in the other case the event stood alone—that is, no such match occurred.

And that was Dembski's key insight. It wasn't the improbability of an event alone that justified the elimination of chance. It was the *match* between one event and a pattern that we recognize from another. In detecting plagiarism, I had recognized not only an event I knew intuitively to be incredibly improbable (a specific sequence of English words and characters), but a pattern in that event that I knew from a completely separate event, namely, my reading of another student's essay. It was the match, the convergence of an improbable event with another pattern of letters and words that made me aware of plagiarism. Pattern recognition helped trigger the rejection of chance.

Dembski later developed his ideas about pattern recognition in more detail. He argued that some kinds of patterns we recognize can, in conjunction with an improbable event, reliably indicate the activity of a designing intelligence. Other types of patterns—highly repetitive patterns of the same physical outcome, such as a coin turning up heads 100 times in a row—might indicate the presence of some physical cause responsible for producing that outcome by necessity. In still other cases, we fabricate or imagine patterns rather than recognize them, and when this occurs the patterns may indicate nothing at all. (In Chapter 16, I discuss how Dembski developed a test to identify and exclude such phony patterns from consideration.) In any case, I came to realize, based on my discussions with Dembski, that pattern recognition played a key role in the rejection of chance.

I also realized that this made sense in light of my own earlier philosophical reflections about what scientists mean by "chance." If a substantive chance hypothesis necessarily negates or nullifies explanations involving physical-chemical necessity and design, then the presence of a pattern necessarily negates chance. The chance hypothesis in effect says, "There is nothing going on in this event to indicate any regular or discernible causal factors." Since patterns signal the presence of deeper causal factors or regularities at work, the presence of patterns negates chance. Because patterns negate the null hypothesis (that "nothing is going on") and the null hypothesis is the chance hypothesis, patterns negate the chance hypothesis. Patterns negate the negation—the negation entailed in a substantive chance hypothesis.

Ex Post Facto Pattern Recognition and Singular Events

As Dembski was thinking about the role that pattern recognition plays in chance elimination—and the way rejection regions define significant patterns—he had another key insight. He realized that it wasn't necessary to anticipate or prespecify a pattern (or rejection region) in order to justify rejecting chance, as Fisher's method required. Dembski regarded Fisher's condition as too restrictive, since he could think of many examples in which recognizing a pattern after the fact, rather than specifying a rejection region before it, obviously justified the elimination of the chance hypothesis.

For example, Dembski cites the case of a spy listening to and recording the communication of an adversary in wartime. Before an encoded message is decrypted, the spy can't tell whether the combinations of sounds are occurring by chance. Nor can he specify in advance the exact combination of sounds that would induce him to reject the chance hypothesis—the hypothesis that he has been listening to random noise or static. Once the spy uncovers the cryptographic key and is able to decode the signals as meaningful text, however, he realizes—after the fact—that his intercepted signals were not random. As Dembski notes, "In contrast to statistics, which always identifies its patterns before an experiment is performed, cryptanalysis must discover its patterns after the fact."[8] This same phenomenon occurs daily for most people as they reflexively reject chance as an explanation for patterns they perceive in events that have already happened, whether a mechanical device or a pattern evident in natural objects such as crystals or sand dunes.

As the result of his deliberations, Dembski developed a method of chance elimination that made the importance of pattern recognition more explicit than it had been in Fisher's. Just as historical scientists can often successfully explain events after the fact that they could not have predicted before the fact, Dembski argued that it was often possible to discern a pattern *after* the fact that could not have been specified *beforehand* (i.e., that was not prespecified). This allowed historical scientists to use his method to evaluate chance explanations of past events—events that had already occurred and would not recur. Since the origin of life constituted precisely such an event, his method could be used to evaluate the chance hypotheses for the origin of life and biological information.

Moreover, as I explain in the next chapter, although the scientists working on the origin of life rarely articulate specific criteria for evaluating chance hypotheses, they typically employ (if implicitly) Dembski's criteria in assessing chance explanations for the origin of life. Clearly, Dembski was on to something.

One Final Factor: Probabilistic Resources

It was now clear to me that the occurrence of an improbable event alone does not justify eliminating the chance hypothesis. Even so, it was also clear that judgments about probability did play an important role in our reasoning about whether something beyond chance is at work. My colleagues and I would not have accused our students of cheating had they written two essays that just happened to use a few of the same phrases. Nor would a gambler suspect that a roulette table had a physical defect just because the ball landed in the same pocket twice in a row. Even though patterns are recognizable in these events, the events in question are simply not improbable enough to convince us of wrongdoing.

Pattern recognition can lead us to suspect that something more than chance is at work. But the presence of a pattern alone does not justify rejecting chance, any more than an improbable event alone, in the absence of a pattern, justifies chance elimination. Instead, both the occurrence of an improbable event and the recognition of a pattern are necessary. Or to make the point using a logical distinction:

the presence of a pattern is a necessary, but not by itself a sufficient condition of chance elimination. Instead, a quantitative judgment of improbability is also involved.

But this raises another question. How improbable does an event have to be to justify the elimination of a chance hypothesis? If we begin to detect a pattern in an improbable sequence of events, at what point should our doubts about the chance hypothesis lead us to reject it as untenable or unreasonable? As a blackjack player repeatedly wins against the house, or a ball repeatedly falls in the same hole, or as a die repeatedly comes up on the same side, or as we observe an event that implies that great odds were overcome in the past—at what point do we finally conclude that enough is enough, that something besides chance must be in play? How improbable is too improbable?

The answer is, "That depends." It depends on the improbability of the event in question. But it also depends upon how many opportunities there are to generate the event. Here again my discussions with Dembski helped me think this through. He noted that there was another common fallacy in reasoning about chance and probability. He had already shown me that some people overlook the importance of pattern recognition in our reasoning about chance. He now explained that in the case where an improbable event manifests a pattern, there is another consideration. What matters, he argued, is not the probability of a particular event or outcome occurring on one occasion (or in one trial), but the probability of generating that event given all the opportunities (or trials) that it has to occur. As Christian de Duve, a Nobel Prize–winning biochemist who has written several books about the origin of life, explains, "What counts [in assessing the probability of a chance hypothesis] is the number of opportunities provided for a given event to happen, relative to the event's probability."[9]

Let's go back to the roulette table to illustrate this point. It is moderately improbable that the roulette ball will land in the red 16 pocket twice in a row. In two spins of the wheel there are 38 × 38, or 1,444, possible outcomes. Thus, there is just one chance in 1,444 of the ball landing on the red 16 twice in a row. Nevertheless, if the croupier has 1,444 opportunities to spin the wheel in a week, he will quite likely witness this otherwise unlikely event sometime during that time. Though the probability of getting two red 16s in just two spins of the roulette wheel is small, the probability of the event occurring by chance within the week is much higher. The croupier has what Dembski calls the "probabilistic resources"—the time and number of trials—necessary to render this event probable.

Dembski uses a coin-tossing scenario to illustrate the concept of probabilistic resources and to point up the human tendency to overestimate the power of our probabilistic resources in overcoming astronomically long odds. He tells a story of a man who had been arrested and convicted for the crime of running a crooked gambling operation. The convict is brought before a judge for sentencing. The judge, a gambling man himself, decides to offer the convict a choice. The judge offers the criminal either ten years in prison or a term in prison that is determined by the outcome of a game of chance. Specifically, the judge tells the criminal that he can leave prison as soon as he flips a coin and it turns up heads 100 times in a row—with no tails interrupting the sequence. The judge stipulates that the coin must be fair, biased toward neither heads nor tails. Dembski asks, "What's the smart criminal to do?" Should he take the coin-flipping option or the ten years?

Here the criminal's intuition might lead him astray. He might think, "Even if it takes two or three years, I'll be better off than sitting in prison for a guaranteed ten years." But imagine that before he takes the coin option, his accountant nudges him from the second row, hands him a calculator, and, whispering in his ear, reminds him that the judge is one of the people he fleeced with his crooked casino. The criminal stares at the calculator, perplexed. What's he supposed to do with that? "Two to the 100th power," his accountant whispers. Then it all comes back to him. As a casino operator, he knows a little about calculating probabilities. He punches 2^{100} into the calculator and is greeted by the figure 1.2676506×10^{30}. The figure, he suddenly realizes, is a life sentence without any realistic possibility of parole. It means that for any given 100-toss series, the man would have a paltry 1 chance in 1,267,650,600,000,000,000,000,000,000,000 of getting all heads. That ten-year window isn't enough to give him a realistic chance of flipping 100 heads in a row. A hundred years isn't enough. A trillion years wouldn't be enough. The criminal should ignore his initial intuition and trust his accountant's calculator.

Dembski explained that if a prisoner flipped a coin at the reasonable rate of once every five seconds for eight hours a day, six days a week, for ten years he would generate only 17,797,120 (~1.7 x 10^7) trials.[10]

Yet this number represents a minuscule portion of the total number of outcomes possible (roughly 10^7 trials out of 10^{30} total possibilities). Even taking the number of trials into account, the odds of generating 100 heads in a row in ten years increases to only roughly 1 chance in 10^{23} (i.e., $10^7/10^{30}$), or 1 in 100,000,000,000,000,000,000,000. Yet this number is obviously far smaller than 1 in 2 or the 50 percent probability that we usually use to gauge a "reasonable chance of success." Thus, it is vastly more likely than not that the criminal will *not* generate the necessary sequence of 100 heads in a row by chance within ten years. He doesn't have the necessary "probabilistic resources."[11]

My discussions with Dembski revealed that there are situations in which a chance hypothesis can be reasonably rejected. Sometimes there are sufficient probabilistic resources to make an unlikely event probable (e.g., two red 16s in a row at the roulette table over the course of a week). And sometimes there are not (as illustrated by Dembski's coin-flipping convict). Scientists can never absolutely prove that some astronomically improbable event could *not* have taken place by chance. But they can determine when it is more likely than not—even vastly more likely than not—that such an event won't or didn't happen by chance alone.

To see this, turn the case of the coin-flipping criminal around. Imagine the convict arrives in prison and, having accepted the judge's challenge, begins to flip a quarter in front of a surveillance camera as required by the court to verify any winning result. After two days, the criminal calls for a prison guard and asks to have the tape reviewed. After viewing the tape of the criminal's last 100 flips, the guard and the warden of the prison verify that, yes indeed, the prisoner flipped 100 heads in a row. What should the court conclude about the prisoner's unexpected success? That the prisoner got lucky on one of his first attempts? This is, of course, a possible, though incredibly improbable, outcome on the chance hypothesis. But because the court sees a pattern (or functional significance) in the event, and because it knows something about the odds of this event and the prisoner's probabilistic resources—that the prisoner had few opportunities to beat the odds—the court suspects that the prisoner cheated. So they investigate before granting release. Sure enough, they find that the prisoner snuck a biased quarter into his cell.

Tools for Further Investigation

So what about the origin of life? Could the chance hypothesis explain the origin of the specified information necessary to produce life in the first place?

As I reflected on Fisher's method of statistical hypothesis testing and Dembski's insights about pattern recognition and probabilistic resources, I realized that I had some analytical tools with which to address the specific questions that interested me. My investigation of how chance hypotheses are generally evaluated suggested a specific line of inquiry and raised several questions that I would need to answer. Is the chance hypothesis for the origin of biological information a substantive hypothesis or merely a vacuous cover for ignorance? If substantive, are there reasons to affirm the adequacy of the chance hypothesis? Or are there grounds for eliminating chance from consideration, at least as the best explanation?

I also wondered: Is there any pattern in the sequence of bases in DNA of the kind that routinely leads us to doubt chance hypotheses in other situations? If so, how improbable is the origin of the information necessary to produce the first life? And finally, what are the relevant probabilistic resources for generating the information in DNA, RNA, and proteins, and are they sufficient to render these presumably improbable events probable? The next two chapters describe what I discovered as I addressed these questions.

9

Ends and Odds

As the result of my work with Bill Dembski in the summer of 1992, I now had a grid by which to evaluate appeals to chance. After I returned from Cambridge that summer to resume my teaching duties at Whitworth College, I began to reexamine specific proposals for how the information necessary for the origin of life might have arisen by chance. Nevertheless, as an assistant professor with a heavy teaching load and a father of young children, I didn't make as much progress on my research over the next few years as I had optimistically projected during my heady summer abroad. That might have turned out for the best, however. I didn't realize it at the time, but I would be in a much better position to evaluate chance as an explanation for the origin of life after Dembski and another scientist I met in Cambridge that summer had completed their own pieces of the research puzzle.

Even so, I began to evaluate the chance hypothesis as best I could. As I did, I rediscovered something that had surprised me when I first encountered it during my graduate studies in England. Most of the leading scientists investigating the origin of life—none sympathetic to the design hypothesis—were deeply skeptical of chance alone. Moreover, I now realized that though these scientists rarely articulated explicit criteria for rejecting the chance hypothesis, they intuitively presupposed criteria that closely resembled those expressed in Fisher's method of chance elimination, or Dembski's, or both. Some also questioned the legitimacy of "chance" because they suspected that it functioned only as a cover for ignorance.

This skepticism had been, at first, surprising to me, because informal discussions of the origin of life and even some college-level textbooks often gave the impression that "chance" was the major factor in the origin of life. For example, in a memorable passage of his popular college textbook, *Biochemistry,* Albert L. Lehninger described how inanimate matter crossed the great divide to become alive. "We now come to the critical moment in evolution," he wrote in 1970, "in which the first semblance of 'life' appeared, through the *chance association* of a number of abiotically formed macromolecular components."[1] Earlier, in 1954, biochemist George Wald argued for the causal efficacy of chance in conjunction with vast expanses of time. As he explained, "Time is in fact the hero of the plot. . . . Given so much time, the impossible becomes possible, the possible probable, and the probable virtually certain."[2] Similarly, in 1968, Francis Crick suggested that the origin of the genetic code might be a "frozen accident."[3]

Most chance hypotheses assumed, sensibly enough, that life could not originate without biological information first arising in some form. In extant cells, DNA stores the information for producing the proteins that perform most critical functions. It follows then that scientists must explain either (a) where the information in DNA (or perhaps some other equivalent information-bearing molecule) came from or (b) how proteins might have arisen directly without DNA (or some equivalent source). Chance-based models have taken both these approaches. In some scenarios, scientists proposed that DNA first arose by chance and then later came into a functional association with the protein molecules that are needed to transcribe and translate genetic information into proteins. In other scenarios, scientists proposed that protein molecules arose first by chance and then later came into association with DNA, RNA, and the other molecules that are now part of the modern system of gene expression. More recently, a third approach suggests that information first arose in RNA molecules and that RNA then functioned initially both as a carrier of genetic information (as DNA does now) and as an enzymatic catalyst (as some proteins do now), although in this model chance is often coupled with other mechanisms (see Chapter 14).

In nearly all these scenarios, origin-of-life theorists first envisioned some favorable prebiotic environment rich in the building blocks out of which DNA, RNA, and proteins are made. Following Oparin's early theory proposing that these chemical building blocks collected in the earth's early ocean, this hypothesized environment acquired the whimsical name "prebiotic soup." Scientists typically imagined subunits of the DNA, RNA, and protein molecules floating freely in this soup or in some other favorable environment. The prebiotic soup would not only provide a rich source of the necessary chemical constituents for building proteins and nucleic acids, but it would also afford many opportunities for these building blocks to combine and recombine to form larger biologically relevant molecules.

Thus, chance theories for the origin of biological information typically envision a process of chemical shuffling. This process eventually would produce large specifically sequenced biopolymers (DNA, RNA, and proteins) by chance starting from an assortment of smaller molecules (such as amino acids in the case of proteins, or bases, sugars, and phosphates in the case of DNA and RNA). Some have likened this process to a "cosmic jackpot." Just as a series of lucky bets might win a gambler a lot of money, so too, perhaps, did a series of fortuitous collisions and connections between chemical building blocks produce the first functional proteins or information-rich nucleic acids.

But does chance provide an adequate explanation for the origin of biological information? Or does it merely conceal ignorance of what actually happened? Do chance-based theories cite processes that are known to generate information-rich structures with some known or regular frequency? Or are there good reasons to reject chance as the best explanation for the origin of the information necessary to life?

These were the questions that animated my investigation of the chance hypothesis in the years that followed my initial conversations with Bill Dembski. I knew that many origin-of-life research scientists were deeply skeptical about explanations that invoked chance alone to explain the origin of life. As I investigated this question, I became convinced that this skepticism was justified—and for several significant reasons.

Functionally Significant Patterns

First, the bases in DNA and RNA and the sequence of amino acids in proteins do not contain mere Shannon information. Rather, these molecules store information that is also functionally specified. As such, they manifest one of the kinds of patterns—a functionally significant pattern—that routinely lead statisticians to reject chance hypotheses, at least in the case of extremely improbable events.

To see why the presence of *specified* information in DNA poses a difficulty for the efficacy of chance processes, recall again our hapless bureaucrat Mr. Jones. When Mr. Jones arrived at the airport in Washington, D.C., having lost Mr. Smith's phone number, he did not despair. He knew that by randomly arranging the numbers 0, 1, 2, 3, 4, 5, 6, 7, 8, and 9 he could produce a large amount of Shannon information. Of course, he quickly discovered that he did not need just unspecified information; he needed specified information. Being naïve about the difference, however, Jones thought that he could arrange and rearrange the ten digits at random to solve the problem by making other numbers. And so he did. He tried 258–197–7292. He tried 414–883–6545. He tried 279–876–2982 and 867–415–5170 and 736–842–3301. He tried hundreds of numbers. In nearly every case, he succeeded only in getting a recorded message from the phone company. Twice a person answered only to tell him that he had a wrong number. Jones kept trying, but he eventually exhausted himself. He simply did not have enough time to have a reasonable chance of finding the correct (i.e., *specified*) phone number by chance alone.

Had Jones been more perceptive, he would have learned an important lesson. Arranging and rearranging characters at random—by chance—can produce lots of unspecified information. It is not, however, an effective way to produce specified information, especially if the time available in which to do so is limited.

During my first few years teaching, I developed another illustration to help students see why many origin-of-life researchers had come to reject the idea that the specified information necessary to the first life could have arisen by chance alone. I would take a bag of Scrabble letters and walk through the aisles of my classroom asking each student to pull out one letter at random from the bag and then walk to the chalkboard and copy that letter on the board. When all had done so, I pointed out that what we had produced was a sequence rich in Shannon information, but no *specified* information. The sequence produced at random was complete gibberish. I would then challenge the class to produce a meaningful sentence, any meaningful sentence, using this random method.

Occasionally, as if to resist the point of my demonstration, the students would persist in their efforts until they had produced a very short sequence of letters that showed some potential for spelling something meaningful, like "ple" or "nes" or even a short word such as "ran." On these occasions, my students would begin to hoot as if they had shown me up—which was exactly what I was hoping they would do. When this happened, I would stop and quickly calculate the amount of specified information they had produced. I would then compare that amount to the amount of information in a short sentence (or functional gene) and ask them to continue selecting letters at random until they had produced an equivalent amount of specified information—a complete sentence or even a long word or two. Invariably the same thing would happen.

After several more selections, whatever promise the students had detected in their initial string of characters was quickly diluted by more gibberish. After their failure became apparent, I would challenge them to start over. No matter how many times they did, they could not produce a long word, let alone a meaningful sentence of any considerable length. By that point, my students saw the moral of the story: a random search might produce a small amount of specified information, but not very much, especially, again, if the time available to search the possible combinations was limited.

Many origin-of-life scientists have similarly recognized how difficult it is to generate *specified* biological information by chance alone in the time available on the early earth (or even in the time available since the beginning of the universe). As one leading biochemist, Alexander Cairns-Smith, wrote in 1971: "Blind chance . . . is very limited. Low levels of cooperation he [blind chance] can produce exceedingly easily (the equivalent of letters and small words), but he becomes very quickly incompetent as the amount of organization [information] increases. Very soon indeed long waiting periods and massive material resources become irrelevant."[4]

In saying things like this, scientists have recognized that the crucial problem is not just generating an improbable sequence of chemical constituents (e.g., an improbable arrangement of nucleotide bases in DNA). Scientists understand that improbable events happen by chance. Instead, the problem is relying on a random search or shuffling to generate one of the rare arrangements of molecules that also *performs a biological function*. It's like the difference between shooting an arrow blindfolded and having it land anywhere and shooting an arrow blindfolded and having it hit the target. It's much harder to do the latter than the former.

Christian de Duve has devised an illustration that underscores the importance of pattern recognition (and the recognition of functional specificity) in our reasoning about chance hypotheses. He points out: "A single, freak, highly improbable event can conceivably happen. Many highly improbable events—drawing a winning lottery number or the distribution of playing cards in a hand of bridge—happen all the time. But a string of [such] improbable events—drawing the same lottery number twice, or the same bridge hand twice in a row—does not happen naturally."[5]

Here de Duve suggests, like Fisher and Dembski, that an improbable event or set of conditions does not alone provide grounds for rejecting a chance hypothesis. Instead, his illustration indicates that pattern recognition also plays a key role. As the same person repeatedly wins the lottery or as the same hand of bridge recurs, a pattern emerges—indeed, one that I have been calling "a functionally significant pattern." In de Duve's examples, an improbable sequence of events has occurred that also produces a significant objective—winning a game or winning money. De Duve suggests that when we recognize that a specific sequence of events has produced a significant outcome—one that achieves a goal or performs a function—we rightly suspect something more than chance at work.

For this reason, he suggests that we should doubt the chance hypothesis for the origin of biological information. According to the chance hypothesis, as the molecular building blocks of DNA, RNA, and proteins interact in a prebiotic environment, these molecular subunits would have attached to one another to form long chainlike molecules. Amino acids would have attached to other amino acids to form a growing polypeptide chain. Sugar and phosphate molecules would have attached to each other. Nucleotide bases would have attached to sugar-phosphate backbones in various arrangements. Thus, the advocates of the chance hypothesis envision such chains growing molecule by molecule (or "letter by letter").

But to form functional genes and proteins, not just any arrangement of bases or amino acids will do. Not by any means. The overwhelming majority of arrangements of bases and amino acids perform no biological function at all. So the assembly of working proteins, for example, in the prebiotic soup would have required functionally appropriate amino acids to attach themselves by chance to a growing chain of other amino acids—time after time after time—despite many opportunities for other less fortuitous outcomes to have occurred. De Duve seems to liken this process to a gambler repeatedly drawing the winning card or the high hand. In both cases he suggests that an astute observer should detect a pattern—a pattern of events that produces a significant outcome.[6]

De Duve insists that as unlikely patterns of fortuitous events emerge, our doubt about the chance hypothesis should increase. Just as we doubt that a card player could draw the same improbable winning hand time after time unless he is cheating, so too should we doubt that appropriate amino acids or bases would happen by chance to attach themselves repeatedly to a growing protein or DNA molecule. In other words, we should be skeptical that the random interaction of molecular building blocks will produce functionally *specified* sequences in DNA (or proteins) by chance alone.

This makes sense. We know from experience that random searches are incredibly inefficient means of generating specified sequences of characters or events—as my Scrabble-bag challenge demonstrated to my students. Thus, when we encounter functionally specified sequences of characters (or molecules that function equivalently)—or when we encounter a functionally significant pattern in a sequence of events (as we do in de Duve's gambling examples)—we are correctly skeptical that chance alone is at work. DNA contains specified sequences of chemicals that function just like digital or alphabetic characters: it contains functionally *specified information,* not just Shannon information. As scientists such as de Duve have recognized this, they understandably have become more skeptical about chance as an explanation for the origin of biological information.[7]

The Improbability of Producing Genes and Proteins by Chance

There is a second reason for doubting the chance hypothesis as an explanation for the origin of biological information. Building a living cell not only requires specified information; it requires a *vast amount* of it—and the probability of this amount of specified information arising by chance is "vanishingly small." But how small? What exactly is the probability that the information necessary to build the first living cell would arise by chance alone?

It turns out that it was initially hard to quantify the answer to this question, because biologists didn't know exactly how much information was necessary to build and maintain the simplest living cell. For one thing, they didn't know how many of the parts of existing cells were necessary to maintain life and how many weren't. But beginning in the 1990s, scientists began to do "minimal complexity" experiments in which they tried to reduce cellular function to its simplest form. Since then, biologists have been able to make increasingly informed estimates of the minimum number of proteins and genes that a hypothetical protocell might have needed to survive.

The simplest extant cell, *Mycoplasma genitalium*—a tiny bacterium that inhabits the human urinary tract—requires "only" 482 proteins to perform its necessary functions and 562,000 bases of DNA (just under 1,200 base pairs per gene) to assemble those proteins. In minimal-complexity experiments scientists attempt to locate unnecessary genes or proteins in such simple life-forms. Scientists use various experimental techniques to "knock out" certain genes and then examine the effect on the bacterial cell to see if it can survive without the protein products of the disabled genes. Based upon minimal-complexity experiments, some scientists speculate (but have not demonstrated) that a simple one-celled organism might have been able to survive with as few as 250–400 genes.

Of course building a functioning cell—at least one that in some way resembles the cells we actually observe today—would have required more than just the genetic information that directs protein synthesis. It would have also required, at the very least, a suite of preexisting proteins and RNA molecules—polymerases, transfer RNAs, ribosomal RNAs, synthetases, and ribosomal proteins, for example—to process and express the information stored in DNA. In fact, there are over 100 specific proteins involved in a simple bacterial translation system; roughly 20 more are involved in transcription and over 30 in DNA replication.[8] Indeed, although the information in DNA is necessary for building proteins, many proteins are necessary for expressing and processing the information in DNA (see Chapter 5). Extant cells, therefore, need *both* types of information-rich molecules—nucleic acids and proteins—functioning together in a tightly integrated way. Therefore, any minimally complex protocell resembling cells we have today would have required not only genetic information, but a sizable preexisting suite of proteins for processing that information.

Building such a cell also would have required other preexisting components. For example, both proteins and DNA are necessary for building and maintaining the energy-production and -processing system of the cell, the ten-enzyme glycolysis pathway that produces high-energy ATP from its low-energy precursor ADP. But information-processing and protein synthesis—and just about everything else in the cell—depends upon a preexisting supply of ATP or one of the closely related energy-carrying molecules. Thus, for life to arise in the first place, ATP (or related molecules) must have also been present along with genes and proteins.

Beyond that, the first cell would have required some kind of semipermeable membrane and a cell wall to protect itself and the chemical reactions taking place inside it. In modern bacteria, the protective barrier is constructed by proteins and enzymes that polymerize (link together) the smaller molecules out of which both the phospholipid bilayer and the cell wall are composed. Cells require a variety of these small molecules—phosphates, lipids, sugars, vitamins, metals, ATP—in addition to proteins and nucleic acids. Many of these small molecules are synthesized with the help of proteins in the cell, but proteins also need these molecules in order to accomplish many of their enzymatic functions. Thus, any protocell likely would have required a preexisting supply of these smaller molecules to establish and maintain itself as well.

The integrated complexity of even a "minimally complex cell" has made it difficult to calculate the odds of all the necessary components of such a system arising in close association with one another by chance alone. Nevertheless, as a first-order approximation, many theorists have contented themselves with calculating the probability of producing just the information necessary to build the suite of necessary proteins, while recognizing that many more components would also be required and that the probability thus calculated vastly underestimates the difficulty of the task at hand.

To calculate this probability, scientists typically use a slightly indirect method. First they calculate the probability of a single functional protein of average length arising by chance alone. Then they multiply that probability by the probability of each of the other necessary proteins arising by chance. The product of these probabilities determines the probability that all the proteins necessary to service a minimally complex cell would come together by chance.

Scientists could, of course, analyze DNA to calculate the probability of the corresponding genetic information arising by chance. But since the information in each gene directs the synthesis of a particular protein, the two sequences (the gene sequence and its corresponding protein sequence) carry the same amount of information.[9] Proteins represent, in the parlance of the discipline, the "expressed function" of the coding region of a DNA molecule. Further, since measures of probability and information are related (albeit inversely), the probability that a particular gene would arise by chance is roughly the same as the probability that its corresponding gene product (the protein that the gene encodes) would do so.

For that reason, the relevant probability calculation can be made either by analyzing the odds of arranging amino acids into a functional protein or by analyzing the odds of arranging nucleotide bases into a gene that encodes that protein. Because it turns out to be simpler to make the calculation using proteins, that's what most origin-of-life scientists have done.[10] In any case by the 1960s, as scientists began to appreciate the complexity of both DNA and proteins and the amount of specified information they contain, many began to suspect that producing these molecules by chance alone would prove to be quite difficult.

Symposium at the Wistar Institute

The first public inkling of concern about the chance explanation surfaced at a now famous conference of mathematicians and biologists held in Philadelphia in 1966. The conference was titled "Mathematical Challenges to Neo-Darwinism."[11] It was held at the Wistar Institute, a prestigious medical research center, and chaired by Sir Peter Medawar, a Nobel laureate from England. The conference was called to discuss the growing doubts of many mathematicians, physicists, and engineers about the ability of random mutations to generate the information needed to produce new forms of life. Though the skeptics mainly expressed doubt about the role of random mutations in biological evolution, the questions they raised had equally important implications for assessing the role of chance in chemical evolutionary theories about the first life.

These doubts were first discussed at an informal private gathering in Geneva in the mid-1960s at the home of MIT physicist Victor Weisskopf. During a picnic lunch the discussion turned to evolution. Several of the MIT math, physics, and engineering professors present expressed surprise at the biologists' confidence in the power of mutations to produce new forms of life in the time available to the evolutionary process. A vigorous argument ensued, but was not resolved. Instead, plans were made for a conference to discuss the concerns of the skeptics.

According to some of these MIT professors, the neo-Darwinian mechanism faced what they called a "combinatorial problem." In mathematics, the term "combinatorial" refers to the number of possible ways that a set of objects can be arranged or combined (the relevant branch of mathematics is known as "combinatorics"). Some simple bike locks, for example, have four *dials* with ten *settings* on each dial. A bike thief encountering one of these locks (and lacking bolt cutters) faces a combinatorial problem because there are $10 \times 10 \times 10 \times 10$, or 10,000 possible ways of combining the possible settings on each of the four dials and only one combination that will open the lock. Guessing at random is unlikely to yield the correct combination unless the thief has a truly generous amount of time to do so.

The Wistar scientists explained that a similar difficulty confronts the Darwinian mechanism. According to neo-Darwinian theory, new genetic information arises first as random mutations occur in the DNA of existing organisms. When mutations arise that confer a survival advantage on the organisms that possess them, the resulting genetic changes are passed on by natural selection to the next generation. As these changes accumulate, the features of a population begin to change over time. Nevertheless, natural selection can "select" only what random mutations first produce. And for the evolutionary process to produce new forms of life, random mutations must first have produced new genetic information for building novel proteins. That, for the mathematicians, physicists, and engineers at Wistar, was the problem. Why?

The skeptics at Wistar argued that it is extremely difficult to assemble a new gene or protein by chance because of the sheer number of possible base or amino-acid sequences. For every combination of amino acids that produces a functional protein there exists a vast number of other possible combinations that do not. And as the length of the required protein grows, the number of possible amino-acid sequence combinations of that length grows exponentially, so that the odds of finding a functional sequence—that is, a working protein—diminish precipitously.

To see this, consider the following. Whereas there are four ways to combine the letters A and B to make a two-letter combination (AB, BA, AA, and BB), there are eight ways to make three-letter combinations (AAA, AAB, ABB, ABA, BAA, BBA, BAB, BBB), and sixteen ways to make four-letter combinations, and so on. The number of combinations grows geometrically, 2^2, 2^3, 2^4, and so on. And this growth becomes more pronounced when the set of letters is larger. For protein chains, there are 20^2, or 400, ways to make a two-amino-acid combination, since each position could be any one of 20 different alphabetic characters. Similarly, there are 20^3, or 8,000, ways to make a three-amino-acid sequence, and 20^4, or 160,000, ways to make a sequence four amino acids long, and so on. As the number of possible combinations rises, the odds of finding a correct sequence diminishes correspondingly. But most functional proteins are made of hundreds of amino acids. Therefore, even a relatively short protein of, say, 150 amino acids represents one sequence among an astronomically large number of other possible sequence combinations (approximately 10^{195}).

Consider the way this combinatorial problem might play itself out in the case of proteins in a hypothetical prebiotic soup. To construct even one short protein molecule of 150 amino acids by chance within the prebiotic soup there are several combinatorial problems—probabilistic hurdles—to overcome. First, all amino acids must form a chemical bond known as a peptide bond when joining with other amino acids in the protein chain (see Fig. 9.1). If the amino acids do not link up with one another via a peptide bond, the resulting molecule will not fold into a protein. In nature many other types of chemical bonds are possible between amino acids. In fact, when amino-acid mixtures are allowed to react in a test tube, they form peptide and nonpeptide bonds with roughly equal probability. Thus, with each amino-acid addition, the probability of it forming a peptide bond is roughly $1/2$. Once four amino acids have become linked, the likelihood that they are joined exclusively by peptide bonds is roughly $1/2 \times 1/2 \times 1/2 \times 1/2 = 1/16$, or $(1/2)^4$. The probability of building a chain of 150 amino acids in which all linkages are peptide linkages is $(1/2)^{149}$, or roughly 1 chance in 10^{45}.

Second, in nature every amino acid found in proteins (with one exception) has a distinct mirror image of itself; there is one left-handed version, or L-form, and one right-handed version, or D-form. These mirror-image forms are called optical isomers (see Fig. 9.2). Function-

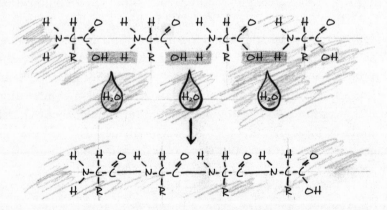

Figure 9.1. Amino acids link together when the amino group of one amino acid bonds to the carboxyl group of another. Notice that water is a by-product of the reaction (called a *condensation* reaction).

ing proteins tolerate only left-handed amino acids, yet in abiotic amino-acid production the right-handed and left-handed isomers are produced with roughly equal frequency. Taking this into consideration further compounds the improbability of attaining a biologically functioning protein. The probability of attaining, at random, only L–amino acids in a hypothetical peptide chain 150 amino acids long is $(1/2)^{150}$, or again roughly 1 chance in 10^{45}. Starting from mixtures of D-forms and L-forms, the probability of building a 150-amino-acid chain at random in which all bonds are peptide bonds and all amino acids are L-form is, therefore, roughly 1 chance in 10^{90}.

Functioning proteins have a third independent requirement, the most important of all: their amino acids, like letters in a meaningful sentence, must link up in functionally specified sequential arrangements. In some cases, changing even one amino acid at a given site results in the loss of protein function. Moreover, because there are 20 biologically occurring amino acids, the probability of getting a specific amino acid at a given site is small—1/20. (Actually the probability is even lower because, in nature, there are also many nonprotein-forming amino acids.) On the assumption that each site in a protein chain requires a particular amino acid, the probability of attaining a particular protein 150 amino acids long would be $(1/20)^{150}$, or roughly 1 chance in 10^{195}.

Nevertheless, molecular biologists have known for a while that most sites along the chain can tolerate several of the twenty different amino acids commonly found in proteins without destroying the function of the protein, though some cannot. At the Wistar conference this tolerance was cited by some of the evolutionary biologists as evidence that random mutations might be sufficient to produce new functional proteins. Most of the mathematicians at the conference had assumed that the functional sequences were incredibly rare within the space of all the

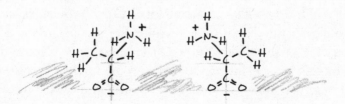

Figure 9.2. Two optical isomers of the same amino acid.

possible combinations. But some of the evolutionary biologists asked: What if functional sequences are more common than the mathematicians are supposing? This raised an important question, one that could not be answered in 1966: How rare, or common, are the *functional* sequences of amino acids among all the possible sequences of amino acids in a chain of any given length?

In the late 1980s, several important studies were conducted in the laboratory of MIT biochemist Robert Sauer in order to investigate this question. His research team used a sampling technique known as "cassette mutagenesis" to determine how much variance among amino acids can be tolerated at any given site in several proteins. This technique would help resolve an important issue. If proteins can tolerate a lot of variance, then that would increase the probability that a random search through the space of possibilities would find a functional sequence. If proteins were more finicky—if the requirements of functionality imposed more rigid or restricted constraints on sequencing—then that would decrease the probability of a random process successfully producing a functional protein. Thus, whatever Sauer's team discovered would be extremely significant. The results of their experiments could help determine the probability that a functional protein would arise by chance from a prebiotic soup.

So what did they find? Their most clear-cut experiments[12] seemed to indicate that, even taking the possibility of variance into account, the probability of achieving a functional sequence of amino acids in several known (roughly 100 amino acid) proteins at random is still "exceedingly small," about 1 chance in 10^{63} (to put this in perspective, there are 10^{65} atoms in our galaxy).[13] Using a variety of mutagenesis techniques, they and other scientists showed that proteins (and thus the genes that produce them) are highly specified relative to biological function.[14] Earlier studies indicated that amino-acid residues at many sites cannot vary without functional loss.[15] Now Sauer and others demonstrated that even for sites that do admit some variance, not just any amino acid will do. Instead, they showed that functional requirements place significant constraints on sequencing at sites where some variance is allowed. By quantifying that allowable variance, they made it possible to calculate the probability of finding a protein with a functional sequence among the larger ensemble of combinatorial possibilities.

I first learned about the work of Robert Sauer and its relevance to assessing the chance hypothesis in 1992 from a postdoctoral researcher at Cambridge University named Douglas Axe. Axe and I met in the summer of that year as I was conducting research with Bill Dembski. Axe had come to Cambridge to perform mutagenesis experiments that were similar to Sauer's. During his Ph.D. work at Caltech, Axe learned about the structure of proteins and the intricate folds they need in order to perform their functions. He began to wonder how difficult it was to produce these folds by random mutations or random molecular interactions. He began to ask a very similar question to the one that had been asked at the Wistar conference: How rare, or common, are the amino-acid sequences that produce the stable folds that make it possible for proteins to perform their biological functions?

As Axe began to examine Sauer's experimental method, he asked whether Sauer might have underestimated how much protein sequences can vary and still maintain function. To test this possibility, he developed a more rigorous method of estimating this allowable variability in order to eliminate possible estimation error. Axe liked what Sauer had done, but wanted to produce a more definitive answer. He wanted to settle the key question from the Wistar conference beyond any reasonable doubt. His interest in the subject eventually led him to the Cambridge University laboratory of Alan Fersht, and then to the Center for Protein Engineering at the famous Medical Research Council Centre in Cambridge, where the likes of Francis Crick, Max Perutz, John Kendrew, and Fred Sanger

Figure 9.3. Douglas Axe.
Printed by permission from Douglas Axe.

had worked in the early 1960s. Axe's work was published in a series of papers in the *Journal of Molecular Biology, Biochemistry,* and the *Proceedings of the National Academy of Sciences* between 1996 and 2004.

The results of a paper he published in 2004 were particularly telling.[16] Axe performed a mutagenesis experiment using his refined method on a functionally significant 150-amino-acid section of a protein called beta-lactamase, an enzyme that confers antibiotic resistance upon bacteria. On the basis of his experiments, Axe was able to make a careful estimate of the ratio of (a) the number of 150-amino-acid sequences that can perform that particular function to (b) the whole set of possible amino-acid sequences of this length. Axe estimated this ratio to be 1 to 10^{77}.

This was a staggering number, and it suggested that a random process would have great difficulty generating a protein with that particular function by chance. But I didn't want to know just the likelihood of finding a protein with a particular function within a space of combinatorial possibilities. I wanted to know the odds of finding *any* functional protein whatsoever within such a space. That number would make it possible to evaluate chance-based origin-of-life scenarios, to assess the probability that a single protein—*any working protein*—would have arisen by chance on the early earth.

Fortunately, Axe's work provided this number as well.[17] Axe knew that in nature proteins perform many specific functions. He also knew that in order to perform these functions their amino-acid chains must first fold into stable three-dimensional structures. Thus, before he estimated the frequency of sequences performing a specific (beta-lactamase) function, he first performed experiments that enabled him to estimate the frequency of sequences that will produce stable folds. On the basis of his experimental results, he calculated the ratio of (a) the number of 150-amino-acid sequences capable of folding into stable "function-ready" structures to (b) the whole set of possible amino-acid sequences of that length. He determined that ratio to be 1 to 10^{74}.

Since proteins can't perform functions unless they first fold into stable structures, Axe's measure of the frequency of folded sequences also provided a measure of the frequency of functional proteins—*any* functional proteins—within that space of possibilities (or "sequence space"). Indeed, by taking what he knew about protein folding into account, Axe estimated the ratio of (a) the number of 150-amino-acid sequences that produce *any functional protein whatsoever* to (b) the whole set of possible amino-acid sequences of that length. Axe's estimated ratio of 1 to 10^{74} implied that the probability of producing any properly sequenced 150-amino-acid protein at random is also about 1 in 10^{74}. In other words, a random process producing amino-acid chains of this length would stumble onto a functional protein only about once in every 10^{74} attempts.

When one considers that Robert Sauer was working on a shorter protein of 100 amino acids, Axe's number might seem a bit less prohibitively improbable. Nevertheless, it still represents a startlingly small probability. In conversations with me, Axe has compared the odds of producing a functional protein sequence of modest (150-amino-acid) length at random to the odds of finding a single marked atom out of all the atoms in our galaxy via a blind and undirected search. Believe it or not, the odds of finding the marked atom in our galaxy are markedly better (about a billion times better) than those of finding a functional protein among all the sequences of corresponding length.

This was a very significant result. Building on the work of Robert Sauer and others, Axe established a reliable quantitative estimate of the rarity of functional sequences within the corresponding space of possible amino-acid sequences, thereby providing an answer to the question first posed at the Wistar symposium. Though proteins tolerate a range of possible amino acids at some sites, functional proteins are still extremely rare within the whole set of possible amino-acid sequences.

In June 2007, Axe had a chance to present his findings at a symposium commemorating the publication of the proceedings from the original Wistar symposium forty years earlier. In attendance at this symposium in Boston was retired MIT engineering professor Murray Eden, with a still incisive mind at the age of eighty-seven. Eden had been one of the original conveners of the Wistar conference and was the one who had most forcefully explained the combinatorial problem facing neo-Darwinism. Forty years later, Axe's experimental work had now confirmed Eden's initial intuition: the odds are prohibitively stacked against a random process producing functional proteins. Functional proteins are exceedingly rare among all the possible combinations of amino acids.

Axe's improved estimate of how rare functional proteins are within "sequence space" has now made it possible to calculate the probability that a 150-amino-acid compound assembled by random interactions in a prebiotic soup would be a functional protein. This calculation can be made by multiplying the three independent probabilities by one another: the probability of incorporating only peptide bonds (1 in 10^{45}), the probability of incorporating only left-handed amino acids (1 in 10^{45}), and the probability of achieving correct amino-acid sequencing (using Axe's 1 in 10^{74} estimate). Making that calculation (multiplying the separate probabilities by adding their exponents: $10^{45 + 45 + 74}$) gives a dramatic answer. The odds of getting even one functional protein of modest length (150 amino acids) by chance from a prebiotic soup is no better than 1 chance in 10^{164}.

It is almost impossible to convey what this number represents, but let me try. We have a colloquial expression in English, "That's like looking for a needle in a haystack." We understand from this expression that finding the needle will be difficult because the needle—the thing we want—is mixed in with a great number of other things we don't want. To have a realistic chance of finding the needle, we will need to search for a long, long time. Now consider that there are only 10^{80} protons, neutrons, and electrons in the observable universe. Thus, if the odds of finding a functional protein by chance on the first attempt had been 1 in 10^{80}, we could have said that's like finding a marked particle—proton, neutron, or electron (a much smaller needle)—among all the particles in the universe (a much larger haystack). Unfortunately, the problem is much worse than that. With odds standing at 1 chance in 10^{164} of finding a functional protein among the possible 150-amino-acid compounds, the probability is 84 orders of magnitude (or powers of ten) *smaller* than the probability of finding the marked particle in the whole universe. Another way to say that is the probability of finding a functional protein by chance alone is a trillion, trillion, trillion, trillion, trillion, trillion, trillion times smaller than the odds of finding a single specified particle among all the particles in the universe.

And the problem is even worse than this for at least two reasons. First, Axe's experiments calculated the odds of finding a relatively *short* protein by chance alone. More typical proteins have hundreds of amino acids, and in many cases their function requires close association with other protein chains. For example, the typical RNA polymerase—the large molecular machine the cell uses to copy genetic information during transcription (discussed in Chapter 5)—has over 3,000 functionally specified amino acids.[18] The probability of producing such a protein and many other necessary proteins by chance would be far smaller than the odds of producing a 150-amino-acid protein.

Second, as discussed, a minimally complex cell would require many more proteins than just one. Taking this into account only causes the improbability of generating the necessary proteins by chance—or the genetic information to produce them—to balloon beyond comprehension. In 1983 distinguished British cosmologist Sir Fred Hoyle calculated the odds of producing the proteins necessary to service a simple one-celled organism by chance at 1 in $10^{40,000}$.[19] At that time scientists could have questioned his figure. Scientists knew how long proteins were and roughly how many protein types there were in simple cells. But since the amount of functionally specified information in each protein had not yet been measured, probability calculations like Hoyle's required some guesswork.

Axe's experimental findings suggest that Hoyle's guesses were pretty good. If we assume that a minimally complex cell needs at least 250 proteins of, on average, 150 amino acids and that the probability of producing just one such protein is 1 in 10^{164} as calculated above, then the probability of producing all the necessary proteins needed to service a minimally complex cell is 1 in 10^{164} multiplied by itself 250 times, or 1 in $10^{41,000}$. This kind of number allows a great amount of quibbling about the accuracy of various estimates without altering the conclusion. The probability of producing the proteins necessary to build a minimally complex cell—or the genetic information necessary to produce those proteins—by chance is unimaginably small.

Conclusion

Axe's work confirmed the intuitions of an older generation of origin-of-life researchers and other scientists—such as Harold Morowitz, Fred Hoyle, Chandra Wickramasinghe, Alexander Cairns-Smith, Ilya Prigogine, Hubert Yockey, Christian de Duve, Robert Shapiro, and Francis Crick (after 1980)[20]—who were deeply skeptical about the chance hypothesis. Many of these scientists had performed their own calculations in which they assumed extremely favorable prebiotic conditions, more time than was actually available on the early earth, and maximally fast reaction rates between the chemical constituents of proteins, DNA, and RNA. Invariably, such calculations have fueled greater skepticism about the chance hypothesis, especially since origin-of-life researchers also recognize that DNA and proteins possess functionally specified, rather than just Shannon, information.

But I knew from my discussions with Dembski and from my own research on the origin-of-life problem that there were other reasons to doubt the chance hypothesis. In the next chapter, I discuss why these other considerations, in conjunction with the factors discussed here, justify eliminating chance from consideration as the best explanation for the origin of the information necessary to produce the first life.

10

Beyond the Reach of Chance

As I investigated the question of whether biological information might have arisen by chance, it became abundantly clear to me that the probability of the necessary events is exceedingly small. Nevertheless I realized, based on my previous conversations with Bill Dembski, that the probability of an event by itself does not alone determine whether the event could be reasonably explained by chance. The probabilities, as small as they were, were not by themselves conclusive. I remembered that I also had to consider the number of opportunities that the event in question might have had to occur. I had to take into account what Dembski called the *probabilistic resources.*

But what were those resources—how many opportunities did the necessary proteins or genes have to arise by chance? The advocates of the chance hypothesis envisioned amino acids or nucleotide bases, phosphates, and sugars knocking into each other in an ocean-sized soup until the correct arrangements of these building blocks arose by chance somewhere. Surely, such an environment would have generated many opportunities for the assembly of functional proteins and DNA molecules. But how many? And were there enough such opportunities to render these otherwise exceedingly improbable events probable?

Here again Bill Dembski's work gave me a way to answer this question. Dembski had calculated the maximum number of events that could actually have taken place during the history of the observable universe.[1] He did this to establish an upper boundary on the probabilistic resources that might be available to produce any event by chance.[2]

Dembski's calculation was elegantly simple and yet made a powerful point. He noted that there were about 10^{80} elementary particles[3] in the observable universe. (Because there is an upper limit on the speed of light, only those parts of the universe that are observable to us can affect events on earth. Thus, the observable universe is the only part of the universe with probabilistic resources relevant to explaining events on earth.) Dembski also noted that there had been roughly 10^{16} seconds since the big bang. (A few more have transpired since he made the calculation, but not enough to make a difference!)

He then introduced another parameter that enabled him to calculate the maximum number of opportunities that any particular event would have to take place since the origin of the universe. Due to the properties of gravity, matter, and electromagnetic radiation, physicists have determined that there is a limit to the number of physical transitions that can occur from one state to another within a given unit of time. According to physicists, a physical transition from one state to another cannot take place faster than light can traverse the smallest physically significant unit of distance (an indivisible "quantum" of space). That unit of distance is the so-called Planck length of 10^{-33} centimeters. Therefore, the time it takes light to traverse this smallest distance determines the shortest time in which any physical effect can occur. This unit of time is the Planck time of 10^{-43} seconds.

Knowing this, Dembski was able to calculate the largest number of opportunities that any material event had to occur in the observable universe since the big bang. Physically speaking, an event occurs when an elementary particle does something or interacts with other elementary particles. But since elementary particles can interact with each other only so many times per second (at most 10^{43} times), since there are a limited number (10^{80}) of elementary particles, and since there has been a limited amount of time since the big bang (10^{17} seconds), there are a limited number of opportunities for any given event to occur in the entire history of the universe.

Dembski was able to calculate this number by simply multiplying the three relevant factors together: the number of elementary particles (10^{80}) times the number of seconds since the big bang (10^{17}) times the number of possible interactions per second (10^{43}). His calculation fixed the total number of events that could have taken place in the observable universe since the origin of the universe at 10^{140}.[4] This then provided a measure of the probabilistic resources of the entire observable universe.

Other mathematicians and scientists have made similar calculations.[5] During the 1930s, the French mathematician Emile Borel made a much less conservative estimate of the probabilistic resources of the universe, which he set at 10^{50}.[6] More recently, University of Pittsburgh physicist Bret Van de Sande has calculated the probabilistic resources of the universe at a more restrictive 2.6×10^{92}.[7] MIT computer scientist Seth Lloyd has calculated that the most bit operations the universe could have performed in its history (assuming the entire universe were given over to this single-minded task) is 10^{120}, meaning that a specific bit operation with an improbability significantly greater than 1 chance in 10^{120} will likely never occur by chance.[8] None of these probabilistic resources is sufficient to render the chance hypothesis plausible. Dembski's calculation is the most conservative and gives chance its "best chance" to succeed. But even his calculation confirms the implausibility of the chance hypothesis, whether chance is invoked to explain the information necessary to build a single protein or the information necessary to build the suite of proteins needed to service a minimally complex cell.

Recall that the probability of producing a single 150-amino-acid functional protein by chance stands at about 1 in 10^{164}. Thus, for each functional sequence of 150 amino acids, there are at least 10^{164} other possible nonfunctional sequences of the same length. Therefore, to have a good (i.e., better than 50–50) chance of producing a single functional protein of this length by chance, a random process would have to generate (or sample) more than one-half of the 10^{164} nonfunctional sequences corresponding to each functional sequence of that length. Unfortunately, that number vastly exceeds the most optimistic estimate of the probabilistic resources of the entire universe—that is, the number of events that could have occurred since the beginning of its existence.

To see this, notice again that to have a good (better than 50–50) chance of generating a functional protein by chance, more than half of the 10^{164} sequences would have to be produced. Now compare that number (call it $.5 \times 10^{164}$) to the maximum number of opportunities—10^{140}—for that event to occur in the history of the universe. Notice that the first number ($.5 \times 10^{164}$) exceeds the second (10^{140})

by roughly twenty-four orders of magnitude, by roughly a trillion trillion times.

What does this mean? It means that if every event in the universe over its entire history were devoted to producing combinations of amino acids of the correct length in a prebiotic soup (an extravagantly generous and even absurd assumption), the number of combinations thus produced would still represent a tiny fraction—roughly 1 out of a trillion trillion—of the total number of events needed to have a 50 percent chance of generating a functional protein—*any* functional protein of modest length by chance alone.

In other words, even if the theoretically maximum number (10^{140}) of amino-acid sequences possible were generated, the number of candidate sequences would still represent a minuscule portion of the total possible number of sequences (of a given length). For this reason, it would be vastly more probable than not that a functional protein of modest length would *not* have arisen by chance—simply too few of the possible sequences would have been sampled to provide a realistic opportunity for this to occur. Even taking the probabilistic resources of the whole universe into account, it is extremely unlikely that even a single protein of that length would have arisen by chance on the early earth. (And, as explained in the accompanying endnote, proteins capable of performing many necessary features of a minimally complex cell often have to be *at least* 150 amino acids in length. Moreover, there are good reasons to think that these large necessary proteins could *not* evolve from simpler proteins or peptide chains.)[9]

To see this probabilistic reasoning in everyday terms, imagine that Slick performs a blind search for a single red marble in a huge gunnysack filled with 10,000 marbles, the remainder of which are blue. To have a better than 50 percent chance of finding the one red marble, Slick must select more than 5,000 of the marbles. But Slick has only ten seconds in which to sample the marbles. Further, it takes one second to find and put each marble aside in another jar. Thus, he can hope to sample only 10 out of the 10,000 marbles in the time available. Is it likely that Slick would find the red marble? Clearly not. Given his probabilistic resources, he has just 1 chance in 1,000 of finding the red marble, which is much less than 1 in 2, or 50 percent.[10] Thus, it is much more likely than not that he will *not* find the red marble by chance in the time available.

In the same way, it is much more likely than not that a random process would *not* produce (or find) even one functional protein (of modest length) in the whole history of the universe. Given the number of possible sequences that need to be sampled and the number of opportunities available to do so, the odds of success are much smaller than $1/2$—the point at which the chance hypothesis becomes reasonable (see below). Indeed, the odds of producing a single functional protein by chance in the whole history of the universe are actually much smaller than the odds of Slick finding the one red marble in my illustration. And beyond that, of course, the odds of producing the suite of proteins necessary to service a minimally complex cell by chance alone are almost unimaginably smaller. Indeed, the improbability of that event—calculated conservatively (see Chapter 9) at 1 chance in $10^{41,000}$—completely dwarfs the probabilistic resources of the whole universe. Taking all those resources—10^{140} possible events—into account only increases the probability of producing a minimally complex cell by chance alone to, at best, 1 chance in $10^{40,861}$, again, an unimaginably small probability.

And that is the third reason why origin-of-life researchers have rejected the chance hypothesis. The complexity of the events that origin-of-life researchers need to explain exceeds the probabilistic resources of the entire universe. In other words, the universe itself does not possess the probabilistic resources necessary to render probable the origin of biological information by chance alone.

The "Chance" of Chance

I knew from my conversations with Bill Dembski and my study of statistical hypothesis testing that the occurrence of an improbable event alone does not justify rejecting a chance hypothesis. To justify eliminating chance, I knew that it was also necessary to recognize a pattern in an event and to consider the available probabilistic resources. The calculations presented in the previous section meet both of these conditions. DNA and proteins manifest functionally significant patterns. And, given all the probabilistic resources of the universe, the odds of producing a functional protein of modest length stands at less than one chance in trillion trillion.

In making this calculation, I have computed what statisticians call a "conditional probability." A conditional probability measures the probability of one thing being true on the assumption that another thing is true. In this case, I have calculated the probability of a functional protein occurring by chance *given,* or "conditioned on," a best-case estimate of the relevant probabilistic resources.

But this calculation has also been "conditioned" on something else. Recall that all along I had been attempting to determine the odds of a functional protein occurring *by chance.* That itself was another kind of conditional or "given." I had not just been asking: What are the odds that a functional protein would arise? I had been asking: What are the odds that a functional protein or a minimally complex cell would arise *by chance,* given the available probabilistic resources? In other words, I had been asking: What are the odds that a functional protein or a cell would arise *given the chance hypothesis* (i.e., given *the truth* of the chance hypothesis)? Recall that the chance hypothesis in this case asserts that functional proteins or information-rich DNA molecules arose from the random interactions of molecular building blocks in a prebiotic soup. Framing the question this way—as a question about the probability of the origin of proteins *given the chance hypothesis*—provided grounds for evaluating whether it was more reasonable or not to accept the chance hypothesis.

This was important, because I often encounter people who think that it makes sense to cling to the chance hypothesis as long as there was some chance—any chance, however small—that life might have arisen by some specified or even unspecified random process. They often say things like, "Sure, the origin of life is overwhelmingly improbable, but as long as there is at least some chance of life arising by chance, then we shouldn't reject the possibility that it did."

This way of reasoning turns out to be fallacious, however, because it fails to recognize what probabilistic resources can tell us about whether it is more reasonable to accept or reject chance. Consider a case in which all probabilistic resources have been considered and the conditional probability of an event occurring *given the chance hypothesis* is greater than 1/2. That means that it is *more likely* than not that the event in question *would have* occurred by chance (if every opportunity for it to occur had been realized). If Slick is given not ten seconds, but, say, two hours (7,200 seconds) to sample the bag of marbles, it is more likely than not that he would find the red marble by sampling randomly. In two hours Slick can sample 7,200 marbles, which gives him a better than 50 percent chance of finding the red marble.

Conversely, if after all probabilistic resources have been considered and the conditional probability of an event occurring by chance is less than 1/2, then it is *less likely* than not that the event will occur by chance. In the case that such an event has already occurred and we have no direct knowledge of how it occurred, it makes more sense to reject the chance hypothesis than to accept it.

My earlier illustrations made this point. When our hypothetical criminal claimed to have flipped 100 heads in a row on his second day of jail, suspicions were rightly raised. Given the improbability of the required outcome (producing 100 heads in a row by flipping a fair coin) and the limited probabilistic resources available to the prisoner, the court assumed—reasonably—that the prisoner had cheated, that something other than chance had been at work.

Similarly, imagine that after starting my demonstration with Scrabble letters, I left the classroom for a few minutes and instructed my students to continue picking letters at random and writing the results on the board in my absence. Now imagine that upon my return they showed me a detailed message on the blackboard such as Einstein's famous dictum: "God does not play dice with the universe." Would it be more reasonable for me to suppose that they had cheated (perhaps, as a gag) or that they had gotten lucky? Clearly, I should suspect (strongly) that they had cheated. I should reject the chance hypothesis. Why?

I should reject chance as the best explanation not only because of the improbability of the sequence my students had generated, but also because of what I knew about the probabilistic resources available to them. If I had made a prediction before I had left the room, I would have predicted that they could not have generated a sequence of that length by chance alone *in the time available*. But even after seeing the sequence on the board, I still should have rejected the chance hypothesis.

Indeed, I should know that my students did not have anything like the probabilistic resources to have a realistic chance of generating a sequence of that improbability by chance alone. In one hour my students could not have generated anything but a minuscule fraction of the total possible number of 40-character sequences corresponding to the length of the message they had written on the board. The odds that they could have produced that sequence—or any meaningful sequence at all of that length—in the time available by choosing letters at random was exceedingly low—much less than 1/2. They simply did not have time to sample anything close to the number of 40-character sequences that they would have needed to have a 50 percent chance of generating a meaningful sequence of that length. Thus, it would be much more likely than not that they would *not* produce a meaningful sequence of that length by chance in the time available and, therefore, it was also vastly more likely than not that something other than chance had been in play.

Decision Time: Assessing the Chance Hypothesis

Following many leading origin-of-life researchers, I came to the same conclusion about the first life and even the first genes and proteins: it is much more likely than not that chance alone did *not* produce these phenomena. Life, of course, does exist. So do the information-rich biological macromolecules upon which living cells depend. But the probability that even one of these information-rich molecules arose by chance, let alone the suite of such molecules necessary to maintain or build a minimally complex cell, is so small as to dwarf the probabilistic resources of the entire universe. The conditional probability that just one of these information-rich molecules arose by chance—in effect, the chance that chance is true—is much less than one-half. It is less than one in a trillion trillion. Thus, I concluded that it is more reasonable to reject the chance hypothesis than to accept it.

This was an intellectually liberating conclusion. Anyone can claim that a fantastically improbable event might have occurred by chance. Chance, in that sense, is always a possible explanation. But it doesn't follow that chance necessarily constitutes the best explanation. And following what I knew about the historical scientific method, I wanted to find the *best* explanation for the origin of biological information. When I realized that I did not need to absolutely *disprove* the chance hypothesis in order to make an objective determination about its merits, clarity came. By assessing the probability of an event in light of the available probabilistic resources, I could determine whether it was more reasonable to affirm or to reject the chance hypothesis for that event. When I realized that it was far more reasonable to reject the chance hypothesis for the origin of functional genes and proteins, I concluded that chance was not a terribly promising candidate for "best explanation" of the DNA enigma. Chance was certainly not a more reasonable explanation than its negation, namely, that something other than chance had been at work in the origin of biological information. Indeed, when I remembered that the chance hypothesis implicitly negated both design and lawlike necessity, and that rejecting chance, therefore, meant affirming "something other than chance" at work, I began to evaluate alternative explanations.

My own reasoning about the chance hypothesis mirrored that of many origin-of-life researchers, many of whom expressed exasperation at the way some scientists used "chance" as a catchall explanation or a cover for ignorance.[11] For example, after I first began reading the scientific articles about the origin of life, I found a spirited critique of chance published in *Nature* in 1963, just about the time molecular biologists were first coming to grips with the complexity of DNA and proteins. The paper was written by P. T. Mora, a senior research biologist at the National Institutes of Health. Here's what he wrote:

> *To invoke statistical concepts, probability and complexity to account for the origin and the continuance of life is not felicitous or sufficient. As the complexity of a molecular aggregate increases, and indeed very complex arrangements and interrelationships of molecules are necessary for the simplest living unit, the probability of its existence under the disruptive and random influence of physico-chemical forces decreases; the probability that it will continue to function in a certain way, for example, to absorb and to repair, will be even lower; and* the probability that it will reproduce, *still lower. Statistical considerations, probability, complexity, etc., followed to their logical implications suggest that the origin and continuance of life is not controlled by such principles. An admission of this is the use of a period of practically infinite time to obtain the derived result. Using such logic, however, we*

can prove anything.... When in statistical processes, the probability is so low that for practical purposes infinite time must elapse for the occurrence of an event, statistical explanation is not helpful.[12]

I had come to much the same conclusion. Not only were the odds overwhelmingly against life arising by chance even considering all available probabilistic resources, but the chance hypothesis was usually invoked in a way that didn't explain anything. To say that "given infinite time, life might have arisen by chance" was, in essence, a tautology. Given infinite time, anything might happen. But that doesn't explain why life originated here or what actually caused it to do so.

Environmental Factors

There were some additional reasons to reject the chance hypothesis. The chance hypothesis for the origin of information-rich biological molecules assumes the existence of a favorable prebiotic soup in which an abundant supply of the chemical building blocks of proteins and nucleic acids could interact randomly over vast expanses of time. These chemical building blocks were thought to have been produced by the kinds of chemical reactions that Stanley Miller simulated in his famous 1953 experiment. Yet when Stanley Miller conducted his experiment simulating the production of amino acids on the early earth, he had presupposed that the earth's atmosphere was composed of a mixture of what chemists call reducing gases, such as methane (CH_4), ammonia (NH_3), and hydrogen (H_2). He also assumed that the earth's atmosphere contained virtually no free oxygen.[13] In the years following Miller's experiment, however, new geochemical evidence showed that the assumptions Miller had made about the early atmosphere were incorrect. Instead, evidence strongly suggested that neutral gases such as carbon dioxide, nitrogen, and water vapor[14]—not methane, ammonia, and hydrogen—predominated in the early atmosphere. Moreover, a number of geochemical studies showed that significant amounts of free oxygen were also present even before the advent of plant life, probably as the result of the photo-dissociation of water vapor.[15]

This new geological and geochemical evidence implied that prebiotic atmospheric conditions were hostile, not friendly, to the production of amino acids and other essential building blocks of life. As had been well known even before Miller's experiment, amino acids will form readily in a mixture of reducing gases. In a chemically neutral atmosphere, however, reactions among atmospheric gases will not take place readily, and those reactions that do take place will produce extremely low yields of biological building blocks.[16] Further, even a small amount of atmospheric oxygen will quench the production of biologically significant building blocks and cause biomolecules otherwise present to degrade rapidly.[17]

An analogy may help to illustrate. Making amino acids in a reducing atmosphere is like getting vinegar and baking soda to react. Because the reaction releases stored chemical energy as heat, it occurs easily. (It is an example of what chemists call an "exothermic" reaction.) Trying to make biological building blocks in a neutral atmosphere, however, is more like trying to get oil and water to mix.[18]

Scientists investigating the origin of life haven't tried to adjust their probability calculations in light of these developments. But they have recognized that these developments do complicate matters further for the chance hypothesis. To make matters worse, an accumulating body of geochemical evidence has shown—perhaps, not surprisingly, in light of the previous discussion—that there likely never was a prebiotic soup. Two leading geochemists, James Brooks and Gordon Shaw, argued that if an ocean rich in amino and nucleic acids had existed, it would have left large deposits of nitrogen-rich minerals (nitrogenous cokes) in metamorphosed Precambrian sedimentary rocks. No evidence of such deposits exists, however. In the words of Brooks: "The nitrogen content of early Pre-Cambrian organic matter is relatively low (less than .015%). From this we can be reasonably certain that: there never was any substantial amount of 'primitive soup' on earth when Pre-Cambrian sediments were formed; if such a soup ever existed it was only for a brief period of time."[19]

Given my own deliberations about what constituted a substantive rather than a vacuous chance hypothesis, this seemed significant. A substantive chance hypothesis must invoke a definite process that produces the outcome in question with some regular or statistically predictable frequency—just as a roulette wheel produces various outcomes with a predictable frequency. The chance hypothesis envisioned DNA and proteins arising from a random process of chemical "roulette" in a favorable nitrogen-rich prebiotic ocean. If no such environment had ever existed, then whatever specificity the chance hypothesis might have once had was now lost. If there was no "chemical roulette" in which life would emerge as an inevitable if improbable outcome, chance could no longer be considered a substantive hypothesis; it would instead be just a vacuous notion that at best concealed ignorance of the true cause of life's origin.

Additionally, I knew from my Ph.D. work that there were other significant, if less quantifiable, problems with the idea that information-rich biomolecules had arisen by chance from a prebiotic soup. Most origin-of-life researchers recognized that, even if there had been a favorable prebiotic soup, many destructive chemical processes would have necessarily been at work at the same time.[20] Simulation experiments of the type performed by Stanley Miller had repeatedly demonstrated this. They have invariably produced nonbiological substances in addition to biological building blocks such as amino acids. Without intelligent intervention, these other substances will react readily with biologically relevant building blocks to form biologically irrelevant compounds—chemically insoluble sludge.[21] To prevent this from happening and to move the simulation of chemical evolution along a biologically promising trajectory, experimenters often remove those chemicals[22] that degrade or transform amino acids into nonbiologically relevant compounds. They also must artificially manipulate the initial conditions in their experiments. For example, rather than using both short- and long-wavelength ultraviolet light, which would have been present in any realistic early atmosphere, they use only short-wavelength UV. Why? The presence of the long-wavelength UV light quickly degrades amino acids.[23]

Zero Hour

I began to wonder if the odds of life arising by chance alone, at least under the circumstances envisioned by advocates of the chance hypothesis, weren't actually zero. Imagine that a casino owner invents a game in which the object is to roll 777 consecutive "sevens" with a set of dice.

He asks the odds makers to calculate the chances of any one contestant winning. They are, of course, infinitesimally small. But now he gives the odds makers some additional information. The dice are made of white chocolate with dark chocolate spots on the faces, both of which will melt as the result of the glare of the lights over the game table and repeated handling by the game players. Now what are the odds of turning up 777 sevens in a row? Clearly, they have diminished further. In fact, they are not just effectively zero, but under these circumstances with destructive processes inevitably predominating they are actually zero. Seven hundred seventy-seven "sevens" will never appear, because the faces of the dice will be destroyed in the attempt to generate them. The destructive processes will ensure that the desired outcome will never occur. I wondered if the same problem didn't afflict the chance hypothesis for the origin of life.

In the face of these and other difficulties, most origin-of-life researchers have decided to consider other theories that do not rely heavily on chance. In the next chapter, I examine one of the other main contending approaches: self-organization. Since the odds of a purely random process producing life are "vanishingly small," many scientists have concluded that some nonrandom, lawlike process must have been at work to help overcome these odds. If chance is insufficient, then perhaps "necessity" will do the job. Christian de Duve expresses the reasoning of researchers who favor this approach: "A string of improbable events—drawing the same lottery number twice, or the same bridge hand twice in a row—does not happen naturally. All of which lead me to conclude that life is *an obligatory manifestation of matter, bound to arise* where conditions are appropriate."[24]

Nevertheless, I should note that a few theorists have attempted to either retain or resuscitate a role for chance. They have done so in one of two ways. They have attempted either to lower the complexity threshold that random processes must first produce or to postulate an increase in the probabilistic resources available to such processes.

Some theorists, notably those proposing an initial "RNA world," have sought to retain a role for chance by suggesting that natural selection might have played a key role in the origin of life, even before the origin of a fully functioning cell. They propose combining chance with natural selection (or other lawlike processes) as a way of explaining how the first cell arose. In doing so, they suggest that random processes would have had to produce much less biological information by chance alone. Once a self-replicating molecule or small system of molecules had arisen, natural selection would "kick in" to help produce the additional necessary information. In Chapter 14, I evaluate theories that have adopted this strategy.

Other theorists have attempted to resuscitate chance theories altogether by postulating the existence of other possible universes beyond our own.[25] In doing so, they have attempted to increase the probabilistic resources available for producing biological information and life itself. These other universes would presumably provide more opportunities to generate favorable environments in which random processes could generate the vast number of combinatorial possibilities necessary to give life a realistic chance of arising somewhere. Since this idea depends upon highly speculative and technical cosmological models, I have decided to evaluate it separately, in Appendix B. There I argue that although it is impossible to disprove the existence of other universes, postulating such universes does not actually provide either a satisfying explanation or, still less, the best explanation of the origin of biological information. As I explain, there are several reasons that such models fail to do so. For now, however, it is fair to say that most serious origin-of-life researchers have not found speculations about other possible universes terribly relevant to understanding what actually happened on the early earth to produce the first life on our planet. So in the next chapters, I turn my attention to evaluating the other leading contenders I have encountered in my investigation of the DNA enigma.

11

Self-Organization and Biochemical Predestination

By 1968, the chance hypothesis was already suspect. The preceding fifteen years had revealed a cellular realm of stunning complexity, and though it would take several more decades to pinpoint the exact odds of a functioning protein or DNA molecule arising by chance alone, many scientists already had grown skeptical of chance as an explanation for the origin of life. Among them was a young biophysicist at San Francisco State University named Dean Kenyon. As he wrote a year later: "It is sometimes argued in speculative papers on the origin of life that highly improbable events (such as the spontaneous formation of a molecule of DNA *and* a molecule of DNA polymerase in the same region of space at the same time) become virtually inevitable over the vast stretches of geological time. No serious quantitative arguments, however, are given in support of such conclusions." Instead, he argued, "such hypotheses are contrary to most of the available evidence."[1] To emphasize the point Kenyon noted: "If the association of amino acids were a completely random event . . . there would not be enough mass in the entire earth, assuming it was composed exclusively of amino acids, to make even one molecule of every possible sequence of . . . a low-molecular-weight protein."[2]

Kenyon began exploring a different approach. With a Ph.D. from Stanford and having worked under Melvin Calvin, a leading biochemist and origin-of-life researcher at the University of California, Berkeley, Kenyon was steeped in the scientific culture of the late 1960s. He was well aware of Jacques Monod's conceptual dichotomy between chance and necessity. He understood that the logical alternative to chance was what Monod called necessity, the lawlike forces of physics and chemistry. If chance events couldn't explain the origin of biological information, then, Kenyon thought, perhaps necessity could. Eventually he and a colleague, Gary Steinman, proposed just such a theory; except that they didn't call it "necessity." They called it "predestination"—biochemical predestination. In 1969 Kenyon and Steinman wrote a book by this title. Through the 1970s and early 1980s it became the bestselling graduate-level text on the origin of life and established Kenyon as a leading researcher in the field.

In *Biochemical Predestination,* Kenyon and Steinman not only presented a new theory about the origin of biological information; they also inaugurated a fundamentally new approach to the origin-of-life problem, one that came to be called "self-organization." In physics, the term "self-organization" refers to a spontaneous increase in the order of a system due to some natural process, force, or law. For example, as a bathtub drains, a highly ordered structure, a vortex, forms as the water swirls down the drain under the influence of gravity and other physical forces such as the Coriolis force. In this system, a force (gravity) acts on a system (the water in the draining tub) to generate a structure (the vortex), thus increasing the order of the system. Self-organizational theories of the origin of life try to attribute the organization in living things to physical or chemical forces or processes—ones that can be described mathematically as laws of nature.

Of course, the self-organizational theorists recognized that they needed to do more than simply assert that life arose by physical or chemical necessity. They needed to identify a specific lawlike process (or series of processes) that could generate life or critical components of living cells starting from some specific set of conditions. In other words, self-organizational theorists needed to identify deterministic processes that could help overcome the otherwise long odds against the origin of life occurring by chance alone. And that is what Kenyon and those who followed in his wake set out to do. In the process Kenyon and other self-organizational theorists formulated theories that either tried to explain, or circumvented the need to explain, the DNA enigma.

More Than the Sum

What accounts for the difference between a chunk of marble and a great sculpture? What explains the difference between a cloud and a message created from steam by a skywriter? What accounts for the difference between a functioning cell and the various molecules that jointly constitute it? These questions raise the classical philosophical issue of reductionism. Does the whole reduce to the sum of its parts? Or conversely, do the properties of the parts explain the structure and organization of the whole? As with many such questions, the best answer is, "It depends."

Consider a silver knife, fork, and spoon. The metal in all three has a definite set of chemical properties describable by physical and chemical laws. Silver atoms will react with certain atoms, under certain conditions, but not with others. Silver will melt at certain temperatures and pressures, but not at others. Silver has a certain strength and resistance to shearing. Insofar as these properties reliably manifest themselves under certain specified conditions, they can be described with general laws. Nevertheless, none of these properties of silver accounts for the shape of the various utensils and, therefore, the functions they perform.

From a purely chemical point of view nothing discriminates the silver in a spoon from that in a knife or fork. Nor does the chemistry of these items explain their arrangement in a standard place setting with the fork on the left and the knife and spoon on the right. The chemistry of the silverware is indifferent to how the silver is arranged on the table. The arrangement is

Figure 11.1. Dean Kenyon.
Printed by permission from Dean Kenyon.

determined not by the properties of the silver or by chemical laws describing them, but by the choice of a rational agent to follow a human convention.

Yet one cannot say the same of every structure. Many objects display a structure and order that results from the chemical properties of their ingredients. For example, a crystal of salt has a lattice structure that exhibits a striking, highly repetitive pattern. This structure results from forces of mutual electrostatic attraction (describable by natural law) among the atoms in the lattice. Similarly, a spiral galaxy has a definite structure that results largely from lawlike forces of gravitational attraction among the stars in the galaxies. Structure does not always defy reduction to chemistry and physics.

What about living cells and the complex molecules they contain? Does their organization derive from the physical and chemical properties of their parts? Do the chemical constituents of proteins or DNA molecules have properties that could cause them to self-organize? Are there physical or chemical forces that make the production of information-rich molecules inevitable under plausible prebiotic conditions? Dean Kenyon thought the answer to these questions might well be yes.

Biochemical Predestination

Like Aleksandr Oparin, whom he credited with inspiration, Kenyon sought to account for the origin of the first life in a series of gradual steps. Like Oparin, Kenyon constructed a scenario describing how life might have arisen through a series of chemical transformations in which more complex chemical structures arose from simpler ones. In Kenyon's model, simple monomers (e.g., amino acids, bases, and sugars) arose from simpler atmospheric gases and energy; polymers (proteins and DNA) arose from monomers; primitive membranes formed around these polymers; and a primitive metabolism emerged inside these membranes as various polymers interacted chemically with one another. Unlike Oparin, however, who relied on chance variations to achieve some of the chemical transformations, Kenyon relied more exclusively on deterministic chemical reactions. In his scenario, each stage along the way to the origin of life was driven by deterministic chemical processes, including the most important stage, the origin of biological information.

Whereas Oparin had suggested that the process by which monomers arranged themselves into polymers in the prebiotic soup was essentially random, Kenyon and Steinman suggested that forces of chemical necessity had performed the work. Specifically, they suggested that just as electrostatic forces draw sodium ($Na+$) and chloride ($Cl-$) ions together into highly ordered patterns within a crystal of salt, amino acids with special affinities for each other might have arranged themselves to form proteins. As the two scientists explained, "In the same fashion that the difference in the nature of reactivity of the units of a growing inorganic crystal determines the final constitution of the three-dimensional crystal array, so differences in reactivities of the various amino acids with one another could possibly serve to promote a defined ordering of sequence in a growing peptide chain."[3]

Kenyon and Steinman first came to this idea while performing "di-mer" bonding experiments in the laboratory of Melvin Calvin at the University of California, Berkeley. They wanted to see whether specific amino acids bond more readily with some amino acids than others. Their experimental results suggested that they do. For example, they discovered that glycine forms linkages with alanine twice as frequently as glycine with valine. Moreover, they discovered that these differences in chemical affinity seemed related to differences in chemical structure. Amino acids with longer side chains bond less frequently to a given amino acid than do amino acids with shorter side chains.[4] Kenyon and Steinman summarized their findings in a table showing the various differential bonding affinities they discovered.

In the wake of these findings, Kenyon and Steinman proposed that these differences in affinity imposed constraints on the sequencing of amino acids, rendering certain sequences more likely than others. As they put it, "It would appear that the unique nature of each type of amino acid as determined by its side chain could introduce nonrandom constraints into the sequencing process."[5] They further suggested that these differences in affinity might correlate with the specific sequencing motifs typical in functional proteins. If so, the properties of individual amino acids could themselves have helped to sequence the amino-acid chains that gave rise to functional proteins, thereby producing the information the chains contain. Biochemical necessity could thus explain the seeming improbability of the functional sequences that exist in living organisms today. Sequences that would be vastly improbable in the absence of differing bonding affinities might have been very probable when such affinities are taken into account.

Kenyon and Steinman did not attempt to extend this approach to explain the information in DNA, since they favored a protein-first model.[6] They knew that proteins perform most of the important enzymatic and structural functions in the cell. They thought that if functional proteins could arise without the help of nucleic acids, then initially they need not explain the origin of DNA and RNA and the information they contained. Instead, they envisioned proteins arising directly from amino acids in a prebiotic soup. They then envisioned some of these proteins (or "proteinoids") forming membranes surrounding other proteins. Thus, the two were convinced that only later, once primitive metabolic function had arisen, did DNA and RNA need to come on the scene. Kenyon and Steinman did not speculate as to how this had happened, content as they were to think they might have solved the more fundamental question of the origin of biological information by showing that it had arisen first in proteins.

Doubts About Self-Organization

Did Kenyon and Steinman solve the DNA enigma? Was the information necessary to produce the first life "biochemically predestined" to arise on the early earth? Surprisingly, even as their bold self-organizational model grew in popularity among origin-of-life researchers, Kenyon himself began to doubt his own theory.[7]

Kenyon's doubts first surfaced in discussions with one of his students at San Francisco State University. In the spring of 1975 near the end of a semester-long upper-division course on evolution, a student began to raise questions about the plausibility of chemical evolution.[8] The student—ironically named Solomon Darwin—pressed Kenyon to examine whether his self-organizational model could explain the origin of the information in DNA. Kenyon might have deflected this criticism by asserting that his protein-first model of self-organization had circumvented the need to explain the information in DNA. But by this point he found himself disinclined to make that defense.

For some time Kenyon himself had suspected that DNA needed to play a more central role in his account of the origin of life. He realized that whether functional proteins had arisen before DNA or not, the origin of information-rich DNA molecules still needed explanation, if for no other reason than because information-rich DNA molecules exist in all extant cells. At some point, DNA must have arisen as a carrier of the information for building proteins and then come into association with functional proteins. One way or another, the origin of genetic information still needed to be explained.

Now he faced a dilemma. Having opted for a self-organizational approach, he had only two options for explaining the information in DNA. Either (a) the specific sequences of amino acids in proteins had somehow provided a template for sequencing the bases in newly forming DNA molecules or (b) DNA itself had self-organized in much the same way he and Steinman supposed proteins had. As he reflected more on Solomon Darwin's challenge, Kenyon realized that neither option was very promising. First, Kenyon knew that to propose that the information in proteins had somehow directed the construction of DNA would be to contradict everything then known about molecular biology. In extant cells, DNA provides the template of information for building proteins and not the reverse. Information flows from DNA to proteins. Moreover, there are several good reasons for this asymmetry. Each triplet of DNA bases (and corresponding RNA codons) specifies exactly one amino acid during transcription and translation. Yet most amino acids correspond to more than one nucleotide triplet or RNA codon. This feature of the genetic code ensures that information can flow without "degeneracy," or loss of specificity, in only one direction, from DNA to proteins and not the reverse.

Additionally, Kenyon realized that for structural and chemical reasons, proteins made poor candidates for replicators—molecules that can function as easily copied informational templates. Unlike DNA, proteins do not possess two antiparallel strands of identical information and thus cannot be unwound and copied in the way DNA can. Further, proteins are highly reactive once they are unwound (due to exposed amino and carboxl groups and exposed side chains). For this reason, most "denatured" (unwound) proteins tend to cross-link and aggregate. Others are quickly destroyed in the cell. Either way, denatured proteins tend to lose their structural stability and function. Moreover, they do not regain their original three-dimensional shape or activity once they lose it. By contrast, DNA is a stable, chemically inert molecule that easily maintains its chemical structure and composition while other molecules copy its information. For all these reasons, it seemed difficult to envision proteins serving as replicators of their own stored information. Indeed, as Kenyon later told me, "getting the information out of proteins and into DNA" would pose an insuperable conceptual hurdle.

And there was another difficulty. By the late 1980s new empirical findings challenged the idea that amino-acid bonding affinities had produced the biological information in proteins. Although Kenyon and Steinman had shown that certain amino acids form linkages more readily with some amino acids than with others,[9] new studies showed that these differential affinities do not correlate with actual sequencing patterns in large classes of known proteins.[10] In other words, differential bonding affinities exist, but they don't seem to explain (or to have determined) the specific sequences of amino acids that actual proteins now possess. Instead, there was a much more plausible explanation of these new findings, one consistent with Crick's central dogma. The amino-acid sequences of known proteins had been generated by the information encoded in DNA—specifically, by the genetic information carried in the DNA of the organisms in which these proteins reside. After all, proteins in extant cells are produced by the gene-expression system. Of course, someone could argue that the first proteins arose directly from amino acids, but now it was clear there was, at the very least, no evidence of that in the sequences of amino acids in known proteins.

All this would later reinforce Kenyon's conviction that he could not circumvent the need to explain the origin of information in DNA by positing a protein-first model. He would have to confront the DNA enigma.

Kenyon realized that if the information in DNA did not arise first in proteins, it must have arisen independently of them. This meant that the base sequences of DNA (and the molecule itself) must have self-organized under the influence of chemical laws or forces of attraction between the constituent monomers. Yet based upon his knowledge of the chemical structure of the DNA molecule, Kenyon doubted that DNA possessed any self-organizational properties analogous to those he had identified in amino acids and proteins. He was strengthened in this conclusion by reading an essay about DNA by a distinguished Hungarian-born physical chemist and philosopher of science named Michael Polanyi. As it happened, this same essay would shape my own thinking about the viability of self-organizational theories as well.

Life Transcending Physics and Chemistry

I first encountered Dean Kenyon in 1985, sixteen years after the publication of *Biochemical Predestination*. I was in the audience in Dallas during the presentation in which Kenyon announced that he had come to doubt all current chemical evolutionary theories of the origin of life—including his own. Thus, when I arrived in Cambridge, I already knew that one of the leading proponents of "self-organization" had repudiated his own work. Nevertheless, I did not yet fully appreciate why he had done so. I learned from Charles Thaxton that Kenyon had begun to doubt that his protein-first models eliminated the need to explain the origin of information in DNA. But I didn't yet see why explaining the DNA enigma by reference to self-organizing chemical laws or forces would prove so difficult.

That began to change as I encountered the work of Michael Polanyi. In 1968, Polanyi published a seminal essay about DNA in the journal *Science,* "Life's Irreducible Structure," and another essay in the *Chemical and Engineering News* the year before, "Life Transcending Physics and Chemistry." During my first year of graduate study, I wrote a paper about the two essays and discussed them at length with my supervisor, Harmke Kamminga, since it seemed to me that Polanyi's insights had profound implications for self-organizational models of the origin of life.

Polanyi's essays did not address the origin of life question directly, but rather a classical philosophical argument between the "vitalists" and "reductionists" (discussed in Chapter 2). Recall, that for centuries, vitalists and reductionists had argued about whether a *qualitative* distinction existed between living and nonliving matter. Vitalists maintained that living organisms contained some kind of immaterial "vital force" or spirit, an *elan vital* that distinguished them qualitatively from nonliving chemicals. Reductionists, for their part, held that life represented merely a *quantitatively* more complex form of chemistry. Thus, in their view, living organisms, like complex machines, functioned as the result of processes that could be "reduced" or explained solely by reference to the laws of physics and chemistry.

The molecular biological revolution of the 1950s and 1960s seemed to confirm the reductionist perspective.[11] The newly discovered molecular mechanisms for storing and transmitting information in the cells confirmed for many biologists that the distinctive properties of life could, as Francis Crick put it in 1966, "be explained in terms of the ordinary concepts of physics and chemistry or rather simple extensions of them."[12] As Richard Dawkins later wrote, the discovery of DNA's role in heredity "dealt the final, killing blow to the belief that living material is deeply distinct from nonliving material."[13]

But had it really? Even if biochemists were no longer looking for some mysterious life force, was it really clear that living things could be explained solely by reference to the laws of physics and chemistry?

Polanyi's answer turned the classical reductionism-vitalism debate on its head. He did this by challenging an assumption held by reductionists and vitalists alike, namely, that "so far as life can be represented as a mechanism, it [can be] explained by the laws of inanimate nature."[14] Whereas vitalists had argued against reductionism by contesting that life can be understood mechanistically, Polanyi showed that reductionism fails even if one grants that living organisms depend upon many mechanisms and machines. To show this, Polanyi argued that even if living organisms function like machines, they cannot be fully explained by reference to the laws of physics and chemistry.[15]

Consider an illustration. A 1960s vintage computer has many parts, including transistors, resistors, and capacitors. The electricity flowing through these various parts conforms to the laws of electromagnetism, for example, Ohm's law ($E = IR$, or voltage equals current times resistance). Nevertheless, the specific structure of the computer, the configuration of its parts, does not result from Ohm's or any other law. Ohm's law (and, indeed, the laws of physics generally) allows a vast ensemble of possible configurations of the same parts. Given the fundamental physical laws and the same parts, an engineer could build many other machines and structures: different model computers, radios, or quirky pieces of experimental art made from electrical components. The physical and chemical laws that govern the flow of current in electrical machines do not determine how the parts of the machine are arranged and assembled. The flow of electricity *obeys* the laws of physics, but *where* the electricity flows in any particular machine depends upon the arrangement of its parts—which, in turn, depends on the design of an electrical engineer working according to engineering principles. And these engineering principles, Polanyi insisted, are distinct from the laws of physics and chemistry that they harness.

Polanyi demonstrated that the same thing was true of living things. He did this by showing that communications systems, like machines, defy reduction to physical and chemical law and by showing further that living organisms contain a communications system, namely, the gene-expression system in which DNA stores information for building proteins.

Polanyi argued that, in the case of communications systems, the laws of physics and chemistry do not determine the arrangements of the characters that convey information. The laws of acoustics and the properties of air do not determine which sounds are conveyed by speakers of natural languages. Neither do the chemical properties of ink determine the arrangements of letters on a printed page. Instead, the laws of physics and chemistry allow a vast array of possible sequences of sounds, characters, or symbols in any code or language. Which sequence of characters is used to convey a message is not determined by physical law, but by the choice of the users of the communications system in accord with the established conventions of vocabulary and grammar—just as engineers determine the arrangement of the parts of machines in accord with the principles of engineering.

Thus, Polanyi concluded, communications systems defy reduction to physics and chemistry for much the same reasons that machines do. Then he took a step that made his work directly relevant to the DNA enigma: he insisted that living things defy reduction to the laws of physics and chemistry because they also contain a system of communications—in particular, the DNA molecule and the whole gene-expression system. Polanyi argued that, as with other systems of communication, the lower-level laws of physics and chemistry cannot explain the higher-level properties of DNA. DNA base sequencing cannot be explained by lower-level chemical laws or properties any more than the information in a newspaper headline can be explained by reference to the chemical properties of ink.[16] Nor can the conventions of the genetic code that determine the assignments between nucleotide triplets and amino acids during translation be explained in this manner. Instead, the genetic code functions as a higher-level constraint distinct from the laws of physics and chemistry, much like a grammatical convention in a human language.

Polanyi went further, arguing that DNA's capacity to convey information actually requires a freedom from chemical determinism or constraint, in particular, in the arrangement of the nucleotide bases. He argued that if the bonding properties of nucleotides determined their arrangement, the capacity of DNA to convey information would be destroyed.[17] In that case, the bonding properties of each nucleotide would determine each subsequent nucleotide and thus, in turn, the sequence of the molecular chain. Under these conditions, a rigidly ordered pattern would emerge as required by their bonding properties and then repeat endlessly, forming something like a crystal. If DNA manifested such redundancy, it would be impossible for it to store or convey much information. As Polanyi concluded, "Whatever may be the origin of a DNA configuration, it can function as a code only if its order is not due to the forces of potential energy. It must be as physically indeterminate as the sequence of words is on a printed page."[18]

DNA and Self-Organization

Polanyi's argument made sense to me. DNA, like other communication systems, conveys information because of very precise configurations of matter. Was there a law of physics or chemistry that determined these exact arrangements? Were there chemical forces dictating that only biologically functional base sequences and no others could exist between the strands of the double helix? After reading Polanyi's essays, I doubted this.

I realized that his argument also had profound implications for self-organizational theories of the origin of life. To say that the information in DNA does not reduce to or derive from physical and chemical forces implied that the information in DNA did not *originate* from such forces. If so, then there was nothing Kenyon could do to salvage his self-organizational model.

But was Polanyi correct? How did we know that the constituent parts of DNA did not possess some specific bonding affinities of the kind that Kenyon and Steinman had discovered in amino acids? Polanyi argued largely on theoretical grounds that the nucleotide bases could not possess such deterministic affinities and still allow DNA to store and transmit information. How do we know that such affinities do not exist?

In 1987 when I first encountered Polanyi's argument, I did not yet understand the chemical structure of DNA well enough to answer this question. And Polanyi, who no doubt did, did not bother to explain it in his articles. So the question lingered in the back of my mind for several years after completing my Ph.D. Not until the mid-1990s, when I began to make a more systematic evaluation of self-organizational theories, did I find an answer to it.

I remember vividly the day the breakthrough came. I was listening to a colleague, a biologist at Whitworth College, teach a college class about the discovery of the double helix when I noticed something about the chemical structure of DNA on the slide that she had projected on the screen. What I noticed wasn't anything I hadn't seen before, but somehow its significance had previously escaped me. It not only confirmed for me Polanyi's conclusion about the information in DNA transcending physics and chemistry, but it also convinced me that self-organizational theories invoking bonding affinities or forces of attraction would never explain the origin of the information that DNA contains.

Figure 11.2 shows what I saw on the slide that suddenly seized my attention. It portrays the chemical structure of DNA, including the chemical bonds that hold the molecule together. There are bonds, for example, between the sugar and the phosphate groups forming the two twisting backbones of the DNA molecule. There are bonds fixing individual nucleotide bases to the sugar-phosphate backbones on each side

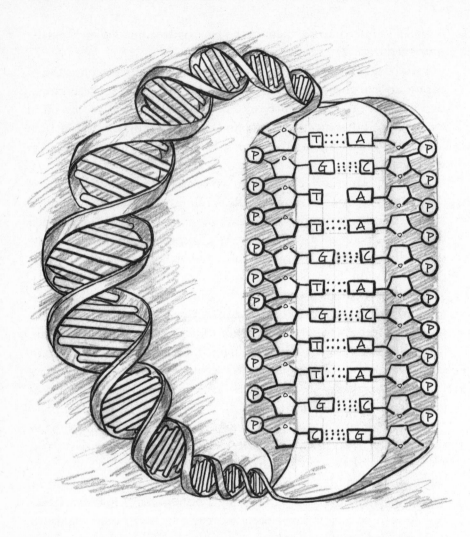

Figure 11.2. Model of the chemical structure of the DNA molecule depicting the main chemical bonds between its constituent molecules. Note that no chemical bonds link the nucleotide bases (designated by the letters in boxes) in the longitudinal message-bearing axis of the molecule. Note also that the same kind of chemical bonds link the different nucleotide bases to the sugar-phosphate backbone of the molecule (denoted by pentagons and circles). These two features of the molecule ensure that any nucleotide base can attach to the backbone at any site with equal ease, thus showing that the properties of the chemical constituents of DNA do not determine its base sequences. *Adapted by permission from an original drawing by Fred Heeren.*

of the molecule. There are also hydrogen bonds stretching horizontally across the molecule between nucleotide bases, forming complementary pairs. The individually weak hydrogen bonds, which in concert hold two complementary copies of the DNA message text together, make replication of the genetic instructions possible. But notice too that there are *no* chemical bonds *between the bases* along the longitudinal axis in the center of the helix. Yet it is precisely along this axis of the DNA molecule that the genetic information is stored.

There in the classroom this elementary fact of DNA chemistry leaped out at me. I realized that explaining DNA's information-rich sequences by appealing to differential bonding affinities meant that there had to be chemical bonds of differing strength between the different bases along the information-bearing axis of the DNA molecule. Yet, as it turns out, there are no differential bonding affinities there. Indeed, there is not just an absence of *differing* bonding affinities; there are no bonds *at all* between the critical information-bearing bases in DNA.[19] In the lecture hall the point suddenly struck me as embarrassingly simple: there are neither bonds nor bonding affinities—differing in strength or otherwise—that can explain the origin of the base sequencing that constitutes the information in the DNA molecule. A force has to exist before it can cause something. And the relevant kind of force in this case (differing chemical attractions between nucleotide bases) does not exist within the DNA molecule.

Of course it might be argued that although there are no bonds between the bases that explain the arrangement of bases, there might be either something else about the different sites on the DNA molecule that inclines one base rather than another to attach at one site rather than another. I investigated the possibility with a colleague of mine named Tony Mega, an organic chemist. I asked him if there were any physical or chemical differences between the bases or attachment sites on DNA that could account for base sequencing. We considered this question together for a while, but soon realized that there are no significant differential affinities between any of the four bases and the binding sites along the sugar-phosphate backbone. Instead, the same type of chemical bond (an N-glycosidic bond) occurs between the base and the backbone regardless of which base attaches. All four bases are acceptable; none is chemically favored.

This meant there was nothing about either the backbone of the molecule or the way any of the four bases attached to it that made any sequence more likely to form than another. Later I found that the noted origin-of-life biochemist Bernd-Olaf Küppers had concluded much the same thing. As he explained, "The properties of nucleic acids indicate that all the combinatorially possible nucleotide patterns of a DNA are, from a chemical point of view, equivalent."[20] In sum, two features of DNA ensure that "self-organizing" bonding affinities cannot explain the specific arrangement of nucleotide bases in the molecule: (1) there are *no* bonds between bases along the information-bearing axis of the molecule and (2) there are no *differential* affinities between the backbone and the specific bases that could account for variations in sequence.

While I was teaching I developed a visual analogy to help my students understand why these two features of the chemical structure of DNA have such devastating implications for self-organizational models, at least those that invoke bonding affinities to explain the DNA enigma. When my children were young, they liked to spell messages on the metallic surface of our refrigerator using plastic letters with little magnets on the inside. One day I realized that the communication system formed by the refrigerator and magnetic letters had something in common with DNA—something that could help me explain why the information in DNA is irreducible to physical or chemical forces of attraction.

To demonstrate this to my students, I would bring a small magnetic "chalkboard" to class with a message spelled on it, such as "Biology Rocks!" using the same kind of plastic letters that my kids used at home. I would point out that the magnetic forces between the letters and the metallic surface of the chalkboard explain why the letters stick to the board, just as forces of chemical attraction explain why the nucleotides stick to the sugar-phosphate backbone of the DNA. But I would also point out that there are no significant forces of attraction between the individual letters that determine their arrangement, just as there are no significant forces of attraction between the bases on the information-bearing axis of the DNA molecule. Instead, the magnetic forces between the letters and the chalkboard allow numerous possible letter combinations, some of which convey functional or meaningful information and most of which do not.

To demonstrate that the magnetic forces do not dictate any specific letter sequence, I arranged and rearranged the letters on the board to show that they have perfect physical freedom to spell other messages or mere gibberish. I would further note that there are no *differential* forces of attraction at work to explain why one letter sticks to the chalkboard at one location rather than another. I then pointed out that the same is true of DNA: there are no differential forces of attraction between the DNA bases and the sites on the sugar-phosphate backbone. In the case of the magnetic board nothing about the magnetic force acting on the "B" inclines it to attach to the board at the front of the "Biology Rocks!" sequence rather than at the back or middle. Instead, each letter can attach to the chalkboard at any location, just as any one of the nucleotide bases can attach to any sugar molecule along the sugar-phosphate backbone of DNA. Forces of attraction (N-glycosidic bonds) do explain why the bases in DNA attach to the backbone of the molecules, but they do not explain why any given nucleotide base attaches to the molecule at one site rather than another. Nor, given the absence of chemical bonds between the bases, do any other bonding affinities or chemical forces internal to the molecule explain the origin of DNA base sequencing.

I later learned that Kenyon already had reckoned with these same stubborn facts of molecular biology and reached the same conclusion about the futility of explaining the arrangement of DNA bases by reference to internal bonding affinities. The properties of the building blocks of DNA simply do not make a particular gene, let alone life as we know it, inevitable. Yet the opposite claim is often made by self-organizational theorists, albeit without much specificity. De Duve states, for example, that "the processes that generated life" were "highly deterministic," making life as we know it "inevitable," given "the conditions that existed on the prebiotic earth."[21] Yet imagine the most favorable prebiotic conditions. Imagine a pool of all four DNA bases and all necessary sugars and phosphates. Would any particular genetic sequence inevitably arise? Given all necessary monomers, would any particular functional protein or gene, let alone a specific genetic code, replication system, or signal transduction circuitry, inevitably arise? Clearly not.

The most obvious place to look for self-organizing properties to explain the origin of genetic information is in the constituent parts (the monomers) of the molecules that carry that information; but biochemistry makes clear that forces of attraction between the bases in DNA do not explain the specific sequences in these large information-bearing molecules. Because the same is true of RNA, researchers who speculate that life began in a self-organizing RNA world also must confront another sequencing problem,[22] in particular, the problem of explaining how information in a functioning RNA molecule could have first arisen in the absence of differential forces of chemical attraction between the bases in RNA.[23]

The Mystery of the Code

Recall from Chapter 5 that, in addition to the specified information in nucleic acids and proteins, origin-of-life researchers need to account for the origin of the integrated complexity of the gene-expression and -translation system. Recall also that the gene-expression system not only utilizes the digital information inscribed along the spine of the DNA molecule (the genetic text); it also depends on the genetic code or translation system imbedded in the tRNA molecule (along with its associated synthetase proteins). (The genetic code is to the genetic information on a strand of DNA as the Morse code is to a specific message received by a telegraph operator.)

It turns out that it is just as difficult to explain the origin of the genetic code by reference to self-organizational bonding affinities as it is to explain the origin of the genetic "text" (the specific sequencing of the DNA bases). The next several paragraphs describe why in some detail. Those unfamiliar with the relevant facts may find it useful to consult the accompanying figure (Figure 11.3) or to skip ahead and pick up the main thread of the argument beginning with the heading "The Necessity of Freedom."

Self-organizational theories have failed to explain the origin of the genetic code for several reasons. First, to explain the origin of the genetic code, scientists need to explain the origin of the precise set of correspondences between specific nucleotide triplets in DNA (or codons on the messenger RNA) and specific amino acids (carried by transfer RNA). Yet molecular biologists have failed to find any significant

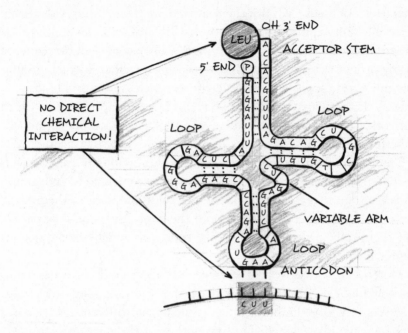

Figure 11.3. The transfer-RNA molecule showing the anticodon on one end of the molecule and an attached amino acid on the other. Notice that there is no direct chemical interaction between the amino acid and the nucleotide codon that specifies it.

chemical interaction between the codons on mRNA (or the anticodons on tRNA) and the amino acids on the acceptor arm of tRNA to which the codons correspond. This means that forces of chemical attraction between amino acids and these groups of bases do not explain the correspondences that constitute the genetic code.

Instead, the mRNA codon binds not to the amino acid directly, but to the anticodon triplet on the tRNA molecule. Moreover, the anticodon triplet and amino acid are situated at opposite ends of tRNA. They do not interact chemically in any direct way. Although amino acids do interact chemically with a nucleotide triplet at the 3' acceptor end of the tRNA molecule, the triplet remains *the same* in all twenty tRNA molecules, meaning the bonds there do not display any differential affinities that could explain the differing amino-acid assignments that constitute the code. All twenty tRNA molecules have the same final triplet of bases (ACC) at the 3' arm where their amino acids attach. Since all twenty amino acids in all twenty tRNA molecules attach to the same nucleotide sequence, the properties of that nucleotide sequence clearly do not determine which amino acids attach and which do not. The nucleotide sequence is indifferent to which amino acid binds to it (just as the sugar-phosphate backbone in DNA is indifferent to which nucleotide base binds to it). All twenty triplets are acceptable; none is preferred.

Thus, chemical affinities between nucleotide codons and amino acids do not determine the correspondences between codons and amino acids that define the genetic code. From the standpoint of the properties of the constituents that compose the code, the code is physically and chemically arbitrary. All possible codes are equally likely; none is favored chemically.

Moreover, the discovery of seventeen variant genetic codes has put to rest any doubt about this.[24] The existence of many separate codes (multiple sets of codon–amino acid assignments) in different microorganisms indicates that the chemical properties of the relevant monomers allow more than a single set of codon–amino acid assignments. The conclusion is straightforward: the chemical properties of amino acids and nucleotides do not determine a single universal genetic code; since there is not just one code, "it" cannot be inevitable.

Instead, scientists now know the codon–amino acid relationships that define the code are established and mediated by the catalytic action of some twenty separate proteins, the so-called aminoacyl-tRNA synthetases (one for each tRNA anticodon and amino-acid pair). Each of these proteins recognizes a specific amino acid and the specific tRNA with its corresponding anticodon and helps attach the appropriate amino acid to that tRNA molecule. Thus, instead of the code reducing to a simple set of chemical affinities between a small number of monomers, biochemists have found a functionally interdependent system of highly specific biopolymers, including mRNA, twenty specific tRNAs, and twenty specific synthetase proteins, each of which is itself constructed via information encoded on the very DNA that it helps to decode. But such integrated complexity was just what needed explanation in the first place. The attempt to explain one part of the integrated complexity of the gene-expression system, namely, the genetic code, by reference to simple chemical affinities leads not to simple rules of chemical attraction, but instead to an integrated system of large molecular components. One aspect of the DNA enigma leads to another.

Certainly, the chemical interactions between the functional polymers in this complex translation system proceed deterministically. But to explain how a windmill or an operating system or a genetic code works is one thing; to explain how any of them originated is quite another. To claim that deterministic chemical affinities made the origin of this system inevitable lacks empirical foundation. Given a pool of the bases necessary to tRNA and mRNA, given all necessary sugars and phosphates and all twenty amino acids used in proteins, would the molecules composing the current translation system, let alone any particular genetic code, have had to arise? Indeed, would even a single synthetase have had to arise from a pool of all the necessary amino acids? Again, clearly not.

As origin-of-life biochemists have taught us, monomers are "building blocks." And like the building blocks that masons use, molecular building blocks can be arranged and rearranged in innumerable ways. The properties of stone blocks do not determine their arrangement in the construction of buildings. Similarly, the properties of *biological* building blocks do not determine the arrangement of monomers in functional DNA, RNA, or proteins. Nor do they determine the correspondences between DNA bases and the amino acids that constitute the genetic code. Instead, the chemical properties of the building blocks of these molecules allow a vast ensemble of possible configurations and associations, the overwhelming majority of which would have no biological function. Thus, functional genes and proteins are no more inevitable, given the properties of their "building blocks," than the palace of Versailles was inevitable, given the properties of the bricks and stone used to construct it.

The Necessity of Freedom

As I thought more about the chemical structure of DNA and tRNA, Polanyi's deeper point about the expression of information requiring chemical indeterminacy, or freedom, came into clearer focus. Polanyi had argued that if "forces of potential energy" determined the arrangement of the bases, "the code-like character" of the molecule "would be effaced by an overwhelming redundancy."[25] I now understood why.

Consider, for example, what would happen if the individual nucleotide bases (A, C, G, T) in the DNA molecule *did* interact by *chemical* necessity (along the information-bearing axis of DNA). Suppose that every time adenine (A) occurred in a growing genetic sequence, it attracted cytosine (C) to it,[26] which attracted guanine (G), which attracted thymine (T), which attracted adenine (A), and so on. If this were the case, the longitudinal axis of DNA would be peppered with repetitive sequences of ACGT. Rather than being a genetic molecule capable of virtually unlimited novelty and characterized by unpredictable and aperiodic sequences, DNA would contain sequences awash in repetition or redundancy—much like the arrangement of atoms in crystals.

To see why, imagine that a group of captive soldiers are told they can type letters to their families at home. The only condition is that they have to begin with the lower case letter *a*, follow it with the next letter in the alphabet, *b*, then the next and the next, moving through the entire alphabet, circling back to *a* at the end and then continuing the process until they have filled in the sheet of typing paper. They are instructed to follow the same process for filling out the front of the envelope. Finally, if they don't feel up to the task or if they accidentally strike a wrong key, the prison guards will take over the task for them and "do it properly."

This would, of course, be nothing more than a cruel joke, for the soldiers couldn't communicate a single bit of information with their letters or even mail the letters to the appropriate addresses—all because the content of the letters was inevitable, strictly governed by the lawlike algorithm forced on them by the prison guards. In the same way, the lawlike forces of chemical necessity produce redundancy (repetition), which reduces the capacity to convey information and express novelty. Instead, information emerges from within an environment marked by indeterminacy, by the freedom to arrange parts in many different ways. As the MIT philosopher Robert Stalnaker puts it, information content "requires contingency."[27]

Information theory reveals a deeper reason for this. Recall that classical information theory equates the reduction of uncertainty with the transmission of information, whether specified or unspecified. It also equates *im*probability and information—the more improbable an event, the more information its occurrence conveys. In the case that a law-like physical or chemical process determines that one kind of event will necessarily and predictably follow another, then no uncertainty will be reduced by the occurrence of such a high-probability event. Thus, no information will be conveyed. Philosopher of science Fred Dretske, the author of an influential book on information theory, explains it this way: "As p(si) [the probability of a condition or state of affairs] approaches 1 [i.e., certainty], the amount of information associated with the occurrence of si goes to 0. In the limiting case when the probability of a condition or state of affairs is unity [p(si) = 1], no information is associated with, or generated by, the occurrence of si. This is merely another way to say that no information is generated by the occurrence of events for which there are no possible alternatives."[28]

Dretske's and Polanyi's observations were decisive: to the extent that forces of attraction among the members of a sequence determine the arrangement of the sequence, the information-carrying capacity of the system will be diminished or effaced by redundancy.[29] Bonding affinities, to the extent they exist, inhibit the production of information, because they determine that specific outcomes will follow specific conditions with high probability.[30] Information-carrying capacity is maximized, however, when the opposite situation occurs, namely, when chemical conditions allow many improbable outcomes.

Polanyi appreciated this point, but also its converse. He knew that it was precisely because the sequences of bases in DNA were not biochemically determined (or *predestined*) that the molecule could store and transmit information. Because any nucleotide can follow any other, a vast array of sequences is possible, which allows DNA to encode a vast number of protein functions. As he explains, "It is this physical indeterminacy of the sequence that produces the improbability of occurrence of any particular sequence and thereby enables it to have a meaning—a meaning that has a mathematically determinate information content equal to the numerical improbability of the arrangement."[31]

As noted in Chapter 4, the base sequences in DNA not only possess information-carrying capacity as measured by classical Shannon information theory, they also store functionally specified information; they are specified as well as complex. Clearly, however, a sequence cannot be both specified and complex if it is not at least complex. Therefore, self-organizational forces of chemical necessity, which produce redundant order and *preclude* complexity, preclude the generation of specified complexity (or specified information) as well. Lawlike chemical forces do not generate complex sequences. Thus, they cannot be invoked to explain the origin of information, whether specified or otherwise.

Conclusion

At a small private conference in 1993, I had a chance to meet and talk with Kenyon personally for the first time. I learned that he had come to the same conclusion about his theory of "biochemical predestination" as I had. It now seemed clear to both of us that there was a significant, in principle, objection to the very idea that chemical attractions could produce information as opposed to simple redundant order. Indeed, if Kenyon had found that bonding affinities between nucleotide bases in DNA determined their sequencing, he would have also discovered that chemists had been mistaken about DNA's information-bearing properties. Yet no one doubted DNA's capacity to store information.

But were there other such theories that could explain what Kenyon's could not? Self-organizational models emphasizing internal bonding affinities had failed, but perhaps models emphasizing lawlike forces external to DNA would succeed. And although simple algorithms like "repeat ATCG" clearly lacked the capacity to convey biological information, perhaps some far more sophisticated algorithm or some dance of forces between a large group of molecules could cause life (or the information it required) to self-organize. By the early 1990s, some researchers were proposing models of just this kind.

12

Thinking Outside the Bonds

There's an oft-told tale, apocryphal but nonetheless instructive, of Christopher Columbus dining with several men who are unimpressed by his discovery of the New World. Columbus, not surprisingly, objects to their disdain. Certainly the way to the New World is obvious now, but if finding the way had been so easy, why hadn't someone done it before? To illustrate what he'd achieved, he asks for an egg and challenges the men at the table to balance it on its head. Each man tries and fails. Columbus then takes the egg and offers a simple solution: he sets the egg down just hard enough to break the end, allowing it to remain upright.

Today we have an expression to describe innovative thinking that transcends unnecessary, but established constraints. We say that innovative problem solvers "think outside the box." Following Kenyon and Polanyi, I had come to see that differences in internal bonding affinity, either between nucleotide bases or amino acids, did not and could not solve the DNA enigma. But if the internal affinities were inside the box, both proverbially and literally, perhaps the failure of models relying on such forces merely signaled the need for researchers to journey outside the box. And that's precisely what subsequent self-organizational researchers tried to do.

External Self-Organizational Forces: Just Add Energy?

If internal bonding affinities did not explain the information in DNA and proteins, might there be some ubiquitous external force that caused the bases in DNA (or amino acids in proteins) to align themselves into information-rich sequences? Magnetic forces cause iron filings to align themselves into orderly "lines of force" around the magnet. Gravitational forces create vortices in draining bathtubs. Perhaps some pervasive self-organizing forces external to DNA and proteins could explain the origin of the information-rich biomolecules or other forms of biological organization.

In 1977 the Russian-born Belgian physicist Ilya Prigogine wrote a book with a colleague, Grégoire Nicolis, exploring this possibility. Prigogine specialized in thermodynamics, the science of energy and heat. He became interested in how energy flowing into a system could cause order to arise spontaneously. His work documenting this phenomenon won him the Nobel Prize in Chemistry in 1977, the same year he published his book with Nicolis.

In their book *Self-Organization in Nonequilibrium Systems,* Prigogine and Nicolis suggested that energy flowing into primitive living systems might have played a role in the origin of biological organization. They characterized living organisms as open systems that maintain their particular form of organization by utilizing large quantities of energy and matter from the environment (and by "dissipating" large quantities of energy and matter into the environment).[1] An open system is one that interacts with the environment and whose behavior or structure is altered by that interaction. Prigogine demonstrated that open systems driven far from equilibrium (i.e., driven far from the normal state they would occupy in the absence of the environmental input) often display self-ordering tendencies as they receive an input of energy. For example, thermal energy flowing through a heat sink will generate distinctive convection currents or "spiral wave activity."

In their book, Prigogine and Nicolis suggested that the organized structures observed in living systems might have similarly "self-originated" with the aid of an energy source. They conceded the improbability of simple building blocks arranging themselves into highly ordered structures under normal equilibrium conditions. Indeed, Prigogine previously had characterized the probability of living systems arising by chance alone as "vanishingly small."[2] But now he and Nicolis suggested that under nonequilibrium conditions, where an external source of energy is supplied, biochemical systems might arrange themselves into highly ordered patterns and primitive biological structures.

Order Versus Information

When I first learned about Prigogine and Nicolis's theory and the analogies by which they justified it, it did seem plausible. But as I considered the merits of their proposal, I discovered that it had an obvious defect, one that the prominent information theorist Hubert Yockey described to me in an interview in 1986. Yockey pointed out that Prigogine and Nicolis invoked external self-organizational forces to explain the origin of *order* in living systems. But, as Yockey noted, what needs explaining in biological systems is not order (in the sense of a symmetrical or repeating pattern), but information, the kind of *specified* digital information found in software, written languages, and DNA.

Energy flowing through a system may produce highly ordered patterns. Strong winds form swirling tornados and the "eyes" of hurricanes; Prigogine's thermal baths develop interesting convection currents; and chemical elements coalesce to form crystals. But Yockey insisted that this kind of symmetric order has little to do with the specified complexity or information in DNA, RNA, and proteins. To say otherwise conflates two distinct types of patterns or sequences.

As was my habit, I developed a visual illustration to convey this point to my college students. Actually, I borrowed the visual aid from the children of a professor friend who lived in my neighborhood. The homemade toy his children played with was meant to entertain, but I realized that it perfectly illustrated Yockey's distinction between order and specified complexity and his critique of Prigogine's self-organizational model.

The toy was made of two one-liter soda bottles that were sealed and fastened together at each opening by a red plastic coupling. The two bottles together made one large hourglass shape. The device also contained a turquoise liquid (probably water with food coloring) and some silver flecks that would sparkle as the liquid swirled around. Liquid from one bottle could flow into the other bottle. The children liked to hold the bottles upright until all the liquid from the top bottle flowed into the bottom one. Then they would quickly turn the whole apparatus over and give it a sudden shake by the narrow neck. Next they would watch as the blue liquid would organize into a swirling vortex in the top bottle and begin to drain into the bottom bottle.

After convincing the children to lend me their toy, I used it in class to illustrate how an infusion of energy could spontaneously induce order in a system. This was an important point to establish with my students, because some of them had heard creationist arguments about how the second law of thermodynamics dictates that order in nature always dissipates into disorder over time. Prigogine had shown that, although disorder will ultimately increase over time in a closed system such as our whole universe, order may arise from disorder spontaneously when energy enters into smaller (than the universe) open systems. I used the big blue vortex maker to demonstrate that order can, indeed, arise from an infusion of energy (in this case, a sudden shake and flipping of the apparatus) into an open system.

Nevertheless, I also wanted my students to understand that there was a difference between order and specified complexity and why that distinction called into question the ultimate relevance of Prigogine's ideas. To illustrate this, I would ask them to focus on the individual flecks sparkling within the swirling blue liquid. Could they see any interesting arrangements of these flecks that performed a communication function? Did the flecks spell any messages or encode any digital information? Obviously, the answer was no. Students could see a highly random arrangement of sparkling flecks. They also could see an orderly pattern in the motion of the liquid as a whole as the swirling blue water formed the familiar funnel shape of a vortex. Nevertheless, they could not detect any specified or functional information, no interesting patterns forming sequences of alphabetic or digital characters.

My students had no trouble comprehending the point of my somewhat crude illustration. Energy flowing through an open system will readily produce *order*. But it does not produce much *specified complexity* or *information*.

The astrophysicist Fred Hoyle had a similar way of making the same point. He famously compared the problem of getting life to arise spontaneously from its constituent parts to the problem of getting a 747 airplane to come together from a tornado swirling through a junkyard. An undifferentiated external force is simply too blunt an instrument to accomplish such a task. Energy might scatter parts around randomly. Energy might sweep parts into an orderly structure such as a vortex or funnel cloud. But energy alone will not assemble a group of parts into a highly differentiated or functionally specified system such as an airplane or cell (or into the informational sequences necessary to build one).

Kenyon's self-organizational model had already encountered this problem. He came to realize that, although internal chemical affinities might produce highly repetitive or ordered sequences, they certainly did not produce the information-rich sequences in DNA. Now a similar problem reemerged as scientists considered whether lawlike external forces could have produced the information in DNA. Prigogine's work showed that energy in an open system *can* create patterns of symmetrical order. But it provided no evidence that energy alone can encode functionally specified information-rich sequences—whether biochemical or otherwise. Self-organizational processes explain well what doesn't need explaining in life.

It's actually hard to imagine how such self-organizing forces could generate or explain the specificity of arrangement that characterizes information-rich living systems. In my vortex maker, an externally induced force infused energy through the system, sweeping all the constituents of the system along basically the same path. In Prigogine's convection baths, an energy source established a pattern of motion throughout the system that affected all the molecules in a similar way, rather than arranging them individually and specifically to accomplish a function or convey a message. Yet character-by-character variability and specificity of arrangement are hallmarks of functional information-rich sequences. Thus, as Yockey notes: "Attempts to relate the idea of order . . . with biological organization or specificity must be regarded as a play on words that cannot stand careful scrutiny. Informational macromolecules can code genetic messages and therefore can carry information because the sequence of bases or residues is affected very little, if at all, by [self-organizing] physicochemical factors."[3]

The Limits of the Algorithm

As a result of these difficulties, few, if any, scientists now maintain that Prigogine and Nicolis solved the problem of the origin of biological information. Nevertheless, some scientists continued to hope that further research would identify a specific self-organizational process capable of producing biological information. For example, biophysicist Manfred Eigen suggested in 1992 that "Our task is to find an algorithm, a natural law that leads to the origin of information."[4]

This sounded good, but I began to wonder whether any lawlike process could produce information. Laws, by definition, describe events that repeatedly and predictably recur under the same conditions. One version of the law of gravity states that "all unsuspended bodies will fall." If I lift a ball above the earth and let it go, it will fall. Every time. Repeatedly. Another law states that "water heated to 212 degrees Fahrenheit at sea level will boil and produce steam." Apply heat to a pan of water. Watch and wait. Bubbles and steam will appear. Predictably. Laws describe highly predictable and regular conjunctions of events—repetitive patterns, redundant order. They do not describe the kind of complexity necessary to convey information.

Here's another way to think of it. Scientific laws often describe predictable relationships between antecedent conditions and consequent events. Many scientific laws take the form, "If A occurs, then B will follow, given conditions C." If the conditions C are present, and an event of type A occurs, then an event of type B will follow, predictably and "of necessity." Thus, scientific laws describe patterns in which the probability of each successive event (given the previous event) approaches one, meaning the consequent must happen if the antecedents are present. Yet, as noted previously, events that occur predictably and "of necessity" do not convey information. Instead, information arises in a context of contingency. Information mounts as *improbabilities* multiply. Thus, to say that scientific laws generate complex informational patterns is essentially a contradiction in terms. If a process is orderly enough to be described by a law, it does not, *by definition*, produce events complex enough to convey information.

Of course, lawlike processes might *transmit* information that already exists in some other form, but such processes do not generate specified information. To see why, imagine that a group of small radio-controlled helicopters hovers in tight formation over a football stadium, the Rose Bowl in Pasadena, California. From below, the helicopters appear to be spelling a message: "Go USC." At halftime with the field cleared, each helicopter releases either a red or gold paint ball, one of the two University of Southern California colors. The law of gravity takes over and the paint balls fall to the earth, splattering paint on the field after they hit the turf. Now on the field below, a somewhat messier but still legible message appears: "Go USC."

Did the law of gravity, or the force described by the law, produce this information? Clearly, it did not. The information that appeared on the field already existed in the arrangement of the helicopters above the stadium—in what physicists call the "initial conditions." Neither the force of gravity nor the law that describes it caused the information on the field to self-organize. Instead, gravitational forces merely transmitted preexisting information from the helicopter formation—the initial conditions—to the field below.

Sometimes when I've used these illustrations I've been asked: "But couldn't we discover a very particular configuration of *initial conditions* that generates biological information? If we can't hope to find a law that produces information, isn't it still possible to find a very particular set of initial conditions that generates information in a predictable law-like way?" But this objection just restates the basic self-organizational proposal in new words. It also again begs the question of the ultimate origin of information, since "a very particular set of initial conditions" sounds precisely like an information-rich—a highly complex and specified—state. As I would later discover, however, this wasn't the only proposal to beg the question about the ultimate origin of information. Indeed, attempts to explain the origin of information by reference to some prior set of conditions invariably shifted—or displaced—the problem someplace else. My first inkling of this problem came as I reflected on perhaps the most innovative—"outside the box"—self-organizational proposal of all.

The Kauffman Model

If neither *internal* bonding affinities between the constituents of DNA nor ubiquitous *external* forces acting upon those constituents can account for the specific sequence of the DNA bases, then what was left? It didn't seem to me that there could be much left, since "forces external" and "forces internal" to the molecule seemed to exhaust the set of possibilities. Yet, even so, I knew that another player was about to step onto the stage, Stuart Kauffman.

Kauffman is a brilliant scientist who trained as a physician at the University of California, San Francisco, and later worked as professor of biochemistry and biophysics at the University of Pennsylvania before leaving to help head up the Santa Fe Institute, a research institute dedicated to the study of complex systems. During the early 1990s, as I was beginning to examine the claims of self-organizational theories, I learned that Kauffman was planning to publish a treatise advancing a new self-organizational approach. His new book promised to bring the progress made at Santa Fe to bear on the problem of the origin of life and indeed to make significant steps toward solving that problem within a self-organizational framework. His long-anticipated book was titled *The Origins of Order: Self-Organization and Selection in Evolution.*

I remember the day Kauffman's book finally arrived at the college where I taught. I quickly opened the package in my office only to discover a rather imposing seven-hundred-page behemoth of a book. After I scanned the table of contents and read the opening chapters, it became

Figure 12.1. Dr. Kauffman's primary work has been as a theoretical biologist studying the origin of life and molecular organization. *Courtesy of the University of Calgary.*

clear that much of Kauffman's book provided a rather generalized discussion of the mathematical properties of complex systems. His specific proposal for explaining the origin of life occupied less than eighty pages of the book and was mainly confined to one chapter. My curiosity got the better of me. I decided to read this section first and return to the rest of the book later.

Kauffman had, in keeping with his reputation as an "outside the box" thinker, made a bold and innovative proposal for explaining the origin of life. Rather than invoking either external lawlike forces or internal bonding affinities to explain how biological information had self-organized, his hypothesis sought to transcend the problem altogether. He proposed a self-organizational process that could bypass the need to generate genetic information.

Kauffman attempted to leapfrog the "specificity" (or information) problem by proposing a means by which a self-reproducing metabolic system might emerge directly from a set of "low-specificity" catalytic peptides and RNA molecules in a prebiotic soup, or what he called a "chemical minestrone."[5] A metabolic system is a system of molecules that react with each other inside a living cell in order to sustain a vital function. Some metabolic systems break down molecules to release energy; others build molecules to store energy (as occurs during ATP synthesis) or information (as occurs in DNA replication or protein synthesis). In extant forms of life, these reactions usually involve or are mediated by a number of highly specific enzyme catalysts and other proteins. As such, Kauffman's proposal represents a kind of protein-first theory similar in some respects to Kenyon's earlier model.

Nevertheless, Kauffman suggests, unlike Kenyon, that the first metabolic system might have arisen directly from a *group* of *low-specificity* polypeptides. He proposes that once a sufficiently diverse set of catalytic molecules had assembled (in which the different peptides performed enough different catalytic functions, albeit inefficiently), the ensemble of individual molecules spontaneously underwent a kind of phase transition (akin to crystallization) resulting in a self-reproducing metabolic system. Kauffman envisions, as the historian of biology Iris Fry puts it, "a set of catalytic polymers in which no single molecule reproduces itself but the system as a whole does."[6] In this way, Kauffman argues that metabolism (and the proteins necessary to it) could have arisen directly without genetic information encoded in DNA.[7]

Kauffman's model was clearly innovative and arguably more sophisticated than many previous self-organizational models. Unlike previous models, Kauffman's claimed that an ensemble of relatively short and "low-specificity" catalytic peptides and RNA molecules would together be enough to establish a metabolic system. He defends the biochemical plausibility of his scenario on the grounds that some proteins can perform enzymatic functions with low specificity and complexity. To support his claim, he cites a class of proteins known as proteases (including one in particular, called trypsin) that cleave peptide bonds at single amino-acid sites.[8] But is he right?

As I thought more about Kauffman's proposal and researched the properties of proteases, I became convinced that his proposal did not solve or successfully bypass the problem of the origin of biological information. Kauffman himself acknowledges that, as yet, there is no experimental evidence showing that such autocatalysis could occur. But, beyond that, I realized that Kauffman had either presupposed the existence of unexplained sequence specificity or transferred the need for specified information out of view. In fact, Kauffman's model has at least three significant information-related problems.

First, it does not follow, nor is it the case biochemically, that just because *some* enzymes might function with low specificity, that *all* the catalytic peptides (or enzymes) needed to establish a self-reproducing metabolic cycle could function with similarly low levels of specificity and complexity. Instead, modern biochemistry shows that most and usually all of the molecules in a closed interdependent metabolic system of the type that Kauffman envisions require high-complexity and -specificity proteins. Enzymatic catalysis (which his scenario would surely require) needs molecules long enough to form tertiary structures. Tertiary structures are the three-dimensional shapes of proteins, which provide the spatial positioning of critical amino-acid residues to convey particular or specialized functions.[9] How long a protein chain needs to be to form one of those shapes depends on its complexity. For the very simplest shapes, something like forty or fifty amino acids are needed. More complicated shapes may require several hundred amino acids. Further, these long polymers require very specific three-dimensional geometries (which in turn derive from sequence-specific arrangements of monomers) in order to catalyze necessary reactions. How do these molecules acquire their specificity of sequencing? Kauffman does not address this question, because his model incorrectly suggests that he does not need to do so.

Second, I discovered that even the allegedly low-specificity molecules (the proteases) that Kauffman cites to illustrate the plausibility of his scenario are actually very complex and highly specific in their sequencing. I also discovered that Kauffman confuses the specificity and complexity of the parts of the polypeptides upon which the proteases act with the specificity and complexity of the proteins (the proteases) that do the enzymatic acting. Though trypsin, for example, acts upon—cleaves—peptide bonds at a relatively simple target (the carboxyl end of two separate amino acids, arginine and lysine), trypsin itself is a highly complex and specifically sequenced molecule. Indeed, trypsin is a non-repeating 247-amino-acid protein that possesses significant sequence specificity as a condition of function.[10]

Further, trypsin has to manifest significant three-dimensional (geometric) specificity in order to recognize the specific amino acids arginine and lysine, at which sites it cleaves peptide bonds. By equivocating in his discussion of specificity, Kauffman obscures from view the considerable specificity and complexity requirements of the proteases he cites to justify his claim that low-specificity catalytic peptides will suffice to establish a metabolic cycle. Thus, Kauffman's own illustration, properly understood (i.e., without equivocating about the relevant locus of specificity), shows that for his scenario to have biochemical plausibility, it must *presuppose* the existence of many high-complexity and -specificity polypeptides and polynucleotides. Where does this information in these molecules come from? Kauffman again does not say.

Third, Kauffman acknowledges that for autocatalysis to occur, the molecules in the chemical minestrone must be held in a very specific spatial-temporal relationship to one another.[11] In other words, for the direct autocatalysis of integrated metabolic complexity to occur, a system of catalytic peptide molecules must first achieve a very specific molecular configuration (or what chemists call a "low-configurational entropy state").[12] This requirement is equivalent to saying that the system must start with a large amount of specified information or specified complexity. By Shannon's theory of information, information is conveyed every time one possibility is actualized and others are excluded. By admitting that the autocatalysis of a metabolic system *requires* a *specific* arrangement of polypeptides (only one or a few of the possible molecular arrangements, rather than any one of the many possibilities) Kauffman tacitly concedes that such an arrangement has a high-information content. Thus, to explain the origin of specified biological complexity at the systems level, Kauffman has to presuppose a highly specific arrangement of those molecules at the molecular level as well as the existence of many highly specific and complex protein and RNA molecules (see above). In short, Kauffman merely transfers the information problem from the molecules into the soup.

In addition to these problems, Kauffman's model encounters some of the same problems that Kenyon's protein-first model and other metabolism-first models encounter. It does not explain (a) how the proteins in various metabolic pathways came into association with DNA and RNA or any other molecular replicator or (b) how the information in the metabolic system of proteins was transferred from the proteins into the DNA or RNA. And it gives no account of (c) how the sequence specificity of functional polypeptides arose (given that the bonding affinities that exist among amino acids don't correlate to actual amino-acid sequences in known proteins).

Robert Shapiro, a leading chemist at New York University, has recently proposed that origin-of-life researchers begin to investigate metabolism-first models of the kind that Kauffman proposed. Shapiro argues that these models have several advantages that other popular origin-of-life scenarios (particularly RNA-first models, see Chapter 14) don't.[13] Though Shapiro favors these metabolism-first approaches, he acknowledges that researchers have not yet identified what he calls a "driver reaction" that can convert small molecules into products that increase or "mobilize" the organization of the system as a whole. He also notes that researchers on metabolism-first models "have not yet demonstrated the operation of a complete [metabolic] cycle or its ability to sustain itself and undergo further evolution."[14] In short, these approaches remain speculative and do not yet offer a way to solve the fundamental problem of the origin of biologically relevant organization (or information).

In any case, I concluded that Kauffman's self-organization model—to the extent it had relevance to the behavior of actual molecules—presupposes or transfers, rather than explains, the ultimate origin of the specified information necessary to a self-reproducing metabolic cycle. I wasn't the only one to find Kauffman's self-organizational model insufficient. Other scientists and origin-of-life researchers made similar criticisms.[15] Though many origin-of-life researchers have expressed their admiration for Kauffman's innovative new approach, few, if any, think that his model actually solves the problem of the origin of information or the origin of life. Perhaps for this reason, after 1993 Kauffman proposed some new self-organizational models for the origin of biological organization. His subsequent proposals lacked the biological specificity of his bold, if ill-fated, original proposal. Nevertheless, Kauffman's later models did illustrate just how difficult it is to explain the origin of information without presupposing other preexisting sources of information.

Buttons and Strings

In 1995 Kauffman published another book, *At Home in the Universe,* in which he attempted to illustrate how self-organizational processes might work using various mechanical or electrical systems, some of which could be simulated in a computer environment.[16] In one, he conceives a system of buttons connected by strings. The buttons represent novel genes or gene products, and the strings represent the lawlike forces of interaction between the gene products, namely, the proteins. Kauffman suggests that when the complexity of the system (as represented by the number of buttons and strings) reaches a critical threshold, new modes of organization can arise in the system "for free"—that is, without intelligent guidance—after the manner of a phase transition in chemistry, such as water turning to ice or the emergence of superconductivity in some metals when cooled below a certain temperature.

Another Kauffman model involves a system of interconnected lights. Each light can flash in a variety of states—on, off, twinkling, and so forth. Since there is more than one possible state for each light and many lights, there are many possible states the system can adopt. Further, in his system, rules determine how past states will influence future states.

Kauffman asserts that, as a result of these rules, the system will, if properly tuned, eventually produce a kind of order in which a few basic patterns of light activity recur with greater than random frequency. Since these patterns represent a small portion of the total number of possible states for the system, Kauffman suggests that self-organizational laws might similarly produce a set of highly improbable biological outcomes within a much larger space of possibilities.

But do these simulations accurately model the origin of biological information? It's hard to think so. Kauffman's model systems are not constrained by functional considerations and, thus, are not analogous to biological systems. A system of interconnected lights governed by preprogrammed rules may well settle into a small number of patterns within a much larger space of possibilities. But since these patterns need not meet any functional requirements, they fail at a fundamental level to model biological organisms. A system of lights flashing "Warning: Mudslide Ahead" would model a biologically relevant self-organizational process, at least if such a message arose without agents previously programming the system with equivalent amounts of functional information. Kauffman's arrangements of flashing lights are not of this sort. They serve no function, and certainly no function comparable to the information-rich molecules found in the biological realm.

Kauffman's model systems differ from biological information in another striking way. The series of information-bearing symbols we find in the protein-coding regions of DNA, in sophisticated software programs, or in the sentences on this page are aperiodic. The sequences of characters do not repeat in a rigid or monotonous way. Kauffman's model, in contrast, is characterized by large amounts of symmetrical order or internal redundancy interspersed with aperiodic sequences (mere complexity) lacking function.[17] Getting a law-governed system to generate repetitive patterns of flashing lights, even with a certain amount of variation, is clearly interesting, but not biologically relevant. Since Kauffman's models do not produce functional structures marked by specified aperiodic symbol sequences such as we find in DNA, they do not serve as promising models for explaining the origin of biological information.

But there is another fundamental problem with Kauffman's model systems, one he tacitly acknowledges. To the extent that Kauffman's systems do succeed in producing interesting nonrandom patterns, they do so only because of an unexplained intervention of information. For example, Kauffman notes that if his system of flashing lights is properly "tuned,"[18] then it will shift from a chaotic regime to an orderly regime that will produce the outcomes or patterns he regards as analogous to processes or structures within living systems. By "tuning," Kauffman means the careful setting of a particular "bias parameter" to make his system shift from a chaotic regime into one in which order is produced. In other words, the tuning of this parameter ensures that certain kinds of outcomes are actualized and others precluded. Such an act constitutes nothing less than an infusion of information. When someone tunes a radio dial or a musical instrument, he or she selects a certain frequency and excludes many others. Yet Shannon defined an informative intervention as precisely electing one option and excluding others.

In his system of flashing lights, Kauffman briefly mentions that two of his collaborators—physicists Bernard Derrida and Gerard Weisbuch—were responsible for the "tuning" that produced the patterns he thinks analogous to order in living systems. Nevertheless, he does not think that any such agency played a role in the origin of life. Thus, even assuming for the sake of argument that his system of flashing lights manifests features analogous to those in living systems, his model still falls prey to what Dembski calls the "displacement problem"—the problem of explaining the origin of one (putatively) information-rich system only by introducing another unexplained source of information.

Conclusion: The Displacement Problem

As I examined Kauffman's model, it occurred to me that I was beginning to see a pattern. Self-organizational models for the origin of biological organization were becoming increasingly abstract and disconnected from biological reality. Model systems such as Prigogine's or Kauffman's did not even claim to identify actual chemical processes that led in a life-friendly direction. Instead, these models claimed to describe processes that produced phenomena with some limited similarity to the organization found in living systems. Yet upon closer inspection these allegedly analogous phenomena actually lacked important similarities to life, in particular, the presence of specified complexity, or information.

But beyond that, I realized that self-organizational models either failed to solve the problem of the origin of specified information, or they "solved" the problem only at the expense of introducing other unexplained sources of information. Kauffman's models provided only the best illustration of this latter "displacement problem." In addition, the earlier models such as Kenyon's or Prigogine's, relying as they did on processes that produced order rather than complexity, each fell prey to both empirical and conceptual difficulties. They not only failed to explain the origin of information; they did so in a highly instructive way—one that helped to clarify our understanding of the nature of information and why it stands conceptually distinct from redundant order and the lawful processes that produce such order.

Thus, despite the cachet associated with self-organizational theories as the "new wave" of thinking in evolutionary biology, I came to reject them as complete nonstarters, as theories that were unlikely ever to succeed regardless of the outcome of future empirical studies. In my view, these models either begged the question or invoked a logical contradiction. Proposals that merely transfer the information problem elsewhere necessarily fail because they assume the existence of the very entity—specified information—they are trying to explain. And new laws will *never* explain the origin of information, because the processes that laws describe necessarily lack the complexity that informative sequences require. To say otherwise betrays confusion about the nature of scientific laws, the nature of information, or both.

As I reflected on the failure of these models, my interest in the design hypothesis increased. But the reason for this was not just that self-organizational scenarios had failed. Instead, it was that self-organizational theories failed in a way that exposed the need for an intelligent cause to explain the relevant phenomena. Remember my magnetic chalkboard demonstration? I had used that demonstration to show that chemical forces of attraction don't explain the specific arrangements of bases in DNA any more than magnetic forces of attraction explained the arrangement of letters on my letter board. But what *did* explain the arrangement of magnetic letters on that board?

Obviously, an intelligent agent. Might the arrangement of bases in DNA have required such a cause as well?

Recall Kauffman's model systems. In each case, they explained the origin of information by reference to an unexplained source of information. These scenarios too lacked just what the design hypothesis provided: a cause known to be capable of generating information in the first place. In one case, Kauffman presupposes that his system will work only once it had been "tuned." But how? Was intelligence necessary to do what self-organizational processes alone could not?

A similar thought had occurred to me earlier when reflecting on chance elimination. Many events that we would not credit to chance—in particular, highly improbable events that matched independent patterns—were actually best explained by intelligent design. The improbable match between the two college papers that led my colleagues and me to exclude the chance hypothesis also led us to conclude plagiarism, a kind of design.

Even Christian de Duve, in explaining why the origin of life could not have occurred by chance, acknowledged (if inadvertently) that the kind of events that lead us to reject chance also suggest design. Recall that de Duve pointed out that a "string of improbable events" such as someone winning the lottery twice in a row (a kind of pattern match) *"do not happen naturally."*[19] Of course, de Duve went on to state that the failure of the chance hypothesis implied, for him, that life must be "an obligatory manifestation of matter"—one that had self-organized when the correct conditions arose.[20]

Having examined the leading self-organizational theories in detail, I now doubted this. I also later learned from de Duve himself that he felt compelled to elect a self-organizational model, because he was unwilling to consider design as an alternative to chance. He thought invoking design violated the rules of science. As he explains: "Cells are so obviously programmed to develop according to certain lines . . . that the word *design* almost unavoidably comes to mind, . . . [but] life is increasingly explained strictly in terms of the laws of physics and chemistry. Its origin *must* be accounted for in similar terms."[21] As I explain later (see Chapters 18 and 19), I saw no reason to accept this prohibition against considering the design hypothesis. To me, it seemed like an unnecessary restriction on rational thought. So the failure of chance and self-organizational models—as well as the way these models failed—only made me more open to intelligent design as a possible hypothesis.

Indeed, the design hypothesis now seemed more plausible to me than when I first encountered it and certainly more plausible than it had before I had investigated the two most prominent alternative categories of explanation, chance and necessity. But I knew that there was another category of explanation that I needed to investigate more fully, one that combined chance and necessity. Some of the most creative proposals for explaining the origin of biological information relied on the interplay of lawlike processes of necessity with the randomizing effects of "chance" variations. So I needed to examine this class of theories as well. The next two chapters describe what I discovered about them.

13

Chance and Necessity, or
The Cat in the Hat Comes Back

In the Dr. Seuss children's book *The Cat in the Hat Comes Back,* a tall bipedal cat makes a return visit to the home of two children, Sally and her unnamed little brother. On the Cat's first visit, he made a colossal mess, so the children are less than delighted to see him a second time. As before, the Cat makes himself right at home, this time by eating a slice of pink cake in the bathtub and leaving behind a pink ring. Ever helpful, the Cat immediately begins cleaning up his own mess. Unfortunately, every attempt he makes to remove the pink ring from the bathtub only results in spreading parts of the pink ring to various other household objects. Next the Cat pulls a series of smaller cats from his enormous hat to lend a hand, but none of them succeeds in solving the problem. Instead, now the pink stuff is everywhere. Finally the last and littlest cat produces a mysterious device called Voom, which suddenly and dramatically cleans up every shred of pink.

What does this story have to do with the origin of biological information? As I examined more and more theories about the origin of life, I discovered a pattern, one curiously reminiscent of Dr. Seuss's Cat and his indissoluble pink icing. Origin-of-life researchers have been looking for their "Voom" for over fifty years, a process that can, once and for all, clean up the problem of explaining the origin of information. I discovered that every attempt to explain the origin of biological information either failed or transferred the problem elsewhere—either by presupposing some other unexplained sources of information or by overlooking the indispensable role of an intelligence in the generation of the information in question. I first noticed this pattern as I was examining self-organizational theories. As I began to examine theories that invoked both chance and necessity in combination, I noticed this same pattern emerging in spades.

The Classical Oparin Model: Chance and Necessity

The strategy of combining chance with necessity to explain the origin of life began with Aleksandr Oparin, who modeled his approach on Darwin's. Just as Darwin sought to explain the origin of new biological forms by reference to the interplay of random variations and the lawlike process of natural selection, Oparin sought to explain the origin of the first life by combining chance and various types of necessity, including both deterministic chemical reactions and a kind of prebiotic natural selection.

In his original model, Oparin asserted that a series of chemical reactions between smaller molecules produced the first amino acids and other building blocks (see Chapter 2).[1] These reactions took place almost exclusively by chemical necessity.[2] Next Oparin envisioned a series of chance interactions between chemical building blocks, such as amino acids and sugars, eventually resulting in the first proteins and other complex polymers, such as carbohydrates, that are needed for establishing metabolic processes.[3] After these molecules arose, and after some were enclosed in coacervate bodies, competition for nutrients ensued. Then "necessity" asserted itself again. The coacervate protocells containing the most complex metabolic processes and means of absorbing nutrients would have grown fastest and multiplied most often. As the availability of nutrient substances diminished, the more numerous, highly organized protocells would have overwhelmed simpler structures in the "struggle for existence."[4] Oparin envisioned natural selection acting on random changes in the metabolic processes of coacervate bodies. More metabolically complex and efficient coacervates would generally outcompete the less complex ones. In this way, the interplay between chance variation and the imperatives of competition (necessity) gradually produced coacervates containing more complex metabolic processes and, eventually, a living cell with the features we see today.

Oparin Revised: The Natural Selection of Unspecified Polymers

The displacement problem was not evident in Oparin's original scenario. But Oparin's original theory did not explain the origin of the information in DNA. When Oparin published his theory in the 1920s and 1930s, he did not yet know about the information encoded in DNA. He therefore, understandably, made no attempt to explain its origin.

By the late 1950s, developments in molecular biology began to cast doubt on Oparin's original scenario. Not only did his model give no explanation for the emerging DNA enigma; his scenario relied heavily on chance to explain the initial formation of the proteins that made cellular metabolism possible. The discovery of the extreme complexity and specificity of protein molecules undermined the plausibility of this proposal for the reasons discussed in Chapters 9 and 10. Protein function depends upon hundreds of specifically sequenced amino acids, and the odds of a single functional protein arising by chance alone are prohibitively low, given the probabilistic resources of the entire universe.

As the complexity of DNA and proteins became apparent, Oparin published a revised version of his theory in 1968 that envisioned a role for natural selection earlier in the process of abiogenesis. The new version of his theory claimed that natural selection acted on unspecified polymers as they formed and changed within his coacervate protocells.[5] Instead of natural selection acting on fully functional proteins in order to maximize the effectiveness of primitive metabolic processes at work within the protocells, Oparin proposed that natural selection might work on less than fully functional polypeptides, which would gradually cause them to increase their specificity and function, eventually making metabolism possible. He envisioned natural selection acting on "primitive proteins" rather than on primitive metabolic processes in which fully functional proteins had already arisen. By claiming that the first polymers need not have been highly sequence-specific to be preserved by natural selection, Oparin attempted to circumvent, at least partially, the information problem.

I say "partially," because Oparin's revised scenario also attempted to explain the origin of genetic information in DNA. It proposed that both unspecified polypeptides *and* unspecified polynucleotides arose by chance and were later enclosed within coacervates. Natural selection then began to act on both types of molecules by favoring the coacervates that gradually developed functionally specified proteins and genes over those that did not.[6] As the specificity of the DNA and protein molecules within the coacervates increased, they came into association with each other, gradually producing metabolic processes and mechanisms of information storage and transmission. Thus Oparin sought to take into account the importance and apparent primacy of the genetic material to living systems. In doing so, he also sought to counter criticisms from advocates of a competing theory of abiogenesis known as the "gene theory."[7]

Gene theorists such as Hermann Muller believed that life had evolved first in the form of a single ("living") genetic molecule as the result of a fortuitous combination of chemicals. To such theorists, the primacy of DNA in information storage and its importance in initiating the process of protein synthesis suggested an inadequacy in Oparin's original theory. Muller attacked Oparin's view that life evolved first in a metabolizing multimolecular system (i.e., in coacervates) "on the grounds that while changes in genes lead to changes in metabolism, the reverse was known not to be the case."[8] Oparin's silence about a specifically genetic molecule in his 1936 work did leave his scenario vulnerable to the criticism that it was inadequate to explain the DNA enigma. But during the 1960s, Oparin adapted his theory to accommodate new information about the function and structure of DNA and the mechanisms of self-replication and protein synthesis.[9] This revision enabled it to weather this initial challenge and to remain in currency for another decade or so.[10]

Oparin's revised theory nevertheless encountered almost immediate criticism. First, many scientists recognized that Oparin's concept of *prebiotic* natural selection begged the question. Natural selection occurs only in organisms capable of reproducing or replicating themselves. Yet, in all extant cells, self-replication depends on functional and, therefore, sequence-specific DNA and protein molecules. As theoretical biologist Howard Pattee explains, "There is no evidence that hereditary evolution [natural selection] occurs except in cells which already have . . . the DNA, the replicating and translating enzymes, and all the control systems and structures necessary to reproduce themselves."[11] But this fact of molecular biology posed an obvious difficulty for Oparin's theory of prebiotic natural selection. In order to explain the origin of specified information in DNA, Oparin invoked a process that depends upon pre-existing sequence-specific (i.e., information-rich) DNA molecules. Yet, the origin of these molecules is precisely what his theory needed to explain. As Christian de Duve explains, theories of prebiotic natural selection necessarily fail because they "need information which implies they have to presuppose what is to be explained in the first place."[12]

I tell a story in lectures to illustrate how the prebiotic natural selection hypothesis begs the question. My story is about a man who has fallen into a thirty-foot pit. To climb out of the pit, the man knows he needs a ladder. Unmoved by the severity of his difficulty, the man goes home and gets a ladder out of his garage, walks back to the pit, climbs back in, sets up the ladder, and climbs out of the hole. The ladder from his garage was just the ticket, but my story begs an obvious question: How did the man get out of the pit in the first place in order to fetch the ladder? In the same way, the concept of prebiotic natural selection begs the question of how nature generated the sequence-specific information-rich DNA and proteins that are needed to make self-replication, and thus natural selection, possible. Indeed, for this reason, Theodosius Dobzhansky, one of the leading evolutionary biologists of the twentieth century, insisted in 1968 that "Prebiological natural selection is a contradiction in terms."[13]

In fairness, Oparin did attempt to circumvent this problem. He proposed that prebiotic natural selection initially would act on unspecified strings of nucleotides and amino acids. But this created another problem for his scenario. Researchers pointed out that any system of molecules for copying information would be subject to a phenomenon known as "error catastrophe" unless those molecules are specified enough to ensure an error-free transmission of information. An error catastrophe occurs when small errors—deviations from functionally necessary sequences—are amplified in successive replications.[14] Since the evidence of molecular biology shows that unspecified polypeptides will not replicate genetic information accurately, Oparin's proposed system of initially unspecified polymers would have been highly vulnerable to such an error catastrophe.

Thus, the need to explain the origin of specified information created an intractable dilemma for Oparin. If, on the one hand, Oparin invoked natural selection early in the process of chemical evolution (i.e., before functional specificity in amino acids or nucleotide strings had arisen), accurate replication would have been impossible. But in the absence of such replication, differential reproduction cannot proceed and the concept of natural selection is incoherent.

On the other hand, if Oparin invoked natural selection late in his scenario, he would need to rely on chance alone to produce the sequence-specific molecules necessary for accurate self-replication. But even by the late 1960s, many scientists regarded that as implausible given the complexity and specificity of the molecules in question.

Oparin's dilemma was made all the more acute by the logic of natural selection itself.[15] As proposed by Darwin, the process of natural selection favors or "selects" functionally advantageous variations. Thus, it can "select" only what random variations first produce. Yet it is extremely unlikely that random molecular interactions would produce the information present in even a single functioning protein or DNA molecule of modest length (see Chapters 9 and 10). Indeed, the improbability of biological information arising by chance undermines the plausibility of prebiotic natural selection because it implies an exceedingly high (i.e., improbable) initial threshold of selectable function. Given the probabilistic resources of the whole universe, it is extremely unlikely that even one functional protein or DNA molecule—to say nothing of the suite of such molecules necessary to establish natural selection—would arise by chance. Yet the hypothesis of prebiotic natural selection presupposes that a series of such improbable events occurred before natural selection played any role at all.

The work of John von Neumann, one of the leading mathematicians of the twentieth century, made this dilemma more acute. Von Neumann showed that any system capable of self-replication would require subsystems that were functionally equivalent to the information storage, replicating, and processing systems found in extant cells.[16] His calculations established an extremely high threshold of minimal biological function, a conclusion that was confirmed by later experimental work.[17] On the basis of minimal complexity and related considerations, several scientists during the late 1960s (for example, physicist Eugene Wigner, biophysicist Harold Morowitz) made calculations showing that random fluctuations of molecules were extremely unlikely to produce the minimal complexity needed for a primitive replication system.[18] Indeed, as we saw in Chapters 9 and 10, the improbability of developing a replication system vastly exceeds the improbability of developing the protein or DNA components of such a system.

As a result, by the late 1960s many scientists came to regard the hypothesis of prebiotic natural selection as indistinguishable from the pure chance hypothesis, since random molecular interactions were still needed to generate the initial complement of biological information that would make natural selection possible. Prebiotic natural selection could add nothing to the process of information generation until *after* vast amounts of functionally specified information had first arisen by chance. Oparin's idea of prebiotic natural selection succeeded only in pushing the pink ring back into the murky realm of small probabilities.

DNA First

As I investigated various other models combining chance and necessity, I found some highly creative proposals. One of the most creative was devised in 1964 by Henry Quastler, an early pioneer in the application of information theory to molecular biology. Quastler proposed a DNA-first model for the origin of life in which chance processes create a system of unspecified polynucleotides that can self-replicate via complementary base pairing (by chemical necessity). Like Oparin, Quastler thought these initial polynucleotides would have arisen without the specificity of sequencing necessary to build specific proteins.[19] Quastler considered this an advantage for his model. He acknowledged that it was extremely improbable that chance alone would produce a *specific* sequence of nucleotide bases long enough to function as a gene. Nevertheless, he knew that if there were no constraints on how DNA was sequenced—if any sequence would do—then chance alone might do the job.

But then how did biological specificity and functional information arise? Quastler thought that specificity in these molecules arose later, after his system of polynucleotides had—by chance—come to associate with a set of proteins and ribosomes capable of producing proteins from the particular nucleotide sequences that happened to have arisen.

At some point, the previously unspecified polynucleotide sequences *acquired* specificity and functional significance by their association with a system of other molecules for producing proteins. Thus, Quastler characterized the origin of information in polynucleotides as an "accidental choice remembered."[20]

Quastler developed an illustration to convey what he had in mind using a combination lock. He asked his readers to imagine someone blindly choosing a combination—any combination—at random. Quastler suggested that after this combination was selected, someone could set the tumblers to match it. As a result, the combination would thereafter open the lock. Quastler used this illustration to show how an initially unspecified sequence that had arisen by chance could later acquire functional specificity.

Quastler's scenario possessed one overriding difficulty, however. It did not account for the origin of the specificity of the molecular system that conferred functional significance and specificity on the initial sequence of nucleotides. In Quastler's combination lock example, a conscious agent chose the tumbler settings that made the initial combination functionally significant. Further, engineers designed the lock that made it possible to specify the tumbler settings. Yet Quastler expressly precluded conscious design as a possibility for explaining the origin of life.[21] Instead, he seemed to suggest that the origin of the set of functionally specific proteins (and the translation system) necessary to create a "symbiotic association" between polynucleotides and proteins would arise by chance.[22] He offered some rough calculations to show that a multimolecular system could have originated by chance in the prebiotic soup. But these calculations, which were performed in 1964, are no longer credible. They vastly underestimated the improbability of generating a single protein by chance alone, to say nothing of the whole gene-expression system.[23]

Moreover, Quastler's approach to the problem of the origin of biological information provided another striking example of the displacement problem. Quastler "solved" the problem of the origin of complex specificity in nucleic acids only by transferring the problem to an equally complex and specified system of proteins, RNA molecules, and ribosomes. In his scenario, any polynucleotide sequence would suffice at first, but the proteins and ribosomal material that made up the transla-

would have to exhibit an extreme specificity
...cleotide sequence. Each of these molecules
...a specific shape and sequence in order to
...t Quastler expected it to play in convert-
...f nucleotide bases into proteins. The pro-
...s the correct sequences of amino acids in
...d copy genetic information; the ribosomal
...eed to be sequenced precisely in order to
...gether to form a functional ribosome; the
...o mediate specific associations in order to
...s of bases on the polynucleotides into spe-
...nd the sequences of amino acids thus pro-
...nged precisely in order to fold into stable
...Like other scenarios I had encountered,
...uence-specificity problem merely shifted it

...laining the origin of life—one that com-
...essity[24]—was proposed in the late 1970s
by Manfred Eigen, a German biophysicist who won the Nobel Prize
in Chemistry in 1967. Eigen proposed a system called a hypercycle to
explain how new biological information and structure might have accumulated over time as life developed.[25] A hypercycle is a hypothetical
self-reproducing system made of many enzymes and RNA molecules.
In the hypercycle, enzymes and RNA molecules react with each other
chemically to make structural improvements in the system as a whole.
The system develops as one group of enzymes increases the replication
rate of a sequence-specific RNA molecule, which in turn increases the
rate of production of other enzymes, which in turn increases the replication rate of a different RNA molecule, and so on. Eventually, a group
of enzymes increases the replication rate of the original RNA in the reaction sequence, forming a cycle that repeats indefinitely (see Fig. 13.1).
Theoretically, the RNA and protein molecules of a hypercycle can be
organized in an interdependent way to give the hypercycle, as a whole, a

kind of structural stability. Because Eigen proposed this as a mechanism for increasing the molecular information and structure of a developing protocell, some have claimed that he solved—at least in theory—the information problem associated with life's origin.

Critics of his model, such as evolutionary biologist John Maynard Smith, physicist Freeman Dyson, and chemist Robert Shapiro, have contested this assessment, showing that Eigen's hypothetical cycles are more likely to lose or degrade genetic information over time.[26] They point out that hypercycles, as conceived by Eigen, lack an error-free mechanism of self-replication. As a result, his proposed mechanism would succumb to various "error catastrophes" that would diminish, rather than increase, the specified information content of the system over time.

In any case, hypercycles have a more obvious limitation. In formulating the idea, Eigen presupposed a large initial complement of information in the form of sequence-specific RNA molecules and some forty functional proteins. Thus, his model does not solve, even hypothetically, the problem of the ultimate origin of biological information. At best, it models how information might have developed from a preexisting source, though even that is doubtful for the reasons his critics have described. Eigen himself has apparently acknowledged that hypercycles presuppose, but do not explain, the ultimate origin of biological

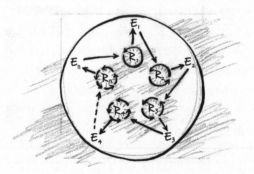

Figure 13.1. A schematic representation of Manfred Eigen's concept of a hypercycle. The E's represent enzymes capable of replicating RNA molecules, the R's in the circles represent RNA molecules. Notice that hypercycles require an initial complement of both sequence-specific enzymes and information-rich RNA molecules.

information. He notes that he devised hypercycles to model how information and structure might *develop*, rather than *originate*.[27]

"Life" in a Computer: Genetic Algorithms

As it became more difficult to envision how life arose by chance and necessity from the molecular constituents of actual living cells, evolutionary biologists sought to simulate the production of biological information using computer programs. The first such attempts occurred during the late 1980s. At that time, Richard Dawkins and Bernd-Olaf Küppers attempted to simulate how natural selection acting on random mutations could explain the origin of biological information. Each tried to do this by developing software programs that simulated the production of genetic information.[28] Both accepted the futility of naked appeals to chance and invoked what Küppers calls a "Darwinian optimization principle." Both used computers to demonstrate the efficacy of natural selection. Thus, both combined chance and necessity in a classically Darwinian way.

In order to show how chance and selection can produce information, Dawkins and Küppers first provided the computer program with a target sequence to represent a desired functional polymer. In their programs, the target sequences were English phrases, such as the line from Shakespeare's play *Hamlet,* "Methinks it is like a weasel." Next the computer program generated another alphabetic sequence the same length as the target. Then the program generated a crop of variant sequences at random. After that, the computer selected the sequences from among the crop that matched the target sequence most closely. The computer then preserved those sequences and amplified the production of them, eliminated the others (to simulate differential reproduction), and repeated the process. As Küppers puts it, "Every mutant sequence that agrees one bit better with the meaningful or reference sequence . . . will be allowed to reproduce more rapidly."[29] After a mere thirty-five generations, his computer succeeded in spelling his target sequence, "Natural selection." Dawkins's simulation took only a bit longer. In a mere forty-three generations, his program produced: "Methinks it is like a weasel."

Such simulations are impressive from a distance, but they conceal an obvious flaw: molecules in situ do not have a target sequence "in mind." In biology, where differential survival depends upon maintaining function, selection cannot occur until functional sequences arise. Nor will sequences confer any selective advantage on a cell and thus differentially reproduce until they combine in a functionally advantageous arrangement. Nothing in nature (biology or chemistry) corresponds to the role that the computer plays in selecting functionally nonadvantageous sequences that happen to agree "one bit better" than others with a target sequence. The simulation doesn't even make sense on its own terms. The sequence "normal election" may agree more with "natural selection" than does the sequence "mistress defection," but neither of the two yields any advantage in communication over the other in trying to communicate something about natural selection. If that is the goal, both are equally ineffectual.

More to the point, a completely nonfunctional polypeptide would confer no selective advantage on a hypothetical protocell, even if its sequence happened to "agree one bit better" with an unrealized target protein than some other nonfunctional polypeptide. Such a molecule would not be preserved but eliminated, stopping the evolutionary process in its tracks. Yet both Küppers's and Dawkins's published results of their simulations show the early generations of variant phrases awash in nonfunctional gibberish.[30] In Dawkins's simulation, not a single functional English word appears until after the tenth iteration (unlike the more generous example above that starts with the actual, albeit incorrect, words). Clearly it is impossible to distinguish sequences based on considerations of comparative functional advantage when the sequences in question have no function whatsoever. Such determinations can be made only if considerations of proximity to possible future function are allowed, but that requires foresight, the very thing *natural* selection lacks.

As philosopher and mathematician David Berlinski has argued, genetic algorithms need something akin to a "forward-looking memory" to succeed. Yet such foresighted selection has no analogue in nature.[31] In biology, where differential survival depends upon maintaining function, selection cannot occur before new functional sequences arise. Natural selection lacks foresight. A computer programmed by a human being can make selections based upon relative proximity to distant targets or goals, but to imply that molecules can do so illicitly *personifies* nature. If computer simulations demonstrate anything, they subtly demonstrate the need for an intelligent agent to elect some options and exclude others—that is, to create information.

Ev

Since the publication of Küppers's and Dawkins's work, other biologists have devised more sophisticated evolutionary algorithms to simulate how mutation and selection—chance and necessity—can generate new information. One of the most famous examples of such an algorithm is called *Ev*. *Ev* was created by Thomas Schneider, a research biologist at the National Institutes of Health.[32] Schneider cites the success of *Ev* in simulating the production of information "from scratch" to support his claim that the undirected material processes of "replication, mutation and selection are necessary and sufficient for information gain to occur."[33]

In particular, Schneider claims *Ev* can simulate the production of nucleotide binding sites. Nucleotide binding sites are very short sequences of nucleotides on the genome located upstream of specific genes. These binding sites enable RNA polymerase to bind to the correct place on a genome, thus allowing it to read and copy the genes in question. The binding sites typically represent between six and twenty specifically sequenced nucleotides, a small (but specified) amount of information in relation to an entire gene. Schneider claims that "the program simulates the process of evolution of new binding sites *from scratch*."[34] He further indicates that the *Ev* program has thus created 131 bits of information.[35]

Though the *Ev* program uses a more complex set of subroutines than either of the early genetic algorithms devised by Küppers or Dawkins, it uses essentially the same strategy to ensure that the program will generate an information-rich sequence. Like Küppers and Dawkins, Schneider supplies *Ev* with a target sequence, in this case a particular sequence of nucleotide bases that function as a binding site. He then has the program generate a random crop of sequences of equal length.

After favoring sequences that manifest the general profile of a binding site, *Ev* applies a fitness function to the remaining sequences. The fitness function then assesses the degree of divergence between the mutated sequences and the target sequence and applies an error value to each mutated sequence. Then the *Ev* program preserves the sequence(s) with the least error (i.e., the degree of difference from the target) and permits them to replicate and mutate.[36] Then he repeats the process again and again, until finally *Ev* converges on the target sequence.

The target sequence involves foresight. So too does the program's fitness function, which makes selections based upon proximity to future function. Thus, it again simulates a goal-directed foresight that natural selection does not possess. It makes use of information about a functional state (nucleotide binding sites) in a way that natural selection cannot.

Ev incorporates one additional step that Dawkins's and Küppers's simulations lack. Before *Ev* applies its fitness function, it applies a filter to the crop of mutated sequences. The filter favors sequences that have the general profile of a binding site. Like the fitness function, this coarser filter makes use of information about the functional requirements of binding sites to favor some sequences over others. As such, it imparts information based on knowledge that Thomas Schneider, not natural selection or the environment, has imparted into the *Ev* simulation. *Ev* exhibits the genius of its designer.

Informational Accounting

Recently, the senior engineering professor Robert Marks, formerly of the University of Washington in Seattle and now at Baylor University in Texas, analyzed evolutionary algorithms such as *Ev*. Marks shows that despite claims to the contrary by their sometimes overly enthusiastic creators, algorithms such as *Ev* do not produce large amounts of functionally specified information "from scratch." Marks shows that, instead, such algorithms succeed in generating the information they seek either by providing information about the desired outcome (the target) from the outset, or by adding information incrementally during the computer program's search for the target. To demonstrate this, Marks distinguishes and defines three distinct kinds of information: exogenous information, endogenous information, and active information.

"Endogenous information" represents the information present in the target. It also provides a measure of the difficulty of the search for that target—that is, the improbability of finding the specific sequence, or target, among the exponentially large space of alternative possibilities. Recall that the amount of information present in a sequence or system is inversely proportional to the probability of the sequence or system arising by chance. If the probability of finding the target is small the information required to find the target is correspondingly large. By calculating the size of the space of alternative possibilities in which the target resides, the computer scientist can determine both the probability of finding the target in a random search and the information content of the target in question. Marks's analysis of evolutionary algorithms shows that, in order to produce or find the (endogenous) information present in the target, a programmer must design a search algorithm that reduces the information requirements of the search to a manageable level. The information added by the programmer to reduce the difficulty of the search he dubs "active information." The "exogenous information" is what is left after the active information is subtracted from the endogenous information. It measures the difficulty of the residual search problem.[37]

In his critique of *Ev* as well as other evolutionary algorithms, Marks shows that each of these putatively successful simulations of undirected mutation and selection actually depends upon several sources of active information. The *Ev* program, for example, uses active information by applying a filter to favor sequences with the general profile of a nucleotide binding site. And it uses active information in each iteration of its evaluation algorithm or fitness function. The fitness function in *Ev* uses information about the target sequence to assess degrees of difference between a prespecified target and the mutated sequences produced by the program. Those sequences that have the lowest error values— greatest proximity to the prespecified functional sequence—are selected to replicate and mutate. Marks shows that each time the program uses knowledge of the target sequence to exclude some sequences and preserve others, it imparts a quantifiable amount of information in its selection. Marks quantifies these sources of active information and shows that they reduce the difficulty well below the 131 bits Schneider claims that *Ev* can produce "from scratch."[38] He also shows that the endogenous information in even modestly difficult search problems usually cannot be generated (or the search problem solved) without added or "active" information to assist the search.

Avida

Since the release of *Ev* in 2000, another evolutionary algorithm has attracted widespread interest. A program called *Avida*—*A* for artificial, and *vida* for life—was designed in 1993 by three computer scientists: Christopher Adami and Titus Brown, of Caltech, and Charles Ofria, of Michigan State University.[39] In 2003, Adami and Ofria teamed up with biologist Richard Lenksi and philosopher Robert Pennock, both of Michigan State, to publish an article in *Nature* describing the results of an evolutionary simulation they had conducted with *Avida*.[40] In the article, the authors claimed that *Avida* demonstrates that "digital organisms" capable of replication could generate complex features and functions "by random mutations and natural selection."[41] *Avida* is now widely cited to show that biological complexity (and presumably the information necessary to produce it) could have arisen from the twin forces of chance and necessity.[42] Unlike the earlier simulations of Dawkins, Küppers, and Schneider, the *Avida* program does not provide the computer with a specific target sequence, nor does it select sequences of characters on the basis of proximity to possible future function. Does it therefore demonstrate the efficacy of chance and necessity in a way that earlier simulations did not?

Avida consists of two main parts, the *Avida* world and the *Avida* digital organisms that populate this world. *Avida* organisms have two basic parts: software and hardware. Each *Avida* organism has a small software program consisting of a loop of instructions composed from a set of twenty-six predefined commands. Additionally, each digital organism also consists of virtual hardware that can execute each of the twenty-six commands among the set of possibilities.

These commands direct *Avida* organisms to perform various operations on two fixed input strings of binary digits, thus producing various output strings. The loop of instructions contains a program that has been written to replicate the loop (initially this is all the instructions in the loop do). Upon replication, the loop of instructions is mutated as the *Avida* world makes random changes to the loop by inserting, deleting, or switching commands, thus making it possible for *Avida* organisms to generate new output strings from the two fixed input strings (see Fig. 13.2).

These *Avida* organisms exist within a larger *Avida* "world." The *Avida* world represents the rules by which the digital organisms operate and compete with each other for survival. The *Avida* world contains an evaluation algorithm—what a colleague of mine calls the "sniffer"— that determines whether one of nine logical relationships exists between the input strings given to each digital organism and the output strings they produce. Another way to think of this is that the sniffer is evaluating whether a "logic function" has been performed on the input strings

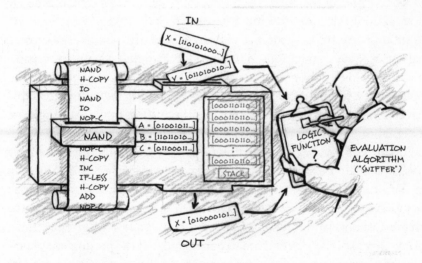

Figure 13.2. Visual representation of the *Avida* organism within *Avida* world. The loop of instructions (*left*) operates on the input sequences (*top*) while stored in the registers (*center*) to produce an output sequence (*bottom*). The evaluation algorithm (depicted in personified form but only for effect) compares the input sequence to the output sequence to see if one of nine specific logical relationships exists between the sequences. If so, the *Avida* world rewards the *Avida* organism with resources that improve the organism's ability to replicate itself.

to produce the output string. For example, one of the logic functions the sniffer is looking for is OR. If the OR function is applied to a pair of binary characters, then it will generate a "1" if at least one *or* the other of the characters is a "1." Otherwise it will generate a "0."[43] If all the binary digits in an output string are consistent with any of the nine logic functions having been applied to each pair of binary digits along the two input strings, then the sniffer will recognize that a logic function has been performed. (See Fig. 13.2.)

In that case, the *Avida* organism that performed the logic function is rewarded with resources that improve its ability to replicate itself. Replication is imperfect, meaning that the "offspring" of an *Avida* organism may carry random changes to the instructions in their program loops. These changes involve replacing a command with another (from the set of twenty-six), inserting a command (making the loop longer), or deleting a command (making the loop shorter).

The authors of the *Nature* paper claim to have used *Avida* to simulate how natural selection and mutation can produce more complex functions by rearranging genomic instructions. Their simulation showed that after many generations *Avida* produced digital organisms capable of performing both simple and compound logic functions. (A compound logic function consists of a series of discrete logic functions that have been performed in sequence.)[44] As the authors put it, starting from a digital organism that could only replicate itself, *Avida* evolved "digital organisms capable of performing multiple logic functions requiring the coordinated execution of many genomic instructions."[45] Since, in actual organisms, biological functions invariably result from specified sequences of bases—that is, from specified information—and since the *Avida* authors claim to have simulated the production of complex functions, they also effectively claim to have simulated the production of new biological information. So, does this state-of-the-art algorithm solve or at least simulate a solution to the DNA enigma? Well, not exactly.

Avida Presupposes Information

First, even as a demonstration of how biological evolution might generate new biological information starting from a preexisting organism,

Avida leaves much to be desired. As I discuss in more detail in an accompanying note,[46] *Avida* lacks realism as a simulation of biological evolution because the program selects functionally significant logic functions possessing too little complexity to represent the actual information content of functional proteins or genes. By allowing it to choose logic functions that are far simpler than any functional gene or protein, the program diminishes the probabilistic task that nature would face in "trying" to evolve the first self-reproducing cell. Thus, the program does not simulate how a random search through a relevantly sized combinatorial space could generate new information—information and structure that must arise first by random mutation before natural selection can act to preserve it. Yet any complete theory of biological evolution must explain precisely the origin of the specified information present in new genes and proteins.

In any case, *Avida* does not simulate how the information necessary to produce the first organism might have originated. Much like Eigen's hypercycles, *Avida* begins with a sizable complement of *preexisting* information, including virtual "organisms" that have been programmed with the capacity to self-replicate. In an *Avida* organism much of this information is stored in the loop of discrete instructions drawn from the set of twenty-six. These instructions express discrete commands such as, "Calculate the difference between the values in two registers." These instructions direct the *Avida* organism to perform specific tasks on the strings of binary digits (the inputs) described above. These commands, whether expressed in the C++ programming language, a machine code, or translated into English, represent a significant amount of functionally specified information.

In addition, the instructions on the initial loop in the digital organisms were specifically arranged and programmed to replicate themselves. The sequencing of these discrete instructions on the loop constitutes another significant source of preexisting information beyond the information in the instructions themselves. Still further, an intricately designed computing device interprets the instruction sets. This computer hardware constitutes a highly improbable and precisely configured arrangement of parts. These devices represent a sizable amount of structural information that would have required a large (and, in principle, calculable) infusion of digital information to manufacture. Therefore, if the *Avida* organisms tell us anything about the origin of life, they tell us something about how much information must be supplied before a self-replicating organism capable of selection could arise.

In fairness, *Avida*'s advocates do not explicitly claim that their program simulates the origin of the information necessary to build a cell in the first place. Instead, they claim that *Avida* shows how new complex functions and features (and the instructions necessary to produce them) might have arisen from *preexisting* organisms. Even so, much of the hype surrounding *Avida* fails to distinguish these two cases. In a recent conference presentation at Oxford, Robert Pennock, one of the four authors of the *Avida* article in *Nature*, claimed—rather categorically—that *Avida* simulates how natural selection and random mutation produce "design without a designer."[47]

Even if one were to ignore the intelligent design of the *Avida* program itself and grant that *Avida* simulates how some appearances of design might have arisen without a designer starting from a preexisting self-replicating organism, it would not follow that *Avida* has simulated the origin of the most salient appearance of design: the original "machine code of the genes." Instead, *Avida* presupposes, but does not explain, the origin of the information necessary to produce the first self-replicating organism. *Avida* "solves" the problem of the origin of biological information by presupposing its solution at the outset. It dissolves the pink stuff only by using more of it as a solvent.

The Conservation of Information

The failure of *Ev* and *Avida* to simulate a naturalistic solution to the DNA enigma illustrates a more general problem with evolutionary algorithms and, indeed, a more general principle about information itself. Invariably, evolutionary algorithms succeed in producing specified information (or its functional equivalent) as the result of preexisting information, or the programming skill of computer scientists, or both. With the exception of a small and quantifiable amount of information that may arise as the result of random effects, the information produced in these algorithms does not exceed the information that was provided to them by the programmers who designed them (and the engineers who designed the hardware). In some programs, such as Richard Dawkins's rather simple simulation, the role of the programmer in providing the information necessary to the outcome is obvious. In other programs, such as *Avida,* programmers play a more subtle role, though not one that escapes a careful accounting of the source of critical informational inputs. Either way, information in a computational context does not magically arise without the assistance of the computer scientist.

My office is in Redmond, Washington, and I have friends who are computer programmers for some of the software companies in the area. One of my friends is a retired forty-something programmer, who was formerly one of Microsoft's elite architect-level programmers. He also has a special interest in the origin of life and evolutionary algorithms. He said something interesting to me about these programs: "There is absolutely nothing surprising about the results of these algorithms. The computer is programmed from the outset to converge on the solution. The programmer designed the code to do that. What would be surprising is if the program didn't converge on the solution. That would reflect badly on the skill of the programmer. Everything interesting in the output of the program came as a result of the programmer's skill—the information input. There are no mysterious outputs."

Computer science has two principles that codify this insight. Indeed, these principles can be or, in some cases, were derived from a careful analysis of evolutionary algorithms themselves. The first principle is called the "no free lunch" (NFL) theorem. The theorem was first developed by David Wolpert and William Macready, two computer scientists at NASA's Ames Research Center.[48] It describes a constraint that evolutionary algorithms invariably face as they attempt to find information-rich targets—such as a meaningful sequence of letters or a functional arrangement of nucleotide bases—in large combinatorial spaces. The NFL theorem states that an evolutionary algorithm will, on average, perform no better than a blind search in finding a target within a large space of possibilities unless external sources of information are provided to the algorithm to point it toward the target. In other words, finding an information-rich target usually requires an enormous amount of "active information" to reduce the difficulty of a random search to a manageable level (i.e., to a level commensurate with the probabilistic resources available to the evolutionary algorithm).

The game Twenty Questions illustrates this principle. In the game, one person has information about a person, place, or thing that another person must guess. If the person guessing is clever, he or she can acquire that information by asking a series of yes or no questions that gradually narrows the field of possibilities. But each time the person guessing receives a yes or no to a question, he or she is also receiving information. The NFL theorem simply states that such information is usually— barring a rare and very lucky search—indispensable to the success of the search. Richard Dawkins's simulation provided an obvious illustration of this principle. His evolutionary algorithm succeeded, but only because he provided it with an external source of information—in particular, a target sequence and fitness function—which the computer uses to guide its selection of sequences in each successive iteration of its search.

The NFL theorem affirms what a careful analysis of individual evolutionary simulations reveals: the information produced by an evolutionary simulation does not arise "for free," that is, without an input from the programmer. Large informational outputs require (roughly) equivalent informational inputs.

The second principle relevant to the assessment of evolutionary algorithms is closely related to the first. It is called the law of the "conservation of information" (COI). Leon Brillouin, a French-American physicist and innovator in computer science and information theory, states the law in its most basic form: "The computing machine does not create any new information."[49] Robert Marks has a more colorful way of expressing the same idea. "Computers," he says, "are no more able to create information than iPods are capable of creating music."[50] Computer scientists have formulated various mathematically precise laws of conservation of information to express this basic principle. Most state that within certain quantifiable limits the amount of information in a computer in its initial state (considering both its hardware and software) equals or exceeds the amount of information in its final state. A careful analysis of evolutionary algorithms confirms this principle. Genetic algorithms can "solve" the information problem, but only if programmers first supply information about proximity to target sequences, selection criteria, or loops of precisely sequenced instructions.

But I noticed something else. The idea of the conservation of information also seems to apply beyond the computer domain. Indeed, most of us know from our ordinary experience that information typically degrades over time unless intelligent agents generate (or regenerate) it. The sands of time have erased some inscriptions on Egyptian monuments. The leak in the attic roof smudged the ink in the stack of old newspapers, making some illegible. In the game of Telephone, the message received by the last child in line bears little resemblance to the one the first child whispered. Common experience confirms this general trend—and so do prebiotic simulation experiments and origin-of-life research. Simulation experiments produce biologically relevant molecules, but only if experimentalists manipulate initial conditions and guide outcomes—that is, only if they first provide specified information themselves. Those origin-of-life theories "succeed" in accounting for the origin of information only by presupposing an ample supply of information in some preexisting form.

In light of this, I formulated a conservation law of my own as a working hypothesis to distill my experience and what I had discovered about origin-of-life research. Since I was not principally concerned with whether biological evolution could generate specified information, I decided to formulate a "conservative" conservation law—one that applied only to a nonbiological context (and thus not to an information-rich initial state). My statement of the law does not say anything about whether undirected natural processes could produce an increase in specified information starting from preexisting forms of life. But it does encapsulate what repeated experience has demonstrated about the flow of information starting from chemistry and physics alone.

Here's my version of the law of conservation of information: "In a nonbiological context, the amount of specified information initially present in a system, S_i, will generally equal or exceed the specified information content of the final system, S_f." This rule admits only two exceptions. First, the information content of the final state may exceed that of the initial state, S_i, if intelligent agents have elected to actualize certain potential states while excluding others, thus increasing the specified information content of the system. Second, the information content of the final system may exceed that of the initial system if random processes have, by chance, increased the specified information content of the system. In this latter case, the potential increase in the information content of the system is limited by the "probabilistic resources" available to the system.

As noted in Chapter 10, the probabilistic resources of the entire universe equal 10^{139} trials, which, in turn, corresponds to an informational measure of less than 500 bits. This represents the maximum information increase that could be reasonably expected to occur by chance from the big-bang singularity to the present—without assistance from an intelligent agent. Systems that exist over a shorter period of time will have correspondingly smaller envelopes of maximal information increase and will, in any case, usually experience informational loss or degradation without input from an agent. Taking these caveats into account allows a more general statement of the law as follows: "In a nonbiological context and absent intelligent input, the amount of specified information of a final system, S_f, will not exceed the specified information content of the initial system, S_i, by more than the number of bits of information the system's probabilistic resources can generate, with 500 bits representing an upper bound for the entire observable universe."[51]

Conclusion

The law of the conservation of information as variously articulated provides another way of describing the displacement problem that I had seen repeatedly demonstrated by attempts to solve the DNA enigma. The failure of genetic algorithms to simulate the production of specified information within the obviously artificial domain of a computer only provided another illustration of this apparently ubiquitous problem. These attempts to simulate how purely undirected processes might have produced information only pushed the information problem back to a decidedly directing entity—the human mind. Thus, in an oddly unexpected way, evolutionary algorithms and the conservation principle derived in part from analyzing them pointed to an external source of specified information. For this reason, the law of conservation of information seemed to have profound implications for the origin-of-life debate.

But is the conservation law true? It was certainly consistent with everything I knew from my ordinary prescientific experience. And I had discovered nothing in my study of origin-of-life research to contradict the law. Quite the opposite was the case. Model after model failed precisely to explain the origin of biological information, the DNA enigma. Those theories that appeared to solve the problem of the origin of information—the metaphorical "pink stuff"—did so only by displacing or transferring it elsewhere.

Yet I also knew that there was another major hypothesis about the origin of life that was attracting attention. It too combined chance and necessity, but envisioned a role for natural selection much earlier in the process of abiogenesis. This hypothesis also held out the possibility of explaining the classical "chicken and egg" problem—the origin of the interdependence of DNA and proteins—by starting the process of abiogenesis in a different place—or rather, with a different molecule.

The "RNA world" hypothesis, as it is called, is the big one—the idea that has currently captivated the wider scientific community and a host of working molecular biologists "at the bench." Could this model solve the problem of the origin of biological information without displacing it elsewhere? Does it refute the law of information conservation—or at least provide a documented exception to it? Before I could establish that biological information pointed decisively to an intelligent cause, I knew I would need to examine this theory in more detail. The idea was too big and too popular to ignore.

14

The RNA World

By the mid-1980s many researchers concluded that both DNA-first and protein-first origin-of-life models were beset with many difficulties. As a result, they sought a third way to explain the mystery of life's origin. Instead of proposing that the first informational molecules were proteins or DNA, these scientists argued that the earliest stages of abiogenesis unfolded in a chemical environment dominated by RNA molecules. The first scientist to propose this idea was Carl Woese, a microbiologist at the University of Illinois. Walter Gilbert, a Harvard biophysicist, later developed the proposal and coined the term by which it is now popularly known, the "RNA world."[1]

The RNA world is now probably the most popular theory of how life began. Scientists in some of the most prestigious labs around the world have performed experiments on RNA molecules in an attempt to demonstrate its plausibility, and in the opinion of many scientists, the RNA-world hypothesis establishes a promising framework for explaining how life on earth might have originated.

I had an encounter with one such scientist in the spring of 2000. I had just written an article about DNA and the origin of life in the April issue of a prominent New York journal of opinion.[2] When the letters to the editor came in, I initially blanched when I saw one from a fierce critic named Kenneth R. Miller, a biology professor at Brown University and a skilled debater. Had I made a mistake in reporting some biological detail in my argument? When I saw his objection, however, I was relieved. Miller claimed that my critique of attempts to explain the

Figure 14.1. Walter Gilbert, photographed in front of a chalkboard in his office at Harvard. *Courtesy of Peter Menzel/Science Photo Library.*

origin of biological information had failed to address the "RNA first" hypothesis. Miller asserted that I had ignored "nearly two decades of research on this very subject" and failed to tell my "readers of experiments showing that very simple RNA sequences can serve as biological catalysts and even self-replicate."[3]

Miller was half right. I hadn't told my readers about these experiments. But I knew that two decades of research on this topic had not solved the problem of the origin of biological information. Because of space constraints and the format of the journal, I had decided not to address this issue in my original article. But now Miller's letter gave me a chance to do so.

At the time I had been studying research articles from origin-of-life specialists who were highly critical of the RNA-world hypothesis, and in my response to Miller I cited and summarized many of their arguments. I heard nothing more from Miller on the matter, but as I attended various conferences over the next several years, I discovered that he was far from alone. Despite the pervasive skepticism about the RNA world among leading origin-of-life researchers, many practicing molecular biologists, including some very prominent scientists at famous labs, continued to share Miller's enthusiasm. Moreover, I discovered that many of these molecular biologists had recently initiated new experimental work inspired by their confidence in the viability of the RNA-world approach. Had they solved the information problem?

Second Things First

The RNA world is a world in which the chicken and egg no longer confound each other. At least that has been the hope. Building proteins requires genetic information in DNA, but information in DNA cannot be processed without many specific proteins and protein complexes. This problem has dogged origin-of-life research for decades. The discovery that certain molecules of RNA possess some of the catalytic properties seen in proteins suggested a way to solve the problem. RNA-first advocates proposed an early stage in the development of life in which RNA performed *both* the enzymatic functions of modern proteins and the information-storage function of modern DNA, thus sidestepping the need for an interdependent system of DNA and proteins in the earliest living system.

Typically RNA-first models have combined chance events and a law-like process of necessity, in particular, the process of natural selection. As Gilbert and others envision it, a molecule of RNA capable of copying itself (or copying a copy of itself) first arose by the chance association of nucleotide bases, sugars, and phosphates in a prebiotic soup (see Fig. 14.2). Then because that RNA enzyme could self-replicate, natural selection ensued, making possible a gradual increase in the complexity of the primitive self-replicating RNA system, eventually resulting in a cell with the features we observe today. Along the way, a simple membrane, itself capable of self-reproduction, enclosed the initial RNA enzymes along with some amino acids from the prebiotic soup.[4]

According to this model, these RNA enzymes eventually were replaced by the more efficient proteins that perform enzymatic functions in modern cells. For that to occur, the RNA-replicating system first had to begin producing a set of RNA enzymes that could synthesize

Figure 14.2. The RNA World Scenario in Seven Steps. Step 1: The building blocks of RNA arise on the early earth. Step 2: RNA building blocks link up to form RNA oligonucleotide chains. Step 3: An RNA replicase arises by chance and selective pressures ensue favoring more complex forms of molecular organization. Step 4: RNA enzymes begin to synthesize proteins from RNA templates. Step 5: Protein-based protein synthesis replaces RNA-based protein synthesis. Step 6: Reverse transcriptase transfers genetic information from RNA molecules into DNA molecules. Step 7: The modern gene expression system arises within a proto-membrane.

proteins. As Gilbert has explained, in this step RNA molecules began "to synthesize proteins, first by developing RNA adapter molecules that can bind activated amino acids and then by arranging them according to an RNA template using other RNA molecules such as the RNA core of the ribosome."[5] Finally, DNA emerged for the first time by a process called reverse transcription. In this process, DNA received the information stored in the original RNA molecules, and eventually these more stable DNA molecules took over the information-storage role that RNA had performed in the RNA world. At that point, RNA was, as Gilbert put it, "relegated to the intermediate role it has today—no longer the center of the stage, displaced by DNA and the more effective protein enzymes."[6]

I knew that origin-of-life theories that sound plausible when stated in a few sentences often conceal a host of practical problems. And so it was with the RNA world. As I investigated this hypothesis, both before and after my exchange with Professor Miller, I found that many crucial problems lurked in the shadows, including the one I had seen before: the theory did not solve the problem of biological information—it merely displaced it.

Because so many scientists assume that the RNA world has solved the problem of the origin of life, this chapter will provide a detailed and, in some places, technical critique of this hypothesis. My critique details five crucial problems with the RNA world, culminating in a discussion of the information problem. To assist nontechnical readers, I have placed some of this critique in notes for the scientifically trained. I would ask technically minded readers to read these notes in full, because in some cases they provide important additional support for, or qualifications to, my arguments.

Each element of this critique stands mostly on its own. So if you find that the technical material under one subheading presupposes unfamiliar scientific concepts or terminology, take note of the heading, which summarizes the take-home message of the section, and skip ahead to the next one, or even the final two, which address the theory's greatest weakness: its inability to explain the origin of biological information.

Problem 1: RNA Building Blocks Are Hard to Synthesize and Easy to Destroy

Before the first RNA molecule could have come together, smaller constituent molecules needed to arise on the primitive earth. These include a sugar known as ribose, phosphate molecules, and the four RNA nucleotide bases (adenine, cytosine, guanine, and uracil). It turns out, however, that both synthesizing and maintaining these essential RNA building blocks, particularly ribose (the sugar incorporated into nucleotides) and the nucleotide bases, has proven either extremely difficult or impossible to do under realistic prebiotic conditions.[7] (See Fig. 14.3.)

Consider first the problems with synthesizing the nucleotide bases. In the years since the RNA world was proposed, chemist Robert Shapiro has made a careful study of the chemical properties of the four nucleotide bases to assess whether they could have arisen on the early earth under

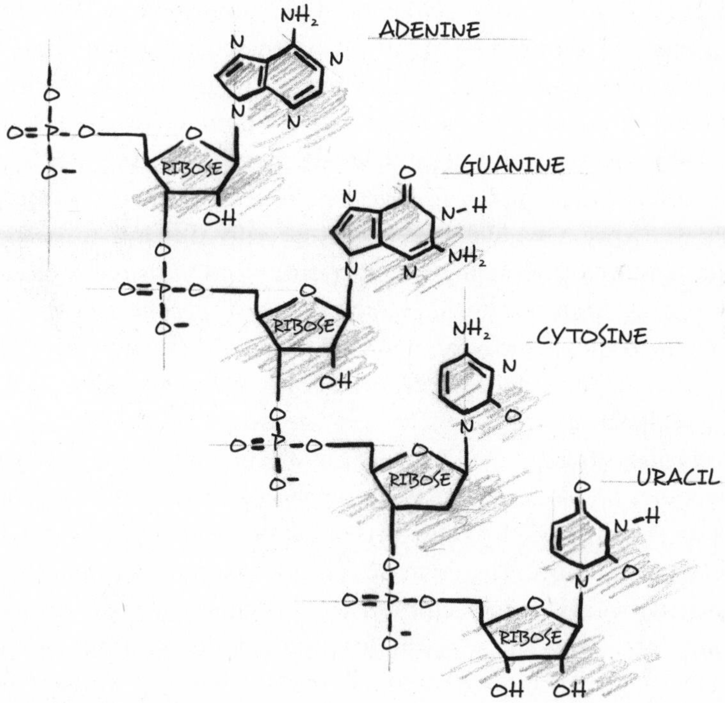

Figure 14.3. The chemical structure and constituents of RNA.

realistic conditions. He notes first that "no nucleotides of any kind have been reported as products of spark-discharge experiments or in studies of meteorites." Stanley Miller, who performed the original prebiotic simulation experiment, published a similar study in 1998.[8] Moreover, even if they did somehow form on the early earth, nucleotide bases are too chemically fragile to have allowed life enough time to evolve in the manner Gilbert and other RNA-first theorists envision. Shapiro and Miller have noted that the bases of RNA are unstable at temperatures required by currently popular high-temperature origin-of-life scenarios. The bases are subject to a chemical process known as "deamination," in which they lose their essential amine groups (NH_2). At 100 degrees C, adenine and guanine have chemical half-lives of only about one year; uracil has a half-life of twelve years; and cytosine a half-life of just nineteen days. Because these half-lives are so short, and because the evolutionary process envisioned by Gilbert would take so long—especially for natural selection to find functional ribozymes (RNA molecules with catalytic activity) by trial and error—Stanley Miller concluded in 1998 that "a high temperature origin of life involving these compounds [the RNA bases] therefore is unlikely."[9] Miller further noted that, of the four required bases, cytosine has a short half-life even at low temperatures, thus raising the possibility that "the GC pair" (and thus RNA) "may not have been used in the first genetic material." Shapiro concurred. He showed that it would have been especially difficult to synthesize adenine and cytosine at high temperatures and cytosine even at low temperatures. Thus he concluded that the presumption that "the bases, adenine, cytosine, guanine and uracil were readily available on the early earth" is "not supported by existing knowledge of the basic chemistry of these substances."[10]

Producing ribose under realistic conditions has proven even more problematic. Prebiotic chemists have proposed that ribose could have arisen on the early earth as the by-product of a chemical reaction called the formose reaction. The formose reaction is a multistep chemical reaction that begins as molecules of formaldehyde in water react with one another. Along the way, the formose reaction produces a host of different sugars, including ribose, as intermediate by-products in the sequence of reactions. But, as Shapiro has pointed out, the formose reaction will not produce sugars in the presence of nitrogenous substances.[11] These include peptides, amino acids, and amines, a category of molecules that includes the nucleotide bases.

This obviously poses a couple of difficulties. First, it creates a dilemma for scenarios that envision proteins and nucleic acids arising out of a prebiotic soup rich in amino acids. Either the prebiotic environment contained amino acids, which would have prevented sugars (and thus DNA and RNA) from forming, or the prebiotic soup contained no amino acids, making protein synthesis impossible. Of course, RNA-first advocates might try to circumvent this difficulty by proposing that proteins arose well after RNA. Yet since the RNA-world hypothesis envisions RNA molecules coming into contact with amino acids early on within the first protocellular membranes (see above), choreographing the origin of RNA and amino acids to ensure that the two events occur separately becomes a considerable problem.

The RNA-world hypothesis faces an even more acute, but related, obstacle—a kind of catch-22. The presence of the nitrogen-rich chemicals necessary for the production of nucleotide bases prevents the production of ribose sugars. Yet both ribose and the nucleotide bases are needed to build RNA. (See note for details).[12] As Dean Kenyon explains, "The chemical conditions proposed for the prebiotic synthesis of purines and pyrimidines [the bases] are sharply incompatible with those proposed for the synthesis of ribose."[13] Or as Shapiro concludes: "The evidence that is currently available does not support the availability of ribose on the prebiotic earth, except perhaps for brief periods of time, in low concentration as part of a complex mixture, and under conditions unsuitable for nucleoside synthesis."[14]

Beyond that, both the constituent building blocks of RNA and whole RNA molecules would have reacted readily with the other chemicals present in the prebiotic ocean or environment. These "interfering cross-reactions" would have inhibited the assembly of RNA from its constituent monomers and inhibited any movement from RNA molecules toward more complex biochemistry, since the products of these reactions typically produce biologically inert (or irrelevant) substances.

Furthermore, in many cases, reactions (such as the formose reaction) that produce desirable by-products such as ribose also produce many undesirable chemical by-products. Unless chemists actively intervene, undesirable and desirable chemical by-products of the same reaction react with each other to alter the composition of the desired chemicals in ways that would inhibit the origin of life. In sum, synthesizing the building blocks of the RNA molecule under realistic prebiotic conditions has proven formidably difficult.

Problem 2: Ribozymes Are Poor Substitutes for Proteins

Another major problem with the RNA world is that naturally occurring RNA molecules possess very few of the specific enzymatic properties of proteins. To date, scientists have shown that RNA catalysts or "ribozymes" can perform a small handful of the thousands of functions performed by modern proteins. Scientists have shown that some RNA molecules can cleave other RNA molecules (at the phosphodiester bond) in a process known as hydrolysis. Biochemists also have found RNAs that can link (ligate) separate strands of RNA (by catalyzing the formation of phosphodiester bonds). Other studies have shown that the RNA in ribosomes (rRNA) promotes peptide-bond formation within the ribosome[15] and can promote peptide bonding outside the ribosome, though only in association with an additional chemical catalyst.[16] Beyond that, RNA can perform only a few minor functional roles and then usually as the result of scientists intentionally "engineering" or "directing" the RNA catalyst (or ribozyme) in question.[17]

For this reason, claiming that catalytic RNA could replace proteins in the earliest stages of chemical evolution is extremely problematic. To say otherwise would be like asserting that a carpenter wouldn't need any tools besides a hammer to build a house, because the hammer performed two or three carpentry functions. True, a hammer does perform some carpentry functions, but building a house requires many specialized tools that can perform a great variety of specific carpentry functions. In the same way, RNA molecules can perform a few of the thousands of different functions proteins perform in "simple" single cells (e.g., in the *E. coli* bacterium), but that does not mean that RNA molecules can perform all necessary cellular functions.

Problem 3: An RNA-Based Translation and Coding System Is Implausible

The inability of RNA molecules to perform many of the functions of protein enzymes raises a third and related concern about the plausibility of the RNA world. RNA-world advocates offer no plausible explanation for how primitive self-replicating RNA molecules might have evolved into modern cells that rely on a variety of proteins to process genetic information and regulate metabolism.[18]

To evolve beyond the RNA world, an RNA-based replication system eventually would have to begin to produce proteins, and not just any proteins, but proteins capable of template-directed protein manufacture. But for that to occur, the RNA replicator first would need to produce machinery for building proteins. In modern cells it takes many proteins to build proteins. So, as a first step toward building proteins, the primitive replicator would need to produce RNA molecules capable of performing the functions of the modern proteins involved in translation. (Recall from Chapter 5 that translation is the process of building proteins from the instructions encoded on an mRNA transcript.) Presumably, these RNA molecules would need to perform the functions of the twenty specific tRNA synthetases and the fifty ribosomal proteins, among the many others involved in translation. At the same time, the RNA replicator would need to produce tRNAs and the many mRNAs carrying the information for building the first proteins. These mRNAs would need to be able to direct protein synthesis using, at first, the transitional ribozyme-based protein-synthesis machinery and then, later, the permanent and predominantly protein-based protein-synthesis machinery. In short, the evolving RNA world would need to develop a coding and translation system based entirely on RNA *and* also generate the information necessary to build the proteins that later would be needed to replace it.

This is a tall order. The cell builds proteins from the information stored on the mRNA transcript (i.e., the copy) of the original DNA molecule. To do this, a bacterial cell depends upon a translation and coding system consisting of 106 distinct but functionally integrated proteins as well as several distinct types of RNA molecules (tRNAs, mRNAs, and rRNAs).[19] This system includes the ribosome (consisting of fifty distinct protein parts), the twenty distinct tRNA synthetases, twenty distinct tRNA molecules with their specific anticodons (all of which jointly embody the genetic code), various other proteins, free-floating amino acids, ATP molecules (for energy), and—last, but not least—information-rich mRNA transcripts for directing protein synthesis. Furthermore, many of the proteins in the translation system perform multiple functions and catalyze coordinated multistep chemical transformations (see Fig. 14.4).

Is it possible that a similar translation and coding system capable of producing genetically encoded proteins might first have arisen using only RNA catalysts (ribozymes)? Advocates of the RNA-world hypothesis have defended the possibility because of the demonstrated catalytic properties of some RNA molecules. Eugene Koonin and Yuri Wolf, two prominent scientists at the National Center for Biotechnology Information, recently reviewed the results of research on the capacities of RNA catalysts in an important article assessing the plausibility of an RNA-based translation system.[20] They note that in the last twenty years, molecular biologists have documented, or engineered, ribozymes that can catalyze "all three elementary reactions"[21] required for translation, including aminoacylation (the formation of a bond between an amino acid and an RNA), the peptidyl-transferase reaction (which forms the peptide bond between amino acids), and amino-acid activation (in which adenosine monophosphate is attached to an amino acid).

At first glance, these results may seem to support the feasibility of an RNA-based translation system. Nevertheless, significant reasons to doubt this aspect of the RNA-world hypothesis remain, as Koonin and Wolf note. First, though ribozymes have demonstrated the capacity to catalyze representative examples of the three main types of chemical reactions involved in translation, they have not demonstrated the ability to catalyze anywhere near all the necessary reactions that fall within these general classifications. Moreover, the gap between "some" and "all" necessary reactions of a given type remains significant. For example, ribozyme engineers have successfully designed an RNA molecule that will catalyze the formation of an aminoacyl bond between itself and the amino acids leucine and phenylalanine.[22] But no one has yet demonstrated that RNA can catalyze aminoacyl bonds

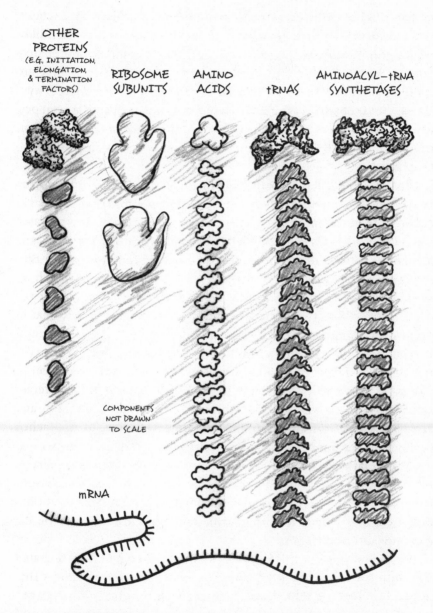

OTHER
PROTEINS
(E.G, INITIATION,
ELONGATION,
& TERMINATION
FACTORS)

RIBOSOME
SUBUNITS

AMINO
ACIDS

tRNAS

AMINOACYL—tRNA
SYNTHETASES

COMPONENTS
NOT DRAWN
TO SCALE

mRNA

Figure 14.4. The main molecular components of the translation system: twenty specific transfer-RNA molecules, twenty specific aminoacyl tRNA synthetases, the ribosome with its two main subunits composed of fifty proteins and ribosomal RNA, the messenger-RNA transcript, and a supply of amino acids.

with the other eighteen protein-forming amino acids, still less with the specificity required to make the resulting molecules useful for translation. Yet establishing a genetic code requires molecules that can catalyze highly specific aminoacylation for each of the twenty protein-forming amino acids. To say that RNA can catalyze "aminoacylation" is true, but it obscures the distinction between part of a group and the whole group, where having the whole group of molecules is necessary to the function in question. Again, it takes more than a hammer to build a house.

Second, unlike RNA catalysts (ribozymes), the protein-based enzymes involved in translation perform multiple functions, often in closely integrated or choreographed ways. Ribozymes, however, are the one-trick ponies of the molecular world. Typically, they can perform one subfunction of the several coordinated functions that a corresponding enzyme can perform. But they cannot perform the entire range of necessary functions, nor can they do so with the specificity needed to execute the many sequentially coordinated reactions that occur during translation.

Consider what ribozymes must do to rival the capacities of the synthetases that catalyze aminoacylation, which occurs between tRNA molecules and their "conjugate" amino acids during translation in actual cells. Researchers have demonstrated that certain RNA molecules can bind a protein-forming amino acid, phenylalanine, to itself, thus performing the function of aminoacylation. They have even isolated a version of the RNA catalyst that binds only phenylalanine, achieving a specificity of sorts. But the synthetase enzymes responsible for aminoacylation in life must catalyze a complex two-stage chemical reaction involving three kinds of molecules: amino acids, ATP (adenosine triphosphate), and tRNAs.

In the first stage of this reaction, synthetases couple ATP to a specific amino acid, giving it the stored energy (in the form of adenosine monophosphate, AMP) needed to establish a bond with a tRNA molecule. Next, synthetases couple specific tRNA molecules to specific activated (AMP-charged) amino acids. These tRNAs have specific shapes and anticodon sites that enable them to bond to mRNA at the ribosome. Thus, synthetases help form molecular complexes with a specificity of fit and with specific binding sites that enable translation to occur in the context of a whole system of associated molecules.

The RNA catalyst proposed as a precursor to the synthetase cannot do this. It does not couple ATP to amino acids as a precursor to catalyzing aminoacylation. Instead, the ribozyme engineer provides "preadenylated" amino acids (amino acids already linked to AMP molecules). Nor does the RNA catalyst couple an amino acid to a specific tRNA with a specific anticodon. The more limited specificity it achieves only ensures that the RNA catalyst will bind a particular amino acid to itself, a molecule that does not possess the specific cloverleaf shape or structure of a tRNA. Moreover, this RNA does not carry an anticodon binding site corresponding to a specific codon on a separate mRNA transcript. Thus, *it has no functional significance within a system of molecules for performing translation*. Indeed, no other system of molecules has even been proposed that could confer functional significance or specificity on the amino acid–RNA complexes catalyzed by the aminoacyl ribozyme.

Thus, even in the one case where ribozyme engineers have produced an RNA-aminoacyl catalyst, the ribozyme in question will not produce a molecule with a functional specificity, or capacity to perform coordinated reactions, equivalent to that of the synthetases used in modern cells. Yet without this specificity and capacity to coordinate reactions, translation—the construction of a sequence-specific arrangement of amino acids from the specific RNA transcript—will not occur.[23]

Similar limitations affect the RNA catalysts that have been shown to be capable of peptidyl-transferase activity (i.e., catalyzing peptide bonds between amino acids). These ribozymes (made of free-standing ribosomal RNA) compare quite unfavorably with the capacities of the protein-dominated *ribosomes* that perform this function in extant cells. For example, researchers have found that free-standing ribosomal RNA can only catalyze peptide-bond formation in the presence of another catalyst. More important, apart from the proteins of the ribosome, free-standing ribosomal RNA does not force amino acids to link together into *linear* chains, which is essential to protein function. (For more details, see the note.)[24]

Why RNA Catalysts Can't Do What True Enzymes Can

There is a fundamental chemical reason for the limited functionality of RNA catalysts—one that casts still further doubt on the RNA-world hypothesis and specifically on its account of the origin of the translation system. Because of the inherent limitations of RNA chemistry,[25] single RNA molecules do not catalyze the coordinated multistep reactions that enzymes, such as synthetases, catalyze. Even if separate RNA catalysts can be found that catalyze each of the specific reactions involved in translation (which is by no means certain), that would leave us very far short of a translation system. Each pony of the RNA world does only its one trick. And even if all the ponies were present together, each one would do only its particular trick separately, decoupled from the others. That's a problem, because producing the molecular complexes necessary for translation requires coupling multiple tricks—multiple crucial reactions—in a closely integrated (and virtually simultaneous) way. True enzyme catalysts do this. RNA and small-molecule catalysts do not.

Here's the chemical backstory. Enzymes couple energetically favorable and unfavorable reactions together into a series of reactions that are energetically favorable overall. As a result, they can drive forward two reactions where ordinarily only one would occur with any appreciable frequency. Water runs downhill because of favorable energetics provided by gravitational force. Water does not run uphill, however, unless there is so much of it that it accumulates and slowly rises up the bank. Whether chemical reactions will occur readily depends upon whether there is enough energy to make them occur. Molecules with enough stored energy to establish new chemical bonds will react readily with one another. Molecules with insufficient stored energy will not react readily with each other unless vast amounts of the reactants are provided (the equivalent of the rising water flooding the banks).

Enzymes use a reaction that liberates energy to drive forward a reaction that requires energy, coupling energetically favorable and unfavorable reactions together. Enzymes can do this because they have a complex three-dimensional geometry that enables them to hold all the molecules involved in each step of the reaction together and to coordinate their interactions. But two independent catalysts cannot accomplish what a compound catalyst (i.e., an enzyme) can. And so far RNA ribozymes have demonstrated the capacity to act only as independent catalysts, not true enzyme catalysts. RNA catalysts might catalyze some energetically favorable reactions, but without the sophisticated active sites of enzymes, they can't couple those favorable reactions to energetically unfavorable reactions.[26] (See Fig. 14.5.)

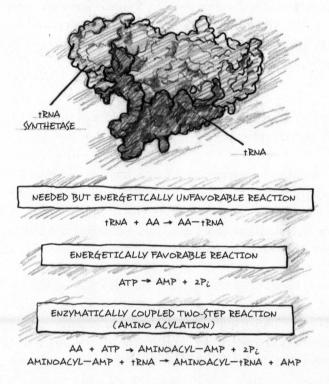

tRNA
SYNTHETASE

tRNA

NEEDED BUT ENERGETICALLY UNFAVORABLE REACTION

tRNA + AA → AA—tRNA

ENERGETICALLY FAVORABLE REACTION

ATP → AMP + 2P$_i$

ENZYMATICALLY COUPLED TWO-STEP REACTION
(AMINO ACYLATION)

AA + ATP → AMINOACYL—AMP + 2P$_i$
AMINOACYL—AMP + tRNA → AMINOACYL—tRNA + AMP

Figure 14.5. Enzymes couple energetically favorable and unfavorable reactions together into a series of reactions that are energetically favorable overall. Enzymes can accomplish this because they have a three-dimensional specificity that allows them to sequester and correctly position all the molecules involved in a series of such reactions. RNA catalysts cannot do this. The figure above shows an enzymatically mediated reaction called aminoacylation. The diagram shows the specificity of fit between a tRNA synthetase and a tRNA molecule during this two-stage chemical reaction. The synthetase links the tRNA to a specific amino acid (AA) using energy from ATP, thus coupling energetically favorable and unfavorable reactions. Amino acids and ATP molecules are not pictured. They would be enveloped by the synthetase during the reactions represented by the chemical equations.

Thus, the demonstration that RNA can catalyze "all the elementary reactions" of translation, but neither the suite of functions nor the co-ordinated functions performed by the necessary enzyme catalysts of the extant translation system, does little to establish the plausibility of ribozyme-based protein synthesis, let alone the transition to enzyme-based protein synthesis, that the RNA-world scenario requires. The inability to account for the origin of the translation system and genetic code, therefore, remains a formidable barrier to the success of the RNA-world hypothesis.

Problem 4: The RNA World Doesn't Explain the Origin of Genetic Information

As I sifted through the primary scientific literature on the RNA-world hypothesis, it did not take me long to realize that the hypothesis faced significant problems quite apart from the central sequencing problem that most interested me. Yet I also realized that it did not resolve the mystery of the origin of biological information—which I had, here-tofore, called the DNA enigma. Indeed, I now realized that I might just as easily have called that mystery the "RNA enigma," because the information problem looms just as large in a hypothetical RNA world as it does in a DNA world. This is not actually surprising. The RNA world was proposed not as an explanation for the origin of biological information, but as an explanation for the origin of the interdependence of nucleic acids and proteins in the cell's information-processing system. And as I studied the hypothesis more carefully, I realized that it presupposed or ignored, rather than explained, the origin of sequence specificity—information—in various RNA molecules.

Consider the step in the RNA-world scenario that I just examined—getting from a primitive replicator to a system for building the first proteins. Even if a system of ribozymes for building proteins had arisen from an RNA replicator, that system of molecules would still need information-rich templates for building specific proteins. RNA-world advocates give no account of the origin of that information beyond vague appeals to chance. But as I argued in Chapters 8–10, chance is not a plausible explanation for the information necessary for building even one protein of modest length, let alone a set of RNA templates for building the proteins needed to establish a protein-based translation system and genetic code.

The need to account for these templates of information stands as a formidable challenge to the RNA world. Nevertheless, the hypothesis faces an even more basic information problem: the first self-replicating RNA molecules themselves would have needed to be sequence-specific in order to perform the function of replication, which is a prerequisite of both natural selection and any further evolution toward cellular complexity.

Though the RNA world was originally proposed as an explanation for the "chicken and egg" functional interdependence problem, not the information problem, some RNA-world advocates nevertheless appear to think that it can somehow leapfrog the sequence-specificity requirement. They imagine short chains (oligomers) of RNA arising by chance on the prebiotic earth. Then, after a sufficiently large pool of these molecules had arisen, some would have acquired the ability to self-replicate. In such a scenario the capacity to self-replicate would then favor the survival of those RNA molecules that could do so and thus would favor the specific sequences that the first self-replicating molecules happened to have. Thus, self-replication arose again as a kind of "accidental choice remembered."[27]

But like Quastler's DNA-first model discussed in the last chapter, this scenario merely shifts the specificity problem out of view. First, for strands of RNA to perform catalytic functions (including self-replication), they, like proteins, must display specific arrangements of their constituent building blocks (nucleotides in the RNA case). In other words, not just any sequence of RNA bases will be capable of self-replication. Indeed, experimental studies indicate that RNA molecules with the capacity to replicate themselves, if they exist at all, are extremely rare among possible RNA base sequences. Although no one has yet produced a fully self-replicating RNA molecule,[28] some researchers have engineered a molecule that can copy a part of itself—though only about 10 percent of itself and then only if a complementary primer strand is provided to the ribozyme by the investigator. Significantly, the scientists selected this partial self-replicator out of an engineered pool of 1,000 trillion (10^{15}) other RNA molecules, almost all of which lack even this limited capacity for self-replication.[29] This suggests that sequences with this capacity are extremely rare and would be especially so within a random (nonengineered) sample.

Further, for an RNA molecule to self-replicate, the RNA strand must be long enough to form a complex structure. Gerald Joyce and the late Leslie Orgel are two prominent origin-of-life researchers who have evaluated the RNA-world scenario in detail. They consider, for the sake of argument, that a replicase could form in a 50-base RNA strand, though they are clearly skeptical that an RNA sequence of this length would really do the job.[30] Experimental results have confirmed their skepticism. Jack Szostak, a prominent ribozyme engineer, and his colleagues have found that it typically takes at least 100 bases to form structures capable of catalyzing simple ligation (linking) reactions. He estimates that getting a ligase capable of performing the other functions that polymerases must perform—"proper template binding, fidelity and strand separation"—may require between 200 and 300 nucleotides.[31] The ribozyme mentioned above—the one that can partially copy itself—required 189 nucleotide bases.[32] It is presently unclear how many bases would be needed to generate enough structural complexity to allow true polymerase function, since no molecule capable of both complete and unassisted self-replication has yet been engineered. It may be as low as 189 bases, but it may be much higher, or it may simply be impossible.[33] Moreover, the problem may be more basic than length. RNA, with its limited alphabet of four bases, may not even have the capacity to form the complex three-dimensional shapes and distributions of charge necessary to perform polymerase or replicase function.

In any case, even if we suppose that RNA-based RNA polymerases (replicases) are possible, experimental evidence indicates that they would have to be information-rich—both complex and specified—just like modern DNA and proteins. Yet explaining how the building blocks of RNA might have arranged themselves into information-rich sequences has proven no easier than explaining how the parts of DNA might have done so, given the requisite length and specificity of these molecules. As Christian de Duve has noted in critique of the RNA-world hypothesis, "Hitching the components together in the right manner raises additional problems of such magnitude that no one has yet attempted to do so in a prebiotic context."[34]

Certainly, appeals to chance alone have not solved the RNA information problem. A 100-base RNA molecule corresponds to a space of possibilities equal to 4^{100} (or 10^{60}). A 200-base RNA molecule corresponds to 4^{200} (or 10^{120}) possibilities. Given this and the experiments mentioned above showing the rarity of functional ribozymes (to say nothing of polymerases) within RNA sequence space, the odds of a functional, self-replicating RNA sequence arising by chance are exceedingly small. Moreover, the odds against such an event occurring are only compounded by the likely presence of destructive cross-reactions between desirable and undesirable molecules within any realistic prebiotic environment.

To make matters worse, as Gerald Joyce and Leslie Orgel note, for a single-stranded RNA catalyst to produce an RNA identical to itself (i.e., to "self-replicate"), it must find an appropriate RNA molecule nearby to function as a template, since a single-stranded RNA cannot function as both replicase and template. Moreover, as they observe, this RNA template would have to be the precise complement of the replicase. Once this chance encounter occurred, the replicase molecule could make a copy of itself by making a complement of its complement (i.e., by transcribing the template), using the physics of nucleotide base pairing.[35]

This requirement, of course, compounds the informational problem facing this crucial step in the RNA-world scenario. Even if an RNA sequence could acquire the replicase function by chance, it could perform that function only if another RNA molecule—one with a highly specific sequence relative to the original—arose close by. (See Fig. 14.6.) Thus, in addition to the specificity required to give the first RNA molecule self-replicating capability, a second RNA molecule with an extremely specific sequence—one with essentially the same specificity as the original—would also have to arise. RNA-world theorists do not explain the origin of the requisite specificity in either the original molecule or its complement. Orgel and Joyce have calculated that to have a reasonable chance of finding two such complementary RNA molecules of a length sufficient to perform catalytic functions would require an RNA library of some 10^{48} RNA molecules.[36] The mass of such a library vastly exceeds the mass of the earth, suggesting the extreme implausibility of the chance origin of a primitive replicator system. They no doubt vastly underestimate the necessary size of this library and the actual improbability of a self-replicating couplet of

RNAs arising, because, as noted, they assume that a 50-base RNA might be capable of self-replication. (See note for qualifying details.)[37]

Given these odds, the chance origin of even a primitive self-replicating system—one involving a pair of sequence-specific (i.e., information-rich) replicases—seems extremely implausible. And, yet, invoking natural selection doesn't reduce the odds or help explain the origin of the necessary replicators since natural selection ensues only *after* self-replication has arisen. As Orgel and Joyce explain, "Without evolution [i.e., prebiotic natural selection] it appears unlikely that a self-replicating ribozyme could arise, but without some form of self-replication there is no way to conduct an evolutionary search for the first primitive self-replicating ribozyme."[38]

Robert Shapiro has resorted to one of my old standbys—Scrabble letters—to illustrate why neither chance, nor chance and natural selection combined, can solve the sequencing problem in the RNA world. While speaking in 2007 at a private conference on the origin of life, he asked an elite scientific audience to imagine an enormous pile of Scrabble letters. Then he said, "If you scooped into that heap [of letters], and you flung them on the lawn there, and the letters fell into a line which contained the words, 'To be or not to be, that is the question,' that is roughly the odds of an RNA molecule, given no feedback [natural selection]—and there would be no feedback, because it [the RNA molecule] wouldn't be functional until it attained a certain length and could copy itself—appearing on earth."[39]

If neither chance, nor chance and selection, can solve the RNA sequencing problem, can self-organization do the trick? It can't. RNA bases, like DNA bases, do not manifest bonding affinities that can explain their specific arrangements. Thus, no one has even attempted to solve the RNA sequencing problem by proposing a "self-organizational RNA world scenario." Instead, the same kind of evidentiary and theoretical problems emerge whether one proposes that genetic information arose first in RNA or DNA molecules. And every attempt to leapfrog the sequencing problem by starting with supposedly "information-generating" RNA replicators has only shifted the problem to the specific sequences that would be needed to make such replicators functional.

In addition, not only does the origin of RNA self-replication depend upon sequence specificity (information), but the transition from the

RNA-based translation system to the current protein-based translation system would have required at some point the production of more than 100 different proteins, each of which would have in turn required an information-rich nucleic acid to guide its construction.

Once again, the pink stuff was spreading.

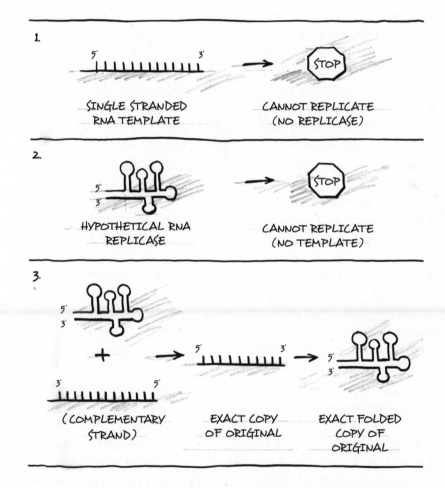

Figure 14.6. The minimal requirements for template-directed RNA self-replication as envisioned by Joyce and Orgel. They insist that any RNA replicase would need to come into close proximity to an exact complementary strand, thus increasing the needed sequence specificity associated with getting such self-replication (and natural selection) started.

Problem 5: Ribozyme Engineering Does Not Simulate Undirected Chemical Evolution

Because of the difficulties with the RNA-world hypothesis and the limited number of enzymatic functions that naturally occurring ribozymes can perform, a new cottage industry has sprung up in molecular biology. Scientists sympathetic to the RNA world have sought to design new RNA molecules with heretofore unobserved functions. In doing so, these scientists have hoped not only to learn more about RNA chemistry, but also to demonstrate the plausibility of the RNA-world hypothesis and possibly even to synthesize an artificial form of life.[40]

These ribozyme-engineering experiments typically deploy one of two approaches: the "rational design" approach or the "directed evolution" approach. In both approaches, biologists try to generate either more efficient versions of existing ribozymes or altogether new ribozymes capable of performing some of the other functions of proteins. In the rational-design approach, the chemists do this by directly modifying the sequences of naturally occurring RNA catalysts. In the directed-evolution (or "irrational design") approach, scientists seek to simulate a form of prebiotic natural selection in the process of producing ribozymes with enhanced functional capacities. To manage this they screen pools of RNA molecules using chemical traps to isolate molecules that perform particular functions. After they have selected these molecules out of the pool, they generate variant versions of these molecules by randomly altering (mutating) some part of the sequence of the original molecule. Then they select the most functional molecules in this new crop and repeat the process several times until a discernible increase in the desired function has been produced.

Most ribozyme-engineering procedures have been performed on ligases, ribozymes that can link together two RNA chains (oligomers) by forming a single (phosphodiester) bond between them. Ribozyme engineers want to demonstrate that these ligases can be transformed into true polymerases or "replicases." These polymerases would not only link nucleotide bases together (by phosphodiester bonds), but also would stabilize the exposed template strands, and use the exposed bases as a template to make sequence-specific copies.

Polymerases are the holy grail of ribozyme engineering. According to the RNA-world hypothesis, once a polymerase capable of template-directed self-replication arose, then natural selection could have become a factor in the subsequent chemical evolution of life. Since ligases can perform one, though only one, of the several functions performed by true polymerases, RNA-world theorists have postulated ligases as the ancestral molecular species from which the first self-replicating polymerase arose. They have tried to demonstrate the plausibility of this conjecture by using ribozyme engineering to build polymerases (or replicases) from simpler ligase ribozymes.

To date, no one has succeeded in engineering a fully functional RNA-based RNA polymerase, from either a ligase or anything else.[41] Ribozyme engineers have, however, used directed evolution to enhance the function of some common types of ligases. As noted, they also have produced a molecule that can copy a small portion of itself. Leading ribozyme engineers such as Jack Szostak and David Bartel have presented these results as support for an *undirected* process of chemical evolution starting in an RNA world.[42] Popular scientific publications and textbooks have often heralded these experiments as models for understanding the origin of life on earth and as the leading edge of research establishing the possibility of evolving an artificial form of life in a test tube.

Yet these claims have an obvious flaw. Ribozyme engineers tend to overlook the role that their own intelligence has played in enhancing the functional capacities of their RNA catalysts. The way the engineers use their intelligence to assist the process of directed evolution would have no parallel in a prebiotic setting, at least one in which only undirected processes drove chemical evolution forward. Yet this is the very setting that ribozyme experiments are supposed to simulate.

RNA-world advocates envision ligases evolving via undirected processes into RNA polymerases that can replicate themselves from free-standing bases, thereby establishing the conditions for the beginning of natural selection. In other words, these experiments attempt to simulate a transition that, according to the RNA-world hypothesis, would have taken place *before natural selection had begun to operate*. Yet in order to improve the function of the ligase molecules, the experiments actually simulate what natural selection does. Starting from a pool of random sequences, the investigators create a chemical trap to isolate only those sequences that evince ligase function. Then they select those sequences for further evolution. Next they use a mutagenesis technique to generate a set of variant versions of these original ligases. Then they isolate and select the best sequences—those manifesting evidence of enhanced ligase function or indications of future polymerase function—and repeat the process until some improvement in the desired function has been realized.

But what could have accomplished these tasks before the first replicator molecule had evolved? Szostak and his colleagues do not say. They certainly cannot say that natural selection played this role, since the origin of natural selection as a process depends on the prior origin of the self-replicating molecule that Szostak and his colleagues are working so hard to design. Instead, in their experiment, Szostak and his colleagues play a role that nature cannot play until a self-replicating system, or at least a self-replicating molecule, has arisen. Szostak and his colleagues function as the replicators. They generate the crop of variant sequences. They make the choices about which of these sequences will survive to undergo another round of directed evolution. Moreover, they make these choices with the benefit of a foresight that neither natural selection nor any other undirected or unintelligent process can—by definition—possess.[43] Indeed, the features of the RNA molecules that Szostak and his colleagues isolate and select are not features that would, by themselves in a precellular context, confer any functional advantage.

Of course, ligase enzymes perform functions in the context of modern cells and in that setting might confer a selectable advantage on the cells that possess them. But prior to the origin of the first self-replicating protocell, ligase ribozymes would not have any functional advantage over any other RNAs. At that stage in chemical evolution, no self-reproducing system yet existed upon which any advantage could be conferred.

The ability to link (ligate) nucleotide chains is, at best, a necessary but not a sufficient condition of polymerase or replicase function. Absent a molecule or, what is more likely, a system of molecules possessing all of the features required for self-replication, nature would not favor any RNA molecule over any other. Natural selection as a process selects only functionally advantageous features and only in self-replicating systems. It passes its blind eye over molecules possessing merely necessary conditions or possible indicators of future function. Moreover, "it" does nothing at all when mechanisms for replication and selection do not yet even exist. In ribozyme-engineering experiments, engineers perform a role in simulating natural selection that undirected natural processes cannot play prior to the commencement of natural selection. Thus, even if ribozyme experiments succeed in significantly enhancing the capacities of RNA catalysts, it does not follow they will have demonstrated the plausibility of an *undirected* process of chemical evolution. Insofar as ribozyme-engineering experiments using a rational-design approach (as opposed to a directed-evolution approach) involve an even more overt role for intelligence, they exemplify the same problem (for example, see note 28).

Conclusion

As I have investigated various models that combined chance and necessity, I have noted an increasing sense of futility and frustration arising among the scientists who work on the origin of life. As I surveyed the literature, it became clear that this frustration had been building for many years. In 1980, Francis Crick lamented, "An honest man, armed with all the knowledge available to us now, could only state that in some sense, the origin of life appears at the moment to be almost a miracle, so many are the conditions which would have had to have been satisfied to get it going."[44] In 1988, the German biochemist and origin-of-life researcher Klaus Dose followed suit with an equally critical assessment of the state of the field. Dose explained that research efforts to date had "led to a better perception of the immensity of the problem of the origin of life on earth rather than to its solution. At present, all discussions on principal theories and experiments in the field either end in a stalemate or a confession of ignorance."[45] After attending a scientific conference on the origin of life in 1989, one of my Cambridge supervisors returned to report, "The field is becoming increasingly populated with cranks. Everyone knows everybody else's theory doesn't work, but no one is willing to admit it about his own."

As I reviewed and evaluated the scientific literature over the ensuing years, I found no reason to amend these assessments. It was demonstrably more reasonable to reject the chance hypothesis than to accept it. Theories relying on necessity awaited the discovery of an oxymoron, namely, "a law capable of producing information"—a regularity that could generate specified *irregularity*. Meanwhile, theories combining law and chance repeatedly begged the question as to the origin of the information they sought to explain. Theorists just spread pink stuff from one place to another, hoping in vain to make it disappear.

Moreover, framing the possibilities differently didn't change the situation. Protein-first theories had failed, giving way to DNA-first theories. DNA-first theories then failed, giving way to RNA-first theories. And now RNA-first theories, like their predecessors, had failed to explain the central question of the origin of the information that living cells require. As Orgel and Joyce concluded in 1993, "The de novo appearance of oligonucleotides [i.e., specifically sequenced RNA bases] on the primitive earth" would have been "a near miracle."[46]

But origin-of-life researchers did not give up. Instead, they had increasingly taken matters into their own hands. As it became more difficult to envision how life could have arisen from the molecular constituents of actual living cells, some evolutionary biologists decided to try to simulate the chemical evolutionary processes on a computer. If explaining how information had first arisen in DNA, RNA, and proteins had proven prohibitively difficult, then, perhaps scientists could at least program computers to show how chance and necessity alone might have done the job in some other hypothetical context. Meanwhile, RNA chemists, frustrated by the embarrassing paucity of proteinlike functions associated with the new miracle molecule, decided that it was time to do a little engineering of their own—"ribozyme engineering" to be precise. Ironically, they sought to demonstrate the power of chance and necessity by intervening—by carefully isolating, selecting, and amplifying those specific features of RNA that could conceivably enhance its capacities.

But as I examined these new approaches,[47] I found them no more convincing than those they were seeking to supplement. Even apart from their limited success, the very fact that these experiments required so much intervention seemed significant. By involving "programming" and "engineering" in simulations of the origin of life, these new approaches had introduced an elephant into the room that no one wanted to talk about, especially not in the methods sections of scientific papers. I began to reflect on the failure of these simulations to explain the DNA enigma apart from the guidance of intelligent scientists. I wondered if they hadn't inadvertently provided evidence for a radically different approach to the problem of biological information. This led me back to where I had started—to the idea of intelligent design and to a consideration of the scientific case in its favor—a case that now I knew I could make.

15

The Best Explanation

My family enjoys a detective show called *Monk,* featuring an obsessive-compulsive detective with almost preternatural powers of observation. Near the end of most episodes, there comes a point where Mr. Monk announces, "I just solved the case. Here's what happened." It's the moment when, in a flash of insight, he at last sees how all the pieces fit together, how all the clues can be explained. The various competing explanations fall away, inadequate to the task, and the one true explanation leaps forth.

When I was studying in Cambridge, I was repeatedly struck by the similarity between the historical method of scientific investigation and the method used by sleuthing detectives, both real and fictional. While learning about how historical scientists use abductive reasoning, I even found a book by the Italian scholar Gian Capretti, who used the fictional Sherlock Holmes detective stories to illustrate how this form of reasoning works.[1]

Historical scientists and detectives alike must gather clues, study each one, and then form tentative hypotheses. With detectives, suspects are identified; with historical scientists, possible causes. Both sleuths and scientists must then weigh the competing possibilities, favored and otherwise, judging their plausibility against a variety of clues and against their own knowledge of how the world works.

Did the murder suspect have the power to strangle the victim?
Is the proposed cause known to be adequate?

Did the suspect have opportunity?
Is there evidence that the cause was in fact present?
Which murder scenario best explains the relevant facts?
Which hypothesis best explains the evidence?

Now, historical scientists, like criminal investigators, are acutely aware that they were not at the "scene of the crime." They did not witness the event or its cause. Instead, they must depend upon circumstantial evidence, the lingering indicators of what happened: the clues, the traces, the signs left behind.

As I first set out to investigate the DNA enigma I was intrigued by an outside-the-box hypothesis that scientists either had long since set aside or now considered *verboten*. Though the concept of design played a seminal role in the foundation of modern science, no one in late-twentieth-century science still regarded intelligent design as a live option for explaining the origin of life. Or so I had assumed. But my encounters with Charles Thaxton, Dean Kenyon, and others introduced me to a body of evidence that seemed—at least—to point toward this radical possibility.

From everyday experience and from what I would soon learn about historical scientific reasoning, I knew that circumstantial evidence could be inconclusive. The same effect can sometimes be produced by more than one cause and, consequently, can be explained in several different ways. As philosopher Charles Sanders Peirce explained, abductive reasoning "often inclines our judgment so slightly toward its conclusion that we cannot say that we believe [the conclusion] to be true; we only surmise that *it may be* so."[2]

To address this uncertainty and to prevent premature leaps to judgment, historical scientists developed a method known as the "method of multiple competing hypotheses," in which they seek to infer the best causal explanation of the evidence in question. As noted in Chapter 7, this method requires a thorough search for and evaluation of various possible causes, leaving—in the best of cases—only one that can explain the relevant facts. Historical scientists proceed in the same manner as the fictional Sherlock Holmes, who, according to Capretti, uses a method of "progressively eliminating hypotheses."[3]

As I investigated the DNA enigma, I consciously followed this method of reasoning. Specifically, I followed and tested my ideas in light of a version of the method that philosopher of science Michael Scriven calls "retrospective causal analysis." Scriven's method distills the key insights of Darwin and Lyell about how to identify past causes and about when a specifically *historical* scientific explanation counted as the "best."

Though inferences about the past can be initially fraught with uncertainty, there nevertheless does come a time in many historical and forensic investigations when the investigator realizes that one hypothesis clearly explains the facts better than any other. And so it was in my case. Yet unlike Mr. Monk, the television detective with a savantlike intellect, I had no single "aha" moment where I suddenly said, "I know what happened." Instead, as I learned more about the DNA enigma, I had a series of insights: about the nature of information; about why the information in DNA transcends physics and chemistry; about why it exceeds the reach of chance; about the real take-home message of various simulation experiments; and about the criteria that make historical explanations *best*. As I used the historical scientific method to evaluate the evidence and competing explanations in light of these criteria, I eventually came to realize that the design hypothesis met each of them. One by one the clues fell into place, the list of suspects dwindled, and the pieces fit together.

Characterizing the Effect

In my research I learned that before historical scientists can evaluate competing explanations of an effect, event, or body of evidence, they usually need to define precisely what it is that needs to be explained.[4] Indeed, part of the historical scientific method of reasoning involves first characterizing what philosophers of science call the *explanandum*—the entity in need of explanation.[5] It's no different for origin-of-life researchers. They need to clearly define what it is they are trying to explain the "origin of."[6]

For this reason, the cell and its contents provide perhaps the most important clues about what happened to cause life to arise. Contemporary biology has shown that cells are a repository of information, and that the information they contain is essential for even minimal biological function. Consequently, origin-of-life researchers have realized that if they were to explain the origin of the first life, they would need to explain how the information in DNA or some other alternative molecular precursor arose.

But what kind of information does DNA possess? This was an important question to answer, because the term "information" denotes at least two distinct concepts: Shannon information, on the one hand, and functionally specified information, on the other.[7] As I discussed in Chapter 4, Shannon's equations allow us to compute the information-carrying capacity of a sequence; they do not determine whether the sequence is meaningful or functional.

DNA and other biological molecules do have large and measurable amounts of information-carrying capacity. But they do not contain just Shannon information; they contain functional information. In virtue of their specific arrangements, the bases in coding regions of DNA and RNA and the amino acids in proteins enable these molecules to perform biological functions. Like the information in machine code or written language, biological information is not just complex; it is also functionally specified.[8]

Thus, to avoid confusion and equivocation, I realized that it was necessary to distinguish:

"information content" from *mere* "information-carrying capacity,"
"specified information" from *mere* "Shannon information," and
"specified complexity" from *mere* "complexity."

The first term in each of these pairs refers to the functional kind of information that DNA possesses. *That* was the kind of information I needed to explain the "origin of."

Assessing the Competing Possible Causes

Having defined the effect in question as specified biological information, I was now in a position to consider the competing explanations of this feature of life to see which explanation, if any, was best. Peter Lipton's seminal *Inference to the Best Explanation* had shown me that the best explanation of an event—whether in the past or in the present—was one that cited a causal difference.[9] I had discovered, further, that historical scientists understandably preferred explanations that posited causes that were known to be capable of producing the effects in question over explanations that posited either no causes or causes lacking such power. Both an earthquake and a bomb can explain the destruction of a building, but only a bomb can explain the presence of charring and shrapnel in the rubble at the scene. Thus, a forensic scientist would likely conclude, in the absence of other evidence, that the bomb best explains the pattern of destruction at the building site. Entities, conditions, or processes that have the capability (or causal powers) to produce the evidence in question constitute better explanations of that evidence than those that do not.[10]

Determining which, among a set of competing explanations, constitutes the best thus depends on knowledge of the causal powers of competing explanatory entities.[11] It is for this reason that the deliberations of historical scientists, like those of good detectives, are inherently comparative. The process of determining the best explanation in the historical sciences necessarily involves generating a list of possible hypotheses, comparing their known (or theoretically plausible) causal powers with respect to the relevant data, and then progressively eliminating potential but inadequate explanations, leaving, in the more illuminating cases, only a single causally adequate explanation. To infer the *best* explanation necessarily implies the need to examine competing explanations. Moreover, as geologist Charles Lyell explained, historical scientists make these judgments of comparative "causal adequacy" based upon their uniform and repeated experience of cause and effect in the present. Historical scientists explain past events "by reference to causes *now* in operation."[12]

The Causal Adequacy of Intelligent Design

As I noted at the close of Chapter 7, I determined that intelligent design should be considered as, at least, a *possible* scientific explanation for the origin of the functionally specified information present in the cell. By intelligent design I mean "the deliberate choice of a conscious, intelli-

gent agent or person to effect a particular outcome, end, or objective."
Clearly, intelligent agents, by their powers of choice, can make a causal
difference in the outcome of events. Moreover, intelligent design qualifies as a presently acting cause—a "cause now in operation"—of the
origin of specified information—the "effect here under study."

I know by introspection of my own mind—my intentions, perceptions, cognitive powers, and deliberations—that I am a conscious
agent. I know from experience that I can—by my own deliberation and
choice—produce specified information. On the entirely reasonable assumption that other people have conscious minds and that they too can
make choices that affect the material world, I continually see evidence
that minds can generate specified information. Programmers write
code. Students ask questions. My wife leaves me a message about whom
to pick up on the way home from work. Intelligent agents produce,
generate, and transmit information all the time. Experience teaches this
obvious truth. Indeed, uniform and repeated experience confirms that
intelligent agents—conscious, rational beings such as ourselves—can
produce information-rich systems, including systems containing digitally encoded, functionally specified information. Intelligent design is
"causally adequate" to produce this effect.

Since intelligent agency has "demonstrated its capacity to produce"
specified information, the "effect of the sort here under study," I concluded that intelligent design must be considered as—at least—a *possible* explanation for the origin of biological information. But was it the
best?

I knew that in order to establish a cause as the best explanation, the
historical scientist must do more than establish that a proposed cause
could have produced the effect in question. He must also provide "evidence that his candidate [cause] was present" and show via "a thorough
search" that there is an *"absence* of evidence" of "other possible causes."
In other words, in addition to meeting a "causal adequacy" condition,
a best explanation must also meet a "causal existence" and/or "causal
uniqueness" condition.

Recall from Chapter 7 that, in practice, meeting the third condition
(causal uniqueness) effectively ensures that the second condition (causal
existence) will also be met. For this reason, there is a simpler way to
think of the problem of historical reconstruction: to qualify as *best* a

historical explanation must cite a uniquely adequate cause—a cause that has *alone* demonstrated the capacity to produce the evidence in question. In such a case—where there is only one known cause of a given effect or body of evidence—the presence of the effect or evidence establishes the past action (and existence) of the cause. The anthropologists who discovered the ancient cave paintings in Lascaux, France, knew of only one cause capable of producing representational art. Consequently, they inferred the past activity and presence of intelligent agents. Moreover, they could make this inference confidently without any other evidence that intelligent agents had been present, because the presence of the paintings alone established the probable presence of the only known type of cause—intelligence—of such a thing. Could there be a similarly strong basis for concluding that an intelligent cause played a role in the origin of biological information?

To answer this question, I made a thorough search for and evaluation of other possible causes in accord with the established canons of historical scientific reasoning, a search recounted in the preceding chapters. As I evaluated the causal adequacy of the other "suspects," I became progressively more intrigued with the design hypothesis. It eventually became clear to me that intelligent design stood as the only known cause of specified information-rich systems and, therefore, that ID provides the best, most causally adequate explanation for the origin of the information necessary to produce the first life. I came to this conclusion for three main reasons.

Reason 1: No Other Causally Adequate Explanations

Despite the "thorough search" described in Chapters 8–14, I found no other causally adequate explanations for the DNA enigma. In my search, I examined the main theories of the origin of life (and/or biological information) exemplifying each of three mutually exhaustive categories of explanation: chance, necessity, and the combination of the two.[13] Of course, there are many specific theories that exemplify one or the other of these three approaches. Yet my examination of the individual theories representing these approaches did not reveal any cause or process capable of producing biologically relevant amounts of specified information. Moreover, the failure of these specific models to explain the origin of biological information often seemed to reflect deeper conceptual or logical limitations.

For example, I discovered that self-organizational laws or processes of necessity cannot generate—as opposed to merely transmit—new information. Laws of nature, by definition, describe highly regular patterns or order, not the aperiodic complexity that characterizes information-rich digital code. This suggested to me that the problems with self-organizational models do not reside in just the details of a specific model, but instead stem from a deeper conceptual incoherence in the basic approach itself. Theories based upon chance face a different, though possibly equally permanent, kind of obstacle. These theories fail because of an inherent limitation in the probabilistic resources of the universe itself.

Theories that combine chance and necessity invariably face a similar dilemma. Since the lawlike processes of necessity do not generate new information, these combination models invariably rely upon chance events to do most, if not all, of the work of producing new information. This problem arises repeatedly for models invoking prebiotic natural selection in conjunction with random events, whether Oparin's theories or various RNA-world scenarios. Since natural selection "selects" for functional advantage, and since functional advantage ensues only *after* the result of a successful random search for functional information, combination models invariably rely upon chance rather than selection to produce new information. Yet these theories face formidable probabilistic hurdles, just as pure chance–based models do.

Nor do these failures seem to be a function of the particular chemical substances involved in the initial stages of information storage or transmission. Neither protein-first, nor DNA-first, nor RNA-first models solve the problem of the origin of specified information. Even pre-RNA-world models that envision information arising in the specific arrangement of smaller molecules only create other insoluble conceptual dilemmas. Either the proposed small-molecule repositories of information are too small to perform biologically significant functions because they lack the ability to form complex folded structures, or scenarios involving small molecules require highly specific arrangements of those molecules. In the former case, no biologically relevant information or processes arise. In the latter case, the scenarios just transfer the information problem into the chemical soup itself.

As I surveyed the landscape of explanatory failures, a clear pattern emerged. Every attempt to explain the origin of biological information either failed because it transferred the problem elsewhere or "succeeded" only by presupposing unexplained sources of information. This displacement problem was particularly evident in computer simulations where positive results depend so obviously on the input of information from intelligent programmers that computer scientists themselves formulated various "no free lunch" theorems and laws of conservation of information, asserting that the information outputs of computer simulations do not exceed (beyond certain probabilistic limits) the informational inputs of the intelligent programmers who designed them.

Rather than treating these explanatory failures as an invitation to still greater flights of theoretical fancy, I began to consider the possibility that nature was telling us something. Perhaps specified information does not arise for free. Perhaps natural processes tend to degrade information, rather than generate it. Messages written in the sand are eventually erased by the waves; old newspapers yellow and eventually crumble without care from archivists; static on the line inevitably interrupts the flow of conversation. Information-rich sequences or systems may maintain their original fidelity over time, but most will show an overall loss as the arrow of time progresses. Information inputs typically exceed (or at best equal) information outputs, unless, of course, intelligent agents have intervened. Ordinary experience confirms this intuition.

Significantly, so did origin-of-life theories and simulations. Every major origin-of-life scenario—whether based on chance, necessity, or the combination—failed to explain the origin of specified information. Thus, ironically, origin-of-life research itself confirms that undirected chemical processes do not produce large amounts of specified information starting from purely physical or chemical antecedents. For this reason, it seemed entirely sensible to think that the conservation laws that computer scientists had devised to describe the flow of information in computational domains applied equally to the larger domain of nature itself. If so, it seemed plausible to think that the informational repositories of life—such as the DNA molecule—were pointing to a source of information beyond the realm of physics and chemistry.

In any case, my long investigation had turned up nothing in the way of materialistic processes with the demonstrated capacity—the proven causal efficacy—to produce the large amounts of specified information necessary to generate a self-replicating organism. Nor was I alone in this conclusion. Leading scientists—Francis Crick, Fred Hoyle, Paul Davies, Freeman Dyson, Eugene Wigner, Klaus Dose, Robert Shapiro, Dean Kenyon, Leslie Orgel, Gerald Joyce, Hubert Yockey, even Stanley Miller—had all expressed skepticism either about the merits of leading theories, the relevance of prebiotic experiments, or both. Even Richard Dawkins, not known for rhetorical restraint in support of evolutionary orthodoxy, candidly admitted in 2008 that "no one knows" how life arose in the first place.[14]

Reason 2: Experimental Evidence Confirms Causal Adequacy of ID

If attempts to solve the information problem only relocated it, and if neither chance, nor physical-chemical necessity, nor the two acting in combination explains the ultimate origin of specified biological information, what does? Do we know of any entity that has the causal powers to create large amounts of specified information?

As I thought about this question, I realized that there was a more positive reason for considering intelligent design as the best explanation for the origin of biological information. Early on, I recognized, based on ordinary experience, that intelligent human agents—in virtue of their rationality and consciousness—have demonstrated the power to produce specified information. But as I investigated the DNA enigma further, I unexpectedly found *experimental* evidence in support of the causal power of intelligent design.

This evidence did not come from weird attempts to detect the paranormal or supernatural. Instead, the evidence came from experiments that pointed to a normal, at least to us, causal power—in particular, the power of our own minds. These experiments inadvertently demonstrated not only the power of mind over matter, but also the *necessity of* a mind to arrange matter into structures relevant to life. I refer, of course, to the simulation experiments of various kinds that scientists have performed for now over fifty years in an attempt to demonstrate the plausibility of some favored origin-of-life scenario. If these experiments were fables, they would have a moral: minds can produce biologically relevant structures and forms of information, but without mind or intelligence little, if any, information arises. Three separate types of experimental results confirm this lesson.

Prebiotic Simulation Experiments

When Stanley Miller conducted his first experiment attempting to simulate the production of amino acids on the early earth, he inaugurated a new form of scientific inquiry. Since then, for over fifty years prebiotic chemists have attempted to simulate the evolution of biologically relevant building blocks—amino acids, sugars, phosphates, and nucleotide bases—from simpler chemicals. Nevertheless, after geochemical evidence established that the early earth probably did not have the reducing atmosphere that Miller first assumed, the relevance of his and many subsequent experiments came into question. If the early earth did not have a reducing atmosphere, then clearly experiments starting with a mixture of reducing gases did not simulate chemical processes at work on the early earth.

Though these simulation experiments failed to demonstrate the plausibility of chemical evolution under realistic prebiotic conditions, they inadvertently demonstrated something else. Assume for the moment that the reducing gases used by Stanley Miller did actually simulate the conditions on the early earth. Would his experimental results establish the plausibility of an undirected process of chemical evolution? Not necessarily. Prebiotic simulation experiments invariably generate biologically irrelevant substances as well as desirable building blocks such as nucleotide bases, sugars, and amino acids. But without investigator intervention, these undesirable by-products react with desirable building blocks to form inert compounds, such as a tar called melanoidin, the curse of the prebiotic chemist.[15] Simulation experiments have repeatedly shown that such destructive chemical processes would have predominated in any realistic prebiotic chemical environment.

To prevent such "interfering cross-reactions," chemists must intervene using various traps and other techniques to isolate and remove chemicals[16] that alter desirable building blocks. For example, as I noted in Chapter 14, the formose reaction that produces ribose sugar also produces many other undesirable chemical by-products that, absent the intervention of a chemist, will react destructively with ribose.[17]

Investigators also artificially manipulate the initial conditions in their experiments. In so doing, they take into account information about—knowledge of—the properties of their reagents. For example, prebiotic chemists typically choose to radiate their chemical mixtures with short-wavelength ultraviolet light because they know that longer-wavelength light degrades the amino acids they are trying to produce.[18]

In these and many other ways, investigators must routinely manipulate chemical conditions both before and after performing "simulation" experiments in order to protect them from destructive naturally occurring processes. These manipulations constitute "profoundly informative intervention[s]."[19] Every choice the investigator makes to actualize one condition and exclude another—to remove one by-product and not another—imparts information into the system. Therefore, whatever "success" these experiments have achieved in producing biologically relevant compounds occurs as a direct result of the activity of the experimentalist—a conscious, intelligent, deliberative mind—performing the experiments.

Thus, these experiments not only fail to simulate an undirected process of chemical evolution, but they actually provide positive evidence for the powers of a guiding hand; they simulate the power of, if not the need for, an intelligent agent to overcome the influences of natural chemical processes—processes that otherwise lead inexorably to biochemical dead ends. In prebiotic simulation experiments, intelligent agents impart information into chemical systems to produce biologically relevant molecules. Therefore, these experiments actually provide positive evidence for the causal adequacy of intelligent design.

Evolutionary Algorithms Demonstrate the Causal Adequacy of ID

Demonstrations of the causal adequacy of intelligent design have inadvertently come from another type of simulation experiment: computer-based evolutionary algorithms. As discussed in Chapter 13, evolutionary algorithms allegedly simulate the creative power of mutation and selection and their ability to generate functional information "from scratch."[20] Nevertheless, some of these programs succeed by the illicit expedient of providing the computer with an information-rich "target sequence" and then treating relatively greater proximity to *future* function (i.e., the target sequence), not actual present function, as a selection criterion. The more recent *Avida* algorithm produces its results without such obviously goal-directed selection. Yet to the extent it succeeds in modeling a realistically biological process (which is itself questionable—see Chapter 13, n. 46), it too relies on several sources of preexisting information—including an information-rich instruction set—all of which came from intelligent computer programmers.

In Chapter 13, I argued that none of these evolutionary algorithms demonstrated the ability of undirected chance and necessity to produce specified information. But the failure of these programs also provides—however much their creators might have intended otherwise—evidence for the causal power of intelligent design.

By selecting for proximity to future function, the Dawkins, Küppers, and Schneider algorithms all utilized a goal-directed search. Yet such foresighted selection has no analogue in nature. In biology, where differential survival depends upon maintaining function, selection cannot occur until new functional structures or sequences actually arise. Natural selection cannot select a nonfunctional sequence or structure based upon the "knowledge" that it may prove useful in the future pending additional alterations.

Nevertheless, what natural selection and mutation lack, intelligent selection—purposive or goal-directed design—provides. Intelligent agents have foresight. They can fix distant goals and arrange both matter and symbols with those goals in mind. They can devise or select material means to accomplish those ends from among an array of possibilities and then actualize those goals in accord with a *pre*conceived design plan or set of functional requirements. They can also actualize intermediate structures and systems in order to execute such plans. Moreover, they can do so without respect to whether such intermediate forms maintain or perform functions along the way. Thus, insofar as evolutionary algorithms set distant goals (target sequences), actualize them using information about them, and do so in a stepwise fashion that ignores considerations of intermediate function, they simulate not the power of undirected selection and mutation but, instead, the powers of mind.

Beyond that, of course, intelligent agents can produce information. And since all evolutionary algorithms require preexisting sources of information provided by designing minds, they show the power—if not the necessity—of intelligent design for a second reason. Rational agents can constrain combinatorial space with information-rich outcomes in mind. In the process of thought, functional objectives precede and constrain the selection of words, sounds, and symbols to generate functional (or meaningful) sequences from among a vast ensemble of meaningless alternative combinations of sounds or symbols.[21] In so doing, minds produce information.

Neither computers by themselves nor the processes of selection and mutation that computer algorithms simulate can produce large amounts of novel information, at least not unless a large initial complement of information is provided. Even *Avida* shows this. To produce a self-replicating digital organism capable of natural selection and mutation, the *Avida* programmers first provided a large digital genome with individual "genes" specifically arranged to ensure the capacity to self-replicate. This information-rich initial condition corresponds to what living organisms must possess to make natural selection possible. Natural selection depends upon the capacity of the organism to replicate a system of different molecules, and this capacity, in turn, derives from preexisting sources of specified information.

Thus, not only do evolutionary algorithms fail to simulate how undirected mutation and selection produce the information necessary to the first life; they actually simulate the opposite. Indeed, computer simulations of the origin of biological information expose limitations in the causal powers of natural selection and mutation that correspond precisely to powers that intelligent agents are known—uniquely—to possess. The causal powers that natural selection lacks—foresight and creativity—are attributes of consciousness and rationality, of purposive intelligence. Where computer simulations depend on these powers—as they surely do to generate functional information and outcomes—they are provided by the knowledge and intelligence of the programmer, not the computer. Thus, like prebiotic simulation experiments, evolutionary algorithms demonstrate the causal power of intelligent design.

Ribozyme Engineering Demonstrates the Causal Adequacy of ID
Conscious intelligence plays the same essential role in ribozyme engineering. Recall that ribozyme engineers attempt to enhance the capacity of RNA catalysts in order to demonstrate the plausibility of the RNA world. In particular, ribozyme engineers want to show that linking enzymes called RNA ligases can acquire true polymerase function, making possible template-directed self-replication.

Yet, as I noted in the previous chapter, ribozyme engineers using an "irrational-design approach" encounter a crucial lacuna that they must use their intelligence to bridge. The irrational-design approach seeks to model a form of prebiotic natural selection to enhance the function of the ligases. Incremental improvements in, or slight additions to, the function of these enzymes are preserved, replicated, amplified, and then selected for further mutation and selection in hopes of eventually producing a polymerase capable of template-directed self-replication. Yet before the emergence of true polymerases, nothing in nature would perform these critical steps (preservation, replication, amplification), even poorly. Absent an enzyme capable of true self-replication, natural selection is not yet a factor.

So what supplies this gap in ribozyme engineering experiments? What causes a molecule possessing merely possible indicators of a future *selectable* function to be preserved? The investigators themselves—the ribozyme engineers. The ribozyme engineers have the foresight to see that ligase capacity, in conjunction with the other capacities of true polymerases, might enable self-replication to proceed. So they select molecules with slightly enhanced ligase capacity. Then they preserve and optimize these molecules. They "enrich by repeated selection and amplification" as one paper puts it.[22] Moreover, they intervene in this way before any of the other functions that true polymerases perform are fully present. Thus, the investigators anticipate a future function not yet present in the emerging ligase itself. They choose RNA sequences informed by knowledge of the conditions required to actualize that future function of template-directed self-replication. Since nature lacks such foresight, the ribozyme engineer supplies what nature does not. The engineer acts as both replicator and selector—though no molecule capable of acting as a replicator would have yet existed in the early stages of the RNA world.

In the most successful ribozyme engineering experiments, the investigators help their ligases along in other ways. In nature, polymerases have the capacity to unwind double-stranded DNA molecules before copying them. Ligases cannot do this. So ribozyme engineers provide only single-stranded RNA molecules to the ribozyme so that they can catalyze the ligation of two such strands. The investigators also provide purified reagents, remove chemical substances to prevent unwanted cross-reactions, and stabilize and position the molecules upon which the ribozymes must act.

Each of these manipulations again constitutes an "informative intervention,"[23] since at every crucial stage ribozyme engineers select some options or possible states and exclude others. By using their knowledge of the requirements of polymerase function to guide their search and selection process, ribozyme engineers also impart what Robert Marks calls "active information" with each iteration of replication. Thus, ribozyme-engineering experiments demonstrate the power of—if not, again, the need for—intelligence to produce information—in this case, the information necessary to enhance the function of RNA enzymes.

Intelligence plays even more obvious roles in ribozyme experiments exemplifying the rational-design approach. In one such experiment in 2002,[24] investigators claimed to have produced a self-replicating RNA molecule, though upon close inspection, not an actual RNA polymerase. Instead, using the familiar mechanism of complementary base pairing, the researchers found that they could get a ribozyme ligase to close the gap between two single-stranded pieces of RNA once the strands had bonded to the longer complementary RNA strand provided by the ribozyme. Yet to get the ribozyme to copy itself, even in this rather trivial sense, the scientists themselves had to provide the two complementary *sequence-specific* strands of RNA. In other words, the scientists themselves solved the specified-information problem by sequencing two RNA strands to match the complementary sites on a longer piece of RNA.[25]

Certainly, the familiar mechanism of hydrogen bonding ensured that the strands would bind to the correct section on their complements, at least if they didn't fold up on themselves or bind to other molecules first. But the specific sequence—the information—that allowed this bonding to occur was provided by intelligent agents. In other words, to generate even this trivial form of self-replication (in which a single molecule, not a system of different kinds of molecules, makes a complement of itself), intelligent agents had to provide the critical sequence-specific information. Thus, ribozyme engineering—whether exemplifying "irrational" or "rational" design procedures—also demonstrates the causal adequacy of intelligent design.

The Quiet Cause

Of course I am belaboring the argument for the causal powers of intelligent agency. But I do so to underscore a point that is too often overlooked: evidence for the causal adequacy of intelligence is all around us both inside and outside the lab. Clearly, we all know that intelligent agents *can* create specified information and that information comes from minds. A computer user who traces the information on a screen back to its source invariably comes to a *mind,* that of a software engineer or programmer. The information in a book or newspaper column or an ancient inscription ultimately derives from a writer or scribe—from a mental, rather than a strictly material cause. The case for the causal adequacy of intelligent design should be obvious from our ordinary experience. But for those of us trained in the natural sciences, appeals to the mental realm sound perilously vague, immeasurable, and unscientific. We reflexively discount knowledge about what minds can do derived from introspection and ordinary experience and we instead credit only what we have learned through experimental studies.

But dismissing evidence from common experience can be a mistake. Children learn that fire burns without establishing rigorous experimental controls. Archaeologists and forensic scientists can identify the evidence of fire in the past without having first isolated every variable in a separate experiment to establish that fire chars. Common experience often counts. If the best explanations cite causes that are known to produce the effect in question, and if conscious intelligent agency is known by experience to cause particular effects, then agency should qualify as a causally adequate and therefore possible explanation of such effects. Logically, it should not matter whether we learn about the true cause-and-effect structure of the world from experimental or other types of observations. Nevertheless, given modern scientific sensibilities, experimental results will always seem weightier. And so it seemed to me not only significant, but ironic, that a careful analysis of prebiotic simulation experiments of various kinds invariably revealed that the choices of intelligent agents played an indispensable role in whatever success those experiments had achieved.

As Robert Marks has shown, scientists can now even measure the effect that intelligence produces in these experiments. Recall that Marks himself quantified the amount of active information that a computer program imparts into a system with each iteration as the result of the knowledge provided to it *by the programmer*. Clearly, ribozyme engineering and prebiotic simulations were also making use of informational inputs (Marks's "active information") as the experimenters made choices about which molecules to preserve and discard based upon their own knowledge of desired outcomes.

In sum, the case for the causal adequacy of intelligent agency no longer depends solely on our ordinary experience of agents producing information in software codes or by using natural languages. Experiments attempting to synthesize biologically relevant substances and information-rich molecules have now established the power of—and arguably the need for—intelligent design. The fact that the experimenters were striving mightily to establish the opposite point makes the demonstration all the more noteworthy since any experimental bias would run in the opposite direction.

Reason 3: ID Is the Only Known Cause of Specified Information

The inability of genetic algorithms, ribozyme engineering, and prebiotic simulations to generate information without intelligence reinforced what I had discovered in my study of other origin-of-life theories.[26] Undirected materialistic causes have not demonstrated the capacity to generate significant amounts of specified information. At the same time, conscious intelligence has repeatedly shown itself capable of producing such information. It follows that mind—conscious, rational intelligent agency—what philosophers call "agent causation," now stands as the only cause known to be capable of generating large amounts[27] of specified information starting from a nonliving state.

But what reason do we have to suppose that an intelligent agent was around for the origin of life? More specifically, is there any way that *historical scientific reasoning* would lead to such a conclusion. As we saw in Chapter 7, historical scientists must establish not only that a given cause *can* produce the effect in question, but that the cause was actually present in order to do so.[28] A contemporary detective might have eyewitnesses or a security-camera video to establish the presence of a potential suspect (a proposed cause). Historical scientists, of course, must depend on other means of establishing the presence of a cause. In Chapter 7, I described two ways that historical scientists have of meeting this causal-existence criterion.

Of the two, the most direct involves making a thorough evaluation of the possible causal explanations and showing that only one of the competing causes has demonstrated the power to produce the main or salient effect in question. In that case, the historical scientist can infer the presence of the cause in the past.[29] The presence of volcanic ash in the sedimentary record establishes the past presence (and existence) of prior volcanic activity because volcanoes, *and only volcanoes,* are known to produce such ash. When a thorough study of various possible causes turns up only a single adequate cause for a given effect, the candidate cause automatically meets the causal-existence criterion.

Consider two other examples of this form of reasoning, one drawn from forensic science and one from a historical science, planetary geology. Several years ago one of the forensic pathologists from the original Warren Commission, which investigated the assassination of President Kennedy, spoke out to quash rumors about a second gunman firing from "the grassy knoll" in front of the motorcade. The bullet hole in the back of President Kennedy's skull evidenced a distinctive beveling pattern that clearly indicated that the bullet had entered his skull from the rear. The pathologist called the beveling pattern a "distinctive diagnostic," because the pattern indicated a single possible direction of entry. Since a rear entry was the only known cause of the beveling pattern, the pattern allowed the forensic pathologists to diagnose the trajectory of the bullet.[30]

Here's another example of reasoning from an effect to a single known cause. The Martian landscape displays evidence of erosion—trenches and rills—that resemble those produced on earth by moving water. Though

Mars currently has no significant liquid water on its surface, planetary scientists have nevertheless inferred that Mars once had a significant amount of water on its surface in the past. Why? Geologists and planetologists have not observed any cause other than moving water that can produce the kind of erosional features observed on Mars today.

One could fill a bookshelf with such examples, because the inferential procedure of the historical sciences is reasonable and makes intuitive sense: when a thorough search reveals only one type of cause with the power to produce a given effect, investigators can infer that uniquely adequate cause from the effect in question. Logically, one can infer the past existence of a cause from its effect, when the cause is known to be *necessary* to produce the effect in question. If there are no other known causes—if there is only one known cause—of a given effect, then the presence of the effect points unambiguously back to the (uniquely adequate) cause.

For this reason, the specified information in the cell establishes the existence and past action of intelligent activity in the origin of life. Experience shows that large amounts of specified complexity or information (especially codes and languages) *invariably* originate from an intelligent source—from a mind or a personal agent.[31] Since intelligence is the only known cause of specified information (at least starting from a nonbiological source), the presence of specified information-rich sequences in even the simplest living systems points definitely to the past existence and activity of a designing intelligence.[32]

The calculus underlying this inference follows the logically valid method of inference used in the historical and forensic sciences. Since both common experience and experimental evidence affirms intelligent design as a necessary condition (and cause) of information, one can detect (or retrodict) the past action of an intelligence from an information-rich effect—even if the cause itself cannot be directly observed.[33] A pattern of flowers spelling "Welcome to Disneyland" allows visitors to the theme park to detect intelligent activity, even if they did not see the flowers planted or arranged. Similarly, the specified and complex arrangement of nucleotide sequences—the information—in DNA implies the past action and existence of an intelligent cause, even if the past action of the cause cannot be directly observed.

Scientists in many fields recognize the connection between intelligence and information and make inferences accordingly. Archaeologists assume that a scribe produced the inscriptions on the Rosetta Stone. Evolutionary anthropologists establish the intelligence of early hominids from chipped flints that are too improbably specified in form and function to have been produced by natural causes. NASA's search for extraterrestrial intelligence (SETI)[34] presupposes that any specified information embedded in electromagnetic signals coming from space would indicate an intelligent source.[35] As yet, radio astronomers have not found any such information-bearing signals. But closer to home, molecular biologists have identified information-rich sequences and systems in the cell, suggesting, by the same logic, the past existence of an intelligent cause for those effects.

The Cell's Information-Processing System, Revisited

In Chapters 3, 4 and 5, I described the main features of the cell that had to be explained by any theory of the origin of life. In Chapter 4 and throughout the book, I've especially stressed the importance of explaining the origin of the functionally specified information in DNA, RNA, and proteins. But in Chapter 5, I also described another problem facing origin-of-life researchers: the problem of explaining the origin of the functionally integrated information-processing system in the cell.

Recall, for example, that specified information in DNA codes for proteins, but specific proteins are necessary to transcribe and translate the information on the DNA molecule. The proteins in the translation and transcription systems even help to process the genetic information for building other copies of themselves. Proteins are needed for protein synthesis. ATP is needed for ATP synthesis. DNA is needed for ATP synthesis. ATP is needed for DNA synthesis. Figure 5.9 depicted many of these interdependent functional relationships in the cell's information processing system.

In Chapter 14, I described how the RNA world hypothesis had been devised in part to explain just one of the "chicken and egg" problems that confront origin-of-life researchers as a result of such interdependent systems. Since DNA is needed to make proteins and proteins are needed to process the information in DNA, perhaps RNA came first, functioning as both an enzyme catalyst and a nucleic acid. Given its many difficulties, however, it now seems clear that the RNA world has explained neither this nor any other functionally interdependent relationship in the cell's information-processing system. In addition to failing to explain the origin of the specific sequencing necessary to establish self-replication, the RNA world has failed to explain the origin of the translation system and the genetic code, key elements in the cell's overall informational system. Similarly, self-organizational models have not only failed to account for the origin of the genetic text, but also for the associations between nucleotide triplets and their cognate amino acids that are the central feature of the genetic code and, again, an indispensable part of information processing in living systems. Further, as noted in Chapters 9 and 10, the probability of building a self-reproducing system of molecules by chance alone is vastly lower than the already prohibitively small probabilities associated with the chance origin of even a single modest-length protein. Clearly, chance will not suffice to explain the origin of information-processing capability either.

But if neither chance nor necessity (self-organizational forces of attraction) nor the combination of the two (as invoked in the RNA world) can explain the origin of the cell's information-processing system, what does? Is there any other entity, or type of cause, that is known to produce the kind of functionally integrated systems that we see in the cell? Is any other kind of entity known to produce functionally integrated information-processing systems?

Of course there is.

In the early 1990s, I first saw a schematic of the cell's information-processing system that I later simplified for presentation as Figure 5.9. When I saw it, a nickel dropped. It reminded me of a diagram that a systems engineer had placed on my desk years earlier when I worked for the oil company. That diagram was a flow chart choreographing how seismic data (information about the earth's subsurface) needed to be collected, processed, analyzed, and interpreted. At each major step along the way in this information-processing system, there were checkpoints, junctions, and feedback loops determining how information was to be routed and passed back and forth between one party (or computer) and another and what functions were to be performed on it. At the time I called it "a spaghetti diagram" because of all the lines going back and forth between the different boxes, each line representing a functional dependence or interdependence between the different functions, players, and/or computers in the system as a whole.

Of course, electrical wiring diagrams and schematics of computer hardware and software provide even better examples of functionally integrated systems. A typical electronic system has multiple unique components as well as multipart gates, filters, converters, digital signal processors, power supplies, switches, and/or functional "blocks." In a typical electronic system, many of these components or subsystems will depend upon others (directly or indirectly) for their proper function. Electronic systems and circuitry are often characterized by what engineers call "many to one," "one to many," and/or "closed loop" functional interdependencies in which one part of a system is needed by many others and the reverse. Oddly, we find the same kind of integrated functional relationships and coordination inside the cell, especially within its information-processing network.

So why does this matter? Again, it matters because—due to the materialistic sensibilities of our intellectual culture—we often inadvertently overlook obvious sources of knowledge about the cause-and-effect structure of the world. We know from uniform and repeated experience of a kind of cause that is capable of producing functionally integrated multipart systems. Engineers—by virtue of their powers of conscious deliberation and rationality—can produce systems that exemplify multipart functional integration. We also know that intelligent agents can produce complex functionally integrated systems specifically for processing information, whether that information is stored and transmitted in audio or digital form or processed by computers or other forms of electronic circuitry. We also know of no other type of cause that has these capacities. Intelligence is the *only known cause* of complex functionally integrated information-processing systems. It follows, once again, that intelligent design stands as the best—most causally adequate—explanation for this feature of the cell, just as it stands as the best explanation for the origin of the information present in DNA itself.

Conclusion

Since the intelligent design hypothesis meets both the causal-adequacy and causal-existence criteria of a *best* explanation, and since no other competing explanation meets these conditions as well—or at all—it follows that the design hypothesis provides the best, most causally adequate explanation of the origin of the information necessary to produce the first life on earth. Indeed, our uniform experience affirms that specified information—whether inscribed in hieroglyphics, written in a book, encoded in a radio signal, or produced in a simulation experiment—*always* arises from an intelligent source, from a mind and not a strictly material process. So the discovery of the specified digital information in the DNA molecule provides strong grounds for inferring that intelligence played a role in the origin of DNA. Indeed, whenever we find specified information and we know the causal story of how that information arose, we always find that it arose from an intelligent source. It follows that the best, most causally adequate explanation for the origin of the specified, digitally encoded information in DNA is that it too had an intelligent source. Intelligent design best explains the DNA enigma.

By the late 1990s, I had become convinced—at least provisionally—that intelligent design was the only known cause of specified information. As a result, I began to sketch out the case for intelligent design as the best explanation for the DNA enigma. In the years that followed (1998–2003), I published a series of articles arguing that intelligent design provides a better explanation than any competing chemical evolutionary model for the origin of biological information.[36] Since then I have continued to examine additional hypotheses and simulations such as the RNA world and genetic algorithms. The case for intelligent design has grown only stronger. Not only have these new approaches failed to provide an adequate explanation for the origin of biological information; they have strengthened the positive case for design that I had previously formulated.

Though advocates of intelligent design have been labeled by some of their opponents as creationists (even "creationists in cheap tuxedos"!), the case for intelligent design depends, ironically, upon a form of scientific reasoning—namely, uniformitarian reasoning—that creationists have often bitterly opposed. Indeed, the case for intelligent design depends on the uniformitarian method of scientific reasoning that Darwin himself used in formulating his argument in *On the Origin of Species*. In light of this and the evidence considered in the preceding chapters, I eventually answered the question that had first seized my attention back in 1985. I concluded that a rigorous scientific argument for intelligent design could be formulated. This chapter has described exactly how I came to that conclusion and why I think it best.

16

Another Road to Rome

When I was teaching I used to use a combination lock to illustrate something about how we detect intelligent causes, though I did not initially tell my students the point I was trying to make. Instead, I began my illustration by announcing to the class that I wanted to show them that chance alone is not a plausible way of generating specified information and, therefore, that chance is not a good explanation for the origin of the information present in DNA.

In the demonstration I asked students to try to open a lock by guessing the combination. I even told them that they needed to turn the dial on the lock first to the right, then to the left, and then back again to the right past the second number. As I passed the lock around the class, and as student after student failed to find the combination in three random trials, I acted increasingly smug as the demonstration, was apparently, proving my point. Then, as if on cue and just as I was becoming insufferable, a student (say "John") nonchalantly turned the dial three times—right, left, right—and popped the lock open.

The class reacted predictably with laughter and taunting—at least for a while. I feigned shock at the outcome of my demonstration. I had been proven wrong. A random search through a space of possible combinations had produced specified information and had done so rather quickly. Or had it? Invariably, as students had a chance to think about what they had witnessed, someone asked whether the student who opened the lock had really guessed the combination by chance.

Then the accusations started. "Was that for real?" "Was he a plant?" "Are you trying to trick us?"

"Who, me?" I replied. "Why would I do something like that?" As more and more students expressed skepticism, I asked why they suspected me. "After all, even though it was improbable that John would guess the combination, he still could have. There was a chance," I said, protesting my ignorance.

"I understand that," one student said, "but it still seems much more likely that he knew the combination already." A consensus would form as other students began to suspect the same thing.

Eventually, I walked over to the student who had opened the lock and asked him to tell the truth. "Did I tell you the combination before class started?" The student stood up, smiled, and then began to fish a small slip of paper out of his pocket. The class erupted in more laughter as he held up the combination for everyone to see.

After order was restored, I explained the real point of the demonstration. "As you thought about what you saw," I said, "most of you began to suspect something fishy. You rejected the chance hypothesis and, instead, began to suspect that intelligent design had played a role. You suspected that John was able to find the combination only because he already knew it. You suspected that he used information to generate information. But beyond that, you also suspected that he and I had colluded in order to fool you in some way—though for what purpose you may not yet know."

"In other words, you inferred intelligent design—both on John's part in using his knowledge to open the lock and on my part in putting him up to it. Even though I, your eminently trustworthy professor, had assured you that each participant was guessing the combination at random, you still detected intelligent design. Why?"

Of course, by this time my students already knew my answer to this question. They knew that I would say that we infer intelligent design as the best explanation for the origin of specified information, because intelligent activity is the only known cause of it. But I wanted them to understand that there was another way to reach the same conclusion. Specifically, I wanted to introduce my students to the ideas of Bill Dembski, who had recently developed a formal theory of design detection, one that my antics with the combination lock and their own thought processes had—unbeknownst to them—just illustrated. But for them to understand this, they needed to know something of Dembski's theory.

Dembski's Method of Design Detection

In Chapter 8, I described how Dembski's work on chance elimination supplied me with an analytical framework for evaluating chance hypotheses about the origin of biological information. In the years after we first met, Dembski went on to develop a scientific method for detecting designed events—events that not only defied explanation by chance, but that could also properly be attributed to intelligent activity. That is, he developed a formal scientific method for detecting intelligent causes in the echo of their effects. The case that I have developed for intelligent design as the best explanation for the origin of biological information does not depend upon the use of Dembski's method of design detection. Nevertheless, his method does apply to the analysis of biological information, and, when applied rigorously, it reinforces my conclusion. Dembski's road also leads to Rome.

In *The Design Inference*, his groundbreaking book on design detection, Dembski notes that rational agents often infer or detect the prior activity of other minds by the character of the effects they leave behind.[1] (This aspect of his work builds on, but is distinct from, his work on probability, which I discussed and applied in Chapters 8–10.) Archaeologists infer, for example, that rational agents produced the inscriptions on the Rosetta Stone; insurance fraud investigators detect certain "cheating patterns" that suggest intentional manipulation of circumstances rather than natural disasters; cryptographers distinguish between random signals and those that carry intelligently encoded messages. Dembski argues that recognizing the activity of intelligent agents constitutes a common and fully rational mode of inference.[2] "We do it all the time," he used to say to me. In this chapter, I explain Dembski's theory of how we infer design, how it applies to DNA, and show why it reinforces the argument for intelligent design made in the preceding chapters.

The Indicators of Intelligent Design

The Design Inference and subsequent works identify two indicators of intelligent activity by which rational agents recognize the effects of other rational agents and distinguish those effects from the effects of purely undirected material causes. Dembski notes that we invariably attribute events, systems, or sequences that have the joint properties of "complexity" (or small probability) and "specification" to intelligent causes—to design—not to chance or physical-chemical necessity.[3] Complex events or sequences of events are extremely improbable and exhibit an irregular arrangement that defies description by a simple rule, law, or algorithm. A specification is a match or correspondence between an observed event and a pattern or set of functional requirements that we know independently of the event in question. Events or objects are "specified" if they exhibit a pattern that matches another pattern that we know independently. I will return to this idea below.

Though Dembski developed his theory using a specialized terminology and precise mathematical formalism, his theory is actually highly intuitive and can be illustrated with many familiar examples of design detection—including the example I described at the beginning of this chapter. As I explained to my students, their suspicions of design were aroused when they perceived an event that exhibited Dembski's twin requirements of small probability and specification.

After my student "plant" popped open the lock, the other students recognized this as an improbable event. Given that the dial had 40 settings, the probability of his generating the correct combination—R26, L28, R14—by chance in three spins of the dial was just one in 64,000. Though that is not very improbable compared to the probability of producing a protein by chance, it was improbable in relation to the handful of opportunities that I gave my students to find the correct combination.

But beyond that, my students recognized a functional significance in the event—one that other equally improbable spins of the dial did not display. When John (my plant) turned the dial three ways to pop the lock open, the other students realized that the event matched a set of independent functional requirements—the requirements for opening the lock that were set when its tumblers were configured. My students also realized that the sequence of numbers that John had generated *matched* the combination. My students perceived an improbable event that matched an independent pattern and met a set of independent functional requirements. Thus, for two reasons, the event manifested a specification as defined above. Since the event was also improbable, my students correctly suspected intelligent design, just as Dembski's theory implies they should have.

Some other homespun examples of design detection illustrate Dembski's criteria and method. Visitors to Mt. Rushmore in South Dakota infer the past action of intelligent design upon seeing some unusual shapes etched in the rock face. Why? The shapes on the hillside are certainly unusual and irregularly shaped, and thus, in this context, improbable. But beyond that, observers recognize a pattern in the shapes that they know from an independent realm of experience, from seeing the faces of ex-presidents in photographs or paintings. The patterns on the mountain *match* patterns the observers know from elsewhere. Thus, according to Dembski's theory, the faces on Mt. Rushmore exhibit a specification (or are specified). Since these particular shapes are also complex, the faces on Mt. Rushmore point, according to his theory, to intelligent design.

Or consider another example. A few years ago, I entered Victoria Harbor in Canada from the sea and noticed a hillside awash in red and yellow flowers. As I got closer, I naturally and correctly detected intelligent design. Why? I recognized a pattern, an arrangement of flowers spelling "Welcome to Victoria." I thus inferred the past activity of an intelligent cause—the careful planning of gardeners.

According to Dembski's theory, two things triggered and justified my recognition of intelligent design in this case. First, the flowers on the hillside appeared to have been arranged in an exceedingly improbable way. The arrangement of flowers did not, for example, exhibit a simple repetitive pattern that might have revealed it to be the expected or likely expression of an underlying natural process. Further, given all the different ways that the same flowers could have been arranged on the hillside, the specific arrangement represented a highly improbable configuration.

In addition to the improbability of the arrangement, I also recognized a pattern, namely the shapes of English letters, and functional significance in it. Based upon my prior knowledge of English, I realized that the letters had been arranged to meet the independent requirements of

English grammar in order to perform a communication function. Had the flowers been more haphazardly scattered so as to defy pattern recognition, I might justifiably have attributed the arrangement to chance—for example, to random gusts of wind having previously scattered the seed. Had the pattern of colors been less complex, with, for example, the red and yellow flowers being segregated by elevation, the pattern might have been explained by some natural necessity, such as certain types of soils or environments favoring certain types of plants. But since the arrangement was highly improbable and since it also conformed to an independent pattern (it exhibited a "specification"), I inferred intelligent design.

This judgment comports with Dembski's theory. If an object or event is both complex and specified, then we should attribute it to intelligent design. By contrast, Dembski notes that we typically attribute to chance low- or intermediate-probability events that do not exhibit discernible patterns. We typically attribute to necessity highly probable events that recur repeatedly in a regular or lawlike way.

Dembski notes that, as we reason about these different kinds of events, we often engage in a comparative evaluation process that he represents with a schematic he calls the "explanatory filter."[4] The filter outlines a method by which scientists (and others) decide among three different types of attributions or explanations—chance, necessity, and intelligent design—based upon the probabilistic features or "signatures" of various kinds of events. (See Fig. 16.1.) His "explanatory filter" constitutes, in effect, a scientific method for detecting the effects of intelligence. The explanatory filter works, he argues, because it reflects our knowledge of the way the world works. Since experience teaches that complex and specified events or systems invariably arise from intelligent causes, we can infer intelligent design from events that exhibit the joint properties of complexity and specification.

From Chance Elimination to Intelligent Design

Dembski's ideas about design detection developed out of his work on chance elimination. In Chapter 8, I described how Dembski came to recognize that low-probability events by themselves do not necessarily indicate that something other than chance is at work. Improbable events happen all the time and don't necessarily indicate anything other than chance in play—as Dembski had illustrated to me by pointing out that if I flipped a coin one hundred times I would necessarily participate in an extremely improbable event. In that case, any specific sequence that turned up would have a probability of 1 chance in 2^{100}. Yet the improbability of this event did not mean that something other than chance was responsible. To eliminate chance, the observer of the event would need to discern a pattern, such as the coin turning up heads every time.

I made this same point in Chapter 8 with my illustration about "Slick" at the roulette wheel. If Slick spins the roulette wheel 100 times, he necessarily generates a series of outcomes that are highly improbable.

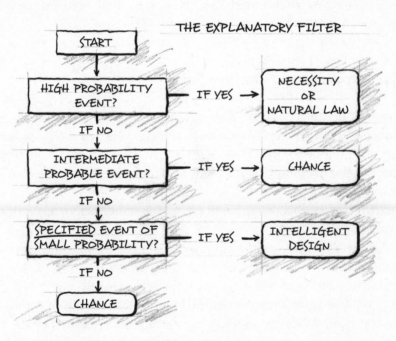

Figure 16.1. Dembski's Original Explanatory Filter. Dembski's filter provides a rational reconstruction of common inferential processes. It suggests that we commonly attribute high-probability events to natural laws (necessity); that we attribute unspecified events of intermediate or low probability to chance; and that we correctly attribute specified events of small probability (complex specified information) to intelligent design. Dembski has since noted that to infer design as the best explanation investigators sometimes must consider explanations that combine chance and law (necessity), as shown in Chapters 13 and 14.

Only if those spins also induced recognition of a pattern—such as the ball landing in the same pocket every time or Slick winning repeatedly—would an observer correctly suspect something other than chance at work.

After Dembski recognized that pattern recognition played a key role in chance elimination, he began to suspect that it also played a key role in the detection of intelligent activity. Though not all of the patterns that helped to justify the elimination of chance also pointed to intelligent design, some did. For example, in my roulette-wheel illustration, the observation that Slick won with every spin of the roulette wheel did suggest that Slick was somehow cheating. Nevertheless, the observation that the ball had landed in the exact same pocket every time did not, by itself, necessarily suggest design.

The Importance of Independence

Dembski began to wonder what it was about some kinds of patterns that helped trigger the recognition of intelligent activity. Why do some patterns indicate a physical or material cause and some an intelligent cause? What distinguishes the patterns that we attribute to lawlike necessity from the patterns that help to indicate intelligent design? As he studied the scientific literature on randomness and pattern recognition, an answer came.

Dembski's answer can be summarized in one word: independence. Patterns that we recognize from an *independent* realm of experience help us to detect design. By contrast, patterns that do not match patterns that we know from another realm of experience indicate nothing about design. When observers see a pattern in an event that exists only in the event itself—or when observers impute patterns to events that do not correspond to patterns they know from elsewhere—the patterns tell us nothing about whether the events in question reflect the activity of another mind. Dembski calls events that match independently known patterns "specifications."[5] He calls patterns that observers merely perceive in the event alone "fabrications."

To see the difference between specifications and fabrications, let's go back to Victoria Harbor. The pattern that I observed in flowers on the hillside did exhibit a pattern that I knew from an independent realm of experience, in particular, from my experience of English vocabulary and grammar. But imagine, instead, that I had fabricated a pattern that did not match any known pattern. Imagine, for example, that I saw several large green bushes randomly scattered in the area just beneath the flower garden. Let's say that I decided to regard the pattern of dots formed by those plants as another way of spelling my name and that I then started to discuss how the gardeners had welcomed me personally to the harbor.

My traveling companions would regard me as irrational, if not crazy. True, I had correctly inferred that the gardeners had intentionally designed the message "Welcome to Victoria." But I had incorrectly inferred that the message was meant for me personally. Why? Rather than detecting an independent pattern in the arrangement of green bushes—one that I recognized from an independent realm of experience—I created a pattern in my mind from my experience of the event—from observing the bushes. Using Dembski's terminology, I fabricated a pattern; I did not detect an independent one.[6]

Dembski devised a test to distinguish these two types of patterns. If observers can recognize, construct, identify, or describe a pattern without observing the event that exemplifies it, then the pattern qualifies as independent from the event. If, however, the observer cannot recognize (or has no knowledge of) the pattern apart from observing the event, then the event does not qualify as independent. Dembski theorized that we reliably detect intelligent activity when we observe events that are both improbable and specified, where events are specified when they conform to a particular kind of pattern, namely, a pattern that can be recognized independently of the occurrence of the event in question.

My roulette-wheel example also illustrates how independent patterns help us to detect design. When the ball repeatedly landed in the same pocket, a pattern did emerge. But in this case the pattern was not complex enough to indicate design. Instead, the croupier observed a highly regular, repetitive pattern in which the same event occurred over and over again with great frequency. This led him to suspect that a physical process or cause was producing the repetitive pattern.[7] And sure enough, he discovered a physical defect in the roulette wheel—an inordinately high lip near the pocket—that was causing the ball to catch and fall into the red 16 pocket every time.

Observing a different type of pattern, however, triggered awareness of a different kind of cause. When Slick repeatedly won the game, despite the ball landing in a variety of different holes, the croupier again detected a pattern. But this time he inferred that Slick was cheating. Why? What was different about this second pattern? The second pattern was different from the first in two ways.

First, rather than the same physical event recurring at the gaming table (with the ball repeatedly landing in the exact same pocket), the croupier observed a complex series of events that defied description by any physical law or mathematical algorithm. Almost every outcome on the game table was different from the one before. The spinning of the wheel generated a complex sequence or pattern, not a repetitive one. The specific sequence he observed was extremely improbable, whereas the repeating sequence of red 16s was highly probable given the defect he discovered in the wheel.

Additionally, the second pattern derived its importance from something independent of the physical events at the roulette table. Observers at the roulette table recognized that the complex sequence of outcomes at the table matched a complex pattern of events that they had noticed independently, in particular, in the sequence of bets that Slick had made. They also recognized a functional significance in the complex pattern of events at the table. They saw that the series of outcomes advanced a *goal* or *objective*—namely, winning money in the game. Because they knew the rules of roulette, and because they observed the bets that Slick made, they recognized a pattern, and indeed a functionally significant one, and detected design. But unlike the other case, in which the pattern was evident simply by observing the event itself, the pattern in this case could be detected only by those who knew something about events and conventions beyond the *physical boundaries* of the roulette wheel.

As Dembski examined cases of pattern recognition that led to the detection of intelligent activity, he began to notice that the same elements were invariably present. A highly improbable event had occurred. Its extreme improbability or complexity defied explanation by reference to a simple lawlike process. And yet, despite the absence of a *regular* pattern, the event clearly seemed significant to the observers because of *something else they knew* from some other realm of experience independent of the event itself.

Often the observers knew of other events that exemplified the same identical pattern as the event in question, in which case, the occurrence of the event made the observer aware of an uncanny coincidence—such as the patterns on Mt. Rushmore matching the exact shape of the ex-presidents' faces from paintings or photographs or such as Slick's bet always matching the hole into which the roulette ball later fell. Other times, observers had knowledge of independent functional requirements or conventions (such as the rules of a game or conventions of grammar) that made them aware of the significance— and, typically, the functional significance—of the complex event in question. For example, the croupier's knowledge of the rules of roulette helped trigger his awareness of the functional significance of the physical outcomes and contributed to his suspicion of design. When I arrived in Victoria Harbor, my knowledge of the conventions of English contributed to my recognition of the pattern and the functional significance of the arrangement on the hillside. Either way, Dembski realized that design detection depended upon observers using knowledge of an independent realm of experience to help them recognize something significant—either functionality or conformity to a pattern—in a complex event.

Two Types of Specifications

Since specifications come in two closely related forms, we detect design in two closely related ways. First, we can detect design when we recognize that a complex pattern of events matches or conforms to a pattern that we know from something else we have witnessed. Tourists at Mt. Rushmore recognize patterns etched in the rock that exactly match patterns that they know from seeing faces of presidents on money or in paintings. Because the shapes carved into the mountain are also highly complex, they correctly infer that they are the product of intelligent activity, rather than undirected processes.

Second, we can detect design when we recognize that a complex pattern of events has a functional significance because of some operational knowledge that we possess about, for example, the functional requirements or conventions of a system. If I observe someone opening a combination lock on the first try, I correctly infer an intelligent cause rather than a chance event. Why? I know that the odds of guessing the combination are extremely low, relative to the probabilistic resources, the single trial available. But beyond that I recognize, based upon my knowledge of how locks work, that the person has produced a functionally significant outcome. And so, I conclude design.

Functional Specifications as Patterns

As Dembski developed his theory, he characterized specification in a general mathematical formalism that subsumed both these cases—the case in which an observer recognizes a pattern in an event and the case in which an observer recognizes a functionally significant outcome in an event. Dembski's mathematical formalism applied to both cases, because there is a sense in which a functionally significant event conforms to a pattern, although one that must be perceived abstractly rather than observed directly.

When I introduced the concept of a specification a few pages back, I described it as a match between an event and an independent pattern *or* as an event that matches or actualizes an independent set of functional requirements. Dembski more commonly uses the first of these two descriptions. Nevertheless, in both his illustrations and his mathematical formalism he treats independent functional requirements as "independent patterns." In Dembski's way of thinking, a functional requirement represents a kind of target and a target is a kind of pattern.

This makes sense. If there are 64,000 three-numeral right-left-right combinations possible for a standard combination lock with 40 dial settings, then the combination that will open the lock represents a tiny target within the abstract space representing all the possibilities. When an event occurs that "hits" such a target, it produces a functionally significant pattern or outcome. Functionally significant outcomes actualize functional requirements, or "hit targets" representing such requirements. And events that actualize independent functional requirements or "hit functional targets" constitute specifications.

Mathematically, Dembski's theory treats events that hit abstract targets in combinatorial space and events that match, or conform to, other kinds of independent patterns as identical. What matters in Dembski's theory is not whether a pattern or target is perceived abstractly or observed directly with the senses. What matters is whether the target or pattern can be known independently of the event that matches or conforms to it. If an abstract target represents a set of outcomes that exemplify real functional requirements, requirements that can be known independently of an event that actualizes them, then the occurrence of such an event is specified.

Here's an illustration that helps to explain why it makes sense to regard functional targets as independent patterns in the sense defined by Dembski. It also shows how Dembski's theory applies to the analysis of specified information and, indeed, identifies it as a reliable indicator of intelligent design. Consider these two strings of characters:

"iuinsdysk]idfawqnzkl,mfdifhs"
"Time and tide wait for no man."

Given the number of possible ways of arranging the letters, spaces, and punctuation marks of the English language for sequences of this length, both these two sequences constitute highly improbable arrangements. Thus, both sequences are complex and meet Dembski's first criterion of a designed system.

Nevertheless, only the second of the two sequences exhibits a specification according to Dembski. To see why, consider that within the set of combinatorially possible sequences, relatively few will convey meaning. This smaller set of meaningful sequences, therefore, delimits a domain (or target or pattern) within the larger set of possibilities. (See Fig. 16.2.) Is this domain or target "independently given" in the way that Dembski's theory requires?

Recall that to qualify as independent, the domain or target must correspond to a known or preexisting pattern or set of functional requirements, not one contrived after the fact of observing the event in question.[8] Notice that the first of the two sequences above —"iuinsdysk]idfawqnzkl,mfdifhs"—does not conform to an "independently given" pattern, or at least any that I recognize. Of course, after observing the sequence "iuinsdysk]idfawqnzkl,mfdifhs" someone simply could declare that it means something such as "It's time to eat lunch." But in that case, the functional significance of the sequence would not have existed independently of the event in question. Instead, the observer would have fabricated, rather than detected, functional significance in the pattern of letters. For that reason, the "meaning" the observer imputed to the sequence would not signify anything about the sequence having originated in another mind.

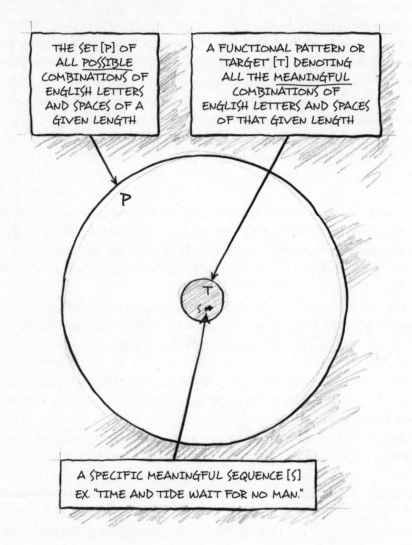

Figure 16.2. A meaningful English phrase or sentence can be thought of as hitting a functional target within a much larger space of possible arrangements and thus is *specified* in the technical sense defined by Dembski's theory of design detection.

But what about the second sequence above? Does it correspond to an independent set of functional requirements or conventions? It does. It conveys meaning. And to convey meaning in English one must employ preexisting or independent conventions of vocabulary (associations of symbol sequences with particular objects, concepts, or ideas) as well as syntax and grammar (such as "Every sentence requires a subject and a verb"). Indeed, when arrangements of symbols utilize, exemplify, or match independent vocabulary and grammatical conventions (functional requirements), meaningful communication can occur in English.

Clearly, all of the English sentences that define the smaller domain or target in Figure 16.2 actualize or utilize existing conventions of vocabulary and requirements of grammar. Since the smaller domain in Figure 16.2 distinguishes functional sequences (English sentences) from nonfunctional sequences, and the functionality of alphabetic sequences depends on independently existing conventions of English vocabulary and grammar, the smaller domain defines an "independently given" target or pattern.

Since the second string of characters ("Time and tide wait . . .") falls within this independently given domain or matches one of the possible meaningful sentences within it (and, therefore, hits a functional target), the second sequence exhibits a specification. Moreover, the functional requirements exemplified by this sequence were not contrived after the fact of observing it in order to impute a meaning or significance to it that it would not possess otherwise. Instead, the second sequence exemplifies a set of known and preexisting requirements and conventions that allow it to convey a message. Thus, this second sequence is specified, according to Dembski's independence condition. Since this second sequence is also complex, it points, according to Dembski's theory, to an intelligent source, just as our intuitions and experience tell us.

In any case, in the most general mathematical expression of Dembski's theory, a specification occurs not only (a) when an object or event matches or exemplifies an independent pattern, but also (b) when an event or object falls within an independently defined domain or hits a functional target, typically, by actualizing an independent set of functional requirements.[9]

Dembski's Criteria Apply to DNA

Dembski's theory of design detection has achieved great prominence in the debate about intelligent design. However, there has long been confusion about whether it applies to biology and, if so, how.

Several years ago, a friend of mine with a background in computer programming was intrigued by the case for intelligent design from DNA. Nevertheless, he was troubled by his inability to make a connection between Dembski's theory of intelligent design and the digital code in DNA. Did DNA manifest Dembski's two criteria of a designed system? He could see that the base sequences in DNA were vastly improbable and, therefore, highly complex. He also understood that the base sequences in DNA performed a critical function by directing protein synthesis in the living cell. But he could not see how DNA was specified in the sense that Dembski had so carefully articulated in *The Design Inference*. Do the base sequences in DNA match a pattern from some other realm of experience? If so, where does that pattern reside?

As I have made my case for intelligent design on university campuses I have encountered this concern more than once. I have even encountered some critics who have rejected the theory because of it. For some, Dembski's work has achieved such prominence that it defines the theory itself. If it does not apply to biology and to DNA specifically, then there is no theory of, or case for, intelligent design in biology. Case closed.

This is a reasonable concern. If Dembski's theory provides a definitive general account of how intelligent design can be detected, as advocates of the theory claim, and if DNA shows evidence of intelligent design as we also claim, then it is certainly reasonable to think that DNA would exemplify Dembski's criteria. So let's look at Dembski's theory in more detail in light of this objection.

Dembski argues that specifications in conjunction with low-probability events reliably indicate design. He describes a specification as a match between an event and a special kind of pattern—patterns that we know independently of, or separately from, the event in question. Does DNA exemplify such a pattern? Do the sequences of bases in DNA match a pattern that we know independently from some other realm of experience? If so, where does that pattern reside?

Based upon a superficial reading of Dembski's work, the answer to these questions might seem to be no. DNA does not exhibit a pattern in its base sequence that matches a pattern scientists have observed from some other realm of experience in the same way, for example, that visitors to Mt. Rushmore recognize a pattern that they have observed elsewhere in photographs or paintings of former presidents.

While certainly we do not *see* any pattern in the DNA molecule that we recognize from having *seen* such a pattern elsewhere, we—or, at least, molecular biologists—do recognize a functional significance in the sequences of bases in DNA based upon something else we know. As discussed in Chapter 4, since Francis Crick articulated the sequence hypothesis in 1957, molecular biologists have recognized that the sequences of bases in DNA produce a functionally significant outcome—the synthesis of proteins. Yet as noted above, events that produce such outcomes are specified, provided they actualize or exemplify independent functional requirements (or "hit" independent functional targets). Because the base sequences in the coding region of DNA do exemplify such independent functional requirements (and produce outcomes that hit independent functional targets in combinatorial space), they are specified in the sense required by Dembski's theory.

To see why, consider the following. The nucleotide base sequences in the coding regions of DNA are highly specific relative to the independent requirements of protein function, protein synthesis, and cellular life. To maintain viability, the cell must regulate its metabolism, pass materials back and forth across its membranes, destroy waste materials, and do many other specific tasks. Each of these functional requirements, in turn, necessitates specific molecular constituents, machines, or systems (usually made of proteins) to accomplish those tasks. As discussed in Chapters 4 and 5, building these proteins with their specific three-dimensional shapes depends upon the existence of specific arrangements of nucleotide bases in the DNA molecule.

For this reason, any nucleotide base sequence that directs the production of proteins hits a functional target within an abstract space of possibilities. As discussed in Chapters 4, 9, and 11, the chemical properties of DNA allow a vast ensemble of possible arrangements of nucleotide bases. Yet within that set of combinatorial possibilities relatively few will—given the way the molecular machinery of the

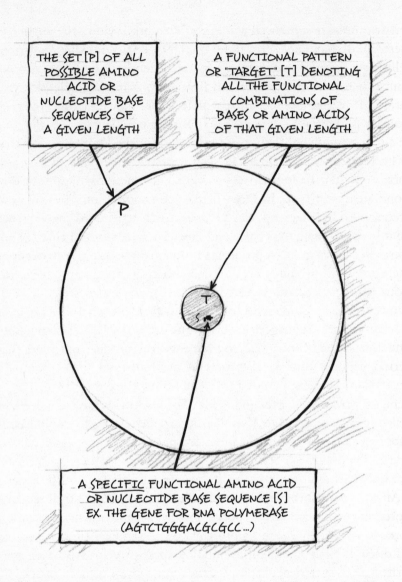

THE SET [P] OF ALL POSSIBLE AMINO ACID OR NUCLEOTIDE BASE SEQUENCES OF A GIVEN LENGTH

A FUNCTIONAL PATTERN OR "TARGET" [T] DENOTING ALL THE FUNCTIONAL COMBINATIONS OF BASES OR AMINO ACIDS OF THAT GIVEN LENGTH

P

T
S

A SPECIFIC FUNCTIONAL AMINO ACID OR NUCLEOTIDE BASE SEQUENCE [S] EX. THE GENE FOR RNA POLYMERASE (AGTCTGGGACGCGCC...)

Figure 16.3. A functionally specified sequence of nucleotide bases or amino acids can be thought of as hitting a functional target within a much larger space of possible arrangements and thus is also *specified* in the technical sense defined by Dembski's theory of design detection. Axe has shown that for a 150-amino-acid-protein domain the ratio of the size of circle P to the size of target T is roughly equal to the ratio of the size of the Milky Way galaxy to the size of a single cotton ball.

gene-expression system works—actually produce functional proteins.[10] (See Fig. 16.3.) This smaller set of functional sequences, therefore, delimits a domain (or target or pattern) within a larger set of possibilities. Moreover, this smaller domain constitutes an independent pattern or target, since it distinguishes functional from nonfunctional sequences, and the functionality of nucleotide base sequences depends on the independent requirements of protein function.

Therefore, any actual nucleotide sequence that falls within this domain or matches one of the possible functional sequences corresponding to it "hits a functional target" and exhibits a specification. Accordingly, the nucleotide sequences in the coding regions of DNA are not only complex, but also specified. Therefore, according to Dembski's theory, the specific arrangements of bases in DNA point to prior intelligent activity, just as I argued in the previous chapter on different, though closely related, grounds.

The Functional Logic of Information Storage and Processing

Dembski's theory applies to the cell's information-processing system as well as to DNA itself. As I noted in Chapters 1 and 4, DNA encodes information by storing it in a digital form. It also uses machinery—indeed, a system of molecular machines—to copy, transmit, and process that information to produce a functional outcome.

All cells use stored digital information to regulate and direct the expression of other digital information (in particular, the information stored in the "protein coding regions" of the DNA molecule). Indeed, as I discuss in more detail in Chapter 18 and in the Epilogue, portions of the genome that many biologists previously regarded as "junk DNA" are now known to perform many important functions, including the regulation and expression of the information for building proteins. In this respect, the nonprotein coding regions of the genome function much like an operating system in a software program, directing and regulating how other information in the system is processed.

In any case, the cell's information-processing system has three key elements: (1) digital storage and encoding of information, (2) machinery for processing that information to produce a functional outcome, and (3) encoding of higher-order (hierarchically arranged) regulatory information. These three key elements for expressing biological information are also found in computer-based information-processing systems. They too (1) encode information digitally, (2) process information with machinery, and (3) use hierarchically organized information to regulate the expression of other information.

These three elements provide a good example of what software designers and other engineers call a design pattern.[11] A design pattern is a general way of solving a design problem. A design pattern distills a functional logic that can be applied and modified in different contexts to solve different but related engineering problems.

DNA and the cell's information-processing system exemplify many design patterns. At the highest level, the way DNA and its information-processing machinery encodes and processes digital information represents a solution to a general design problem: how to store information and transmit it across a communication channel in order to produce a functional outcome. It also represents a solution to many other more specific subproblems such as how to convert one-dimensional digital information into useful three-dimensional structural information, how to copy information with fidelity from one medium to another, how to automate error correction during information processing, how to organize information about related and unrelated functions, and how to access and utilize distributed information.

The design patterns exemplified in the cell's information-processing system also exhibit specifications. Why? Because we recognize design patterns in the cell's information-processing system that match ones we know from an independent realm of experience, in particular, from our own information technology. We recognize a match or, rather, several of them.

We see in the cell's use of nucleotide bases as digital characters a functional motif that we know well from our own information technology. Recall Richard Dawkins's observation that "the machine code of the genes is uncannily computer-like" or Bill Gates's observation that "DNA is like a computer program." We also recognize a design pattern in the way the cell stores, transmits, and processes information. Recall Hubert Yockey's observation that the "genetic code is constructed to confront and solve the problems of communication and recording by the same principles found . . . in modern communication and computer codes."[12] We also recognize a functional pattern in the way the cell uses digital characters to construct three-dimensional mechanical parts. Recall the discussion in Chapter 5 of the airline manufacturing industry's use of CAD-CAM (Computer Assisted Design and Manufacture). And scientists familiar with the logic of software design can recognize many other more specialized design patterns and strategies in the subsystems of the cell's information-processing system.

This point was made to me with particular poignancy one afternoon a couple of years ago. I dropped by to visit a software engineer who was working closely with one of the molecular biologists from the Biologic Institute, a laboratory that we, at the Discovery Institute, helped to start. The software engineer had been studying how the cell processes information in order to write a computer simulation of gene expression. He showed me a book called *Design Patterns,* a standard text for software engineers. The text was full of different design strategies— strategies for processing, storing, copying, organizing, accessing, and correcting digitally encoded strings of information.

My colleague told me that he recognized many of these specific design patterns and strategies at work in the cell. He expressed his awe at the "sophistication of its design logic" and its resemblance to that used in the software industry. He said the cell often employs a functional logic that mirrors our own, but exceeds it in the elegance of its execution. "It's like we are looking at 8.0 or 9.0 versions of design strategies that we have just begun to implement. When I see how the cell processes information," he said, "it gives me an eerie feeling that someone else figured this out before we got here."

According to Dembski's theory, my colleague may have had more than merely an intuitive feeling about the design of the cell's information-processing system. He had two good reasons—specification and low probability—for suspecting design. My colleague recognized in the cell's information-processing system many design patterns that he knew from an independent realm of experience. But he realized that the system of molecules and machines that he had been studying was also extremely complex. Thus, according to Dembski's theory, the digitally encoded information and information-processing system in the cell points to intelligent design, just as my friend's intuition and his expertise had led him to suspect.

Convergence: Complexity and Specification Are Information

When I first met Dembski in the summer of 1992, he and I were attacking the problem of design detection from what we thought were two different directions. He was interested in the logic of chance elimination. As he developed his ideas, he realized first that pattern recognition played a key role in the elimination of chance and then, second, that particular types of patterns, namely, independent patterns, could also help us detect design. His starting point in this analysis was statistical hypothesis testing. Thus, he spoke in the language of probability and kept talking about identifying "small probability specifications" as a way of detecting design.

On the other hand, I had been first impressed in my discussions with Charles Thaxton by the uncanny applicability of information theory to molecular biology. Though I soon realized that Shannon's theory of information could not be equated with meaning or function, I was convinced that Shannon's concept could be supplemented easily to make it useful to biology. Just as English speakers can distinguish meaningful from nonmeaningful arrangements of English letters in virtue of their knowledge of established conventions of grammar and vocabulary, biologists have been able to make determinations about the functional role of DNA base sequences based on knowledge they have acquired of the genetic code and the gene-expression system. As noted in Chapter 4, molecular biologists recognized "specificity"[13] in the arrangement in nucleotide bases and amino acids and defined specificity as what's necessary to produce a particular effect or perform a function.

I suspected that the presence of the extra information-theoretic property of "function" or "specificity" in conjunction with a quantifiable amount of Shannon information might well indicate design. Thus, in my discussions with Dembski, I kept talking about the presence of information content or functional information as possible indicators of design in biology. Nevertheless, Dembski was reluctant to consider information as an indicator of design. He knew that many people confused Shannon information with meaning or function. And he knew that improbability alone, which Shannon's theory measured, did not necessarily indicate design. Thus, Dembski kept proposing the presence of "small probability *specifications*," not information, as the definitive indicator of an intelligent cause.

The next winter, while I was staring at a snowstorm, the obvious resolution to our conflict hit me. Small-probability specification *was* information, at least, the functional kind of information that I was considering as an indicator of design. If an improbable sequence produced a functional outcome, then it was also specified in the sense that Dembski's method required. Dembski and I had been describing the same reality—the same criterion—in different words. Small-probability specification was just another way of saying "Shannon information plus"—either Shannon information plus functionality, or Shannon information plus meaning, or Shannon information plus *specification,* which, in Dembski's theory, subsumed both meaning and function.

This made perfect sense. After all, Shannon's theory of information provided a way to measure the complexity or improbability of a sequence of characters in a communication channel. But since Shannon's equations did not measure or detect meaning or function, determinations about function needed to be made in other ways. I realized that. But I did not initially understand how Dembski's notion of specification related to function. As I came to realize that function constituted a special case of specification—or that specification subsumed function as an instance of it—I realized that we had been talking past each other.

Dembski, who had been previously resistant to using the language of information altogether, now made this equivalence explicit. He began to talk about small-probability specifications and "complex specified information" as the same thing. He used the term "complex" as well as "specified" to modify information, to describe cases where the amount of specified information exceeded available probabilistic resources and thus defied explanation by chance. When I refer to "large amounts" of specified information as I have done in previous chapters, I mean precisely what Dembski means by "*complex* specified information," namely, that the amount of specified information in the system exceeds the amount that can be best explained by chance given the available probabilistic resources.

In any case, the convergence between Dembski's criterion of design detection and mine is evident throughout the examples that I have used to illustrate his method. The flower arrangement in Victoria Harbor not only exhibited complexity and specification, but also contained functional or specified information. The sentence "Time and tide wait for no man" represents not only an improbably specified sequence of characters, but also one that clearly conveys information and performs a communication function.

This equivalence suggests a final way in which Dembski's theory of design applies to biology. Since functionally specified information exhibits complexity and specification, events or objects that possess functionally specified information also necessarily exhibit Dembski's twin indicators of design. Dembski's theory therefore applies to any system that has large amounts of such functional information. Since DNA, RNA, and proteins do have large amounts of functionally specified information, and since even the first simple self-replicating organism would have required large amounts of it, Dembski's theory also implies that the origin of the specified information necessary to build the first living cell is best explained by intelligent design. All roads lead to Rome.

17

But Does It Explain?

Imagine a team of researchers who set out to explore a string of remote islands near Antarctica. After many days at sea, they arrive on an icy, windswept shore. Shouldering their packs, the team hikes inland and eventually takes shelter from the bitter cold in a cave. There, by the light of a small campfire built to cook their freeze-dried rations, they notice a curious series of wedgelike markings vaguely reminiscent of Sumerian cuneiform. It occurs to them that perhaps these scratches in the rock constitute some sort of written language, but dating techniques reveal that the markings are more than five hundred thousand years old, far older than any known human writing and, indeed, far older than anatomically modern human beings.

The researchers investigate other possibilities. Perhaps the markings are animal scratchings. Perhaps they were left by some sort of leeching process or by glacial action, perhaps in conjunction with winds bringing sand through gaps in ice at high speeds. After extensive research by investigators with a broad range of expertise, these and other explanations invoking purely mindless undirected causes fail to explain the evidence. An additional discovery reinforces this conclusion. In a broad cavern farther inside the cave, the explorers find a series of drawings on the walls of various fish, birds, and mammals; geological features such as mountains and valleys; and what appear to be tools of modest sophistication. Each picture has beneath it a sequence of markings, sequences also found among the markings along the walls of the entrance cave. The markings, it seems, are words. The investigators are unable to unravel the function of every word or sequence in the cave writings, and there even appears to be possible misspelled words among them, but eventually it becomes clear that the markings have been arranged in a variety of complex patterns to describe hunting and fishing techniques as well as methods for using the various creatures for food and fuel.

In the process of their painstaking investigation, the explorers make an inference. They note that, although the markings do not reveal the identity of the scribes, they do point to intelligent activity of some kind. The markings reveal a sophisticated system for conveying information, and the only known cause for such a thing is intelligence—conscious rational activity. They conclude that the remote and barren islands were once settled by a group of toolmakers and hunters who employed written, alphabetic language some five hundred thousand years before modern humans were believed to have invented the technique.

The team expects to face skepticism over the dating of the cave markings, and indeed the first wave of doubts focus on this. However, two independent teams, including one using a new dating technique, confirm that the markings are ancient, between four hundred and six hundred and fifty thousand years old. The scientific community soon accepts these findings.

The original researchers think they have now established a solid scientific conclusion. But then some naysayers begin to level a series of philosophical and methodological objections to their work. Some of the critics dispute the conclusion, because they claim it's based merely on our ignorance of any known natural process capable of generating inscriptions. Scientists discover new things all the time, the naysayers point out. Surely it's only a matter of time before one of them discovers a natural cause that can explain the inscriptions. They claim that the research team has made a fallacious "scribe of the gaps" argument.

Another skeptic, a philosopher of science, insists that it's fine to infer that an intelligent agent was at work at times and in places where humans were known to be present, but it's not clear that any human agent was on the scene when these markings were made. Since we don't know of any nonhuman examples of intelligent activity, we can't say anything about the origin of the markings in the cave. Still another skeptic, a famous biologist, poses what he sees as the ultimate stumper. He insists that the team hasn't really explained the origin of the markings at all. "If inscriptions point to an ancient scribe, then who designed the designer of the inscriptions?" he asks. Since presumably the mind who designed the inscriptions was full of information, then invoking an intelligence as the explanation for the information on the cave wall merely pushes the question of the ultimate origin of information back in time, generating an unacceptable and possibly infinite regress.

For these and other reasons, skeptics question the logic of the research team's inference to design. Some insist that the inference to an intelligent cause was not warranted, others that it explained nothing.

As I have presented the argument for intelligent design as the best explanation for the DNA enigma at academic and scientific conferences, I rarely encounter scientists who claim to have a better, or even adequate, explanation for it. In response to the case for design from DNA, I frequently have the experience of debate opponents actually conceding that they do not know how to explain the origin of the information necessary to produce the first life, as indeed leading origin-of-life researchers have often done as well.

Instead, as I have made my case for design as the best explanation for the origin of biological information, I have found that the scientists and philosophers who reject it typically do so on philosophical grounds not unlike the objections raised by the hypothetical naysayers in the preceding parable of discovery. Critics insist that the argument to intelligent design is logically flawed, that it is unwarranted or that it explains nothing.

In this chapter, I defend my case for intelligent design against those who claim that it is, in some way, unwarranted or logically flawed. In the process, I also show how making the case for intelligent design as an inference to the best explanation inoculates it against several common philosophical criticisms.

Argument from Ignorance?

Over the years, I have participated in many debates about the theory of intelligent design at scientific conferences, on university campuses, and on television and radio programs. In nearly every debate, my debate partner has claimed that the case for intelligent design constitutes an argument from ignorance. Arguments from ignorance occur when evidence against a proposition is offered as the sole (and conclusive) grounds for accepting some alternative proposition.

Critics of intelligent design often assert that the case for intelligent design commits this fallacy.[1] They claim that design advocates use our present ignorance of any natural or material cause of specified information as the sole basis for inferring an intelligent cause for the origin of biological information. They accuse ID advocates of arguing for intelligent design based only upon evidence against the adequacy of various natural causes. Since we don't yet know how biological information could have arisen, we invoke the mysterious notion of intelligent design. In this view, intelligent design functions not as an explanation, but as a fig leaf for ignorance.

The inference to design as developed here does not commit this fallacy. True, some of the previous chapters of this book do argue that, at present, all types of material causes and mechanisms fail to account for the origin of biological information from a prebiotic state. And clearly this lack of knowledge of any adequate material cause does provide part of the grounds for inferring design from information in the cell, although it is probably more accurate to characterize this supposed "absence of knowledge" as knowledge of absence, since it derives from a thorough search for alternative materialistic causes and a thorough evaluation of the results of numerous experiments performed over several decades.

In any case, the inadequacy of proposed materialistic causes forms only part of the basis of the argument for intelligent design. We also *know* from broad and repeated experience that intelligent agents can and do produce information-rich systems: we have positive experience-based knowledge of a cause that is sufficient to generate new specified information, namely, intelligence. We are not ignorant of how information arises. We know from experience that conscious intelligent agents can create informational sequences and systems. To quote Quastler again, "The creation of new information is habitually associated with conscious activity."[2] Experience teaches that whenever large amounts of specified complexity or information are present in an artifact or entity whose causal story is known, invariably creative intelligence—intelligent design—played a role in the origin of that entity. Thus, when we encounter such information in the large biological molecules needed for life, we may infer—based on our *knowledge* of established cause-and-effect relationships—that an intelligent cause operated in the past to produce the specified information necessary to the origin of life.

For this reason, the design inference defended here does not constitute an argument from ignorance. Instead, it constitutes an "inference to the best explanation" based upon our best available knowledge.[3] As noted in Chapter 7, to establish an explanation as best, a historical scientist must cite positive evidence for the causal adequacy of a proposed cause. Indeed, unlike an argument from ignorance, an inference to the best explanation does not assert the adequacy of one causal explanation merely on the basis of the inadequacy of some other causal explanation. Instead, it asserts the superior explanatory power of a proposed cause based upon its proven—its *known*—causal adequacy *and* based upon a lack of demonstrated efficacy among the competing proposed causes.

In Chapter 15 I provided evidence for the causal adequacy of intelligent design to account for large amounts of specified information. There I showed that we know from ordinary experience as well as from the results of scientific experiments and computer simulations that intelligent agents do produce large amounts of specified information. Since I had previously shown via a thorough search that no known material process produces this effect, I argued that we can infer design as the best explanation for the origin of information in the cell. The inference to design, therefore, depends on present *knowledge* of the demonstrated causal powers of material entities and processes (inadequate) and intelligence (adequate). It no more constitutes an argument from ignorance than any other well-grounded inference in geology, archaeology, or palaeontology—where present knowledge of cause-and-effect relationships guides the inferences that scientists make about the causes of events in the past.

Formulated as an inference to the best explanation, the argument for design from biological information exemplifies the standard uniformitarian canons of method employed within the historical sciences. The principle of uniformitarianism states that "the present is the key to the past." In particular, it specifies that our *knowledge* of present cause-and-effect relationships should govern how we assess the plausibility of inferences we make about the cause of events in the remote past. Determining which, among a set of competing explanations, constitutes the best depends on *knowledge* of the causal powers of the competing explanatory entities, knowledge that we acquire through our repeated observation and experience of the cause-and-effect patterns of the world.[4] Such knowledge, not ignorance, undergirds the inference to intelligent design from the specified information in DNA.

Arguments from ignorance make an obvious logical error. They omit a necessary kind of premise, a premise providing positive support for the conclusion, not just negative evidence against an alternative conclusion. The case for intelligent design as an inference to the best explanation does not omit that necessary type of premise. Thus, it does not commit the fallacy.

Let's take a closer look. In an explanatory context, arguments from ignorance have the form:

Premise One: Cause X cannot produce or explain evidence E.
Conclusion: Therefore, cause Y produced or explains E.

Critics of intelligent design claim that the argument for intelligent design takes this form as well. As one of my frequent debating partners, Michael Shermer, likes to argue, "Intelligent design . . . argues that life is too specifically complex (complex structures like DNA) . . . to have evolved by natural forces. Therefore, life must have been created by . . . an intelligent designer."[5] In short, critics claim that ID proponents argue as follows:

Premise One: Material causes cannot produce or explain specified information.
Conclusion: Therefore, an intelligent cause produced specified biological information.

If proponents of intelligent design were arguing in the preceding manner, they would be guilty of arguing from ignorance. But the argument made in this book does not assume this form. Instead, it takes the following form:

Premise One: Despite a thorough search, no material causes have been discovered that demonstrate the power to produce large amounts of specified information.

Premise Two: Intelligent causes have demonstrated the power to produce large amounts of specified information.
Conclusion: Intelligent design constitutes the best, most causally adequate, explanation for the information in the cell.

Or to put it more formally, the case for intelligent design made here has the form:

Premise One: Causes A through X do not produce evidence E.
Premise Two: Cause Y can and does produce E.
Conclusion: Y explains E better than A through X.

In addition to a premise about how material causes lack demonstrated causal adequacy, the argument for intelligent design as the best explanation also affirms the demonstrated causal adequacy of an alternative cause, namely, intelligence. This argument does not omit a premise providing positive evidence or reasons for preferring an alternative cause or proposition. Instead, it specifically includes such a premise. Therefore, it does not commit the informal fallacy of arguing from ignorance. It's really as simple as that.

Science and Saying Never

Some might still deny the legitimacy of inferring intelligent design (even as a best explanation), because we are ignorant of what future inquiry may uncover about the causal powers of other natural entities or material processes. Some would characterize the design inference presented here as invalid because it depends on a negative generalization—that is, *purely physical and chemical causes do not generate large amounts of specified information*—a generalization that future discoveries may later falsify. We should "never say never," they say; to do so is a science stopper.[6]

Yet science often says "never," even if it can't say so with absolute certainty. Negative or proscriptive generalizations often play an important role in science. As many scientists and philosophers of science have pointed out, scientific laws often tell us not only what does happen, but also what does not happen.[7] The conservation laws in thermodynamics, for example, proscribe certain outcomes. The first law tells us that energy is never created or destroyed. The second tells us that the entropy of a closed system will never decrease over time. Those who claim that such "proscriptive laws" do not constitute *knowledge* because they are based on past but not future experience will not get far if they try to use their skepticism to justify funding for research on, say, perpetual motion machines.

Further, without proscriptive generalizations, without knowledge of what various possible causes cannot or do not produce, historical scientists could not determine things about the past. As we saw previously, reconstructing the past *requires* making abductive inferences from present effects back to past causal events.[8] Making such inferences requires a progressive elimination of competing causal hypotheses. Deciding which causes can be eliminated from consideration requires knowing what effects a given cause can—and cannot—produce. If historical scientists could never say that particular entities lack particular causal powers, they could never eliminate them from consideration, even provisionally. Thus, they could never infer that a specific cause had acted in the past. Yet historical and forensic scientists make such inferences all the time.

As archaeology, cryptography, and criminal forensics show, we often infer the past activity of an intelligent cause without worrying about committing fallacious arguments from ignorance. And we do so for good reason. A vast amount of human experience shows that intelligent agents have unique causal powers that purely material processes lack. When we observe features or effects that from experience we know only agents produce, we rightly infer the prior activity of intelligence. To determine the best explanation, scientists do not need to say "never" with absolute certainty. They need only say that a postulated cause is best, given what we know at present about the demonstrated causal powers of competing entities. That cause C can produce effect E makes it a better explanation of E than some cause D that has never produced E (especially if D also seems incapable of doing so on theoretical grounds), even if D might later demonstrate causal powers of which we are presently ignorant.[9]

Thus, the objection that the design inference constitutes an argument from ignorance reduces in essence to a restatement of the classical problem of induction, the problem of not knowing whether our gener-alizations about nature based upon past experience will be confirmed by future experience. Yet one could make the same objection against any scientific law or explanation or any historical inference that takes pres-ent knowledge, but not possible future knowledge, of natural laws and causal powers into account. As physicists John Barrow and Frank Tipler note, to criticize design arguments, as Hume did, simply because they assume the uniformity and normative character of natural law cuts just as deeply against "the rational basis of any form of scientific inquiry."[10] Our knowledge of what can and cannot produce large amounts of new information later may have to be revised, but so might the laws of ther-modynamics. Such possibilities do not stop scientists from making gen-eralizations about the causal powers of various entities or from using those generalizations to identify probable or most plausible causes in particular cases. Inferences based on past and present experience con-stitute knowledge (albeit provisional), not ignorance. Thus, those who object to such inferences are objecting not only to the design inference, but to scientific reasoning itself.

The Human Factor

Some critics of intelligent design marshal a more subtle version of the pre-ceding argument. They admit that we may justifiably infer a past human intelligence operating (within human history) from an information-rich artifact or event, but only because we already know that human minds existed during that time. But, they argue, since we do not know (we are ignorant of) whether any intelligent agent existed prior to humans, inferring a designing intelligence that predates humans cannot be justi-fied, even if we observe effects that typically arise only from intelligent agents.[11] This objection asserts, in effect, that since we do not have in-dependent knowledge of the existence of intelligent agents prior to the advent of human beings, the case for intelligent design does not meet the causal-existence requirement of a best explanation.

Though this objection reflects an understanding of the historical sci-entific method, it reflects only a partial understanding of that method. It also overlooks the strong logical basis for the design inference as formulated in Chapter 15. True, historical scientists must meet both adequacy and past-existence conditions in order to establish a causal claim. But, as noted in Chapter 7, one way to meet that causal-existence requirement is to show that there is only one known cause of a given effect. In such a case, the observation of the effect automatically establishes the past existence of the cause and satisfies the causal-existence requirement. Since, as argued in Chapters 8 through 15, intelligence is the only known cause of large amounts of specified information, the presence of such information in the cell points decisively back to the action of a designing intelligence.

In the parable of discovery at the beginning of this chapter, my fictional anthropologists reasoned in this manner. They inferred the existence of a past intelligence prior to the existence of humans because they had discovered an effect that, based upon their repeated experience, had only one known cause. Thus, they inferred the existence and activity of intelligent agents, even though the evidence of that activity predated the origin of anatomically modern humans. Did this evidence provide a basis for affirming an earlier origin for modern humankind? Perhaps. Did it provide the basis for inferring the existence of a nonhuman form of intelligence? They couldn't be sure. But based on their repeated experience—and their introspective awareness of the capacities that we possess that allow us to generate information—they could affirm that a conscious and intelligent mind or minds had acted to produce the specified information—the inscriptions—of interest.

My hypothetical explorers aren't the only scientists to reason in this manner. Actual anthropologists have often revised their estimates for the beginning of human history or particular civilizations, because they have discovered information-rich artifacts dating from times that predated their previous estimates of human or cultural origins. Such inferences to design establish the existence and activity of human agents operating in a time or place in which they were previously unknown. In making such inferences, anthropologists do not initially have *independent* knowledge of the existence of humans from those times or places. Instead, they have only artifacts displaying features that intelligence alone produces. That alone, however, enables them to establish the existence of a prior intelligent cause. The anthropologists do not need independent knowledge of the existence of humans from those earlier times or locales.

Similarly, the scientists searching for extraterrestrial intelligence do not already know that extraterrestrial intelligence exists. Yet they assume that the receipt of specified information or complexity from an extraterrestrial source would indicate the existence of an intelligence in space. In the science-fiction novel *Contact*, scientists detect extraterrestrial intelligence in radio signals carrying the first one hundred prime numbers. In actual SETI research, scientists are looking for more subtle indicators of intelligence, namely, unnaturally modulated and focused radio signals.[12] Either way, SETI does presume that the presence of a complex and specified pattern would provide grounds for suspecting the existence of an intelligence. Moreover, SETI seeks precisely to establish the activity of an intelligent cause in a remote place and from a remote time in which intelligence is currently unknown. If scientific methods can—in principle, at least—detect the presence of an extraterrestrial (and nonhuman) intelligence in a faraway galaxy, why can't methods of design detection be used to establish the activity of nonhuman intelligence in the remote past as the cause of the specified complexity in the cell?

Hume's Objection: A Failed Analogy?

Students of philosophy know about another common objection to intelligent design, since they usually encounter it in their freshmen textbooks. According to many philosophy textbooks, the debate about the design argument was settled by the skeptical philosopher David Hume (1711–76).[13] Hume refuted the classical design argument in biology by showing that it depends on a flawed analogy between living forms and human artifacts. In his *Dialogues Concerning Natural Religion*, Hume admits that artifacts derive from intelligent artificers and that biological organisms have certain similarities to complex human artifacts. Eyes and pocket watches both depend on the functional integration of many precisely configured parts. Nevertheless, he argues, biological organisms also differ from human artifacts—they reproduce themselves, for example—and the advocates of the design argument fail to take these dissimilarities into account. Since uniform experience teaches that organisms always come from other organisms, Hume argues that analogical arguments really ought to suggest that organisms ultimately come from an infinite regress of earlier organisms or from some eternally existent primeval organism (perhaps a giant spider or vegetable), not a transcendent mind.

Despite the assertions of some critics,[14] Hume's objections to the classical design argument fail to refute the argument of this book for several reasons. First, we now know that organisms come from organisms, because organisms possess information-rich macromolecules and a complex information-rich system for processing and replicating the information stored in those molecules. Thus, his argument that uniform experience suggests that organisms necessarily arise from an infinite regress of primeval organisms (or an eternally self-existent one) fails. Repeated experience about the origin of information-rich systems suggests two possibilities, not one. Either information-rich systems arise from preexisting systems of information via a mechanism of replication, or information-rich systems arise from minds. We have repeated experiences of both. Even so, our experience also affirms—based on cases in which we know the cause of such systems—that systems capable of copying and processing other information ultimately arise from intelligent design. After all, the computer hardware that can copy and process information in software originated in the mind of an engineer.

Beyond that, advances in our understanding of planetary and cosmic evolution have ruled out the possibility that biological life has always existed, either on earth or in the cosmos. At some point in the remote past, the conditions on earth and in the larger cosmos were simply incompatible with life. The big-bang theory alone implies that the cosmos itself is finite. Thus, scientifically informed people generally don't argue that biological life always existed or even that it always existed on earth. The question is whether life originated from a purely undirected material process or whether a mind also played a role. Between these two options uniform experience affirms only the latter as an adequate cause for information-rich systems capable of processing and copying information. Since we know that organisms capable of reproduction constitute information-rich systems, a Humean appeal to uniform experience actually suggests intelligent design, not undirected processes, as the explanation for the origin of the first life.

Second, the contemporary case for intelligent design (such as the one made in this book) is not an analogical argument, even though many interesting similarities do exist between living organisms and human information technology. If, as Bill Gates says, "DNA is like a computer program," it makes sense, on analogical grounds, to consider inferring that DNA also had an intelligent source. Nevertheless, although the digitally encoded information in DNA is similar to the information in a computer program, the case for design made here does not depend upon mere similarity. Here's why.

Classical design arguments in biology typically seek to draw analogies between whole organisms and machines based on similar features present in both systems, reasoning from similar effects back to similar causes. These arguments are a bit like those sixth-grade math problems in which students are given a ratio of known quantities on one side of the equation and a ratio of an unknown to a known quantity on the other and then asked to "solve for x," the unknown quantity. In analogical design arguments, two similar effects are compared. In one case, the cause of the effect is known. In the other case, the cause is unknown, but is presumed to be knowable because of the alleged similarity between the two effects. The analogical reasoner "solves for x," in this case, the unknown cause.

The status of such design arguments inevitably turns on the degree of similarity between the systems in question. If the two effects are very similar, then inferring a similar cause will seem more warranted than if the two effects are less similar. Since, however, even advocates of these classical design arguments admit there are dissimilarities as well as similarities between living things and human artifacts, the status of the analogical design argument has always been uncertain. Advocates argued that similarities between organisms and machines outweighed dissimilarities. Critics claimed the opposite.

But the DNA-to-design argument does not have an analogical form. Instead, it constitutes an inference to the best explanation. Such arguments do not compare degrees of similarity between different effects, but instead compare the explanatory power of competing causes with respect to a single kind of effect.

As noted, biological information, such as we find in DNA and proteins, comprises two features: complexity and functional specificity. Computer codes and linguistic texts also manifest this pair of properties ("complexity" and "specificity"), what I have referred to throughout this book as *specified information*. Although a computer program may be similar to DNA in many respects and dissimilar in others, it exhibits a precise identity to DNA insofar as both contain specified complexity or specified information.

Accordingly, the design argument developed here does not rely on a comparison of similar effects, but upon the presence of a single kind of effect—specified information—and an assessment of the ability of competing causes to produce that effect. The argument does not depend upon the *similarity* of DNA to a computer program or human language, but upon the presence of an *identical* feature in both DNA and intelligently designed codes, languages, and artifacts. Because we know intelligent agents can (and do) produce complex and functionally specified sequences of symbols and arrangements of matter, intelligent agency qualifies as an adequate causal explanation for the origin of this effect. Since, in addition, materialistic theories have proven universally inadequate for explaining the origin of such information, intelligence or mind now stands as the only entity with the causal power known to produce this feature of living systems. Therefore, the presence of this feature in living systems points to intelligent design as the best explanation of it, whether such systems resemble human artifacts in other ways or not.

Information as Metaphor: Nothing to Explain?

A related objection is answered in much the same way. Though most molecular biologists see nothing controversial in characterizing DNA and proteins as "information-bearing" molecules, some historians and philosophers of biology have recently challenged that description. The late historian of science Lily Kay characterized the application of information theory to biology as a failure, in particular because classical information theory could not capture the idea of meaning.[15] She suggests that the term "information" as used in biology constitutes nothing more than a metaphor. Since, in Kay's view, the term does not designate anything real, it follows that the origin of "biological information" does not require explanation.[16] Instead, only the origin of the *use* of the term "information" within biology requires explanation. As a social constructivist, Kay explains this usage as the result of various social forces operating within the "Cold War Technoculture."[17] In a different but related vein, philosopher Sahotra Sarkar has argued that the concept of information has little theoretical significance in biology because it lacks predictive or explanatory power.[18] He, like Kay, seems to regard the concept of information as a superfluous metaphor.

Of course, insofar as the term "information" connotes semantic meaning, it does function as a metaphor within biology. That does not mean, however, that the term functions *only* metaphorically or that origin-of-life biologists have nothing to explain. Though information *theory* has a limited application in describing biological systems, it has succeeded in rendering quantitative assessments of the complexity of biomacromolecules. Further, experimental work has established the functional specificity of the base sequences in DNA and amino acids in proteins. Thus, the term "information" as used in biology refers to two real and contingent properties: complexity and functional specificity.

Since scientists began to think seriously about what would be required to explain the phenomenon of heredity, they have recognized the need for some feature or substance in living organisms possessing precisely these two properties together. Thus, Erwin Schrödinger envisioned an "aperiodic crystal";[19] Erwin Chargaff perceived DNA's capacity for "complex sequencing";[20] James Watson and Francis Crick equated complex sequences with "information," which Crick in turn equated with "specificity";[21] Jacques Monod equated irregular specificity in proteins with the need for "a code";[22] and Leslie Orgel characterized life as a "specified complexity."[23] The physicist Paul Davies has more recently argued that the "specific randomness" of DNA base sequences constitutes the central mystery surrounding the origin of life.[24] Whatever the terminology, scientists have recognized the need for, and now know several locations of, complex specificity in the cell, information crucial for transmitting heredity and maintaining biological function. The incorrigibility of these descriptive concepts suggests that specified complexity constitutes a real property of biomacromolecules—indeed, a property that could be otherwise, but only to the detriment of cellular life. Indeed, recall Orgel's observation that "Living organisms are distinguished by their specified complexity. Crystals . . . fail to qualify as living because they lack complexity; mixtures of random polymers fail to qualify because they lack specificity."[25]

The origin of specified complexity, to which the term "information" in biology commonly refers, therefore does require explanation, even if the concept of information connotes only complexity in Shannon information theory, and even if it connotes meaning in common parlance, and even if it has no explanatory or predictive value in itself. Instead, as a descriptive (rather than an explanatory or predictive) concept, the term "information" (understood as specified complexity) helps to define an essential feature of life that origin-of-life researchers must explain "the origin of." So, only where information connotes subjective meaning does it function as a metaphor in biology. Where it refers to complex *functional specificity,* it defines a feature of living systems that calls for explanation every bit as much as, say, a mysterious set of inscriptions on the inside of a cave.

But Who Designed the Designer?

Once when I was explaining the theory of intelligent design on a radio talk show, a caller challenged me with another, now common, objection to the design inference—one that the caller clearly considered a knockdown rebuttal: "If an intelligence designed the information in DNA," he demanded, "then who designed the designer?" I asked for clarification. "Are you arguing that it is illegitimate to infer that an intelligence played a role in the origin of an event unless we can also give a complete explanation of the nature and origin of that intelligence?" Yes, he said, that was exactly what he meant. I then answered as best I could in the available time, but I remember thinking later how facile I thought the objection was. It reminded me of the three-year-old child in the neighborhood where I grew up who used to follow older children around asking them "why" questions. "Why are you going swimming?" "Because it's hot." "Why is it hot?" "Because the sun's out." "Why is the sun out?" "Because there are no clouds today." "Why are there . . ." No matter how you answered, he would ask "why" again, as if in so doing he had somehow invalidated the answer you had just given.

But does the ability to ask about the cause of a cause of an event invalidate a causal explanation? That had always seemed such an obviously flawed idea that I never bothered to refute it in print. Imagine my surprise, then, to learn that Professor Richard Dawkins, holder of the Charles Simony Professorship in the Public Understanding of Science at Oxford University, had advanced precisely that argument as the centerpiece of his case against intelligent design in his bestselling book *The God Delusion*.[26]

There Dawkins argues that the design hypothesis fails to explain anything, because it evokes an infinite regress of questions. If complexity points to the work of a designing intelligence, then who designed the designer? According to Dawkins, the designer would need to be as complex (and presumably as information-rich) as the thing designed. But then he argues, by the logic of the ID advocates, the designer must also be designed. But that would settle nothing, because we then would have to explain the origin of a designing intelligence by reference to a previous designing intelligence, ad infinitum, always leaving unexplained something as mysterious as we started with. Thus, "the design hypothesis" fails, says Dawkins, because it "immediately raises the larger problem of who designed the designer. The whole problem we started out with was the problem of explaining statistical improbability. It is obviously no solution to postulate something even more improbable."[27]

When I read Dawkins's version of this argument, I could see why it sounded plausible to some people. As I thought about it more, I became intrigued by the fascinating philosophical issues it raised. I realized that Dawkins had posed a serious philosophical objection to intelligent design, even though his objection failed for several obvious and fundamental reasons.

Dawkins's objection fails, first, because it does not negate a causal explanation of one event to point out that the cause of that event may also invite a causal explanation. To explain a particular event, historical scientists often cite or infer past events as causes (see Chapter 7). But the events that explain other events presumably also had causes, each of which also invites a causal explanation.[28] Is the original explanation thereby vitiated? Of course not. Pointing out that the past event cited in a causal explanation also has a prior cause—typically, another event—does not render the explanation void, nor does it negate the information it provides about past conditions or circumstances. It merely raises another separate question. Clearly, the young inquisitor in our neighborhood could decide to ask why it was hot after I told him I was going swimming on account of the heat. But his decision to do so did not negate the information he received about my reasons for going swimming or my prior state of mind. A proximate explanation of one event is not negated by learning that it does not supply a comprehensive or ultimate explanation of all the events in the causal chain leading up to it.

Nevertheless, Dawkins's objection to the design hypothesis presupposes precisely the opposite principle, namely, that causal explanations of specific events count as explanations only if there is a separate and comprehensive causal narrative that explains how the cause cited in the explanation itself came into existence from something simpler, and only then if the narrative does not involve an infinite regress of other past causes. Yet Dawkins cannot seriously apply that principle in any other case without absurdity. If applied consistently, Dawkins's principle would, for example, prevent us from inferring design in cases where no one, not even Dawkins, questions the legitimacy of such inferences.[29] One needn't explain who designed the builders of Stonehenge or how they otherwise came into being to infer that this complex and specified structure was clearly the work of intelligent agents. Nor would someone need to know how the scribes responsible for cave markings in my opening parable came into being in order to attribute those inscriptions to a designing intelligence.

Imagine you have traveled to Easter Island to view the famous Moai statues. A child beside you asks no one in particular, "Who carved these statues?" A man standing next to the kid looks over the top of his glasses and asks, "Why do you assume they're sculpted?" Dumbfounded by the question, the kid has no reply, so you rush to his aid. "The carvings manifest a pattern that conforms to the shape of a human face. The match in the patterns is too close and the figures are too intricate, for it to be mere coincidence." The man scoffs. "Don't tell me you've been reading intelligent-design propaganda, all of that rubbish about specified complexity? Let me ask you this: Who sculpted the sculptor? Who designed the designer? Do you see the problem? Your reasoning leads to an infinite regress. Who designed the designer's designer's designer's . . ." The child, appropriately unimpressed by this display of erudition, rolls his eyes and mutters under his breath, "Yeah. But I *know* someone carved these." And, indeed, some*one* did.

The absurdity that results from consistently applying Dawkins's implicit principle of reasoning has invited parody from various quarters. In a popular YouTube mock interview, the fictional Dr. Terry Tommyrot argues that Richard Dawkins is himself a delusion, despite the extensive textual evidence for his existence in his many books. As Dr. Tommyrot asks his interviewer, "If Dawkins designed the books, then who designed the Dawkins? Just tell me that!"[30]

Of course, Dawkins insists that the problem of regress does not afflict properly scientific (read: materialistic) explanations, even explanations involving ordinary human designers. Why? Because as a scientific materialist, Dawkins assumes that physical and chemical processes provide complete materialistic explanations for the thoughts and actions of human agents, and that Darwinian evolution can provide a comprehensive and fully materialistic account of the origin of *Homo sapiens* from earlier and simpler forms of life. Thus, the materialists' answer to the question "Who designed Dawkins?" is "No one." Dawkins descended by material processes from a series of human parents, the first of whom evolved by natural selection and random mutation from lower animal forms, which in turn did the same. Further, the evidence of intelligence in Dawkins's books that points proximately to the activity of a mind, points ultimately to simpler physical and chemical processes in his brain. These processes make his conscious mind—like all human minds—either an illusion or a temporary "epiphenomenon" that has no ability to affect the material world. Material processes can explain everything simply and completely without any appeals to mysterious immaterial minds and without any regress equivalent to that implied in appeals to intelligent design.

But is this true? Is there really such a seamless and fully materialistic account of reality available? Oddly, Dawkins himself has admitted that there is not. As noted in the previous chapter, Dawkins has acknowledged that neither he nor anyone else possesses an adequate explanation for the origin of the first life.

Yet the Darwinian explanation holds that every living thing ultimately evolved from the first self-replicating life-form. Thus, by Dawkins's own logic, one could vitiate the entire edifice of *Darwinian* explanation simply by demanding an explanation for the cause of the cause it cites— that is, the cause of the process of natural selection itself, the origin of the first self-replicating organism. If human life evolved from simpler forms of life, and if biological evolution commences only once a self-replicating organism has arisen, couldn't the skeptic ask, "What evolved the evolver? How did the first self-replicating organisms arise?"

Of course, the lack of a materialistic explanation for the origin of life does not invalidate Darwinian explanations of the origin of higher life-forms. Logically, it's perfectly possible that some unknown, non-Darwinian cause produced the first life, but then natural selection and random mutation produced every living form thereafter.[31] But Dawkins's criterion of a satisfying explanation seems to imply otherwise.

There is an additional problem. Suppose scientists did formulate a completely adequate materialistic explanation for the origin of life. Couldn't a skeptic of Dawkins's materialism still ask for an account of the origin of matter itself? If every material state arose because of the laws of nature acting on a previous material state, then materialistic causal narratives would seem to have their own problems with infinite regress. From whence came the first material state? No physical cosmology now provides a causal explanation of how matter and energy came into being. But suppose one did. How would it do so without invoking a regress of prior material (and/or energetic) states? But what then would become of Dawkins's insistence that causal explanations of particular events fail unless all such regresses are eliminated?

There is still another difficulty with Dawkins's argument. Part of the force of his objection lies in its implicit accusation of inconsistency in the case for intelligent design. If specified complexity always points to intelligent design, then the existence of a designing mind in the past would, by Dawkins's understanding of the logic of the design inference, necessarily point to a still prior designing mind, ad infinitum. In asserting this, Dawkins assumes that designing minds are necessarily complex (and, presumably, specified) entities (itself a questionable proposition).[32] He then argues that advocates of intelligent design can escape the need for an infinite regress only by violating the rule that specified (or irreducible) complexity always points to a prior intelligent cause. Inferring an uncaused designer, he seems to be arguing, would represent an unjustified exception to the principle of cause and effect upon which the inference to design is based.

But positing an uncaused designer would not constitute an *unjustified* exception to this principle, if it constitutes an exception at all.[33] In every worldview or metaphysical system of thought something stands as the ultimate or prime reality, the thing from which everything else comes. All causal explanations either involve an infinitive regress of prior causes, or they must ultimately terminate with explanatory entities that do not themselves require explication by reference to anything more fundamental or primary. If the latter, then something has to stand as the ultimate or primary causal principle at the beginning of each causal chain. If the former—if all explanations inevitably generate regresses—then all explanations fail to meet Dawkins's implicit criterion of explanatory adequacy, including his own. Since, however, most cosmological theories now imply that time itself had a beginning, and further imply that life itself first arose sometime in the finite past, it seems likely that every chain of effect back to cause must terminate at some starting point. Either way, materialistic explanations as well as those involving mind are subject to these same constraints. If so, why couldn't an immaterial mind function as the ultimate starting point for causal explanation just as well as matter and energy?

In Dawkins's worldview, matter and energy must stand as the prime reality from which everything else comes. Thus, Dawkins simply assumes that a material process must function as the fundamental explanatory principle or first cause of biological complexity and information. His "who designed the designer" objection shows this. Why? Dawkins assumes that explanations invoking intelligent design must either generate a regress of designing minds or that such explanations must eventually account for mind by reference to an undirected material process. Either way, Dawkins simply presupposes that mind cannot function as the ultimate explanation of biological complexity and information. For Dawkins and other philosophical materialists, matter alone can play this role. But that begs the fundamental question at issue in the debate about the origin of life.

A more philosophically neutral way to frame the issue would be to simply ask: What is a better candidate to be that fundamental explanatory principle, the thing from which specified complexity or information ultimately comes? What is a better candidate to be the first cause of this phenomenon: mind or matter?

Based upon what we know from our own experience, as opposed to deductions from materialistic philosophical doctrine, the answer to that question would seem to be mind. We have first-person awareness of our own minds. We know from direct introspection what attributes our minds possess and what they can do. Our uniform experience shows that minds have the capacity to produce specified information. Conversely, experience has shown that material processes do not have this capacity. This suggests—with respect to the origin of specified information, at least—that mind is the better candidate to be the fundamental explanatory entity, the thing from which such information comes in the first place.

In any case, explanations invoking intelligent design do not necessarily imply either an infinite regress or the need for further reductionistic accounts of intelligence as a cause. A self-existent immaterial mind might well function as the ultimate cause of biological information, just as prior to the acceptance of the big-bang theory matter functioned as the self-existent entity from which everything else came for philosophical materialists. Intelligent design—defined as a choice of a rational agent to actualize a possibility—might well be a fundamental cause that requires no prior explanatory cause of itself. Agents have the power by their choices to initiate new sequences of cause and effect. Most human agents reflexively assume (and intuitively know) they have this power. Perhaps an uncaused agent with similar powers generated the first biological information. Dawkins cannot foreclose that possibility without first assuming an answer to the question at issue, namely, whether mind or matter stands as the ultimate explanation of biological information. Thus, his "who designed the designer" objection commits, among other errors, the logical fallacy of begging the question. It, therefore, fails as an objection to the design inference based upon DNA.

Conclusion

Over the years, I have encountered each of the objections discussed in this chapter. Initially, I responded to them in great earnest, hoping to persuade the objector. I continue to make every effort to do so, but I'm no longer surprised or disappointed when I don't. Eventually, I realized something odd about all of these objections. None of the alleged logical errors involved in inferring intelligent design from DNA would prevent any reasonable person from inferring or detecting the activity of intelligent agents in any other realm of experience. Even in exotic situations like the one limned in my opening parable or in the Search for Extra-Terrestrial Intelligence, most reasonable people would not dispute the possibility of detecting intelligence. In the case of SETI, many scientists doubt that the program will ever discover extraterrestrial intelligence, since they doubt that our galaxy holds other technological civilizations, but few would question the premise of the search, namely, that we should treat information-rich radio signals as a signature of intelligence. The discovery of specified information alone would suggest antecedent intelligence as the best explanation for the origin of that information, without independent evidence of designing agents existing in the relevant places or times, remote though they might be. Nor when we detect intelligence in more ordinary situations do we worry about making arguments from ignorance, or generating infinite regresses, or running afoul of Hume's critique of analogical reasoning. Neither would we deny that something as interesting as specified digitally encoded information requires explanation.

Instead, in hypothetical and real-world cases, the inference to intelligent design as the best explanation for the origin of specified information is straightforward and unproblematic—except, for some, when considering the origin of life. This suggested to me that perhaps many of these objections to the design inference actually constituted a form of special pleading, perhaps on behalf of a favored idea. Someone once said, "Behind every double standard lies a single hidden agenda." What might that agenda be? As it turns out, I did not have to wait long or do much detective work to find out.

18

But Is It Science?

In December 2005, Judge John E. Jones III ruled that a Dover, Pennsylvania, school district could not tell its biology students about a book in the school library that explained the theory of intelligent design. The judge based his decision on the testimony of expert witnesses—two philosophers, Robert Pennock and Barbara Forrest—who argued that the theory of intelligent design is not scientific *by definition*.[1] Since it is not scientific, the judge reasoned, it must be religious. As he put it in his ruling, "Since ID is not science the conclusion is inescapable that the only real effect of the ID Policy is the advancement of religion."[2] Therefore, he ruled, telling students about the theory of intelligent design would violate the establishment clause of the U.S. Constitution.

Many people have heard about the theory of intelligent design only from news reports about the Dover trial in 2005.[3] Naturally, such reports about the trial and the judge's decision have strongly influenced public perceptions of the theory. For many people, if they know anything at all about the theory, they know—or think they know—that intelligent design is "religion masquerading as science."

I encounter this perception nearly every time I speak about the evidence for intelligent design, whether on university campuses or in the media. As noted previously, as I present the evidence for intelligent design, critics do not typically try to dispute my specific empirical claims. They do not dispute that DNA contains specified information, or that this type of information always comes from a mind, or that competing materialistic theories have failed to account for the DNA enigma. Nor do they even dispute my characterization of the historical scientific method or that I followed it in formulating my case for intelligent design as the best explanation for the evidence. Instead, in addition to the philosophical objections described in the previous chapter, critics simply insist that intelligent design "is just not science," sometimes even citing Judge Jones as their authority.

Since Jones is a lower-level district judge who entered the trial with no apparent background in either science or the history and philosophy of science, and since he made several clear factual errors[4] in his ruling, it would be easy to dismiss his opinion. Jones rendered this response all the more tempting by telling one reporter, apparently in all seriousness, that during the trial he planned to watch the old Hollywood film *Inherit the Wind* for historical background.[5] *Inherit the Wind* is a thinly veiled fictional retelling of the 1925 Scopes "Monkey Trial." But as historian of science Edward Larson has shown in his Pulitzer Prize–winning *Summer for the Gods,* the drama is grossly misleading and historically inaccurate. Clearly, Jones had little, if any, relevant expertise from which to make a judgment about the merits or scientific status of intelligent design.

His opinion, however, reflected a much broader consensus among scientific and academic critics of intelligent design. Indeed, it was later discovered that Jones lifted more than 90 percent of his discussion of "Whether ID Is Science" in his lengthy opinion, virtually verbatim, from an American Civil Liberties Union brief submitted to him before his ruling. The ACLU brief, in turn, recapitulated the most common reasons for challenging the scientific status of intelligent design based upon the testimony of their own expert witnesses.[6] Thus, the Jones opinion and the witnesses who influenced it effectively expressed an entrenched view common not only among members of the media, but within the scientific establishment at large.

I first discovered how deeply entrenched this view is in 1993. I was attending a private conference of scientists interested in the case for intelligent design. In attendance was Dean Kenyon, the evolutionary biologist who had pioneered the self-organizational approach to the origin of life (see Chapter 11). After he repudiated his own theory in the late 1970s, Kenyon eventually came to favor the case for intelligent design. He later announced his change of view in 1985 at the conference in Dallas that first sparked my interest in the origin-of-life debate.

Eight years later, Kenyon was in trouble. He had been removed from teaching introductory biology at San Francisco State University after he explained to his students why he had rejected chemical evolutionary theories of the origin of life—including his own—and why he now favored the design hypothesis. Kenyon explained that the presence of digital information in DNA seemed to provide compelling evidence for the actual, not just apparent, design of the first life. But colleagues insisted that his discussion of intelligent design was inappropriate. Why? Because, in their view, intelligent design did not qualify as a scientific hypothesis.

Reasoning much like Judge Jones in the Dover case, Kenyon's colleagues accused him of forsaking science and bringing religion into the science classroom. They then removed him—a senior, highly published professor with a Stanford Ph.D. in biophysics and a world-class reputation in origin-of-life research—from his introductory biology class and reassigned him to supervising labs.

After hearing about Kenyon's predicament, I wrote an opinion article in hopes of publicizing the attack on his academic freedom. I submitted the article to the *Wall Street Journal* in the fall of 1993. Though the paper accepted it, it held it for nine weeks before deciding to publish it. Then one Friday afternoon in early December, long after I had given up on ever seeing it published, I got a call from an editor at the *Journal*. The article would run on the editorial page the next Monday morning. I called Professor Kenyon with the news. He was shocked. He had just heard that the San Francisco State University faculty senate had decided to hear his case the following Tuesday. On Monday the article was published. On Tuesday, Kenyon won his appeal before the faculty senate by a vote of 25–8—albeit with the representatives of the biology department voting unanimously against him.

Despite the favorable outcome of Kenyon's case, other ID-friendly scientists have experienced similar treatment.[7] Like the judge in the Dover case, many scientists regard the design hypothesis as "unscientific" by definition. In their view, any scientist who advocates the theory of intelligent design has, in their view, broken a fundamental rule of scientific inquiry and may deserve censure, denial of tenure, or termination.

But why isn't the theory of intelligent design scientific? On what basis do critics of the theory make that claim? And is it justified?

In the previous chapters, I argued that intelligent design provided the best explanation for the origin of biological information, and I sought to refute those who claim that intelligent design doesn't provide a logically valid or adequate explanation of this phenomenon. Now I address another objection, perhaps the most pervasive objection of all to intelligent design: the claim that intelligent design doesn't provide a specifically *scientific* explanation—that the theory isn't scientific at all.

A Matter of Definitions?

As a philosopher of science, I've always thought there was something odd and even disingenuous about the objection that intelligent design is not scientific. The argument shifts the focus from an interesting question of truth to a trivial question of definition. To say that an idea or theory does or does not qualify as science implies an accepted definition of the term by which to make that judgment. But to say that a claim about reality "is not science" according to some definition says nothing about whether the claim is true—unless it can be assumed that only scientific theories are true. A definition of science does not, by itself, tell us anything about the truth of competing statements, but only how to classify them (whether as scientific or something else, such as philosophical, historical, or religious statements).

So, at one level, I regarded the debate about whether intelligent design qualifies as science as essentially a semantic dispute, one that distracts attention from significant questions about what actually happened in the past to cause life to arise. Does life exhibit evidence of intelligent design or just apparent design? Did life arise by undirected processes, or did a designing intelligence play a role? Surely such questions are not settled by defining one of the competing hypotheses as "unscientific" and then refusing to consider it.

At another level the debate is tacitly a debate about the basis of the theory itself. Since the term "science" connotes a rigorous experimental or empirical method for studying nature, denying that an idea is scientific implies that rigorous empirical methods played no role in its formulation.

To emphasize this impression, many critics of intelligent design insist that the theory is not testable and, for this reason, is neither rigorous nor scientific.[8] Because many people assume that only "the" scientific method produces justified conclusions, the charge that the theory isn't science seems to justify dismissing it as merely a subjectively based opinion or belief. The objection "ID isn't science" is code for "It isn't true," "It's disreputable," and "There is no evidence for it."

That is why the claim that intelligent design is not science—repeated often and with great presumed authority—has led many to reject it before considering the evidence and arguments for it. I realized that in order to make my case—and open minds to the evidence in favor of it—I needed to defend the theory of intelligent design against this charge. To do so, indeed to defend any theory against this charge and to do so with intellectual integrity, requires one to navigate some treacherous philosophical waters. To claim that intelligent design is science implicitly invokes a definition of science—some understanding of what science is. But which definition?

Because of my background, I knew that historians and philosophers of science—the scholars who study such questions—do not agree about how to define science.[9] Many doubt there is even a single definition that can characterize all the different kinds of science. In the philosophy of science this is known as the "demarcation problem," the problem of defining science and distinguishing (or "demarcating") it from "pseudoscience," metaphysics, history, religion, or other forms of thought or inquiry.

Typically philosophers of science have tried to define science and distinguish it from other types of inquiry (or systems of belief) by studying the methods that scientists use to study nature. But that's where the trouble started. As historians and philosophers of science studied the methods that scientists use, they realized that scientists in different fields use different methods.

This, incidentally, is why historians and philosophers of science are generally better qualified to adjudicate the demarcation question than scientific specialists, such as inorganic chemists, for example. As they say of the catcher in baseball, the philosopher and historian of science has a view of the whole field of play, meaning he or she is less likely to fall into the error of defining all of science by the practices used in one corner of the scientific world. I already had some inkling of this from my work as a geophysicist. I was aware that historical and structural geology use distinct (if partially overlapping) methods. But as I delved into the demarcation question, I discovered that different sciences use a wide variety of methods.

Some sciences perform laboratory experiments. Some do not. Some sciences name, classify, and organize natural objects; some sciences seek to discover natural laws; others seek to reconstruct past events. Some sciences seek to formulate causal explanations of natural phenomena. Some provide mathematical descriptions of natural phenomena. Some sciences construct models. Some explain general or repeatable phenomena by reference to natural laws or general theories. Some study unique or particular events and seek to explain them by reference to past (causal) events.

Some sciences test their theories by making predictions; some test their theories by assessing their explanatory power; some test their theories by assessing both explanatory power and predictive success. Some methods of scientific investigation involve direct verification; some employ more indirect methods of testing. Some test theories in isolation from competing hypotheses. Some test theories by comparing the predictive or explanatory success of competing hypotheses. Some branches of science formulate conjectures that cannot yet be tested at all. Some sciences study only what can be observed. Some sciences make inferences about entities that cannot be observed. Some sciences reason deductively; some inductively; some abductively. Some use all three modes of inference. Some sciences use the hypothetico-deductive method of testing. Some use the method of multiple competing hypotheses.

This diversity of methods has doomed attempts to find a single definition (or set of criteria) that accurately characterizes all types of science by reference to their methodological practices. Thus, philosophers of science now talk openly about the "demise" of attempts to demarcate or define science by reference to a single set of methods.[10]

To say that an idea, theory, concept, inference, or explanation is or isn't scientific requires a particular definition of science. Yet if different scientists and philosophers of science could not agree about what the scientific method is, how could they decide what did and did not qualify as science? And how could I argue that the theory of intelligent design is scientific, if I could not say what I meant by "science"? Conversely, how could critics of intelligent design assert that intelligent design is not science without articulating the standard by which they made this judgment? How could any headway in this debate be made without an agreed-upon definition?

I discovered that though it was difficult to define science by reference to a single definition or set of methodological criteria, it was not difficult to define science in such a way that either acknowledged the diversity of methodological practices or refused to specify which method made a discipline scientific. Such an approach allows science to be defined more broadly as, for instance, "a systematic way of studying nature involving observation, experimentation, and/or reasoning about physical phenomena." So far, so good. The difficulty has come when scholars tried to equate science with *a particular* systematic method of studying nature to the exclusion of other such methods.

The situation was not hopeless, however. I discovered that although it was impossible to describe the rich variety of scientific methods with a single definition, it was possible to characterize the methodological practices of specific disciplines or types of science. This made sense. It was precisely the diversity of scientific methods that made defining science as a whole difficult in the first place. Focusing on a single established scientific method as the relevant standard of judgment eliminated the practical problem of deciding how to assess the scientific status of a theory without an established definition of science. Furthermore, from my own studies, I knew the methodological practices of the sciences directly relevant to the questions I was pursuing—the sciences that investigate the causes of particular events in the remote past. Stephen Jay Gould called these sciences the *historical sciences*.[11] I knew that the inference to design followed from a rigorous application of the logical and methodological guidelines of these disciplines. As shown in Chapter 15, I carefully followed these guidelines in constructing my own case for design, so I concluded that there was a good (if definition-dependent) reason to regard intelligent design as a scientific—and, specifically, historically scientific—theory. In fact, there are several such reasons.

Reason 1: The Case for ID Is Based on Empirical Evidence

The case for intelligent design, like other scientific theories, is based upon empirical evidence, not religious dogma. Contrary to the claims of Robert Pennock,[12] one of the expert witnesses in the Dover trial, design theorists have developed specific empirical arguments to support their theory. In this book, to name just one example, I have developed an argument for intelligent design based on the discovery of digital information in the cell. In addition, other scientists now see evidence of intelligent design in the "irreducible complexity" of molecular machines and circuits in the cell,[13] the pattern of appearance of the major groups of organisms in the fossil record,[14] the fine-tuning of the laws and constants of physics,[15] the fine-tuning of our terrestrial environment,[16] the information-processing system of the cell, and even in the phenomenon known as "homology" (evidence previously thought to provide unequivocal support for neo-Darwinism).[17] Critics may disagree with the conclusions of these design arguments, but they cannot reasonably deny that they are based upon commonly accepted observations of the natural world. Since the term "science" commonly connotes an activity in which theories are developed to explain observations of the natural world, the empirical, observational basis of the theory of intelligent design provides a good reason for regarding intelligent design as a scientific theory.

Reason 2: Advocates of ID Use Established Scientific Methods

The case for intelligent design follows from the application of not one, but two separate systematic methods of scientific reasoning—methods that establish criteria for determining when observed evidence supports a hypothesis. In Chapter 15, I discussed the primary method, the method of multiple competing hypotheses, by which I inferred and justified the inference to intelligent design as the best explanation for the origin of biological information. As noted there, this method is a standard method of scientific reasoning in several well-established scientific disciplines. I discuss in more detail below how this method makes it possible to test intelligent design. Additionally, and as discussed in Chapter 16, advocates of intelligent design have developed another method that complements the method of multiple competing hypotheses.

In *The Design Inference* (and in subsequent works), William Dembski established criteria by which intelligently designed systems can be identified by the kinds of patterns and probabilistic signatures they exhibit. On the basis of these criteria, Dembski developed a comparative evaluation procedure—his explanatory filter[18]—to guide our analysis and reasoning about natural objects and artifacts and to help investigators decide among three different types of explanations: chance, necessity, and design.[19] As such, it constitutes a rigorous, systematic, evidence-based method for detecting the effects of intelligence, again suggesting a good reason to regard intelligent design as scientific in accord with common definitions of the term.

Reason 3: ID Is a Testable Theory

Most scientists and philosophers of science think that the ability to subject theories to empirical tests constitutes an important aspect of any scientific method of study. But for a theory to be testable, there must be some evidential grounds by which it could be shown to be incorrect or inadequate. And, contrary to the repeated claims of its detractors, the theory of intelligent design *is* testable. In fact, it is testable in several interrelated ways.

First, like other scientific theories concerned with explaining events in the remote past, intelligent design is testable by comparing its explanatory power to that of competing theories. Darwin used this method of testing in *On the Origin of Species*. In my presentation of the case for intelligent design in Chapters 8 through 16, I tested the theory in exactly this way by comparing the explanatory power of intelligent design against that of several other classes of explanation. That the theory of intelligent design can explain the origin of biological information (and the origin of the cell's interdependent information-processing system) better than its materialistic competitors shows that it has passed an important scientific test.

This comparative process is not a hall of mirrors, a competition without an external standard of judgment. The theory of intelligent design, like the other historical scientific theories it competes against, is tested against our knowledge of the evidence in need of explanation and our knowledge of the cause-and-effect structure of the world. As noted in Chapters 7 and 15, evaluations of "causal adequacy" guide historical scientific reasoning and help to determine which hypothesis among a competing group of hypotheses has the best explanatory power. Considerations of causal adequacy provide an experience-based criterion by which to test—accept, reject, or prefer—competing historical scientific theories. When such theories cite causes that are known to produce the effect in question, they meet the test of causal adequacy; when they fail to cite such causes, they fail to meet this test. To adapt my example from Chapter 7, the earthquake hypothesis fails the test of causal adequacy because we do not have evidence that earthquakes cause layers of volcanic ash to accumulate, whereas the volcanic eruption hypothesis passes the test of causal adequacy because experience has shown that eruptions do cause this phenomenon.[20]

Since empirical considerations provide grounds for rejecting historical scientific theories or preferring one theory over another, such theories are clearly testable. Like other historical scientific theories, intelligent design makes claims about the cause of past events, thus making it testable against our knowledge of cause and effect. Moreover, because experience shows that an intelligent agent is not only a known, but also the *only* known cause of specified, digitally encoded information, the theory of intelligent design developed in this book has passed two critical tests: the tests of causal adequacy and causal existence (see Chapter 15). Precisely because intelligent design uniquely passed these tests, I argued that it stands as the best explanation of the DNA enigma.

Finally, though *historical* scientific theories typically do not make predictions that can be tested under controlled laboratory conditions, they do sometimes generate discriminating predictions about what we should find in the natural world—predictions that enable scientists to compare them to other historical scientific theories. The theory of intelligent design has generated a number of such discriminating empirical predictions. These predictions not only distinguish the theory of intelligent design from competing evolutionary theories; they also serve to confirm the design hypothesis rather than its competitors.

Consider the case of so-called junk DNA—the DNA that does not code for proteins found in the genomes of both one-celled organisms and multicellular plants and animals. The theory of intelligent design and materialistic evolutionary theories (both chemical and biological) differ in their interpretation of so-called junk DNA. Since neo-Darwinism holds that new biological information arises as the result of a process of mutational trial and error, it predicts that nonfunctional DNA would tend to accumulate in the genomes of eukaryotic organisms (organisms whose cells contain nuclei). Since most chemical evolutionary theories also envision some role for chance interactions in the origin of biological information, they imply that nonfunctional DNA would have similarly accumulated in the first simple (prokaryotic) organisms—as a kind of remnant of whatever undirected process first produced functional information in the cell. For this reason, most evolutionary biologists concluded upon the discovery of nonprotein-coding DNA that such DNA was "junk." In their view, discovery of the nonprotein-coding regions confirmed the prediction or expectation of naturalistic evolutionary theories and disconfirmed an implicit prediction of intelligent design.

As Michael Shermer argues, "Rather than being intelligently designed, the human genome looks more and more like a mosaic of mutations, fragmented copies, borrowed sequences, and discarded strings of DNA that were jerry-built over millions of years of evolution."[21] Or as Ken Miller argues: "The critics of evolution like to say that the complexity of the genome makes it clear that it was designed. . . . But there's a problem with that analysis, and it's a serious one. The problem is the genome itself: it's not perfect. In fact, it's riddled with useless information, mistakes, and broken genes. . . . Molecular biologists actually call some of these regions 'gene deserts,' reflecting their barren nature."[22] Or as philosopher of science Philip Kitcher puts it, "If you were designing the genomes of organisms, you would not fill them up with junk."[23]

ID advocates advance a different view of nonprotein-coding DNA.[24] The theory of intelligent design predicts that most of the nonprotein-coding sequences in the genome should perform some biological function, even if they do not direct protein synthesis. ID theorists do not deny that mutational processes might have degraded or "broken" some previously functional DNA, but we predict that the functional DNA (the signal) should dwarf the nonfunctional DNA (the noise), and not the reverse. As William Dembski explained and *predicted* in 1998: "On an evolutionary view we expect a lot of useless DNA. If, on the other hand, organisms are designed, we expect DNA, as much as possible, to exhibit function."[25] *The discovery in recent years that nonprotein-coding DNA performs a diversity of important biological functions has confirmed this prediction.* It also decisively refutes prominent critics of intelligent design—including Shermer, Miller, and Kitcher—who have continued to argue (each as recently as 2008) that the genome is composed of mostly useless DNA.[26]

Contrary to their claims, recent scientific discoveries have shown that the nonprotein-coding regions of the genome direct the production of RNA molecules that regulate the use of the protein-coding regions of DNA. Cell and genome biologists have also discovered that these supposedly "useless" nonprotein-coding regions of the genome: (1) regulate DNA replication,[27] (2) regulate transcription,[28] (3) mark sites for programmed rearrangements of genetic material,[29] (4) influence the proper folding and maintenance of chromosomes,[30] (5) control the interactions of chromosomes with the nuclear membrane (and matrix),[31] (6) control RNA processing, editing, and splicing,[32] (7) modulate translation,[33] (8) regulate embryological development,[34] (9) repair DNA,[35] and (10) aid in immunodefense or fighting disease[36] among other functions. In some cases, "junk" DNA has even been found to code functional genes.[37] Overall, the nonprotein-coding regions of the genome function much like an operating system in a computer that can direct multiple operations simultaneously.[38] Indeed, far from being "junk," as materialistic theories of evolution assumed, the nonprotein-coding DNA directs the use of other information in the genome, just as an operating system directs the use of the information contained in various application programs stored in a computer. In any case, contrary to the often heard criticism that the theory makes no predictions, intelligent design not only makes a discriminating prediction about the nature of "junk DNA"; recent discoveries about nonprotein-coding DNA confirm the prediction that it makes.[39] Appendix A describes several other discriminating predictions that the theory of intelligent design makes.

Reason 4: The Case for ID Exemplifies Historical Scientific Reasoning

There is another good—if convention-dependent—reason for classifying intelligent design as a scientific theory, one already hinted at in the previous section on testability and explored at length in Chapter 15. Not only do scientists use systematic methods to infer intelligent design; the specific methods they use conform closely to established patterns of inquiry in the historical sciences. Indeed, the theory of intelligent design and the patterns of reasoning used to infer and defend it exemplify each of the key features of a historical science.

During my doctoral studies I discovered several distinctive characteristics of historical scientific disciplines—disciplines that try to reconstruct the past and explain present evidence by reference to past causes rather than trying to classify or explain unchanging laws and properties of nature (see Chapter 7). I found that historical sciences generally can be distinguished from nonhistorical scientific disciplines by reference to four criteria. And the theory of intelligent design (and the modes of inference used to establish and test it) provides a good example of each of the key features of a historical science.

A Distinctive Historical Objective
Historical sciences focus on questions of the form, "What happened?" or "What caused this event or that natural feature to arise?" rather than questions of the form, "How does nature normally operate or function?" or "What causes this general phenomenon to occur?"[40] Those who postulate the past activity of an intelligent designer do so as an answer, or a partial answer, to distinctively historical questions. The theory of intelligent design attempts to answer a question about what caused certain features in the natural world to come into existence—such as the digitally encoded, specified information present in the cell. It attempts to answer questions of the form "How did this natural feature arise?" as opposed to questions of the form "How does nature normally operate or function?"

A Distinctive Form of Inference
The historical sciences use inferences with a distinctive logical form. Unlike many nonhistorical disciplines, which typically infer general-

izations or laws from particular facts (induction), historical sciences employ *abductive* logic to infer a past event from a present fact or clue. Such inferences are also called "retrodictive." As Gould put it, the historical scientist infers "history from its results."[41]

Inferences to intelligent design exemplify this abductive and retrodictive logical structure. They infer a past unobservable cause (in this case, an instance of creative mental action or agency) from present facts or clues in the natural world, such as the specified information in DNA, the irreducible complexity of certain biological systems, and the fine-tuning of the laws and constants of physics.[42]

A Distinctive Type of Explanation

Historical sciences usually offer causal explanations of particular events, not lawlike descriptions or theories describing how certain kinds of phenomena—such as condensation or nuclear fission—generally occur. In historical explanations, past causal events, not laws or general physical properties, do the main explanatory work.[43] To explain a dramatic erosional feature in eastern Washington called the Channeled Scablands, a historical geologist posited an event: the collapse of an ice dam and subsequent massive flooding. This and other historical scientific explanations emphasize past events as causes for subsequent events and/or present features of the world.

The theory of intelligent design offers such a distinctively historical form of explanation. Theories of design invoke the act or acts of an agent and conceptualize those acts as causal events, albeit ones involving mental rather than purely physical entities. Advocates of design postulate past causal events (or a sequence of events) to explain the origin of present evidence or clues, just as proponents of chemical evolutionary theories do.

Use of the Method of Multiple Competing Hypotheses

Historical scientists do not mainly test hypotheses by assessing the accuracy of the predictions they make under controlled laboratory conditions. Using the method of multiple competing hypotheses, historical scientists test hypotheses by comparing their explanatory power against that of their competitors. And advocates of the theory of intelligent design use this method (as I have done in this book). (For more on this, see Chapter 15.)

In sum, the theory of intelligent design seeks to answer characteristically historical questions, it relies upon abductive/retrodictive inferences, it postulates past causal events as explanations of present evidence, and it is tested indirectly by comparing its explanatory power against that of competing theories. Thus, the theory of intelligent design exhibits each of the main features of a historical science, suggesting another reason to regard it as scientific.

Reason 5: ID Addresses a Specific Question in Evolutionary Biology

There is another closely related reason to regard intelligent design as a scientific theory. It addresses a key question that has long been part of historical and evolutionary biology: How did the appearance of design in living systems arise? As noted in Chapter 1, both Darwin and contemporary evolutionary biologists such as Francisco Ayala, Richard Dawkins, and Richard Lewontin acknowledge that biological organisms *appear* to have been designed.[44] Nevertheless, for most evolutionary theorists, the appearance of design is considered illusory, because they are convinced that the mechanism of natural selection acting on random variations (and/or other similarly unguided mechanisms) can fully account for the appearance of design in living organisms.[45]

In *On the Origin of Species,* Darwin sought to show that natural selection has creative powers comparable to those of intelligent human breeders. In doing so, he sought to refute the design hypothesis by providing a materialistic explanation for the origin of the appearance of design in living organisms. Following Aleksandr Oparin, chemical evolutionary theorists have sought to provide similarly materialistic accounts for the appearance of design in the simplest living cells.

Is the appearance of design in biology real or illusory? Clearly, there are two possible answers to this question. Neo-Darwinism and chemical evolutionary theory provide one answer, and competing theories of intelligent design provide an opposite answer. By almost all accounts the classical Darwinian answer to this question—"The appearance of design in biology does not result from actual design"—has long been considered a scientific proposition. But what is the status of the opposite answer? If the proposition "Jupiter is made primarily of methane gas" is a scientific propo-sition, then the proposition "Jupiter is not made primarily of methane gas" would seem to be a scientific proposition as well. The negation of a propo-sition does not make it a different type of claim. Similarly, the claim "The appearance of design in biology does not result from actual design" and the claim "The appearance of design in biology does result from actual design" are not two different kinds of propositions; they are two different answers to the same question, a question that has long been part of evolutionary biology and historical science. If one of these propositions is scientific, then it would seem that the other is scientific as well.[46]

Reason 6: ID Is Supported by Peer-Reviewed Scientific Literature

Critics of the theory of intelligent design often claim that its advocates have failed to publish their work in peer-reviewed scientific publica-tions. For this reason, they say the theory of intelligent design does not qualify as a scientific theory.[47] According to these critics, science is what scientists do. Since ID scientists don't do what other scientists do—namely, publish in peer-reviewed journals—they are not real scien-tists and their theory isn't scientific either.

Critics of the theory of intelligent design made this argument before and during the Dover trial in support of the ACLU's case against the Dover school-board policy. For example, Barbara Forrest, a philosophy professor from Southeastern Louisiana State University and one of the expert witnesses for the ACLU, asserted in a *USA Today* article before the trial that design theorists "aren't published because they don't have any scientific data."[48] In her expert witness report in support of the ACLU, Forrest also claimed that "there are no peer-reviewed ID ar-ticles in which ID is used as a biological theory in mainstream scientific databases such as MEDLine."[49] Judge Jones apparently accepted such assertions at face value. In his decision, he stated not once, but three separate times, that there were no peer-reviewed scientific publications supporting intelligent design.[50]

But Dr. Forrest's carefully qualified statement gave an entirely mis-leading impression. In 2004, a year in advance of the trial, I published a peer-reviewed scientific article advancing the theory of intelligent design in a mainstream scientific journal. As I mentioned in the Prologue, the publication of the article evoked a huge backlash at the Smithsonian Institution, where the journal, *The Proceedings of the Biological Society of Washington,* was published. Moreover, controversy about the editor's decision and his subsequent treatment spilled over into both the scientific and the mainstream press, with articles about it appearing in *Science, Nature,* the *Wall Street Journal,* and the *Washington Post* among other places.[51] Both Dr. Forrest and Judge Jones had every opportunity to inform themselves about the existence of at least one peer-reviewed scientific article in support of intelligent design.

In any case, as my institute informed the court in an *amicus curiae* (friend of the court) brief, my article was by no means the only peer-reviewed or peer-edited scientific publication in support of the theory of intelligent design.[52] By 2005, scientists and philosophers advocating the theory of intelligent design had already developed their theory and the empirical case for it in peer-reviewed scientific books published both by trade presses[53] and by university presses.[54] Michael Behe's groundbreaking *Darwin's Black Box* was published by the Free Press in New York. William Dembski's *The Design Inference* was published by Cambridge University Press. Both were peer-reviewed. In addition, design proponents have also published scientific articles advancing the case for intelligent design in peer-reviewed scientific books and anthologies published by university presses[55] and in scientific conference proceedings published by university presses and trade presses.[56] Advocates of intelligent design have also published work advancing their theory in peer-reviewed philosophy of science journals and other relevant interdisciplinary journals.[57] Moreover, since the publication of my article in 2004, several other scientific articles supporting intelligent design (or describing research guided by an ID perspective) have been published in mainstream peer-reviewed scientific journals.[58]

Of course, critics of intelligent design may still judge that the number of published books and articles supporting the theory does not yet make it sufficiently mainstream to warrant teaching students about it. Perhaps.[59] But that is a judgment about educational policy distinct from deciding the scientific status, or still less, the merits of the theory of intelligent design itself. Clearly, there is no magic number of supporting peer-reviewed publications that suddenly confers the adjective "scientific" on a theory; nor is there a tribunal vested with the authority to make this determination. If there were a hard-and-fast numerical standard as low as even one, no new theory could ever achieve scientific status. Each new theory would face an impossible catch-22: for a new theory to be considered "scientific" it must have appeared in the peer-reviewed scientific literature, but anytime a scientist submitted an article to a peer-reviewed science journal advocating a new theory, it would have to be rejected as "unscientific" on the grounds that no other peer-reviewed scientific publications existed supporting the new theory.

Critics of intelligent design have actually used a similarly circular kind of argument to claim that ID is not science. Before 2004, critics argued that the theory of intelligent design was unscientific, because there were no published articles supporting it in peer-reviewed scientific journals (ignoring the various peer-reviewed books that existed in support of ID). Then once a peer-reviewed scientific journal article was published supporting intelligent design, critics claimed that the article should not have been published, because the theory of intelligent design is inherently unscientific.[60] Indeed, critics accused the editor who published my article of editorial malfeasance, because they thought he should never have considered sending the article out for peer review in the first place.[61] Why? Because, according to these critics, the perspective of the article should have immediately disqualified it from consideration. In short, critics argued that "intelligent design is not scientific because peer-reviewed articles supporting the theory have not been published" and that "peer-reviewed articles supporting intelligent design should not be published because the theory is not scientific," apparently never recognizing the patent circularity of this self-serving, exclusionary logic.

Logically, the issue of peer review is a red herring—a distracting procedural side issue. The truth of a theory is not determined or guaranteed by the place of, or procedures followed in, its publication.[62] As noted in Chapter 6, many great scientific theories were first advanced and published without undergoing formal peer review. Though modern peer-review procedures often do a good job of catching and correcting factual mistakes, they also can enforce ideological conformity, stifle innovation, and resist novel theoretical insights. Scientific experts can make mistakes in judgment and, being human, they sometimes reject good new ideas because of prejudicial attachments to older, more familiar ones. The history of science is replete with examples of established scientists summarily dismissing new theories that later proved able to explain the evidence better than previously established theories. In such situations, proponents of new theories have often found traditional organs of publication closed to them. Thus, it is neither surprising nor damning to intelligent design that currently many scientific journals are implacably opposed to publishing articles supporting the theory.

Yet if science is what scientists do, and if publishing peer-reviewed scientific books and articles is part of what scientists do that makes their theories scientific (as critics of ID assert), then there is another good, convention-dependent reason to regard intelligent design as scientific. The scientists who have developed the case for intelligent design have begun to overcome the prejudice against their ideas and have published their work in peer-reviewed scientific journals, books, conference volumes, and anthologies.[63]

Conclusion

As I examined the question of whether intelligent design qualified as a scientific theory, it was clear to me that the answer to this question depended upon the definition of science chosen to decide the question. But as I considered both common definitions of science and what I had learned about the specialized methodological practice of the historical sciences, it seemed equally clear that there were many good—if definition-dependent—reasons for considering intelligent design as a scientific theory.

But maybe there was some other, better definition of science that I should have considered. Perhaps there was some specific feature of a scientific theory that intelligent design did not possess, or some specific criterion of scientific practice that its advocates did not follow. I knew that the theory of intelligent design met the criterion of testability, despite what many critics of the theory asserted, but perhaps there were other criteria that it could not meet. If so, then perhaps these definitional criteria would establish a good reason for disqualifying intelligent design from consideration as science after all. Certainly, many critics of intelligent design have argued that the theory lacks many key features of a bona fide scientific theory—that it fails to meet criteria by which science could be defined and distinguished from nonscience, metaphysics, or religion. In the next chapter, I examine why critics of the theory— including the judge in the Dover case—have insisted that, despite the arguments developed in this chapter, intelligent design does not qualify as a scientific theory.

19

Sauce for the Goose

In 1992, the year after I received my Ph.D., I was invited to an academic conference to respond to a paper by Michael Ruse, a well-known British philosopher of science. Ruse had long ago made his reputation as a prolific defender of Darwinian evolution and an archnemesis of young-earth creationism, the idea that God created the world roughly in its present form between six and ten thousand years ago. In 1981, in *Mclean v. Arkansas Board of Education,* opponents of creationism, represented by attorneys working for the American Civil Liberties Union (ACLU), sued the state of Arkansas, arguing that a law that required teachers to teach creationism alongside Darwinian evolution in public-school science classrooms was unconstitutional. Ruse testified for the ACLU against the Arkansas law.[1]

In his Arkansas testimony, Ruse did not just argue that creation science (as it was called in the statute)[2] was wrong. Instead, he argued that it did not qualify as a scientific theory *by definition.* To make this case, he offered a fivefold definition of science. According to Ruse, to qualify as scientific a theory must be: (1) guided by natural law, (2) explanatory by reference to natural law, (3) testable against the empirical world, (4) tentative in its conclusions, and (5) falsifiable.[3] He argued that these demarcation criteria could distinguish science from pseudoscience, metaphysics, and religion, and that creation science failed to meet them.

Now, eleven years later, Ruse was turning his attention to intelligent design at a conference convened to assess the merits of this (then) new theory of biological origins. Having just started my academic career, I found the prospect of critiquing such a well-known figure in my field

Figure 19.1. Professor Michael Ruse. *Courtesy of Ray Stanyard, Florida State University and Review Magazine.*

rather daunting. But as I read the advance copy of Ruse's essay coming off the fax machine in our college library, my concern subsided. Ruse was making a demarcation argument based upon a conception of science that I had already refuted in my Ph.D. thesis. I knew his argument didn't work and knew what I needed to say.

Ruse was, at this point, completely unaware of my work characterizing the methods of the historical sciences. So he didn't know why I thought that intelligent design provided a good example of a historical scientific theory. Instead, he had his own definition of science in mind—the one he had promulgated in the Arkansas trial.

In the trial, Ruse had used this definition of science to define creationism as an unscientific or "pseudoscientific" idea. Now, in 1992, he wanted to use one prong of his definition—one of his demarcation criteria—to argue that intelligent design did not qualify as a scientific theory either.

In philosophy, such negative arguments are often called "defeaters," because they are intended to "defeat" or refute a positive argument for a given claim. My encounter with Professor Ruse made me aware that there was a host of such proposed defeaters. I realized that I needed to tackle these negative arguments head-on if I was going to open minds to the strong evidential case for intelligent design.

In the previous chapter, I enumerated several positive—if definition-dependent—reasons for considering intelligent design a scientific theory.

In this chapter, I respond to several of the most common demarcation arguments that have been used to deny or "defeat" its scientific status. I show why these arguments don't do the work that critics of intelligent design suppose—why they don't establish that intelligent design is any more or less scientific than rival evolutionary theories and why they don't provide good reasons for treating intelligent design as inherently *un*scientific.

So, "Is intelligent design scientific?" remains the topic of this chapter, but now the context of the discussion shifts. Think of it as a trial before the court of reason (not to be confused, as we will see, with Judge Jones's court). In Chapter 18, I gave an opening argument *for* the scientific status of intelligent design by appealing to various definitions of science, the most important of which was derived from a study of the methods of the historical sciences. Here I defend intelligent design against various defeaters, in particular, arguments that purport to establish that intelligent design does not qualify as a scientific theory.

At the close of this chapter, I'll return to central Pennsylvania to take a closer look at the trump card—the supposedly ultimate defeater—that Judge Jones played to justify his much publicized 2005 ruling against intelligent design. Before I do, I need to start where my own thinking about these negative demarcation arguments began, back in 1992, at a conference on the campus of Southern Methodist University in an encounter with the eminent Professor Ruse.

Intelligent Design and Explanation by Natural Law

At the conference, Ruse made one of the most common demarcation arguments against intelligent design. He argued that science must assume that "there are no powers, seen or unseen, that interfere with or otherwise make inexplicable the normal working of material objects."[4] Since the scientific enterprise is characterized by a commitment to "unbroken regularity" or "unbroken law,"[5] scientific theories must explain events or phenomena *by reference to natural laws.*[6] And since intelligent design invokes an event—the conscious activity of a designing agent— rather than a law of nature to explain the origin of biological form and information, Ruse argued that it was scientifically "inappropriate."

Ruse also seemed to think that if an intelligent designer had acted during the history of life, then its actions would have necessarily violated the laws of nature, since intelligent agents typically interfere with the otherwise "normal workings of material objects." Since, for Ruse, the activity of an intelligent designer violates the laws of nature, positing such activity—rather than a law—would violate the rules of science.

In response,[7] I pointed out that the activity of a designing intelligence does not necessarily break or violate the laws of nature. Human agents design information-rich structures and otherwise interfere with the "normal workings of material objects" all the time. When they do, they do not violate the laws of nature; they alter the conditions upon which the laws act. When I arranged the magnetic letters on my metallic chalkboard to spell the message "Biology Rocks!" I altered the way in which matter is configured, but I did not alter or violate the laws of electromagnetism. When agents act, they initiate new events within an existing matrix of natural law without violating those laws. (For a quick definition of a law of nature and a primer on the difference between a law and an event, see this note.)[8]

I also pointed out that Ruse's key demarcation criterion, if applied strictly, *cut just as much against Darwinian and chemical evolutionary (and many other scientific) theories* as it did against intelligent design. I showed, for example, that natural laws often describe, but do not explain natural phenomena. Newton's law of universal gravitation described, but did not explain, what caused gravitational attraction. A strict application of Ruse's second criterion would therefore imply that Newton's law of gravity had been "unscientific," since it did not offer an *explanation* by natural law.

I also showed that many historical scientific theories do not offer an explanation *by natural law.* Instead, they postulate past events (or patterns of events) to explain other past events as well as presently observable evidence. Historical theories explain mainly by reference to events or causes, not laws.

For example, if a historical geologist seeks to explain what caused the unusual height of the Himalayas, he or she will cite particular events or factors that were present in the case of the Himalayan mountain-building episode that were not present in other such episodes. Knowing the laws of physics that describe the forces at work in all mountain-building events will not aid the geologist in accounting for the contrast between the Himalayas and other mountain ranges. To explain what caused the Himalayas to rise to such heights, the geologist does not need to cite a general law, but instead evidence of a distinctive set of past events or conditions.[9] Evolutionary theories, in particular, often emphasize the importance of past events in their explanations.[10] For example, Aleksandr Oparin's chemical evolutionary theory postulated a series of events (a scenario), not a general law, in order to explain how the first living cells arose.

Of course, past events and historical scenarios are assumed to take place in a way that obeys the laws of nature. Moreover, our knowledge of cause-and-effect relationships (which we can sometimes formulate as laws) will often guide the inferences that scientists make about what happened in the past and will influence their assessment of the plausibility of competing historical scenarios and explanations. Even so, many historical scientific theories make no mention of laws at all. Laws, at best, play only a secondary role in historical scientific theories. Instead, events play the primary explanatory role.

In my response to Ruse, I pointed out that the theory of intelligent design exemplified the same style of scientific explanation as other historical scientific theories. Intelligent design invoked a past event—albeit a mental event—rather than a law to explain the origin of life and the complexity of the cell. As in other historical scientific theories, our knowledge of cause and effect ("Information habitually arises from conscious activity") supports the inference to design. A law (conservation of information) also helps to justify the inference of an intelligent cause as the best explanation. Advocates of intelligent design use a law ("Complex specified information always arises from an intelligent source in a nonbiological context") to infer a past causal event, the act of a designing mind. But that act or *event* explains the evidence in question. Though laws play a subsidiary role in the theory, a past event (or events) explains the ultimate origin of biological information.

If explaining events primarily by reference to prior events, rather than laws, does not disqualify other historical scientific theories, including evolutionary theories, from consideration as science, then by the same logic it should not disqualify the theory of intelligent design either. Oddly, in a discussion of population genetics—part of the explanatory framework of contemporary Darwinian theory—Ruse himself noted that "it is probably a mistake to think of modern evolutionists as seeking universal laws at work in every situation."[11] But if laws can play no role or only a subsidiary role in other historical theories, then why was it "inappropriate" for a law to play only a supportive role in the theory of intelligent design?

Conversely, if invoking a past event, rather than a law, made intelligent design unscientific, then by the same token it should make materialistic evolutionary theories unscientific as well. Either way, I concluded that Ruse's key criterion for scientific status did not provide a basis for discriminating the scientific status of the two types of theories. Both were equivalent in their capacity to meet Ruse's definitional standard.

Ruse's reaction to my critique of his paper surprised me. On the podium as we discussed our differences, he seemed genuinely interested in how I had come to my position and curious to understand it. When I talked with him privately afterward, I found him to be genial and kindly. He offered me some well-intended career advice and asked me about people that we both knew in Cambridge. During our private conversation he also shocked me by admitting his own reservations about the validity of using demarcation arguments to settle the debate about biological origins. The following year in a much-publicized talk before the American Association for the Advancement of Science, he made some of these doubts public (see below).

Defeaters Defeated

Encouraged by my discussion with Professor Ruse, I began to examine the "intelligent design isn't science" objection in more detail. As I did, I gradually came to a radical conclusion: not only were there many good—if convention-dependent—reasons for classifying intelligent design as a historical scientific theory (as I had concluded based upon my doctoral research), but there were *no* good—non-question begging—reasons to define intelligent design as *un*scientific.

Typically those who argued that "intelligent design isn't science" invoked various demarcation criteria, as Ruse had done in the Arkansas trial. Since my conversation with Michael Ruse, I have encountered numerous such arguments. Critics claim that intelligent design does not qualify as a scientific theory because: (1) it invokes an unobservable entity,[12] (2) it is not testable,[13] (3) it does not explain by reference to natural law,[14] (4) it makes no predictions,[15] (5) it is not falsifiable,[16] (6) it cites no mechanisms,[17] and (7) it is not tentative.[18]

As I studied these arguments I discovered a curious pattern. Invariably, if the critics applied their definitional criteria—such as observability, testability, or "must explain by natural law"—in a strict way, these criteria not only disqualified the design hypothesis from consideration as science; they also disqualified its chief rivals—other historical scientific theories—each of which invoked undirected evolutionary processes.

Conversely, I discovered that if these definitional criteria were applied in a less restrictive way—perhaps one that took into account the distinctive historical aspects of inquiry into the origin of life—then these criteria not only established the scientific bona fides of various rivals of intelligent design; they confirmed the scientific status of the design hypothesis as well. In no case, however, did these demarcation criteria successfully differentiate the scientific status of intelligent design and its competitors. Either science was defined so narrowly that it disqualified both types of theory, or it was defined so broadly that the initial reasons for excluding intelligent design (or its competitors) evaporated. If one theory met a specific criterion, then so did the other; if one theory failed to do so, then the rival theory did so as well—provided the criteria were applied in an evenhanded and non–question begging way. Intelligent design and its materialistic rivals were equivalent in their ability to meet various demarcation criteria or methodological norms. (I later coined the not-so-catchy phrase "methodological equivalence" to describe how these competing theories compared in their ability to measure up to various demarcation criteria.) Given this equivalence, and given that materialistic evolutionary theories were already widely regarded as scientific, I couldn't see any reason to classify intelligent design as unscientific. The defeaters didn't work.

Because these "defeaters" are used against intelligent design all the time, it's important to see why they fail. So, in what follows, I examine some additional demarcation arguments that are commonly used against intelligent design. (I don't provide in these pages an exhaustive demonstration of the equivalence I discovered. That would require a book in itself and a level of detail that only philosophers of science would happily stomach. But those interested in a more detailed analysis of other demarcation arguments against intelligent design might consult the Web site for this book and the references in this note.[19])

Observability

During the controversy over the treatment of Dean Kenyon at San Francisco State University, I noticed his critics using a second common demarcation argument against ID. After the biology department removed Kenyon from teaching his biology class, some of Kenyon's colleagues argued that the theory of intelligent design did not qualify as a scientific theory because it invoked an *unobservable* entity, in particular, an unseen designing intelligence. In making this argument, Kenyon's colleagues assumed that scientific theories must invoke only *observable* entities. Since Kenyon discussed a theory that violated this convention, they insisted that neither the theory he discussed, nor he himself, belonged in the biology classroom.[20]

Others who defended the action of the biology department, such as Eugenie Scott of the National Center for Science Education, used a similar rationale. She insisted that the theory of intelligent design violated the rules of science because, "you can't put an omnipotent deity in a test tube (or keep it out of one)."[21] At the conference at SMU where I had met Ruse the year before, I also encountered this complaint about the theory of intelligent design. There, molecular biologist Fred Grinnell argued that intelligent design can't be a scientific concept, because if something "can't be measured, or counted, or photographed, it can't be science."[22] According to these critics of intelligent design, the unobservable character of a designing intelligence renders it inaccessible to empirical investigation and, therefore, makes it unscientific.

But was that really the case? Does a reference to an unobservable entity provide a good reason for defining a theory as unscientific? Does my postulation of an unobservable intelligence make my case for intelligent design unscientific?

The answer to that question depends, again, upon how science is defined. If scientists (and all other relevant parties) decide to define science as an enterprise in which scientists can posit only observable entities in their theories, then clearly the theory of intelligent design would not qualify as a scientific theory. Advocates of intelligent design infer, rather than directly observe, the designing intelligence responsible for the digital information in DNA.

But it didn't take me long to realize that this definition of science would render many other scientific theories, including many evolutionary theories of biological origins, unscientific by definition as well. Many scientific theories infer or postulate unobservable entities, causes, and events. Theories of chemical evolution invoke past events as part of the scenarios they use to explain how the modern cell arose. Insofar as these events occurred millions of years ago, they are clearly not observable today. Darwinian biologists, for their part, have long defended the putatively unfalsifiable nature of their claims by reminding critics that many of the creative processes to which they refer occur at rates too slow to observe in the present and too fast to have been recorded in the fossil record. Further, the existence of many transitional intermediate forms of life, the forms represented by the nodes on Darwin's famous branching tree diagram, are also unobservable.[23] Instead, unobservable transitional forms of life are *postulated* to explain observable biological evidence—as Darwin himself explained. But how is this different from postulating the past activity of an unobservable designing intelligence to explain observable features of the living cell? Neither Darwinian transitional forms, neo-Darwinian mutational events, the "rapid branching" events of Stephen Jay Gould's theory of punctuated equilibrium, the events comprising chemical evolutionary scenarios, nor the past action of a designing intelligence are directly observable. With respect to direct observability, each of these theories is equivalent.

Thus, if the standard of observability is applied in a strict way, neither intelligent design nor any other theory of biological origins qualifies as a scientific theory. But let's consider the flip side. What if the standard of observability is applied in a more flexible and, perhaps, realistic way? What if science is defined as an enterprise that examines the observable natural world, but does not necessarily explain empirical observations by reference to observable entities?

Does it make sense to define science in this more flexible way? It does. Many entities and events posited in scientific theories cannot be observed directly either in practice, or sometimes even in principle. Instead, scientists often must infer unobservable entities to explain observable events, evidence, or phenomena. Physical forces, electromagnetic or gravitational fields, atoms, quarks, past events, subsurface geological features, biomolecular structures—*all* are unobservable entities inferred from observable evidence. In 2008 under the border between France and Switzerland, European scientists unveiled the Large Hadron Collider. This supercollider will enable physicists to "look" for various elementary particles including the elusive Higgs boson. Yet none of these particles are observable in any direct sense. Instead, physicists try to detect them by the energetic signatures, traces, or decay products they leave behind.

Scientists in many fields detect unobservable entities and events in their effects. They often infer the unseen from the seen. Nevertheless, such entities and events are routinely considered to be part of scientific theories. Those who argue otherwise confuse the event or evidence in need of explanation (which in scientific investigations is nearly always observable in some way) with the event or entity doing the explaining (which is often not).

The presence of unobservable entities in scientific theories creates a problem for those who want to use observability as a demarcation criterion by which to disqualify intelligent design from consideration as scientific. Many theories—theories that are widely acknowledged to be scientific—invoke unobservable entities. But if other scientific theories, including materialistic theories of biological origins, can invoke unobservable entities or events to explain observable evidence and still qualify as scientific, then why can't the theory of intelligent design do so as well?

Testability Revisited

In the previous chapter, I showed that the theory of intelligent design is testable by reference to empirical evidence. I described in general terms how scientists can subject intelligent design to various kinds of empirical tests, and I provided an example of a testable prediction that the theory makes. But for years, I have talked to people—scientists, theologians, philosophers, lawyers, journalists, callers on talk shows—who purport to know that the theory of intelligent design cannot be tested. (Oddly, some of these same people also claim that the theory has been tested and found wanting).[24]

Sometimes critics say that intelligent design is untestable because the designing intelligence is unobservable, thus combining two demarcation criteria, observability and testability. Other critics assert that intelligent design cannot be tested, because they assume that ID advocates are necessarily positing an omnipotent deity. Some critics say that intelligent design is untestable because the actions of intelligent agents (of any kind) are inherently unpredictable, and testability depends upon the ability to make predictions.

These common objections to the testability and thus the scientific status of intelligent design have dissuaded many people from considering evidence for intelligent design. So what should we make of these "defeaters"? Do these specific demarcation arguments provide a good reason for denying that intelligent design is a scientific theory? Do they show, despite my arguments to the contrary in the previous chapter, that intelligent design cannot be tested? Let's take a closer look.

Unobservables and Testability

Robert Pennock, one of the witnesses in the Dover trial, argued that the unobservable character of a designing intelligence precludes the possibility of testing intelligent design scientifically because, as he explained, "science operates by empirical principles of *observational* testing; hypotheses must be confirmed or disconfirmed by reference to . . . accessible empirical data."[25] Eugenie Scott also seemed to argue that intelligent design cannot be tested *because* it invokes an unobservable entity. In the article I cited above, in which she defended the actions of Kenyon's detractors at San Francisco State, Scott also linked the criterion of observability to testability. After saying, "You can't put an omnipotent deity in a test tube," she went on to say: "As soon as creationists invent a 'theo-meter,' maybe then we can test for miraculous intervention. You can't (scientifically) study variables you can't test, directly or indirectly."[26]

In this version of the argument, critics insist that the unobservable character of a designing intelligence renders the theory inaccessible to empirical investigation, making it both untestable and unscientific. Both "observability" and "testability" are asserted as necessary to scientific status and the converse of one (unobservability) is asserted to preclude the possibility of the other (testability). Superficially this version of the argument seems a bit more persuasive than demarcation arguments that simply invoke observability by itself to disqualify design. Yet it does not stand up to close inspection either.

In the first place, there are many testable scientific theories that refer to unobservable entities. For example, as we saw in Chapter 3, during the race to elucidate the structure of the genetic molecule, both double helix and triple helix models were considered, since both could explain the X-ray images of DNA crystals.[27] Although neither structure could be observed directly, the double helix of Watson and Crick eventually won out, because it could explain other observations that the triple helix model could not. The inference to one unobservable structure (the double helix) was accepted because it was judged to possess a greater explanatory power than its competitor.

Claims about unobservables are routinely tested in science indirectly against observable evidence. In many fields the existence of an unobservable entity is established or detected by testing the explanatory power that would result if a given hypothetical entity (i.e., an unobservable) were accepted as actual. Many sciences infer to the best explanation—where the explanation presupposes the reality of an unobservable entity—including theoretical physics, geology, molecular biology, genetics, cosmology, psychology, physical and organic chemistry, and evolutionary biology.

Second, as I showed in Chapters 7 and 15, historical sciences, in particular, commonly use indirect methods of testing, methods that involve assessing the causal powers of competing unobservable events to determine which would, if true, possess the greatest explanatory power. Recall that Darwin defended the scientific status of his theory by pointing out that assessing the relative explanatory power of his theory of common descent—a theory about the unobservable past—was a perfectly legitimate and accepted method of scientific testing.[28]

Third, as I showed in Chapters 15 and 18, intelligent design is testable in precisely this fashion—by examining its explanatory power and comparing it to that of competing hypotheses. The unobservable intelligence referred to in the theory of intelligent design does not preclude testing the theory, if indirect methods of testing hypotheses—such as evaluating comparative explanatory power—are allowed as scientific. If, however, science is defined more narrowly so that only the direct observation of a causal factor counts as a confirmatory test of a causal hypothesis, then neither intelligent design nor a host of other theories would qualify as scientific.

Either way, the theory of intelligent design and various evolutionary theories of origins are equivalent in their ability to meet the joint criteria of observability and testability. If critics of intelligent design construe these criteria to forbid *inferring* the existence of unobservables and indirect testing for them, then both intelligent design and its potential competitors fail. If critics construe these criteria to allow inferences to, and indirect testing of, unobservable entities and events, then both intelligent design and many competing evolutionary theories qualify as scientific theories. Either way, these criteria fail to discriminate between intelligent design and many other theories that are already accepted as scientific. Thus, they fail to provide a good reason for disqualifying intelligent design from consideration as a scientific theory.

Testability, Omnipotence, and the Supernatural

Robert Pennock argues that there is something else about the unobservable designing intelligence posited by intelligent design that makes it untestable. Specifically, Pennock claims that intelligent design is untestable because it invokes an unobservable *supernatural* being with unlimited powers. He argued that since such a being has powers that could be invoked to "explain any result in any situation," all events are consistent with the actions of such a being. Therefore, no conceivable event could disprove the hypothesis of intelligent design. As Ken Miller asserts, "The hypothesis of design is compatible with any conceivable data, [and] makes no testable predictions."[29]

This argument fails for two reasons. First, it misrepresents the theory of intelligent design. The theory of intelligent design does not claim to detect a supernatural intelligence possessing unlimited powers. Though the designing agent responsible for life may well have been an omnipotent deity, the theory of intelligent design does not claim to be able to determine that. Because the inference to design depends upon our uniform experience of cause and effect in this world, the theory cannot determine whether or not the designing intelligence putatively responsible for life has powers beyond those on display in our experience. Nor can the theory of intelligent design determine whether the intelligent agent responsible for information in life acted from the natural or the "supernatural" realm. Instead, the theory of intelligent design merely claims to detect the action of some *intelligent* cause (with power at least equivalent to those we know from experience) and affirms this because we know from experience that only conscious, intelligent agents produce large amounts of specified information. The theory of intelligent design does not claim to be able to determine the identity or any other attributes of that intelligence, even if philosophical deliberation or additional evidence from other disciplines may provide reasons to consider, for example, a specifically theistic design hypothesis.[30] (I discuss the possible *implications* of the theory of intelligent design in the next chapter.)

Pennock's argument also fails because the theory of intelligent design *is* subject to empirical testing and refutation. Indeed, intelligent design actually makes a much stronger claim than the caricature of it he critiqued during the trial. Pennock critiques the hypothesis that "an omnipotent deity *could* explain the origin of life." But the theory of intelligent design developed in this book differs from that hypothesis. My theory of intelligent design does not merely affirm that intelligence constitutes a *possible* explanation of certain features of life. Instead, the design hypothesis developed here asserts that intelligent design constitutes the *best* explanation of a particular feature of life because of *what we know about the cause-and-effect structure of the world*—specifically, because of what we know about what it takes to produce large amounts of specified information. For this reason, the design hypothesis is not "compatible with any conceivable data" or observations whatsoever.

If it were shown, for example, that the cause-and-effect structure of the world were different than what advocates of intelligent design claim—if, for example, someone successfully demonstrated that "large amounts of functionally specified information *do* arise from purely chemical and physical antecedents," then my design hypothesis, with its strong claim to be the best (clearly superior) explanation of such phenomena, would fail. Intelligent design would remain as a possible explanation (much as chance does now). But the claim that intelligent design provides the best (most causally adequate) explanation for the origin of biological information would be refuted. Similarly, if it could be shown that key indicators of intelligence—such as specified information—were not present in living systems, the basis of the design hypothesis in its present strong form would evaporate. Thus, Pennock and Miller incorrectly portray the theory of intelligent design as being consistent with any empirical situation. The theory of intelligent design is, in fact, testable—just as I argued in the previous chapter.

Testability and Predictability

I have learned that when critics of intelligent design are confronted with refutations of a particular demarcation argument, they typically shift their ground and formulate other arguments either by invoking a different demarcation criterion or by applying the original criterion in a more demanding way. For example, after explaining how intelligent design can be tested and how it does make certain kinds of predictions, I commonly hear the objection that the theory of intelligent design is not scientific, because it cannot make other kinds of predictions. Critics correctly point out, for example, that we cannot predict with complete accuracy what intelligent agents will do, since, presumably, intelligent agents possess the capacity to act freely of their own volition. Since ID invokes the action of an unpredictable intelligent agent, and since scientific theories must make predictions, theories invoking the activity of intelligent agents are not scientific—or so the argument goes.

Yet standard materialistic theories of evolution (whether chemical or biological) do not make predictions of this kind either. Specifically, evolutionary theory does not make predictions about the future course of evolution. It makes no prediction about the kind of traits or species that random mutations and natural selection will produce in the future.

As Ken Miller notes, "The outcome of evolution is not predictable."[31] Even so, most evolutionary biologists think that these theories are scientific—and for good (if convention-dependent) reasons. Evolutionary theories provide explanations of past events and present evidence, and they make predictions about the patterns of evidence that scientists should find *in their future investigations* of, for example, the genome or the fossil record.

In the same way, the theory of intelligent design does not make predictions about when (or whether) the designing intelligence responsible for life will act in the future. Yet it does explain past events and present evidence, and it also makes discriminating predictions about the kind of evidence scientists should find in their future investigations (see Chapter 18 and Appendix A).[32] Thus, neither type of origins theory qualifies as scientific if the "ability to generate predictions" is treated as a condition of scientific status and interpreted in a strict way, though both types of theories qualify as scientific if this criterion is equated with scientific status and interpreted in a more flexible way.

As I studied the various demarcation arguments against intelligent design, I repeatedly found this same pattern. Invariably, the criteria that supposedly showed that intelligent design is inherently unscientific either disqualified both intelligent design *and* its materialistic rivals, or, if the criteria were applied more flexibly, legitimated both types of theories—provided, that is, that the criteria were not applied in a question-begging way.

As this pattern became more pronounced with each of the definitional criteria I examined, I became more convinced that there was no good reason to exclude intelligent design from consideration as a scientific explanation for the origin of biological information. Since—by convention—materialistic theories of biological origin were considered scientific, and since the theory of intelligent design met various criteria of scientific status just as well as these rival theories, it seemed clear that the theory of intelligent design, by the same conventions, must be considered scientific as well. Nevertheless, Judge John E. Jones disagreed.

The Demise of Demarcation Arguments

In 2005, before the Dover trial, I knew that the ACLU would try to persuade Judge Jones that "intelligent design isn't science." Yet I wondered how it was going to do this. I knew the long history of attempts to use specific demarcation criteria to discredit various ideas and theories. And I knew that most philosophers of science, some of whom the ACLU would need to use as expert witnesses, did not regard these arguments as valid. Philosophers of science who had studied the history of the demarcation question already knew what I had discovered: it was not only difficult to define science by reference to a single set of methodological practices; it was difficult to find demarcation criteria that could differentiate the scientific status of competing theories without applying a double standard or using question-begging logic.

I also knew that using these arguments in court had a checkered history. Though the ACLU won the *Mclean v. Arkansas* case in 1981, leading philosophers of science,[33] none sympathetic to creationism, later severely criticized Ruse's use of demarcation arguments in his testimony. They pointed out that many of the definitional criteria that Ruse had used to establish creation science as pseudoscience could actually be used to establish creation science as a scientific theory. They also pointed out that the same criteria, if applied strictly, could have the effect of disqualifying Darwinian evolution from that same honorific designation. For more detail on this curious result, see this note.[34]

As a result of such difficulties, several leading philosophers of science, such as Larry Laudan, Philip Quinn, and Philip Kitcher,[35] argued that the question, "What distinguishes science from nonscience?" is both intractable and uninteresting. Instead, they and most other philosophers of science have increasingly realized that the real issue is not whether a theory is "scientific" according to some abstract definition, but whether a theory is true, or supported by the evidence.

Scientists do not decide these questions using abstract criteria that purport to tell in advance how all good scientific theories are constructed or what they will, in general, look like. Instead, scientists look to the evidence to decide the merits of competing theories. Theories are not rejected with definitions, but with evidence. Thus, in a now famous article called "The Demise of the Demarcation Problem," Larry

Laudan shows that demarcation criteria of the kind that Ruse had proposed in 1981 do not do the work Ruse wanted them to do. Specifically, these criteria will not decide the merits of, or discriminate between, competing theories.

Ruse himself later publicly acknowledged this—at least in part. During a talk to the American Association for the Advancement of Science (AAAS) in 1993, Ruse repudiated his previous support for the demarcation principle by admitting that Darwinism (like creationism) "depends upon certain unprovable metaphysical assumptions."[36] In his more recent scholarship Ruse has gone further, arguing that evolutionary theory has often functioned as a kind of "secular religion."[37] In any case, by the early 1990s a consensus had developed in the philosophy of science about the use of demarcation arguments. As one philosopher, Martin Eger, summarizes: "Demarcation arguments have collapsed. Philosophers of science don't hold them anymore. They may still enjoy acceptance in the popular world, but that's a different world."[38]

How Dover Was Decided

In 2005 I wondered, "Could the ACLU pull it off?" Could its representatives convince a judge to rule that intelligent design isn't science by definition? If so, on what basis? What criterion would they invoke to discriminate between the scientific status of intelligent design and Darwinian or chemical evolution, both of which are routinely taught in public high-school science textbooks? Observability? Testability? Falsifiability? Given what I had discovered about the inability of such criteria to discriminate the scientific status of competing origins theories, I wondered: What criterion *could* a judge use to deny that intelligent design qualified as a scientific theory?

As it turned out, Judge Jones did a clever thing. He didn't reject intelligent design as science because it failed to meet a neutral definition of science or methodological norm. At the urging of the ACLU, he circumvented the whole demarcation problem by defining science as the exclusion of intelligent design—only he didn't call it that. Instead, following the ACLU's expert witnesses and brief, he called the exclusionary principle "methodological naturalism."[39] He then equated science with adherence to that principle and rejected intelligent design because it violated it.

But what is the principle of methodological naturalism? Methodological naturalism asserts that to qualify as scientific, a theory must explain all phenomena by reference to purely material—that is, non-intelligent—causes. As philosopher Nancey Murphy explains, methodological naturalism forbids reference "to creative intelligence" in scientific theories.[40]

So, did the judge find a demarcation criterion or methodological norm that could discriminate between intelligent design and materialistic theories of evolution? Clearly, he did. If science is defined as Judge Jones defined it, intelligent design does not qualify as a scientific theory. But should science be defined that way? Did the judge offer a *good* reason for excluding intelligent design from consideration as science?

He did not. Instead, he provided an entirely arbitrary, circular, and question-begging justification for the exclusion of design. I knew, as did many other philosophers of science, that demarcation arguments based upon neutral methodological norms such as testability could not justify a prohibition against intelligent causes in science. The judge in the Dover case supposedly offered a reason for this prohibition, but his reason turned out to be just a restatement of the prohibition by another name. According to Judge Jones, the theory of intelligent design cannot be part of science, because it violates the principle of methodological naturalism. But that principle turns out to be nothing more than the claim that intelligent causes[41]—and thus the theory of intelligent design—must be excluded from science. According to this reasoning, intelligent design isn't science because it violates the principle of methodological naturalism. What is methodological naturalism? A rule prohibiting consideration of intelligent design in scientific theories.

Thus, despite appearances to the contrary, Judge Jones did not offer a good reason—a theoretically neutral norm or definition of science—by which to justify the exclusion of intelligent design "from science." Instead, he simply asserted a prohibition against the consideration of intelligent design, invoked the same prohibition by another name, and then treated it as if it were a reason—a methodological principle—justifying the prohibition itself.

Fortunately we don't look to federal judges to settle great questions of scientific and philosophical import. Did life arise as the result of purely undirected material causes or did intelligence play a role?[42] Surely a court-promulgated definition of science, especially one so logically problematic as methodological naturalism, does not answer that question.[43]

No doubt Judge Jones felt justified in offering such a thin and circular justification for his definition of science because he knew many scientists agreed with him. And, indeed, the majority of scientists may well accept the principle of methodological naturalism. So, if science is what scientists do, and if many or most scientists do not think that hypotheses invoking intelligent causes have a place in their theories, then perhaps intelligent design doesn't qualify as a scientific theory after all. According to this line of thinking, Judge Jones did not impose an arbitrary definition of science. Instead, his ruling merely expressed a preexisting consensus about proper scientific practice from within the scientific community. As the judge himself wrote in the ruling, methodological naturalism is simply a "centuries-old ground rule" of science.[44]

So why shouldn't scientists continue to accept methodological naturalism as a strict rule governing scientific practice? Maybe we should just accept this convention and move on. Of course, some scientists may decide to do exactly that. But if they do, it's important to recognize what that decision would and would not signify about the design hypothesis. Scientists who decide to define explanations involving creative intelligence as unscientific cannot then treat the failure of such hypotheses to meet their definition of science as a tacit refutation of, or reason to reject, such hypotheses. Why? It remains logically possible that an "unscientific" hypothesis (according to methodological naturalism) might constitute a better explanation of the evidence than the currently best "scientific" hypothesis. Based upon the evidence presented in this book, I would contend that, whatever its classification, the design hypothesis provides a better explanation than any of its materialistic rivals for the origin of the specified information necessary to produce the first life. *Reclassifying an argument does not refute it.*

In any case, there is no compelling reason for the currently dominant convention among scientists to continue. Conventions are just that.

Without a good reason for holding them, they may do nothing more than express an unexamined prejudice and block the path of inquiry. When good reasons for rejecting conventions come along, reasonable people will set them aside—and there are now good reasons to set this convention aside.

First, scientists have not always restricted themselves to naturalistic hypotheses, contrary to the claims of one of the expert witnesses in the Dover trial. Newton, for example, made design arguments within his scientific works, most notably in the *Principia* and in the *Opticks*. Louis Agassiz, a distinguished paleontologist and contemporary of Darwin, also made design arguments within his scientific works, insisting that the pattern of appearance in the fossil record strongly suggested "an act of mind." Defenders of methodological naturalism can claim, at best, that it has had normative force during *some* periods of scientific history. But this concedes that canons of scientific method change over time— as, indeed, they do. From Newton until Darwin, design arguments were a common feature of scientific research. After Darwin, more materialistic canons of method came to predominate. Recently, however, this has begun to change as more scientists are becoming interested in the evidence for intelligent design.

Second, many scientific fields currently posit intelligent causes as scientific explanations. Design detection is already part of science. Archaeologists, anthropologists, forensic scientists, cryptographers, and others now routinely infer intelligent causes from the presence of information-rich patterns or structures or artifacts. Further, astrobiologists looking for extraterrestrial intelligence (SETI) do not have a rule against inferring an intelligent cause. Instead, they are open to detecting intelligence, but have not had evidence to justify making such an inference. Thus, the claim that all scientific fields categorically exclude reference to creative intelligence is actually false.

Even some biologists now contest methodological naturalism. Granted, many evolutionary biologists accept methodological naturalism as normative within their discipline. Nevertheless, biologists intrigued by the design hypothesis reject methodological naturalism, because it prevents them from considering a possibly true hypothesis. Indeed, a central aspect of the current debate over design is precisely about whether methodological naturalism should be regarded as normative for biology today. Most evolutionary biologists say it should remain normative; scientists advocating intelligent design disagree. But critics of intelligent design cannot invoke methodological naturalism to settle this debate about the scientific status of intelligent design, because methodological naturalism is itself part of what the debate is about.

Third, defining science as a strictly materialistic enterprise commits scientists to an unjustified—and possibly false—view of biological origins. It is at least logically possible that a personal agent—a conscious goal-directed intelligence—existed before the appearance of the first life on earth. Moreover, as shown in Chapters 15 and 16, there are now rigorous scientific methods by which the activity of intelligent agents can be inferred or detected from certain kinds of effects. Thus, if a personal agent existed before the advent of life on earth, then it is also at least possible that the activity of such an agent could be detected using one of these methods. If so, then prohibitions against the design hypothesis in investigations of the origin of life amount to an assumption that no intelligence of any kind existed or could have acted prior to that event. But this assumption is entirely unjustified, especially given the absence of evidence for a completely materialistic account of abiogenesis.

Finally, allowing methodological naturalism to function as an absolute "ground rule" of method for all of science would have a deleterious effect on the practice of certain scientific disciplines, especially the historical sciences.[45] In origin-of-life research, for example, methodological naturalism artificially restricts inquiry and prevents scientists from exploring and examining some hypotheses that might provide the most likely, best, or causally adequate explanations. To be a truth-seeking endeavor, the question that origin-of-life research must address is not, "Which materialistic scenario seems most adequate?" but rather, "What actually caused life to arise on earth?" Clearly, one possible answer to that latter question is this: "Life was designed by an intelligent agent that existed before the advent of humans." If one accepts methodological naturalism as normative, however, scientists may never consider this possibly true hypothesis. Such an exclusionary logic diminishes the significance of any claim of theoretical superiority for any remaining hypothesis and raises the possibility that the best "scientific" explanation (according to methodological naturalism) may not be the best in fact.

Scientific theory evaluation is an inherently comparative enterprise. Theories that gain acceptance in artificially constrained competitions can claim to be neither "best" nor "most probably true." At most such theories can be considered "the best, or most probably true, among an artificially limited set of options." Openness to the design hypothesis would seem necessary, therefore, to any fully rational historical biology—that is, to one that seeks the truth, "no holds barred."[46] A historical biology committed to following the evidence wherever it leads will not exclude hypotheses a priori because of their possible metaphysical implications. Instead, it will employ only metaphysically neutral criteria—such as causal adequacy—to evaluate competing hypotheses. Yet this more open (and arguably rational) approach would now seem to affirm the theory of intelligent design as the best, most causally adequate, scientific explanation for the origin of the information necessary to build the first living organism.

20

Why It Matters

At the height of the media frenzy surrounding the Dover trial in the fall of 2005, I was asked to appear on an MSNBC program called *The Abrams Report*. As is customary on "talking heads" shows, the host, Dan Abrams, played a short prerecorded "backgrounder" before the interview portion of the program. The report about the trial had been filed by Robert Bazell, a science correspondent for NBC News. After playing the piece and before asking his guests any questions, Abrams took the somewhat unusual step of offering his own opinion about the theory of intelligent design.

What he had to say wasn't too favorable. Abrams explained as how he thought that intelligent design was "dishonest." In his opinion, it was a stealth form of creationism that refused to mention God in order to conceal a religious agenda. He also alleged that the theory wasn't scientific. Not only had advocates of the theory "provided no new evidence"; there were no "peer-reviewed studies" in support of it, or so he claimed. After getting the other guest on the program, my old nemesis Eugenie Scott (see Chapter 6), to confirm this (falsely, as it happens), Abrams initiated a line of questioning to establish that intelligent design was "religion." To do this, he tried to get me to say that I thought the designing intelligence responsible for life was God.

But Abrams was setting a trap, one that, by this time, I knew all too well. If I answered truthfully (which I did) and told him that neither the evidence from biology nor the theory of intelligent design could prove the identity of the designer, he would accuse me of dishonesty and "refusing to come clean" about the religious nature of the theory (which he also did). If, on the other hand, I told him—again truthfully—that I personally thought that God had designed the universe and life, he would seize upon my words as proof that the *theory* of intelligent design was "religion," thus establishing in his mind that it must lack any scientific basis. "Just admit it, it's religion," he kept demanding.

As a Christian, I've never made any secret about my belief in God or even why I think theism makes more sense of the totality of human experience than any other worldview. But I was on Mr. Abrams's show to discuss the theory of intelligent design, and the theory does not make claims about a deity, nor can it. It makes a more modest claim based upon our uniform experience about the kind of cause—namely, an intelligent cause—that was responsible for the origin of biological form and information.

Of course, that modest claim raises a separate question, indeed, an important religious or philosophical question, namely, the very question about the identity of the designing intelligence that Abrams was pressing me to answer. Clearly, his question was legitimate. But I wanted to answer it *after* I had explained what the theory of intelligent design is and *after* I had established that there is scientific evidence for it. Otherwise, I knew the minute I said that I personally thought that God was the designer, he would dismiss the case for intelligent design as "religion" because he, and perhaps many of his viewers, assume that if an idea is religious it has no basis in fact or evidence.

And so a little tug-of-war ensued. To get me to either "admit it" or look evasive, Abrams asked two different questions in rapid succession: "What is intelligent design?" and "Who is the intelligent designer?" As I tried to answer his first question by defining intelligent design and describing some of the evidence that supports it, he kept demanding that I admit the designer is God. He was playing the journalist on the scent of a scandal, and the scandal he wanted to reveal was my belief in God. If I "admitted" that I thought God had designed the universe, then that would invalidate my position by showing intelligent design to be "religion." And so he peppered me with a series of questions: "Is it religion or not?" "You just can't . . . It's religion." "Is it religion or not?" "Just admit it. It's religion."

Religion, Science, or What?

Perhaps more than any other objection, the accusation that intelligent design is religion or "religion masquerading as science" has closed minds to considering the evidence for the design hypothesis. This has occurred partly because the media have successfully portrayed those who advocate intelligent design as having a hidden religious agenda. But there is another, more fundamental reason that this criticism has had the effect of closing minds. Many people assume that science and religion do not interact in any significant way. They assume that scientific theories have nothing to say about religious or philosophical questions and that if they do, then they must not really be scientific.

Abrams was clearly making this assumption as he pursued his "either or" line of questioning. Judge Jones's ruling in the Dover case also betrayed this same way of thinking. Either intelligent design is science or it is religion. Since, as both men noted, "major scientific organizations" say it isn't science, it must be religion. Similarly, since some advocates of "intelligent design" think that life was designed by God, intelligent design must be a religious belief rather than an evidence-based scientific theory.

But does this follow? Is intelligent design religion? And, if so, does that mean that the theory of intelligent design lacks a scientific basis? And what about the beliefs and motives of advocates of intelligent design: Do they invalidate the case for intelligent design, including the case I have developed in this book? There are several reasons to think not.

Not Religion

First, by any reasonable definition of the term, intelligent design is not "religion." When most people think of religion, they think of an institutionalized form of worship or meditation based upon a comprehensive system of beliefs about ultimate reality. Religions also typically involve various formal structures, practices, and ritualistic observances, including "formal services, ceremonial functions, the existence of clergy" and "the observance of holidays."[1]

Though intelligent design, like its materialistic evolutionary counterparts, does address questions about the origin of living things and may, therefore, have *implications* for metaphysical questions about ultimate reality (see Chapter 2), it does not proffer a comprehensive system of belief about that reality. Intelligent design does not answer questions about the nature of God or even make claims about God's existence. The theory of intelligent design does not promulgate a system of morality or affirm a body of doctrines about the afterlife. It doesn't require belief in divine revelation or tell adherents how to achieve higher consciousness or how to get right with God. It simply argues that an intelligent cause of some kind played a role in the origin of life. It is a theory about the origin of biological information and other appearances of design in living systems.

Moreover, the theory of intelligent design does not involve any of the practices or have any of the institutional structures or features typically associated with religions. It does not involve worship or meditation or recommend a system of spiritual disciplines. It does not have sacred texts, ordained ministers, rabbis, or priests; there are no intelligent-design liturgies, prayer meetings, or intelligent-design holidays. Advocates of intelligent design have formed organizations and research institutes,[2] but these resemble other scientific or professional associations rather than churches or religious institutions.

Despite this, some critics, such as Robert Pennock and Gerald Skoog, have gone so far as to characterize the theory of intelligent design as narrowly "sectarian."[3] Yet upon examination, this claim evaporates into nothing more than the observation that the theory of intelligent design is popular with some Christians and not others. In any case, the theory of intelligent design does not affirm sectarian doctrines. It has nothing to say about, for example, the virgin birth, the immaculate conception, predestination, infant baptism, the validity of Islamic law, salvation, original sin, or the reality of reincarnation. Moreover, the belief that a designing intelligence played a role in the origin of the living world is hardly unique to Christians or to religious persons in general. Historically, advocates of design have included not only religious theists, but nonreligious ones, pantheists, polytheistic Greeks, Roman Stoics, and deistic Enlightenment philosophers and now include modern scientists and philosophers who describe themselves as religiously agnostic.[4]

Theistic Implications

To deny that intelligent design is a religion is not to say, however, that the evidence for intelligent design in biology has no religious or metaphysical *implications*. Indeed, there is another option that Mr. Abrams and Judge Jones did not consider in their attempts to *classify* the theory rather than assess its merits. Theories, especially origins theories, needn't be *either* scientific *or* religious. They might be both. Or more precisely, some scientific theories—although not themselves *religions*—might have philosophical or religious *implications*.

There are good reasons to think that intelligent design is a scientific theory of this kind. First, as I've already shown (see Chapters 18 and 19) there are good reasons for thinking that intelligent design is a scientific theory. Second, the theory of intelligent design addresses a major philosophical question that most religious and metaphysical systems of thought also address, namely, "What caused life and/or the universe to come into existence?" Thus, like its materialistic counterparts, the theory of intelligent design inevitably raises questions about the ultimate or prime reality, "the thing from which everything else comes" (see Chapter 2).[5]

Moreover, intelligent design, arguably, has specifically theistic implications because intelligent design confirms a major tenet of a theistic worldview, namely, that life was designed by a conscious and intelligent being, a purposive agent with a mind. If intelligent design is true, it follows that a designing intelligence with *some* of the attributes typically associated with God acted to bring the first living cells into existence. The evidence of intelligent design in biology does not prove that God exists (or that a being with *all* of the attributes of a transcendent God exists), since it is at least logically possible that an immanent (within the universe) intelligence rather than a transcendent intelligence might have designed life. Nevertheless, insofar as a transcendent God (as conceived by theists) does possess conscious awareness and intelligence, it possesses the causal powers necessary to produce (and explain the origin of) specified biological information. Thus, the activity of a theistic God *could* provide an adequate explanation of the evidence of intelligent design in biology, though other entities *could* conceivably do so as well. Further, insofar as the evidence for intelligent design in biology increases the explanatory power of theism (as a kind of metaphysical hypothesis), it makes theism more plausible or more likely to be true than it would have been otherwise in the absence of such evidence.

Those who believe in a transcendent God may, therefore, find support for their belief from the biological evidence that supports the theory of intelligent design. They may cite this and other evidence as a reason to identify the designing intelligence responsible for life's origin with the God of their religious belief. Thus, it's fair to say that intelligent design has theistic *implications,* or implications that are friendly to theistic belief, even though the theory is not itself a religion (or a proof of God's existence).

Metaphysical or Religious Implications?

But if intelligent design makes belief in God more plausible or likely, doesn't that still mean intelligent design is essentially a religious, rather than a scientific, concept of biological origins? And shouldn't that induce some skepticism about its scientific merit?

No. On the contrary, the religious implications of intelligent design are not grounds for dismissing it. To say otherwise confuses the evidence for a theory and its possible implications. It also fails to recognize that intelligent design is not the only theory that has metaphysical or religious implications. Contrary to the popular "just the facts" stereotype of science, many scientific theories have larger ideological, metaphysical, or religious implications. Origins theories in particular have such implications since they make claims about the causes that brought life or humankind or the universe into existence.[6]

For example, many scientists believe that the big-bang theory, with its affirmation that the universe had a temporal beginning,[7] has affirmative implications for a theistic worldview. In fact, many scientists with materialistic philosophical leanings initially rejected the big-bang theory, because they thought it challenged the idea of an eternally self-existent universe and because they thought it pointed to the need for a transcendent cause of matter, space, and time.[8] Nevertheless, scientists eventually accepted the theory despite its (to some) unsavory philosophical implications. They did so because they thought the evidence strongly supported it.

Scientific theories must be evaluated on the basis of the evidence, not on the basis of philosophical preferences or concerns about implications. Antony Flew, the longtime atheistic philosopher who has come to accept the case for intelligent design, insists correctly that we must "follow the evidence wherever it leads," regardless of its implications. Were that not the case, the metaphysical implications of other scientific theories would invalidate them—and yet they do not.

Or consider another example. Some scientists think that Darwinism and other materialistic origins theories have significant metaphysical and religious (or antireligious) implications. Because both classical Darwinism and modern neo-Darwinism deny that the appearance of design in living organisms is real, they affirm that the process that gave rise to that appearance is blind and undirected. Chemical evolutionary theorists likewise insist that the first life arose, without direction, from brute chemistry.[9] Richard Dawkins has dubbed the idea that life arose as the result of an undirected process the "blind watchmaker" thesis.[10] He and other leading evolutionary theorists claim that biological evidence overwhelmingly supports this purposeless and fully materialistic account of creation.[11] As George Gaylord Simpson, the leading neo-Darwinist a generation ago, stated: "Man is the result of a *purposeless* and materialistic process that did not have him in mind. He was not planned."[12]

In light of this, Simpson and a host of prominent Darwinian scientists—from Douglas Futuyma[13] to William Provine[14] to Stephen Jay Gould[15] to Richard Dawkins[16]—have insisted that Darwinism (and the broader blind-watchmaker thesis) has made a materialistic worldview more plausible. They also argue that materialistic evolutionary theories have made traditional religious beliefs about God either untenable or less plausible. As Dawkins stated, "Darwin made it possible to become an intellectually fulfilled atheist."[17] Or as the late Harvard paleontologist Stephen J. Gould argued, Darwin formulated "an evolutionary theory based on chance variation and natural selection . . . a rigidly materialistic (and basically atheistic) version of evolution." Or as Gould explained elsewhere, "Before Darwin, we thought that a benevolent God had created us," but after Darwin, "biology took away our status as paragons created in the image of God."[18]

Similarly, many major biology texts present evolution as a process in which a purposeful intelligence (such as God) plays no detectable role.

As Kenneth Miller and Joseph Levine explained in the fourth edition of their popular textbook, *Biology,* the evolutionary process is "random and undirected" and occurs "without plan or purpose."[19] Or as W. H. Purvis, G. H. Orians, and H. C. Heller tell students in *Life: The Science of Biology,* "The living world is constantly evolving without any goals. Evolutionary change is not directed."[20] Other texts openly state that Darwin's theory has profoundly negative implications for theism. As Douglas Futuyma's biology text puts it, "By coupling undirected, purposeless variation to the blind, uncaring process of natural selection, Darwin made theological or spiritual explanations of the life processes superfluous."[21] For this reason, many people may find support for materialistic metaphysical beliefs in Darwinian and chemical evolutionary theory. Conversely, some scientists, such as Kenneth Miller, believe that evolutionary theory reinforces their religious beliefs.[22] Thus, if he is correct, the study of evolutionary theory may lead a student to "find Darwin's God."

Either way, chemical evolutionary theory and neo-Darwinism raise unavoidable metaphysical and religious questions. Arguably, these theories also have incorrigibly metaphysical and religious (or antireligious) implications. At the very least, many scientists *think* that evolutionary theory has larger metaphysical, religious (or antireligious), or worldview implications. Yet this fact has not prevented Darwinism from being regarded as a scientific theory. Nor does anyone think that the possible implications of the theory should determine its scientific merit or invalidate the evidence in its favor. Yet if the religious (or antireligious) implications of materialistic evolutionary theories do not make these theories religion or invalidate the evidence in support of them, then neither should the religious implications of the theory of intelligent design negate the evidence in its favor or make it a "religion"—with all that implies to the modern mind.

Instead, the content of a scientific theory, not its implications, should determine its merit. Scientific theories must be evaluated by the quality of the evidence and the arguments marshaled in their favor. But if that principle applies generally, and specifically, in the case of materialistic theories of evolution, then it should apply to the assessment of intelligent design as well. If it does, then the metaphysical or religious implications of intelligent design do not invalidate the evidential case in its favor.

Religious Motivations?

Just as the implications of particular theories do not determine their merit or truth, the motivations of the theorists who advance these theories do not invalidate them either. Indeed, there is an obvious distinction between what *advocates* of the theory of intelligent design *think* about the identity of the designing intelligence responsible for life and what the *theory* of intelligent design itself *affirms*. Just because some advocates of intelligent design think that God exists and acted as the designer does not mean that the theory of intelligent design affirms that belief.

Notwithstanding, there is no question that many advocates of the theory of intelligent design do have religious interests and beliefs and that some are motivated by their beliefs. I personally think that the evidence of design in biology, considered in the context of other evidence, strengthens the case for theism and, thus, my personal belief in God. Subjectively, as a Christian theist, I find this implication of intelligent design "intellectually satisfying."

Does that negate the case for intelligent design that I have presented? Some have argued as much. For example, in the Dover trial, Barbara Forrest and Robert Pennock argued that the religious beliefs of advocates of intelligent design delegitimized the theory. But this doesn't follow.

First, it's not what motivates a scientist's theory that determines its merit, status, or standing; it's the quality of the arguments and the relevance of the evidence marshaled in support of a theory. Even if all the scientists who have advocated the theory of intelligent design were motivated by religious belief (and they are not), motives don't matter to science. Evidence does. To say otherwise commits an elementary logical fallacy known as the genetic fallacy, in which an alleged defect in the source or origin of a claim is taken to be evidence that discredits the claim.

Here's an example. Suppose someone argues that because Richard was raised by evil atheists, his arguments against the existence of God are wrong. The reasoning is obviously fallacious. The facts of Richard's upbringing are irrelevant to the soundness of the arguments he makes. The arguments must be considered separately and on their own merits.

Similarly, that many ID advocates have religious beliefs that may increase their openness to considering intelligent design says nothing about the truth or falsity of the theory. Instead, the theory must be assessed by its ability to explain the evidence.

In any case, scientists on both sides of the origins controversy have ideological or metaphysical or religious (or antireligious) motivations. Barbara Forrest, a leading critic of intelligent design, is a board member of the New Orleans Secular Humanist Association. Other prominent critics of intelligent design such as Eugenie Scott and Michael Shermer have signed the American Humanist Manifesto III. Richard Dawkins's sympathies are well-known.[23] Aleksandr Oparin was a committed Marxist. Kenneth Miller takes a different, though no less disinterested tack. He claims that Darwinism illuminates his religious beliefs as a Catholic.[24]

Do the religious or antireligious motives of leading advocates of evolutionary theory disqualify Darwinian evolution or chemical evolutionary theory from consideration as scientific theories or diminish the merit of the theories? Obviously they do not. The motivations of the proponents of a theory don't negate the scientific status, merit, or validity of that theory. But if that general principle applies to the evaluation of materialistic evolutionary theories, then it should apply when considering the merits of intelligent design. In short, the motives of the advocates of intelligent design do not negate the claims of the theory.

It Gets Personal: Why It Matters

In public debates, I've often encountered critics of intelligent design who quote design advocates acknowledging their religious beliefs as a way to discredit the case for the design hypothesis. Though this happens frequently, I'm always a bit surprised that scientists and especially professional philosophers (who have presumably taught logic) would resort to such fallacious motive-mongering. Nevertheless, I suppose it's not surprising that religious motives and worldview implications do surface in the heat of discussion. The issue of biological origins raises deeply personal and philosophical issues. As I have reflected on these issues, I've become convinced that my former philosophy professor, Norman Krebbs, was right. The scientific case for intelligent design is fraught with philosophical significance and poses a serious challenge to the materialistic worldview that has long dominated Western science and much of Western culture.

With the rise of materialistic evolutionary theories in the nineteenth and twentieth centuries, science purported to explain the origin of everything from the solar system to the cell to the longings of the human soul, all by reference to undirected physical processes. Collectively, Laplace, Darwin, Oparin, and others portrayed the universe as an eternal, self-existent, self-creating system. Skinner, Freud, and Marx applied this perspective to understanding human beings by asserting that the same impersonal forces that shaped the material cosmos also determined human behavior, thought, and history. This view of reality, derived as it was from the natural and social sciences, understandably seemed to support the comprehensive philosophy or worldview of scientific materialism.

According to scientific materialism, reality is ultimately impersonal: matter and energy determine all things and, in the end, only matter matters. "In the beginning were the particles. And the particles became complex stuff. And the complex stuff reacted with other stuff and became alive. As the living stuff evolved, it eventually became conscious and self-aware . . . but only for a time." According to the materialist credo, matter and energy are the fundamental realities from which all else comes, but also the entities into which all that exists, including our minds and conscious awareness, ultimately dissolves. Mind and personhood are merely temporary "epiphenomena," a restless foam effervescing for a time atop a deep ocean of impersonality.

Though this view of existence proved initially liberating in that it released humans from any sense of obligation to an externally imposed system of morality, it has also proven profoundly and literally dispiriting. If the conscious realities that constitute our personhood have no lasting existence, if life and mind are nothing more than unintended ephemera of the material cosmos, then, as the existential philosophers have recognized, our lives can have no lasting meaning or ultimate purpose. Without a purpose-driven universe, there can be no "purpose-driven life."

The British analytical philosopher Bertrand Russell understood the connection between the denial of design (or what he called "prevision") and humankind's existential predicament. As he explained in 1918:

That Man is the product of causes which had no prevision of the end they were achieving; that his origin, his growth, his hopes and fears, his loves and his beliefs, are but the outcome of accidental collocations of atoms; that no fire, no heroism, no intensity of thought and feeling, can preserve an individual life beyond the grave; that all the labours of the ages, all the devotion, all the inspiration, all the noonday brightness of human genius, are destined to extinction in the vast death of the solar system, and that the whole temple of Man's achievement must inevitably be buried beneath the debris of a universe in ruins—all these things, if not quite beyond dispute, are yet so nearly certain, that no philosophy which rejects them can hope to stand.[25]

As a teenager in the mid-1970s, I sensed this absence of meaning in modern life. I'm not sure why. Perhaps it was that I had been acutely aware of the distress of the generation coming of age just ahead of me. Perhaps it was that my family had left the church. Perhaps it was because the questions that I kept asking did not seem to have any obvious answers. "What's it going to matter in a hundred years?" And by "it," I meant anything. What heroism, thought or feeling, labor, inspiration, genius, or achievement will last, if impersonal particles are all that ultimately endure?

Though the theory of intelligent design does not identify the agent responsible for the information—the signature—in the cell, it does affirm that the ultimate cause of life is personal. By personal I mean a self-conscious, deliberative mind in possession of thoughts, will, and intentions. Only persons have such minds and only minds of this kind can create complex specified information. If we know anything we certainly know this. Thus, while the theory of intelligent design does not prove the existence of God or answer all of our existential questions, it does reestablish the conditions of a meaningful "search for meaning." The case for intelligent design challenges the premise of the materialist credo and holds out the possibility of reversing the philosophy of despair that flows from it. Life is the product of mind; it was intended, purposed, "previsioned." Hence, there may be a reality behind matter that is worth investigating.

These implications of the theory are not, logically speaking, reasons to affirm or reject it. But they are reasons—very personal and human reasons—for considering its claims carefully and for resisting attempts to define the possibility of agency out of bounds. Is intelligent design science? Is it religion? Perhaps these are not the right questions. How about, "Is there evidence for intelligent design?" "Is the theory of intelligent design true?" And, if so, "What does it imply?"

Indeed, for me, far from wanting to avoid the philosophical or theological questions that naturally arise from a consideration of the evidence for intelligent design, these questions have done much to sustain my long interest in the scientific controversy surrounding the origin of life. And why not? If there is evidence of design or purpose behind life, then surely that does raise deeper philosophical questions. Who is the designer, indeed? Can the mind that evidently lies behind life's digital code be known? Can we as persons know something of the agent responsible for the intricacies of life? Is there a meaning to existence after all? I have asked these questions for many years. What excites me about the theory of intelligent design and the compelling evidence now on display in its favor is not that the theory answers these questions, but instead that it provides a reason for thinking that they are once again worth asking.

Conclusion

For one hundred and fifty years many scientists have insisted that "chance and necessity"—happenstance and law—jointly suffice to explain the origin of life on earth. We now find, however, that orthodox evolutionary thinking—with its reliance upon these twin pillars of materialistic thought—has failed to explain the origin of the central feature of living things: information.

Even so, many scientists insist that to consider another possibility would constitute a departure from science, from reason itself. Yet ordinary reason and much scientific reasoning that passes under the scrutiny of materialist sanction not only recognize but require us to recognize the causal activity of intelligent agents. The sculptures of Michelangelo, the software of the Microsoft Corporation, the inscribed steles of Assyrian kings—each bespeaks prior mental activity rather than merely impersonal processes. Indeed, everywhere in our high-tech environment we observe complex events, artifacts, and systems that impel our minds to recognize the activity of other minds: minds that communicate, plan, and design. But to detect the presence of mind, to detect the activity of intelligence in the echo of its effects, requires a mode of reasoning—indeed, a form of knowledge—that science, or at least official biology, has long excluded. If living things—things that we manifestly did not design ourselves—bear the hallmarks of design, if they exhibit a signature that would lead us to recognize intelligent activity in any other realm of experience, then perhaps it is time to rehabilitate this lost way of knowing and to rekindle our wonder in the intelligibility and design of nature that first inspired the scientific revolution.

Epilogue

A Living Science

John F. Kennedy was assassinated by a single gunman. The plays attributed to William Shakespeare were written by William Shakespeare. Troy was an actual city in the ancient world, not a mere legend. Imagine that a history professor has just asserted these three claims to spark a class discussion about how historians sift evidence and arrive at conclusions. But before the discussion can take its expected course, an administrator steps through the door of the classroom and clears his throat. "The consensus among the university's leadership is that such claims should simply be dismissed," he begins. "Such claims lack utility. History, properly understood, develops theories that allow the science of history to move forward. Unless these historical claims can be shown to be *useful* for making new discoveries, you as students should reject them. Thank you for your attention. Carry on." He takes a seat.

The students respond in various ways—acquiescence, dumbfounded silence, a one-fingered salute hidden behind a textbook. The history professor, after recovering from the unexpected intrusion (and remembering that she has already secured tenure), seizes the teachable moment by immediately responding. She begins by noting that one doesn't assess the truth of a proposition by determining if it's useful for doing or discovering something else. Neptune circles the sun whether or not we can make any further use of the fact. What's more, she continues, some historical conclusions that seem to lack practical utility can eventually lead to useful insights. The knowledge that Shakespeare wrote Shakespeare tells us a great deal about what is and isn't essential for fostering great literature, since, for instance, Shakespeare was raised in an unimportant English village by parents of modest means and received only a grammar-school education. What appears useless to one may prove quite useful to others. The students nod appreciatively.

The administrator is unfazed. "Unless historical claims demonstrate their utility," he repeats, "they should be rejected."

This fictional scenario illustrates an important feature of the current controversy over intelligent design. After making my case for intelligent design as the best explanation for the origin of biological information, I often encounter an objection similar to the one lodged by the intrusive administrator in this imaginary history class. Scientists often respond to my argument not by proposing a better explanation for the origin of information, but by insisting that intelligent design is irrelevant since it doesn't lead anywhere. "What can I *do* with intelligent design?" they ask. "What predictions does it make?" "What research questions does it generate?" These scientists don't necessarily deny the truth of intelligent design. They reject it on the grounds that it lacks utility. Yet a proposition may lack utility for generating predictions and still be true.

Philip Kitcher articulates a more nuanced but related objection. Kitcher is a respected philosopher of science who holds a distinguished professorship in philosophy at Columbia University. In 2008, he published *Living with Darwin,* largely in response to the increasing interest in the theory of intelligent design. In the book, Kitcher rejects intelligent design because he sees it as unable to explain putatively new DNA evidence derived from studies of the genome. Kitcher does not address the argument I make in this book—namely, that intelligent design best explains the origin of the specified information necessary to produce the first living cell. Instead, he argues that there are other features of the genome—in particular, the many supposed "junk" sequences in DNA—that intelligent design can't explain. Because of this failure, he concludes that intelligent design is a "dead science."

Kitcher is a sophisticated philosopher of science, and so he chose his words carefully. Kitcher acknowledges that attempts to use demarcation arguments to define intelligent design as inherently unscientific have failed. To establish that intelligent design does not qualify as a scientific theory, "we must," he says, "explain which rule of proper science has been broken." But, he observes, this "leads into thickets of philosophy from which no clear resolution has yet emerged. For the past half-century, philosophers have tried and failed to produce a precise account of the distinction between science and pseudo-science. We cannot seem to articulate that essential demarcation." Besides this, Kitcher acknowledges that the design hypothesis played an important role in the formation of many sciences and often guided fruitful scientific investigations in centuries past. As he explains, "Intelligent design has deep roots in the history of cosmology, and in the earth and life sciences."[1]

Nevertheless, Kitcher argues, the theory no longer plays that kind of role. Though many earlier natural philosophers used the design hypothesis to guide their research, Kitcher says, intelligent design can no longer do so because it fails to explain the facts of biology today—in particular, the pervasive presence of "junk DNA" within the genome. Thus, Kitcher concludes that intelligent design is a dead science—one that lacks utility as an interpretive framework—because it cannot *explain* the current facts of molecular biology.

Kitcher presupposes, as do I, a connection between the explanatory power of a theory and its probable truth. He does not insist that theories about the past must guide discovery or predict the future. Nor does he say that ID isn't science because it doesn't do these things. Instead, he claims that ID no longer provides a useful interpretive framework *because it lacks explanatory power.* This constitutes a significant objection to the case I have made. Further, since I have based my case for intelligent design without reference to recent developments in the field of genomics, some might charge, echoing Kitcher's argument about "junk" DNA, that my argument for intelligent design is obsolete.

Is Kitcher correct? Does intelligent design fail to explain the current DNA evidence from genomic studies? Are others correct when they say intelligent design no longer has utility for guiding research or making predictions? In short, is intelligent design a dead science or an active, useful, living science? If living, in what way? As an explanatory framework? A guide to discovery? Or as a generator of predictions?

In this Epilogue, I argue that intelligent design is a "living science," one that can: (1) readily explain new facts, including some surprising new discoveries in genomics, and (2) generate many fruitful research questions and lines of inquiry. Indeed, I show that *because* ID explains the most current data about the cell's information storage and processing system (and, arguably, does so uniquely well), it *naturally* suggests many new research questions and lines of inquiry. Then, in Appendix A, I also show that, though we do not generally require historical scientific theories to do so, (3) intelligent design does make a number of discriminating predictions about what we ought to find in future investigations of living systems if intelligent design actually played a role in the origin of these systems.

Before proceeding, a note of warning is probably appropriate. This Epilogue is written primarily for scientists who wonder what they *can* do with intelligent design. This means that it's a bit more technical than most of the previous chapters. It is also longer, since it goes into some detail about how intelligent design provides both an *interpretive* framework for a wider class of facts than I've previously considered (including recent discoveries in genomics) and an *investigative* framework for ongoing research. Certainly curious nonscientists will find in these pages much to fascinate them, but perhaps also some parts to skim over in favor of summarizing material at the beginnings and endings of the various sections.

The Ongoing Explanatory Power of Intelligent Design

A successful scientific research program provides, perhaps first and foremost, an interpretive framework for explaining new evidence, including evidence that competing theories have difficulty explaining. Throughout the history of science new theories or research programs have been established—as "living sciences"—when they can explain anomalies and new facts that are unexpected from the point of view of older established theories. For example, Newton made the case for his theory of universal gravitation in part by showing that it could readily explain many facts that the established vortex theory could not. As he famously asserted at the beginning of the *Principia* before reciting these anomalous facts, "The hypothesis of vortices is beset by many difficulties."[2]

As noted (see Chapters 7, 18, and 19), many theories in the history of science have been accepted mainly because of their ability to explain established facts better than their competitors, irrespective of whether they also successfully predict new facts—though clearly many successful theories have done both. Explanatory power, especially with respect to a body of anomalous data, often establishes a theory as an active scientific research program. Certainly a theory typically also suggests further research questions that scientists can pursue in the lab, in the field, or, increasingly, on the computer, but it typically gains its initial traction by providing a useful framework for understanding and interpreting a broad range of evidence, particularly evidence recently uncovered.

And so it is with the theory of intelligent design. Thus, before I examine the research questions and predictions that the theory of intelligent design generates, I want to look in some detail at the new evidence that it explains, specifically the new evidence about the genome and the cell's information-processing and storage system that has deepened the DNA enigma and proven increasingly puzzling from the standpoint of conventional materialistic theories of evolution.

Though intelligent design and the selection and mutation mechanism can both explain some appearances of design, the two forms of causation work differently. For this reason, there are some features of living systems that we should expect only if mutation and selection had generated them, and other features or outcomes that we should expect if intelligent design had done so. And there are some features of living systems that we should expect to find whether either cause was operative.

We know a lot from experience about how intelligent agents design information-processing and storage systems. We know that intelligent agents regularly use particular design strategies and patterns for such systems. They organize information hierarchically so as to maximize storage density and facilitate its efficient retrieval. We also know that they generally can conceive of outcomes before they exist. Those outcomes are then actualized by arranging the many disparate parts of the system without the need to develop and maintain functional "intermediate forms" or structures along the way to the desired functional end point. We also know that designers often seek to optimize one or more objectives, and where they seek to optimize several, they often do so by balancing competing objectives and by making judicious compromises or "elegant" trade-offs between them. Thus, there are many features of a designed system that we should expect to see if in fact the information-processing and storage system in the cell was intelligently designed. The main body of this book showed that two such features—functionally specified digital information and a functionally integrated system for processing that information—are clearly evident in the cell.

Conversely, we know a lot about natural selection and random mutation and other similarly naturalistic mechanisms. We know how such
mechanisms should work in theory and, therefore, what types of features
in living systems these mechanisms do, or do not, readily explain.

Evolutionary biologists commonly remind us that natural selection
and mutation can produce the appearance of design. But many also
acknowledge that selection and mutations can explain some appearances of design and not others. This mechanism explains appearances
of design that are optimized in a way that is consistent with the results
of a blind trial-and-error process. Yet it wouldn't explain optimized
structures or designs that are inconsistent with the outworking of such
a process. Similarly, since natural selection is by definition mindless and
undirected, it cannot explain structures or systems or informational features (appearances of design) that require *foresight* to assemble, whereas
actual intelligent design can explain such appearances.

Since natural selection "selects" or preserves functionally advantageous mutations or variations, it can explain the origin of systems that
could have arisen through a series of incremental steps, each of which
maintains or confers a functional advantage on a living organism. Nevertheless, by this same logic, selection and mutation face difficulty in
explaining structures or systems that could not have been built through
a close series of functional intermediates. Moreover, since selection operates only on what mutation first produces, mutation and selection do
not readily explain appearances of design that require discrete jumps of
complexity that exceed the reach of chance; that is to say, the available
probabilistic resources.

Since these two different types of causes (mutation and selection,
and intelligent design) generate at least some different types of effects, they also generate different empirical expectations about what
we should find in living systems (or in the fossil record). Scientists can,
therefore, use evidence from living systems to adjudicate the comparative explanatory power of these different modes of causation. (Since
other materialistic mechanisms produce certain types of features and
effects and not others, their explanatory power also can be assessed in
a similar fashion.)

Since we would expect to find different patterns of evidence in living
systems if they had one type of cause rather than the other, the evidence

present in living systems enables us to assess which of the two types of cause provides a better explanation of the evidence in question, making theories invoking these modes of causation eminently testable. Designing agents produce information-rich systems in one way. Mutation and selection are thought to have produced information in another. Therefore, the features of the genome and its surrounding information-processing system provide a critical test of the comparative explanatory power of these two competing causal explanations.

So, given what we know about these prospective causes, what should we expect to find in the genome and in the cell's larger information-processing system, and how does it compare with what we do see?

In fact, we find features that are just what we would expect if the information systems in the cell were the product of a designing intelligence as opposed to an undirected process such as selection and mutation.

The New Genomics

Over the last fifteen years, genetic studies have revolutionized our understanding of how the cell performs operations on genetic and other forms of biological information. Biologists still affirm that DNA contains specified information, but they have discovered that the system for storing and processing this information is even more complex than I described in Chapter 5. Though the mechanisms of transcription and translation described in Chapter 5 are still thought to play a central and essential role in the expression of genetic information, leading researchers have discovered that the information for building a given protein is not always (or even usually) located in just one place along the DNA molecule. They have also discovered that one gene does not always code for just one protein, as George Beadle and Edward Tatum first claimed in the early 1940s.

Instead, leading researchers now realize that, depending upon how the cell processes the information stored in DNA, a single gene may contribute to the production of thousands of proteins and other gene products (such as regulatory and structural RNA molecules).[3] The cell also uses genetic information to produce critical RNA molecules that do not undergo translation, but instead direct the processing of other genetic information.[4] Further, during the translation process, additional processes edit the chains of amino acids produced before they fold into their final functional forms.[5] Equally revolutionary is the discovery that biological information beyond (not resident in) DNA plays a critical role in the development of organisms. As molecular biology and genomics have revealed new features of the cell's information storage and processing system, they have inspired a new conception of the gene—one in which the gene is no longer understood as a singular, linear, and localized entity on a DNA strand, but rather as a distributive set of data files available for retrieval and context-dependent expression by a complex information-processing system.[6]

For this reason, some might argue that the case for intelligent design presented thus far is outdated. Nevertheless, far from making the case for intelligent design obsolete, these recent discoveries strengthen it. The cell's information-processing and storage systems possess many features that we find only in intelligently engineered systems and they are, for this reason, precisely what we should have expected to find if these systems had in fact been designed. Based upon our experience of the cause-and-effect structure of the world, we know of only one type of cause for the origin of these newly discovered features—and that cause is, again, intelligence. That the theory of intelligent design can provide a causally adequate explanation for otherwise anomalous evidence and unexpected discoveries underscores its ability to provide an interpretive framework for scientific research and illustrates a key way that the theory functions as a dynamic scientific research program.

Consider three new discoveries about the cell's informational system that illustrate the ability of intelligent design to explain otherwise unexpected discoveries, including many that the competing explanations do not. First, functionally specified information is densely concentrated in DNA. Second, the genome is hierarchically arranged to optimize access and retrieval of information. And, third, the organism provides an informational context—involving both genomic and extragenomic information—that determines the expression of lower-level genetic modules.

1. The Concentration of Information in DNA

Over the last ten years, scientists have come to realize that genetic information in DNA is organized to maximize the storage of information. Far from containing a preponderance of "junk"—nonprotein-coding regions that supposedly perform no function—the genome is dominated by sequences rich in functional information. Indeed, even the nonprotein-coding regions of DNA serve multiple functions. In Chapter 18, I listed some of the many functions that the nonprotein-coding regions play. I also noted that—overall—these regions of the genome perform many of the same functions as an operating system in a computer. For example, the noncoding regions of DNA (in concert with other sources of information) direct the expression of base sequences in the protein-coding regions in DNA. Thus, far from being dispersed sparsely, haphazardly, and inefficiently within a sea of nonfunctional sequences (ones that supposedly accumulated by mutation), functional genetic information is densely concentrated on the DNA molecule.

This is not to say there is no evidence of mutational accumulation or degradation (of the functional genome) over time. The genome does display evidence of past viral insertions, deletions, transpositions, and the like, much as digital software copied again and again accumulates errors. Nevertheless, the vast majority of base sequences on the genome, and even the many sequences that do not code for proteins, serve essential biological functions. Genetic signal dwarfs noise, just as design advocates would expect and just as they predicted in the early 1990s. (See Chapter 18, backnotes 25 and 39).

Another feature of the cell's information-storage system is reshaping our understanding of the gene itself. Genomic studies reveal that the cell accesses "distributed genetic data sets" and then combines these modules of specified information to direct the production of various proteins during translation—much as a computer operating system retrieves and accesses modular data sets stored in various places on a hard drive and then reassembles them into a single data file. As a result of these recent discoveries, biologists have come to understand the gene less as a string of nucleotide bases stored in just one place along the DNA molecule and more as a distributed data set of specified base sequences stored in various places along the DNA helix (and sometimes

across various chromosomes).[7] These distributed information-rich sequences are assembled by the organism's information-processing system
using—in addition to the genetic code—other higher-level codes that
determine how various modules of genetic information should be re-
trieved and concatenated before translation occurs.

There is another more striking—even eerie—way that the genome
maximizes information storage. In the traditional picture of the gene,
a sequence of bases for building a given protein begins at a start codon
(a triplet of bases designating the beginning of relevant coding instructions) and ends at a stop codon. The sequence in between constitutes *the*
genetic assembly instructions for building that particular protein. Many
proteins are built in this fashion. But recently molecular biologists have
discovered that multiple messages (sets of assembly instructions) can be
stored in the same sequence of bases or region of the genome.

Sometimes these multiple messages overlap along the genome. In such
cases, the RNA polymerase complex accesses and transcribes different
genetic messages by starting and stopping at different places in the same
gene or genome region.[8] Additionally, two large ribonucleoprotein complexes called "spliceosomes" and "editosomes" splice and edit the RNA
transcripts to produce many other genetic messages before their translation at the ribosome.[9] (RNA transcripts from the same genome region
and even RNA transcripts from different chromosomes can be spliced
together to form many new sets of assembly instructions.)[10] At the ribosome, the translation machinery—in concert with various other protein
and RNA factors—then determines how these messages will be read.

The splicing, editing, and reading process can produce more than
one protein from the same RNA message. Indeed, one gene or region
of the genome, in concert with extragenomic codes and machinery,
can produce many thousands of different RNA messages and proteins.
This polypotency results from a highly efficient system of information
storage involving both the DNA molecule and a larger information-
processing system composed of numerous specific RNAs and protein
factors. Indeed, as a result of the overlapping genetic messages and
different modes of information processing, the specified information
stored in DNA is now recognized to be orders of magnitude greater
than was initially thought in the immediate wake of the molecular biological revolution.

Genetic messages are also often embedded within other messages in another way. Genes are culled from regions of DNA that contain both introns and exons. Exons contain coding instructions for specific proteins (and sometimes RNAs). Introns are sections of DNA interspersed between the exons—sections that do not contain coding information for the proteins or RNAs that exons specify. To produce a transcript for translation, the information-processing machinery of the cell must cut out introns from an RNA transcript and splice together individual exons (i.e., those previously separated by introns) to form a single seamless message. At first glance it looks as if the introns are just getting in the way. Nevertheless, scientists recently have discovered that, though introns do not have information that codes for the proteins that exons specify, they often have imbedded within them other genes for building other proteins.[11] Additionally, individual exons and introns have imbedded within them separate messages or coding regions for specifying structural and regulatory RNAs.[12] Thus, like Russian dolls stored within Russian dolls, exons and introns encode multiple genetic messages within themselves and are themselves part of a larger genetic message.

To get a picture of how this works, imagine a coded message in a letter sent back across enemy lines during the Revolutionary War. Read in the standard way, the letter describes problems on the farm, the weather, household challenges, and progress the children are making in their studies, all of it clear and accurate information. Yet the letter also manages to encode another message about enemy troop strength and movements, supplies of ammunition, and an impending enemy attack. The soldier receiving the message can read it because he possesses a key—a cipher—that enables him to identify the location of the second message embedded in the first and then to translate it.

In the same way, the cell has protein machinery and RNA codes that jointly function as a cipher enabling it to access and read the secondary imbedded messages within the primary message of the genome. Within the cell, higher-level RNA codes, protein factors, and cutting-and-splicing enzymes work together to enable the cell to identify, access, and transcribe these genetic messages within messages—messages that are transcribed into RNA and read at the ribosome during translation. The presence of these genes embedded within genes (messages within messages) further enhances the information-storage density of the genome and underscores how the genome is organized to enhance its capacity to store information.

Signs of Design?

This extraordinarily dense concentration of functional information, and the storage system that makes it possible, suggest design for several reasons. First, the amount of information present in even the simplest prokaryotic genome is orders of magnitude greater than previously assumed. Since even the amounts previously known vastly exceeded the probabilistic resources of the universe, the origin of the volume of information now known to be stored in the genome is even more unlikely to have arisen by chance alone. As W.-Y. Chung, a bioinformatician at the Center for Comparative Genomics and Bioinformatics at Penn State University, has noted, the existence of "dual coding" and overlapping protein-coding reading frames, just one of many cellular innovations for concentrating genomic information, is "virtually impossible by chance."[13]

Nor does natural selection acting on random mutations help explain the efficient information-storage density of the genome. Quite the reverse. Natural selection and random mutation is essentially a trial-and-error process. It would, therefore, necessarily generate many mutational errors in the process of producing any functionally specified sequences. It should produce a genome in which genetic noise rivals or dwarfs genetic signal. This is why the discovery of supposed junk-containing regions of the genome was seen as confirming that selection and mutation had shaped the genome. Indeed, the scientists who first reflected upon the discovery of the supposed junk DNA—Susumu Ohno, Richard Dawkins, Francis Crick, Leslie Orgel, W. Ford Doolittle, Carmen Sapienza—all assumed that junk DNA was an expected by-product of mutational processes and that the presence of junk in the genome confirmed that such processes had played a central role in the origin of genetic information.[14] Yet, as noted in Chapter 18, advocates of intelligent design expected early on that the nonprotein-coding regions of the genome should play a functional role in the cell, because the design logic of an information-processing system precludes carrying a preponderance of useless code, especially in biological settings where such excess would impose a burdensome energy cost on the cell.

There are other positive reasons to suspect that intelligent design played a role in the origin of the information-storage system of the genome. Based upon our experience of how intelligent agents design information-processing systems and encrypt coded messages, the features of the genome responsible for its storage density are not at all unexpected. We have experience of information-processing systems that rely on one section of encoded information to regulate and direct the use of another. And we know of systems that direct the retrieval of dispersed data modules in different places and then reassemble them as complete data files. Computer operating systems, undisputed products of intelligent design, perform both these functions.

Operating systems use digitally encoded information stored in one part of the computer hard drive to direct the use of other digitally coded information, in particular, the application programs stored in another part of the hard drive. In the cell, nonprotein-coding regions of the genome provide formatting, bracketing, and indexing codes that enable the cell to locate and express specific modules of stored genetic information, the expression of which may be needed to respond to specific environmental stresses or changing developmental conditions.[15] Operating systems also direct the retrieval of dispersed data modules that are stored on the hard drive in different places and then reassemble them as complete data files, just as the cell's information-processing system directs the retrieval of genetic data modules for expression, assembly, and later concatenation (of gene products).

Operating systems also store code to perform functions ("services") that many application programs need, allowing specific application programs to be more streamlined and store less information than they otherwise would have to do. Similarly, nonprotein-coding DNA provides services and needed functions to the protein-coding DNA during gene expression. For example, nonprotein-coding sections of DNA produce small microRNAs crucial for translational regulation whenever protein-coding regions of the gene are being accessed and expressed. Every protein-coding region also needs promoter sequences and a host of other codes (including some stored in nonprotein-coding regions as far as a million bases upstream from the coding region of the gene).[16] These promoters are necessary to orient the cell's transcriptional machinery correctly.

Though vastly more complex, the cell's information-processing and storage system performs many of the same functions, and does so using functional logic and design patterns that again are reminiscent of those in the operating systems of modern digital computers. Clearly, biological and computer systems use different material media to store and process information. Yet the logic embodied in the two systems exhibits many uncanny similarities and functional mappings.

The discovery of genetic messages encoded within genetic messages also calls to mind systems that have their origin in intelligence. With cipher codes, human agents conceal one message in another. A bit earlier, I offered the fictional scenario of a letter sent to a Revolutionary War soldier containing messages at two levels. The scenario, as it turns out, is only fictional up to a point. During the Revolutionary War, George Washington's men did in fact use simple ciphers in which each English letter in the primary text was to be read as a different English letter in a hidden secondary text. Thus, if the recipient of a letter bearing the primary message knew the code—the set of correspondences—he could translate the primary message and reveal the encrypted message. Clearly, encrypting a message within a message is more difficult than just writing a single message, because the writer must consider two sets of functional constraints. For each letter selected, the writer has to consider how two sequences are affected simultaneously—whether they are both meaningful and whether the second encrypted sequence expresses the meaning intended. The decoded meaning of the secondary sequence of characters has to conform to the conventions of English communication to convey its message, and the surface message also must express some meaning as well—at least, if it is to conceal the presence of the encrypted message.

Clearly, making two sequences that satisfy two sets of functional constraints simultaneously is more difficult than constructing a single sequence that must satisfy only one set of such constraints. Thus, the probability of generating such a meaningful message within another meaningful message by change alone is vastly smaller than the odds of getting a single message to arise by chance on its own. For this reason, the discovery of dual and overlapping messages in genetic texts—messages essential to function—only complicates the information problem for scenarios that rely on chance and/or natural selection. Indeed, a trial-and-error process seems unlikely to produce nested coding of information, since the probability of error increases with each trial when two or more sets of functional constraints have to be satisfied. And many functional outcomes in the cell depend upon satisfying multiple sets of constraints.

Further, since self-organizational affinities fail to explain the sequential arrangements of DNA base sequences generally, they do nothing to account for even more sophisticated forms of sequencing (i.e., those involving dual messaging) in the genome. Instead, this form of encryption seems to point decisively to design, because the use of such encryption techniques are, based upon our experience, the sole province of intelligent agents. We know of no other cause of this effect. The evidence of sophisticated encryption techniques within the genome thus constitutes another distinctive diagnostic—or signature—of intelligence in the cell.

2. Hierarchical Arrangement, Optimized for Access and Retrieval

Another feature of the genome evokes comparisons to information-processing and storage systems that are known to have been intelligently designed. Like the files-within-folders system used to organize data files in a personal computer, the genome is hierarchically organized to make retrieving, manipulating, and expressing information-rich data sets more efficient.

The genome manifests this hierarchical organization in several ways. First, different types of genes and various genomic regions occur in nonrandom groupings along a DNA string.[17] It was long thought that the arrangement of one sequence relative to another is a chance affair, with gene clustering arising only occasionally from duplications or some other type of mutation. But studies of genome sequences in disparate organisms have revealed that genes themselves are ordered into longer functional complexes, just as the bases in various coding sections of a gene are specifically sequenced to enable the production (and efficient regulation) of distinct RNAs and proteins. In the same way that words are ordered into sentences and sentences into paragraphs, nucleotide bases are ordered into genes and genes are ordered into specifically arranged gene clusters. Or think of these individual genes as computer data files and groupings of genes as folders containing several files.

The groupings of DNA "files" that we observe serve several roles. These groupings allow the cell to make longer transcripts that are combinations of different gene messages. In other words, the coding modules of the gene files in a "folder" can be combined in numerous ways—and in both directions—to greatly increase the number of encoded transcripts and protein products from the same genomic region or resources. In addition, gene folders are structured to permit the en masse retrieval of the DNA files needed to make these larger combination transcripts, or they are structured to permit the selective accessing of subfiles.[18]

Second, gene folders are themselves nonrandomly grouped along chromosomes to form higher-order folders.[19] "Housekeeping" gene folders are clustered to form housekeeping gene "superfolders"; tissue-specific gene folders are clustered to form tissue-specific gene superfolders. Also, in the genomes of mammals, specific types of functionally polyvalent DNA elements preferentially associate with superfolders. SINEs (Short Interspersed Nuclear Elements) are densest in housekeeping regions, for instance, where they modulate a host of genomic activities. LINEs (Long Interspersed Nuclear Elements) are by contrast a common feature of tissue-specific chromosome domains, where they commonly act as chromosome "scaffold attachment regions" that determine how strands of DNA will fold and unfold in the nucleus. They also function as "molecular rheostats" that fine-tune gene expression. Even the gene "deserts" that occur between superfolders—long segments devoid of protein-coding genes once thought to be junk and often still cited as evidence against intelligent design[20]—are now known to contain a set of superfolders. Indeed, these gene deserts are extensively transcribed and code for regulatory RNAs.[21] Overall, these clusters of gene clusters perform many other functions that are only now being discerned.

Third, the combinatorial arrays of gene superfolders, including gene deserts, are in turn grouped into yet larger sets of sequences termed "isochores." A way to think about isochores is as DNA megafolders. Moreover, these isochores, millions of bases long, are arranged into triplets to form the band patterns, or "bar codes," visible when scientists stain the chromosomes of mammals.[22] Although the existence of isochore megafolders has been known for well over two decades, only in the past few years have their many functions become clear. One of these functions is to regulate the three-dimensional association of adjacent chromosomes in the nucleus. These isochores also serve as a framework for the formation of nuclear organelle-like compartments.[23] And isochores and chromosome bands are likely only the lower tiers of still higher codes that yet await discovery.

More Signs of Design?

All this is, again, just what we would expect from a design-theoretic perspective based upon our experience of how intelligent agents design information-processing systems. We have extensive experience of systems that organize data sets as files within folders to facilitate access and retrieval of information—and we know what caused these systems to originate. The computer operating system and word-processing program that I am using to write this Epilogue will allow me to store this file within a folder of other drafts of this same chapter. I can store that folder within another folder that contains the folders housing drafts of each of the other chapters of this book. That folder can in turn be stored within another folder that stores folders containing material for other books. Such hierarchical informational filing systems are the undisputed products of intelligent design. Our knowledge of these computer filing systems induces a sense of déjà vu when we encounter the hierarchical informational filing systems in biological organisms—as it has for software engineers I know who have studied molecular biology. This eerie sense of déjà vu is nothing mystical, however. It derives from the quite rational human penchant for pre-theoretic uniformitarian reasoning, registering that we have experience of hierarchical systems for storing information and that we are familiar with a type of cause that can produce them: intelligent design.

On the other hand, the linear and hierarchical arrangement of the genome is not what we would expect if the information in DNA and the chromosomes had developed by undirected mutation and selection. Just as this mechanism should have generated much more genetic noise than signal, because of its reliance upon a random trial-and-error process, so too should we not expect random mutation and selection to produce a nonrandom functionally specified organization of genetic files along the DNA strand, or a similar arrangement of genetic files within folders, or folders within folders. Natural selection and random mutations (recombinations) might generate a few seemingly nonrandom clusters of genes. But we would not expect this mechanism to produce a system of genomic organization in which the location and arrangement of adjacent genes invariably assists, and indeed is essential to, a functional outcome. Nor would we expect the arrangement of genes within a cluster or the arrangement of clusters to always embody a clear and indispensable functional logic.

Mutation might occasionally produce a chance association of, for example, a transcriptional regulatory element next to a protein-coding gene. But we would not expect mutations to produce arrangements and associations of genes such that cells would repeatedly deploy gene products to the right tissues at the right time and in the right amount. And yet the organization of the genome allows the organism to express gene products with precisely such targeted and calibrated efficiency.

As University of Indiana evolutionary biologist Michael Lynch has argued using standard population genetics, the size of breeding populations of multicellular organisms are simply not large enough to have afforded natural selection sufficient opportunity to shape genomes into structures with the kind of hierarchically organized systems of information storage that they exhibit. Lynch instead hopes that the structure of the genome can be explained by a neutralist theory of evolution based mainly on genetic drift.[24] But by removing natural selection as a potential shaping influence on the structure of the genome, Lynch repairs to purely random mutations to explain the incredibly improbable arrangement and structure of the genome as a whole. This scenario seems, if anything, less likely than one dependent on natural selection.

In any case, the hierarchical organization and layering of many types of information onto the same physical medium would seem to require considerable forethought, precisely what natural selection by definition cannot provide. And experience affirms that the hierarchical organization of information evident in the genome constitutes another distinctive hallmark or signature of intelligence, another feature of living systems for which intelligent agency is the only known cause. Thus, intelligent design constitutes the best, most causally adequate explanation for the origin of this and other key features of the cell's information-processing system, underscoring the continuing and growing explanatory power of intelligent design as a scientific theory of the origin of information.

3. Informational Context—Genomic and Extragenomic

In recent years, developmental and evolutionary biologists have discovered that coding sequences in the genome do not by themselves determine the function of gene products during embryological development. Instead, scientists have found that the larger informational context in which genes are expressed often determines the specific function of the proteins they produce. Biologists have shown this in part by taking a gene out of one kind of organism and then expressing it in a radically different kind of organism.[25] Moreover, by comparing genes between groups, biologists discovered that largely identical sequences regulate the development of very different structures in different organisms. In fruit flies, the gene *distal-less* regulates the development of compound limbs with exoskeletons and multiple joints. In sea urchins, however, the homologous gene regulates the development of spines. In vertebrates, by contrast, it regulates the development of another type of limb, with multiple joints but an internal bony skeleton.[26] Except insofar as these structures all exemplify a broad general class, namely, appendages, they have little in common with each other. As Stuart Newman, professor of cell biology at New York Medical College, notes, "Insect and vertebrate appendages have little in common anatomically other than being produced, in part, by outgrowth of the body surface."[27] Yet in each case the same gene plays a critical role in the production of these different anatomical structures. The gene *distal-less* and its homologues function as switches, but in each case a switch that regulates many different downstream genes, leading to different anatomical features, depending upon the large informational context in which the gene finds itself.

This was surprising to many evolutionary biologists because they had long assumed on the basis of orthodox evolutionary theory that (a) genes control the development of organisms and anatomical structures and that (b) homologous genes should, therefore, produce homologous organisms and structures.[28] Yet the anatomical structures produced by *distal-less* and *Pax-6* are so different and they are found in such distantly related organisms, that they are unlikely to have evolved from a single common ancestor possessing a precursor structure to each.[29] There is no conceivable organism whose appendages were ancestral precursors to echinoderm spines, arthropod compound limbs, and vertebrate bony limbs. As biologists Douglas Erwin and Eric Davidson note in reflecting on this general puzzle, "Although the heads, hearts, eyes, etc., of insects, vertebrates and other creatures carry out analogous functions, neither their developmental morphogenesis nor functional anatomies are actually very similar if considered in any detail."[30]

On the other hand, the way genes perform different functions based upon the larger informational context in which they find themselves is not at all unexpected from a design-theoretic point of view. We have extensive experience of informational systems, designed by intelligent agents, that exhibit exactly this feature of context-dependent modularity.

Consider any body of meaningful English text made of paragraphs, sentences, and words. In such a text, the meanings of the lower-level modules—the words—are determined by the context in which the words occur. English words have a range of associated meanings. This range is not infinitely large: words cannot mean anything at all. Nor is it usually completely discrete. Words rarely have only a single, univocal meaning. Instead, the particular meaning a word assumes, among a set of allowable meanings, is determined by the larger context in which the word occurs.

Consider two largely identical sequences of words that have completely different meanings. First, "'My, this cake is delicious,' he said enthusiastically." Second, "'My, this cake is delicious,' he said sarcastically." The first seven words in each eight-word sentence are identical. Yet these two groups of words have completely different meanings. Why? The final word in each case provides important contextual information that determines the meaning of the other modular elements—the first seven words—in the sentence. Notice too that in the larger context of this paragraph, both sentences have another function as well—not to express either my appreciation or disdain for a piece of cake, but to illustrate how context determines the meaning of modular elements in intelligently arranged communications.

A colleague of mine, Paul Nelson, illustrates how the context of a whole informational system can determine the meaning or function of individual modular elements in another way. He shows that the same forty-three words used to write the conclusion of the Gettysburg Address can be rearranged to write "An Anarchist Manifesto," with a meaning diametrically opposite that of Abraham Lincoln's.[31] What's different in the two cases? Not the lower-level modules (i.e., the words). Instead, the difference is the overall arrangement of the words and the context it provides for interpreting the meaning of the individual words, the lower-level modular elements. Same modular elements, completely different meanings.

Increasingly, it appears that modular genomic elements within biological systems manifest this same context dependence. Yet according to standard evolutionary models for the origin of biological form, this should not be. In both chemical and biological evolutionary theory, novel organismal form arises from the "bottom up," first as chemical building blocks and then new genes and proteins determine higher-level biological form and function. Yet the context dependence of these lower-level modular elements (genes and proteins) shows that their functions are determined by a larger informational system and, presumably, only once this system is in place. Indeed, the function of many genes and proteins is determined "top-down," by the larger system-wide informational and organismal context—by the needs of the organism as a whole.

The question is inescapable: How could genes and proteins have survived, much less reproduced with variation, before there existed the extraordinarily complex organismal context in which they alone appear to function? Though this feature of genes seems puzzling given standard materialistic and reductionistic theories of evolution, it is entirely expected from a design-theoretic perspective. Why? Simply this: the function of modular elements in intelligently constructed blocks of text (or software) routinely exhibits such context dependence. In other words, we know of a cause of this feature of information-rich systems and that cause is, again, intelligent design. Thus, design readily explains—provides a cause known to produce—this unexpected feature of life's informational structure.

Extragenomic or Ontogenetic Information

Design can explain another surprising feature of life's informational structure and hierarchy. Much of the information stored in the organism that determines the context and function of genes is stored on other genes. But developmental biologists in particular have learned that the cell stores other kinds of information, so-called extragenomic (not stored in DNA), or ontogenetic, information that plays a critical role in the development of the body plans of organisms.

There is now a wealth of embryological evidence showing that DNA does not wholly determine morphological form in organisms.[32] DNA directs the synthesis of proteins and RNAs. It also helps to regulate the timing and expression of the synthesis of various proteins within cells. Yet DNA alone does not determine how individual proteins assemble themselves into larger systems of proteins; still less does it, by itself, determine how cell types, tissue types, and organs arrange themselves into body plans.[33] Instead, other factors—such as the three-dimensional structure and organization of the cell membrane and cytoskeleton, and the spatial architecture of the fertilized egg—play important roles in determining body-plan formation during embryogenesis.

Two analogies may help clarify what is going on in organisms. At a building site, construction workers make use of many materials: lumber, wire, nails, drywall, piping, and windows. Yet building materials do not determine the floor plan of the house or the arrangement of houses in a neighborhood. Similarly, electronic circuits are composed of many components, such as resistors, capacitors, and transistors. But such lower-level components do not determine their own arrangement in an integrated circuit. Instead, builders have a blueprint or floor plan and electrical engineers have a wiring diagram, each of which determines the arrangement of lower-level parts.

Biological organisms also depend on higher-level information and a hierarchical arrangement of parts. Genes and proteins are made from simple building blocks—nucleotide bases and amino acids—arranged in specific ways. Cell types are made of, among other things, systems of specialized proteins. Organs are made of specialized arrangements of cell types and tissues. And body plans comprise specific arrangements of specialized organs and tissues. Yet the properties of individual proteins (or, indeed, the lower-level parts in the hierarchy generally) do not fully determine the organization of the higher-level structures and organizational patterns.[34] Nor does the genetic information that codes for proteins determine these higher-level structures.

Instead, higher-level structural information appears to play a critical role in the development of organisms. Developmental biologists do not know where all this extragenomic information in the cell resides, but they have located some of it. For example, they know that the structure and location of the cytoskeleton influence the patterning of embryos. Arrays of microtubules help to distribute the essential proteins used during development to their correct locations in the cell. These microtubules themselves are made of many protein subunits. Nevertheless, like bricks that can be used to assemble many different structures, the tubulin subunits in the cell's microtubules are identical to one another. For this reason, it is not possible to predict the structure of the cytoskeleton of the cell from the characteristics of the protein constituents that form that structure.[35] Neither the tubulin subunits nor the genes that produce them account for the different shape of microtubule arrays that distinguish different kinds of embryos and developmental pathways.

Instead, the structure of the microtubule array itself is determined by the location and arrangement of its subunits, not the properties of the subunits themselves. And the location of specified target sites on the interior of the cell membrane helps to determine the shape of the cytoskeleton and also influences the development of organismal form. So does the position of the centrosome, which "nucleates" or spurs the assembly of the microtubules that form the cytoskeleton. Although both the membrane targets and the centrosomes are made of proteins, the location and form of these structures is not wholly determined by the proteins that form them. Instead, centrosome structure and membrane patterns *as a whole* convey three-dimensional structural information that helps determine the structure of the cytoskeleton and the location of its subunits.[36] Moreover, the centrioles that compose the centrosomes replicate independently of DNA replication.[37] The daughter centriole receives its form from the overall structure of the mother centriole, not from the individual gene products that constitute it.[38] In ciliates, microsurgery on cell membranes can produce heritable changes in membrane patterns, even though the DNA of the ciliates has not been altered.[39] This suggests that membrane patterns (as opposed to membrane constituents) are impressed directly on daughter cells. In both cases, form is transmitted from parent three-dimensional structures to daughter three-dimensional structures directly and is not wholly contained in constituent proteins or genetic information.[40]

Thus, in each new generation, the form and structure of the cell arise as the result of *both* gene products *and* preexisting three-dimensional structure and organization. Cellular structures are built from proteins, but proteins find their way to correct locations in part because of preexisting three-dimensional patterns and organization inherent in cellular structures. Preexisting three-dimensional form present in the preceding generation (whether inherent in the cell membrane, the centrosomes, the cytoskeleton, or other features of the fertilized egg) contributes to the production of form in the next generation. Neither structural proteins alone nor the genes that code for them are sufficient to determine the three-dimensional shape and structure of the entities they form. Gene products provide necessary but not sufficient information for the development of three-dimensional structure within cells, organs, and body plans.[41]

This is not expected in standard views of evolutionary theory. Neo-Darwinism has long sought to explain the origin of new information, form, and structure as a result of selection acting on randomly arising variation at a low level within the biological hierarchy, namely, within the genetic text. Yet major morphological innovations depend on a specificity of arrangement at a much higher level of the organizational hierarchy, a level that DNA alone does not determine. Yet if DNA is not wholly responsible for body-plan morphogenesis, then DNA sequences can mutate indefinitely, without regard to realistic probabilistic limits, and still not produce a new body plan, suggesting the possibility of something else at work in the origin of major morphological innovations.

Might intelligent design have played a role? There are reasons to consider this possibility. Organisms contain both information-rich components (such as proteins and genes) and information-rich arrangements of those components forming a rich multilayered informational hierarchy. We know that design engineers can also produce hierarchical systems in which both individual modules and the arrangements of those modules exhibit complexity and specificity—information so defined. Individual transistors, resistors, and capacitors exhibit considerable complexity and specificity of design; at a higher level of organization, their specific arrangement within an integrated circuit represents additional information and reflects further design. Conscious and rational agents have, as part of their powers of purposive intelligence, the capacity to design information-rich parts and to organize those parts into functional information-rich systems and hierarchies. A rich multilayered informational hierarchy in life is, therefore, just what we might have expected from an ID perspective. That biological organisms exhibit such hierarchy further underscores the relevance of intelligent design as an explanatory and interpretive framework for understanding biological systems.

New Research Questions

Clearly, the genome and the cell's information-processing and storage system manifest many features—hierarchical filing, nested coding of information, context dependence of lower-level informational modules, sophisticated strategies for increasing storage density—that we would expect to find if they had been intelligently designed. Conversely, many of these newly discovered features are not readily explained by standard materialistic evolutionary mechanisms.

Moreover, these informational features are found in higher-level multicellular organisms as well as in single-celled prokaryotes. This suggests an intriguing, if radical, possibility. It suggests that intelligent design may have played a role in the origin of complex multicellular organisms, and that mutation and selection, along with other undirected mechanisms of evolutionary change, do not fully account for the origin of these higher forms of life. Might intelligent design have played a role in biological evolution—that is, in the origin or historical development of new living forms from simpler preexisting forms? Given the centrality of information to living systems, and given that all forms of life, including complex multicellular organisms, display distinctive hallmarks of intelligent design in their informational systems, there would now seem to be an increasing reason to consider this possibility.

This possibility in turn suggests a number of research questions that are not being addressed because of the limitations of the neo-Darwinian perspective. We now know that organisms contain information of different types at every organizational level, including ontogenetic or structural information not encoded in DNA. Yet, according to neo-Darwinism, new form and structure arise as the result of information-generating mutations in DNA. In its population-genetics models of evolutionary change, neo-Darwinism has long assumed a number of things about genes that we also now know to be incorrect. For example, these models assume that genetic information is context-independent, that genes independently associate, and that genes can mutate indefinitely with little regard to extragenomic and other functional constraints. In short, neo-Darwinism gives primacy to the gene as the locus of biological change and innovation. In so doing, however, it assumes a one-dimensional conception of biological information.

Consequently, neo-Darwinism provides little reason to consider or investigate (and every reason to ignore) the additional tiers of information and codes that reside beyond the gene. On the other hand, advocates of intelligent design not only acknowledge, but expect to find, sophisticated modes of information storage and processing in the cell. Since the theory of intelligent design treats the hierarchical organization of information as theoretically significant, advocates of the theory have naturally shown intense interest in the cell's informational hierarchies and intricate modes of coding. Therefore, a design-theoretic perspective tends to encourage questions about the hierarchies of information in life that neo-Darwinists tend to ignore.

Such questions include: Where exactly does this ontogenetic information reside? How does it affect the function of lower-level genetic modules? How many types of information are present in the cell? How much ontogenetic information is present in the cell? How do we measure this information, given that it is often structural and dynamic rather than digital and static? And how mutable are various forms of non-DNA-based information—if at all?

We know that animal body plans are static over long periods of time. Is this morphological stasis the result of constraints imposed upon mutability by the interdependence of informational hierarchies? Are there other constraints—even probabilistic constraints operating at the level of the individual gene and proteins—that limit the transformative power of the selection and mutation mechanism? Given the phenomenon of "phenotypic plasticity" (individuals in a population with the same genotype that have different phenotypes) and the recurrence of similar variations in the same species, how much variability in organisms is actually the result of preprogramming as opposed to random mutations? If recurring variations result from preprogramming, where does the requisite information for these programs reside and how is it expressed? How many phenomena currently regarded as examples of so-called neo-Lamarckian processes can be properly attributed to preprogrammed, intelligently designed, adaptive capacity?

All these questions arise naturally from a design-theoretic perspective and have little place in a neo-Darwinian framework. Many are questions about the structure, function, and composition of living systems themselves. Some are questions about the efficacy of various evolutionary mechanisms—questions about whether these mechanisms can explain various appearances of design as well as an actual designing intelligence. What can mutation and selection produce, and what can they not produce? Can selection and mutation produce novel genes and proteins? New anatomical structures? Novel body plans? If not, are there mechanisms or features of life that impose limits on biological change? Or are there perhaps other materialistic mechanisms that have the causal powers to produce novel forms of life? If not, and if the pervasive hallmarks of intelligent design in complex organisms indicate actual design, what does that imply about what else we should find in living systems or the fossil record? Are there perhaps other design patterns known from software design or mechanical engineering that await description or discovery in living systems or molecular machines? Are there patterns in the fossil record indicative of intelligent activity?

If so, do they suggest that intelligent design played a role in the production of novel body plans or smaller innovations in form? Is intelligent design needed to build new forms of life that exemplify higher taxonomic categories—such as orders, classes, and phyla? Is it also needed to build the new forms encompassed by lower taxonomic categories such as genera and species? How much morphological change can undirected mechanisms such as selection and mutation produce, and at what point, if any, would intelligent design need to play a role? Such questions move from whether intelligence played a role in the history of life to *how, when,* and *how often* intelligence acted.

If a designing intelligence generates new form by infusing new information into the biosphere, do we see evidence of that anywhere in the geological time scale? If so, where, and how can we detect it? If intelligent design played a role in the history of life after life's initial origin, did that designing agency act gradually or discretely? Did it effect a gradual transformation of form from simple to more complex organisms? Or did that intelligence effect more sudden transformations or innovations in biological form, thus initiating new and separate lines of genealogical descent? In other words, is the history of life monophyletic or polyphyletic? If the latter, how many separate lines of descent or trees of life have existed during life's history, and what should the fossil record, embryological development, comparative anatomy, and phylogenetic studies show in that case? If life's history is polyphyletic, how wide are the envelopes of variability in the separate trees of life?

Conversely, if undirected evolutionary mechanisms are sufficient to account for the origin of all new forms of life, is it possible that the pervasive signs of design in higher forms of life were preprogrammed to unfold from the origin of life itself? If design was thus "front-loaded" in the first simple cell, what does that imply about the capacity of cells to store information for future adaptations? And what should the structure and organization of the prokaryotic genome look like in that case?

Many of the preceding questions follow from considering the informational signature of intelligence in life. But design arguments, such as Michael Behe's argument from irreducible complexity, suggest other kinds of research questions. Are specific molecular machines irreducibly complex, as Behe argues? If so, is intelligent design the only cause known to produce this feature of systems? Can mechanisms of "co-option" or "exaptation" explain the origin of the flagellar motor and other molecular machines, as his critics argue in recent scientific publications?[42]

Indeed, the controversy over Behe's argument from irreducible complexity has already motivated specific new lines of empirical inquiry and generated a number of new research questions. The argument of this book, based on the informational features of life, raises many others. Many of these lines of inquiry are admittedly radical from an orthodox neo-Darwinian point of view, and some scientists may not want to consider them. Nevertheless, they cannot argue that the scientists who do will have nothing *to do*. Nor can they argue that the new genomics has weakened the case for intelligent design presented in the main body of this book.

Appendix A

Some Predictions of Intelligent Design

Critics of intelligent design often argue that the theory cannot be tested, because it makes no predictions. The charge turns on a fundamental misunderstanding of how historical scientific theories are tested. Primarily, such testing is accomplished by comparing the explanatory power of competing hypotheses against already known facts. The theory of intelligent design, like other theories about the causes of past events, is testable, and has been tested, in just this way. That said, the theory of intelligent design also has predictive consequences. Since the design hypothesis makes claims about what caused life to arise, it has implications for what life should look like. Moreover, the explanatory framework that intelligent design provides leads to new research questions, some of which suggest specific predictions that are testable against observations or by laboratory experiments.

Some of these predictions can help adjudicate proposals that invoke either intelligent causes or materialistic mechanisms as explanations for various features of life or events in life's history. Other predictions can help discriminate between competing ideas of how a designing intelligence influenced the history of life—for instance, between design hypotheses that affirm universal common ancestry and those that envision more discrete or discontinuous intelligent activity in the history of life. Indeed, depending upon how scientists envision intelligent design playing a role in the history of life, they may formulate different kinds of design hypotheses, each entailing different though testable predictions.

Some predictions (those that discriminate between the explanatory power of intelligent causes and materialistic mechanisms) will necessarily function as tests of the causal efficacy of mechanisms of evolutionary change. Since design hypotheses are often formulated as strong claims about intelligence as the best causal explanation of some particular phenomenon, these hypotheses entail counterclaims about the insufficiency of competing materialistic mechanisms. But such claims also entail predictions. The claim that intelligent design constitutes the best explanation of particular informational features of organisms leads inevitably to the claim that other competing causal hypotheses *will not* demonstrate the power to produce these effects—just as they may not have done so to this point. Similarly, the claim that intelligent design constitutes the best explanation of the integrated or "irreducible complexity" of molecular machines entails a prediction about the insufficiency of competing materialistic causes for these systems.

In addition to predictions about what future evidence will show about the causal powers of various processes, intelligent design also generates predictions about what we are likely to find in living systems as we investigate them. We have extensive experience-based knowledge of the kinds of strategies and systems that designing minds devise to solve various kinds of functional problems. We also know a lot about the kinds of phenomena that various natural causes produce. For this reason, the theory of intelligent design makes predictions about the kinds of features we are likely to find in living systems if they were in fact intelligently designed.

Other types of predictions flow from considering the possibility that intelligence influenced, directed, or guided the history of life, either gradually or discretely. Various ID hypotheses generate different predictions about what, for example, the fossil record or phylogenetic studies should show. Depending upon how ID theories conceive of the designing intelligence affecting the history of life over time and what other characteristics they attribute to this intelligence (such as benevolence, for example), design hypotheses may make specific claims about the causes of so-called dysteleology, or bad design. These claims may entail specific empirical predictions as well.

Thus, intelligent-design hypotheses may generate several distinct types of predictions: predictions about causal powers, or lack thereof, of various mechanisms; predictions about the structure, organization, and functional logic of living systems; predictions about what evidence will show about the history of life; and predictions about the causes of putatively bad design. Consider a dozen or so ID-based predictions, each of which exemplifies one or more of these types. I start with two predictions directly relevant to testing the main arguments made in this book.

The Causal Powers of Materialistic Mechanisms

The theory of intelligent design makes predictions about what the evidence will show about the causal powers of various material mechanisms. According to the hypothesis developed in this book, intelligent design is the best explanation of the origin of the information necessary to produce the first life. To make this case, I argued that no purely physical or chemical entity or process had demonstrated the causal powers to produce complex specified information—where "complex" refers to a specific amount of information (roughly 500 bits or more) and the inverse of a probability measure (Dembski's universal probability bound). An obvious prediction follows from this claim—in particular, that large amounts of new functionally specified information (over 500 bits) will not accumulate as a result of random or undirected natural processes and that no such process will be discovered that can produce over 500 bits of new specified information starting from purely physical and chemical antecedents. My theory acknowledges that small amounts of specified information can occasionally arise by random processes, but that the amount of information that can be generated is limited by the probabilistic resources of the universe. This prediction can be clearly falsified by the discovery of an undirected physical or chemical process that can generate over 500 bits of functionally specified information.

This general prediction entails other more specific ones. For example, based upon the considerations just discussed, the theory of intelligent design developed here predicts that genetic algorithms simulating the power of undirected evolutionary processes will not produce more than 500 bits of new complex specified information (information not supplied by the programmer in the algorithm)—unless, that is, an intelligent programmer provides necessary "active information." It further predicts that a close examination of programs that appear to produce novel complex specified information will reveal either (a) inputs from programmers that account for the putative creation of new complex specified information beyond what could have been expected given the available probabilistic resources, or (b) a lack of biological realism in the computer simulation, or (c) both. Recently, William Dembski and Robert Marks have produced peer-reviewed papers performing informational accounting on genetic algorithms that confirm these predictions in specific case studies.[1]

Typically, genetic algorithms may lack realism (1) by providing the program with a target sequence, (2) by programming the computer to select for proximity to future function rather than actual function, or (3) by selecting for changes that fail to model biologically realistic increments of functional change, increments that reflect the extreme rarity of functional sequences of nucleotide bases or amino acids in the relevant sequence space (see Chapter 13).

The theory of intelligent design presented in this book entails specific predictions about RNA-world research as well. For example, it predicts that investigations of the properties of ribozymes (RNA catalysts) will reveal an insufficient number of enzymatic functions to sustain a primitive cell or even an alternative RNA-based system of protein synthesis, demonstrating that RNA could not have performed both the necessary enzymatic functions of modern proteins and the information-storage functions of nucleic acids that the RNA-world scenario envisions. Further, the theory of intelligent design developed in this book also predicts that putatively "successful" ribozyme engineering experiments—in particular, experiments that appear to enhance the replicase activity of ribozymes—will require significant sequence specificity in the RNA bases in any functional ribozyme and that, upon examination, active information provided by ribozyme engineers will account for much of this sequence specificity (in particular, the amount beyond what could have been expected to arise spontaneously given available probabilistic resources).

The Structure, Organization, and Functional Logic of Living Systems

ID also makes predictions about the structure, organization, and functional logic of living systems. In 2005, University of Chicago bacterial geneticist James Shapiro (not an advocate of intelligent design) published a paper describing a regulatory system in the cell called the lac operon system.[2] He showed that the system functions in accord with a clear functional logic that can be readily and accurately represented as an algorithm involving a series of if/then commands. Since algorithms and algorithmic logic are, in our experience, the products of intelligent agency, the theory of intelligent design might expect to find such logic evident in the operation of cellular regulatory and control systems. It also, therefore, expects that as other regulatory and control systems are discovered and elucidated in the cell, many of these also will manifest a logic that can be expressed in algorithmic form.

Molecular biologist Jonathan Wells has also used design reasoning to infer the possible existence of a previously undiscovered molecular machine within the cell. While researching the causes of cancer, Wells became intrigued with the possibility that an organelle in the cell called the centrosome may play a significant role in this disease.

In animal cells, the centrosome sits alongside the nucleus. It serves as the focus of the microtubules that give the cell its shape and provides tracks along which proteins from the nucleus are moved to their proper locations elsewhere in the cell. Centrosomes play a role in the process of cell division, and some cancer researchers—noting that cancer cells usually contain damaged, abnormal, or extra chromosomes—have suggested that centrosomal defects may be the first stage in cancer.

Many cancer researchers disagree. Influenced by neo-Darwinism, they believe that cancer is caused by DNA mutations. As a skeptic of neo-Darwinism, Wells was inclined to doubt this and to consider the centrosomal theory as a possibility. As he studied the literature on cancer, he surmised that there was no consistent pattern of mutations in cancer. As he studied centrosomes, he became even more intrigued with the possibility that they may play a significant role in cancer.

When animal cells divide, they rely on an intracellular apparatus called the "spindle." Looking somewhat like the barrel-shaped pattern formed by iron filings around the two poles of a magnet, the spindle has a centrosome at each pole. The already duplicated chromosomes are contained within it, and before the cell divides, a "polar ejection force" helps to move the chromosomes to the midpoint of the spindle. After the chromosomes are properly aligned at the midpoint, the calcium level inside the cell rises abruptly. Then the chromosomes split into two equal sets and move to the spindle poles—the centrosomes.

Suspecting that centrosomes produce the polar ejection force, Wells turned his attention to them. Each animal-cell centrosome contains two "centrioles," cylindrical structures only half a millionth of a meter long. Each centriole looks like a tiny turbine with nine tilted blades. As an advocate of intelligent design and a critic of undirected evolution, Wells assumed provisionally that these structures actually were designed to be turbines. From there, Wells used reverse engineering to predict other features of centrioles and their action in both normal and cancerous cells.

First, hints from other centriole studies—together with engineering considerations—led Wells to suggest that each centriole contains an Archimedes screw, a helical pump that draws fluid into one end of the turbine and pushes it out through the blades. Second, Wells postulated that dynein-motor molecules inside the centriole would provide the force to turn the helical pump. Third, Wells concluded from engineering considerations that such an arrangement would operate like a laboratory vortexer, a common device that produces a wobble to swirl the contents of a test tube.

Doing the math, Wells deduced that centrioles could rotate tens of thousands of times a second. A pair of centrioles at both ends of the spindle could produce the polar ejection force that moves chromosomes to the midpoint before cell division. The rise in intracellular calcium that accompanies chromosome separation would shut off the dynein motors, thereby turning off the polar ejection force. This would permit the chromosomes to move poleward without being pushed away at the same time. But if the centriole turbines fail to turn off, the continuing polar ejection force would subject the chromosomes to unusual stress and could cause the damage now thought by some researchers to be the first step in cancer. The fact that there is a correlation between calcium deficiency and cancer is consistent with this hypothesis.

Wells is currently testing his hypothesis experimentally. If corroborated, the hypothesis could aid in the prevention and early diagnosis of cancer. Wells's work also shows how an intelligent-design perspective can lead to new hypotheses, testable predictions, and new lines of research. The outcome of his work won't directly confirm or disconfirm intelligent design, or neo-Darwinism for that matter, since the truth of neither theory depends upon whether any specific structure is or is not a turbine. But it illustrates how an ID perspective can prove fruitful for generating new testable hypotheses and predictions about the structure and function of the cell (as well as the causes of cellular malfunctions when they occur).

Wells himself has noted that scientists operating out of a Darwinian framework could have formulated a similar hypothesis. Nevertheless, he also notes that the underlying assumptions of Darwinians (about the role of mutations in DNA) disinclined them to do so. Conversely, Wells's convictions about intelligent design inclined him to suspect that appearances of design might in fact be evidence of real design, which led him to suspect the presence of a molecular machine in the centriole. It also led him to use reverse engineering to develop a testable hypothesis about its structure and function. As he explained in an abstract to a scientific article about his hypothesis, "Instead of viewing centrioles through the spectacles of molecular reductionism and neo-Darwinism, this hypothesis assumes that they are holistically designed to be turbines.... What if centrioles really are tiny turbines? This is much easier to conceive if we adopt a holistic rather than reductionistic approach, and if we regard centrioles as designed structures rather than accidental by-products of neo-Darwinian evolution."[3]

The History of Life

Various ID hypotheses also generate predictions about the history of life. Advocates of design who think that the information necessary to produce new forms of life was front-loaded in the first cell might predict that prokaryotic cells would demonstrate the capacity to carry amounts of genetic information (information in excess of the needs of those cells) or that such cells would retain vestiges of having done so.

Similarly, advocates of design who think that the information necessary to produce new forms of life was front-loaded in the initial conditions of the universe and the fine-tuning of the laws of physics might predict that physical and chemical laws would demonstrate biologically relevant self-organizing tendencies. Since both these design hypotheses favor a monophyletic view of the history of life, they would predict, along with neo-Darwinism, that the traditional evidences for universal common descent (from biogeography, palaeontology, embryology, comparative anatomy, and genomics) would withstand challenge. Indeed, such theories generate a host of specific predictions about what the evidence in each of these subdisciplines of biology should show. These are not, of course, predictions that provide reasons for preferring a front-loaded design hypothesis to a materialistic explanation for the origin of the first life. The evidence and predictions considered already do that. Instead, they are predictions that could help decide the merits of hypotheses of front-loaded design versus those of discrete design.

Conversely, advocates of design who envision a designing intelligence acting discretely at intervals across the geological time scale tend to favor a polyphyletic rather than a monophyletic view of the history of life. Thus, they predict emerging patterns of evidence from these same subdisciplines that contradict a traditional monophyletic view and support a polyphyletic interpretation of the history of life.

Design theorists who expect a polyphyletic rather than a monophyletic geometry for the history of life often do so because of the functional integration and interdependence of parts exhibited in living systems and because of their understanding of how such functionally integrated systems constrain changes in form. The constraints principle of engineering asserts that the more functionally integrated a system is, the more difficult it is to perturb any part of the system without destroying the whole. Because many ID theorists think organisms were designed as functionally integrated systems comprising many parts and subsystems, they think it is difficult to alter these systems significantly without destroying them, particularly when such alterations consist of a series of blind mutations "in search of" increased functionality. Thus, these ID theorists predict that there should be significant and discoverable limits to the amount of change that various organisms can endure and that major body plans should exhibit significant stasis over time in the fossil record.

Design hypotheses envisioning discrete intelligent action also predict a pattern of fossil evidence showing large discontinuous or "quantum" increases in biological form and information at intervals in the history of life. Advocates of this kind of design hypothesis would expect to see a pattern of sudden appearance of major forms of life as well as morphological stasis.[4] Since designing agents are not constrained to produce technological innovations in structure from simpler precursors or to maintain the function of these simpler precursors through a series of intermediate steps, they would also predict a "top-down" pattern of appearance in which large-scale differences in form ("disparity" between many separate body plans) emerge suddenly and prior to the occurrence of lower-level (i.e., species and genus) differences in form. Neo-Darwinism and front-loaded design hypotheses expect the opposite pattern, a "bottom-up" pattern in which small differences in form accumulate first (differentiating species and genera from each other) and then only much later building to the large-scale differences in form that differentiate higher taxonomic categories such as phyla and classes.

Both types of design hypothesis also make distinct predictions about what phylogenetic analyses should show. Those design advocates who (like neo-Darwinists) accept common descent predict that phylogenetic analyses performed on different molecules and structures in two species should yield harmonious trees of life—trees that indicate similar degrees of difference, relatedness, and divergence from a common ancestor regardless of which molecules or anatomical structures are compared. According to the theory of common descent, since all the molecules and subsystems of an organism evolved from the same common ancestor, the phylogenetic trees generated from different molecules and structures in two species should agree regardless of which molecules or subsystems are compared.

For this same reason, design advocates who affirm polyphyly (and with it more discrete modes or infusions of design) predict that phylogenetic analyses would often yield conflicting trees of life—that is, dissimilar measures of difference, relatedness, and divergence, depending upon which molecules or anatomical structures are compared in the same two species. They predict this for another reason. Many design advocates see similarities in functional biomacromolecules and anatomical structures as resulting from functional or engineering considerations rather than common ancestry. Since intelligent agents have the freedom to combine modular elements and subsystems in unique ways from a variety of information sources, we should expect phylogenetic analyses of diverse systems and molecules to generate some conflicting trees.

Clearly, there are several possible hypotheses about how design played a role in the history of life. Since each of these hypotheses has different empirical consequences, design hypotheses can generate different and competing predictions about what different classes of evidence should show. There is nothing unusual about this, however. Philosophers of science have long recognized that hypotheses generate predictions when they are conjoined with so-called auxiliary hypotheses, that is, other claims or suppositions about the world. In this, design advocates are no different from advocates of other scientific theories. Some may conjoin the hypothesis of design with a monophyletic hypothesis about the history of life; some may conjoin it with a polyphyletic view or with other claims about the world or life (such as the constraints principle of engineering) in order to generate different though still specific and testable predictions. These differing predictions do not demonstrate that intelligent design is incoherent. They merely make testing and assessing the relative merits of competing design hypotheses possible.

Dysteleology, or Bad Design

The theory of intelligent design generally affirms that complex biological structures were designed for functional reasons. Thus, it predicts that the study of supposedly "dysteleological" or "poorly designed" structures will reveal either (a) functional reasons for their design features or (b) evidence of degenerative evolution—that is, evidence of decay of an otherwise rational and beneficial original design. Neo-Darwinists have argued, for example, that the backward wiring of the vertebrate retina exhibits a suboptimal or bad design, one ill-befitting an intelligent designer. Design theorists have challenged this and predict that further study of the anatomy of the vertebrate retina will reveal functional reasons for its nonintuitive design parameters. Biologists George Ayoub and Michael Denton have identified a number of functional reasons for the design of the vertebrate retina confirming this prediction.[5] Ayoub has shown, for example, that the vertebrate retina provides an excellent example of what engineers call constrained optimization, in which several competing design objectives are elegantly balanced to achieve an optimal overall design.

The theory of intelligent design also predicts that instances of "bad" design in nature may turn out to be degenerate forms of originally elegant or beneficial designs. Critics of design have pointed to the existence of organisms such as virulent (disease-producing) bacteria to refute the ID hypothesis, arguing that an intelligent and beneficent designer would not have made such organisms. Some design theorists (in this case, those who hold that the designer is both intelligent and benevolent) predict that genetic studies will reveal that virulent bacterial systems are degenerative systems that have resulted from a loss of aboriginal genetic information.[6] University of Idaho microbiologist Scott Minnich, an ID advocate, has specifically predicted that the virulence capacity in *Yersinia pestis*, the bacterium that caused the black plague in medieval Europe, resulted from genetic mutations that stopped it from manufacturing molecules and structures recognized by the human immune system. He is currently conducting experimental tests of this hypothesis. He and his team already have shown that the more limited virulence capacity of *Yersinia pseudotuberculosis* (a bacterium that causes gastroenteritis) resulted from the mutational degradation of genes that produce flagellin, a protein that the human immune system recognizes in flagellar motors of bacteria. Minnich and his team have found that virulence in *Yersinia pseudotuberculosis* can be reduced by restoring its gene for producing flagellin.[7] This plus a growing body of data showing that virulence capacity in bacteria generally results from a loss of genetic information have provided a significant initial confirmation of the ID-generated prediction about the cause of bacterial virulence.[8]

The Efficacy of Causal Mechanisms and the Features of Living Systems

Theories of intelligent design also sometimes link considerations of causal adequacy to predictions about what we should find in the cell or the genome. These predictions test whether intelligent causes or competing materialistic mechanisms are more likely to have caused the origin of some system by explicating how the two modes of causation are different and then showing how each would produce different features in some part of a living system.

Our previous discussions about nonprotein-coding DNA illustrate this. Since neo-Darwinism affirms that biological information arises as the result of an undirected trial-and-error process, it predicts the existence of vast regions of nonfunctional DNA in the genome. By contrast, advocates of intelligent design think that DNA sequences arose mainly as the result of purposeful intelligence. They predict, therefore, that nonprotein-coding DNA should perform important biological functions. Thus, ID and neo-Darwinism affirm competing causes of the origin of biological information, both of which have different implications for what scientists should find in the genome.

The debate about the origin of irreducibly complex molecular machines provides another example of this kind of prediction. In his book *Darwin's Black Box,* biochemist Michael Behe argues that the many miniature machines and circuits that have been discovered in cells provide strong evidence for intelligent design. A crucial part of Behe's argument for intelligent design involves his attempt to show that miniature machines, such as the now infamous bacterial flagellar motor, could not have developed from simpler precursors in a gradual step-by-step fashion. In Behe's view, the coordinated interaction of the many parts of the flagellar motor resulted ultimately from an idea in the mind of a designing intelligence, not from a process of gradual step-by-step evolution from a series of simpler material precursors.

Behe's critics, however, such as biologist Kenneth Miller of Brown University, have suggested that the flagellar motor might have arisen via a different causal pathway. They suggest that the flagellar motor might have arisen from the functional parts of other simpler systems or from simpler subsystems of the motor. They have pointed to a tiny molecular syringe called a type-3 secretory system (or T3SS). They note that the T3SS usually functions as part of the flagellar motor, but it is sometimes found in bacteria that do not have the other parts of the flagellar motor. Since the type-3 secretory system is made of approximately ten proteins that have close homologues in the thirty-protein flagellar motor, and since this tiny pump does perform a function, Miller intimates that the flagellar motor might have arisen from this smaller pump.[9] Behe, however, remains convinced that the flagellar motor is the aboriginal system. To defend Behe, other ID theorists have suggested that when the type-3 secretory system appears in isolation from the other parts of the flagellar motor, it does so as the result of degenerative evolution—that is, as the result of a loss of the genetic information necessary to produce the other parts of the motor.

Thus, these two different views of the T3SS imply something different about the relative age of the genes that produce the flagellar motor and the T3SS, respectively. The co-option theory predicts that the genes that produce the T3SS syringe should be older than the genes that produce the flagellar motor, since the syringe in this view is a precursor system. The hypothesis from design and degenerative evolution predicts the opposite—that the genes of the flagellar motor should be older than those of the T3SS. Thus, the two theories posit causal histories for these molecular machines that contradict each other, and both make testable predictions about features of the systems (the age of the genes) as a result.

As it happens, phylogenetic analyses of the distribution of flagellar systems in bacteria now make it possible to assess the relative age of two suites of genes. These phylogenetic studies suggest that the flagellar motor genes are older than the T3SS genes, thus providing an initial confirmation of the design-theoretic hypothesis about the origin of the flagellar motor.[10]

There is an interesting twist to this story, however—one that underscores how intelligent design leads to many new and unresolved research questions. There is another design hypothesis about the flagellar motor and the T3SS. It envisions both systems as the products of independent instances of design, despite the similarity of the T3SS to parts of the flagellar motor. It predicts that if the T3SS were designed independently of the flagellar motor, then we ought to find many unique (nonhomologous) genes encoding the T3SS—genes that exhibit little similarity to those found in the flagellar system. It is now known that the T3SS does have several such nonhomologous genes, confirming this prediction of a separate ID hypothesis. So did the T3SS devolve from the flagellar motor, or did it arise independently by a separate act of design? This is another research question generated by the theory of intelligent design.

Clearly, further experimental tests are needed to discriminate between these two competing design hypotheses.

Here's another prediction of this type that has also generated an interesting program of experimental research. If intelligence played a role in the origin of new biological forms after the first life, it likely would have done so by generating the biological information necessary to produce these forms. If so, intelligent design would have played a role in either the origin of new genes and proteins or the origin of extragenomic information, or both. Either way, this design hypothesis implies that the undirected mechanism of random mutation and selection is not sufficient to produce the information necessary for such biological innovation. Thus, this particular design hypothesis would predict that mutation and selection lack the capacity to produce fundamentally new genes and proteins. (Again, those who object that this is a merely negative claim against neo-Darwinism rather than a positive prediction of intelligent design misunderstand the fundamentally comparative nature of historical scientific theories, which take the form of arguments to the best explanation.)

Douglas Axe, whom I discussed in Chapter 9, has been intrigued with intelligent design since the early 1990s. He devised a way to test this ID-inspired prediction with a program of rigorous experimental work—work that he first performed at Cambridge University and continues to perform at the Biologic Institute in Redmond, Washington.

In developing this test, Axe reasoned as follows. According to neo-Darwinism new biological information arises as natural selection acts on functionally advantageous mutations in genes. To produce any fundamentally new biological forms these mutations would—at the very least—have to produce a number of new proteins. But natural selection can act only on what mutations first generate. Thus, for mutation and selection to produce new functional proteins or protein folds—the smallest unit of selectable function—new proteins or protein folds must first arise by chance—that is, by random mutation. If the probability of this were extremely low—beyond the reach of available probabilistic resources—then this would undermine the plausibility of the neo-Darwinian mechanism and confirm Axe's own ID-based expectation of its inadequacy.

Thus, as a specific test of the efficacy of the neo-Darwinian mechanism (as well as the chance origin of information in a prebiotic setting), Axe posed the question: How rare or common are functional protein folds within their corresponding amino acid–sequence space? He realized that if functional sequences were common enough for mutations to stumble upon them relatively easily (within the time required for relevant evolutionary transitions), mutation and selection might be able to build otherwise extremely improbable structures in small incremental steps. On the other hand, if functional proteins are extremely rare within sequence space, such that mutations will not have a realistic chance of finding them in the available time, selection will have little or nothing to work on, undermining its ability to produce biological information.

It's important to emphasize that Axe's prediction follows from the premise that intelligent design played a role in the origin of new genes and proteins during biological (or chemical) evolution. Since the case for intelligent design as the best explanation for the origin of biological information necessary to build novel forms of life depends, in part, upon the claim that functional (information-rich) genes and proteins cannot be explained by random mutation and selection, this design hypothesis implies that selection and mutation will not suffice to produce genetic information and that, consequently, functional sequences of amino acids within protein-sequence space will be extremely rare rather than common. Axe's mutagenesis experiments have tested, and continue to test, this prediction of ID theory. As noted in Chapter 9, Axe has shown that the ratio of functional amino-acid sequences to nonfunctional amino-acid sequences is extremely small, 1 in 10^{74} for a protein fold 150 amino acids in length. Since most new proteins exceed 150 amino acids in length, and since fundamentally new life-forms require many new proteins, this experimental result, published in the *Journal of Molecular Biology*, provides an initial confirmation of Axe's ID-inspired prediction.[11]

Of course, predictions about the insufficiency of materialistic mechanisms do not form the whole basis of the case for intelligent design—whether as a theory of chemical or biological evolution. To establish a design hypothesis as a best explanation requires more than a confirmed prediction about the inadequacy of a competing causal explanation. Instead, it requires positive evidence for the efficacy of an intelligent cause and refutations of other relevant causal hypotheses using either predictive methods of testing, evaluations of explanatory power, or both. For this same reason, a single failed prediction rarely falsifies a theory. How well a theory explains (or predicts) the preponderance of relevant data is usually a better gauge of its merit. In Chapter 15 and the Epilogue, I showed that intelligent design exhibits broad explanatory power, at least as a hypothesis about the origin of the first life. Even so, any claim that intelligent design best explains some particular feature of life also entails a prediction about some relevant body of evidence showing the inability of competing hypotheses to explain that same feature. For this reason, and others, design hypotheses do make predictions. Lest there be any doubt about this, I summarize a number of the key predictions of the preceding discussion in the following list:

A Dozen ID-Inspired Predictions

• No undirected process will demonstrate the capacity to generate 500 bits of new information starting from a nonbiological source.

• Informational accounting will reveal that sources of active information are responsible for putatively successful computer-based evolutionary simulations.

• Future experiments will continue to show that RNA catalysts lack the capacities necessary to render the RNA-world scenario plausible.

• Informational accounting will reveal that any improvements in replicase function in ribozymes are the result of active information supplied by ribozyme engineers.

• Investigation of the logic of regulatory and information-processing systems in cells will reveal the use of design strategies and logic that mirror (though possibly exceed in complexity) those used in systems designed by engineers. Cell biologists will find regulatory systems that function in accord with a logic that can be expressed as an algorithm.

• Sophisticated imaging techniques will reveal nanomachines (turbines) in centrioles that play a role in cell division. Other evidence will show that malfunctions in the regulation of these machines are responsible for chromosomal damage.

• If intelligent design played a role in the origin of life, but not subsequently, prokaryotic cells should carry amounts of genetic information that exceed their own needs or retain vestiges of having done so, and molecular biology should provide evidence of information-rich structures that exceed the causal powers of chance, necessity, or the combination of the two.

• If a designing intelligence acted discretely in the history of life, the various subdisciplines of biology should show evidence of polyphyly.

• The fossil record, in particular, should show evidence of discrete infusions of information into the biosphere at episodic intervals as well as a top-down, rather than bottom-up, pattern of appearance of new fossil forms.

• If an intelligent (and benevolent) agent designed life, then studies of putatively bad designs in life—such as the vertebrate retina and virulent bacteria—should reveal either (a) reasons for the designs that show a hidden functional logic or (b) evidence of decay of originally good designs.

• If the flagellar motor was intelligently designed and the type-3 secretory system devolved from it, the genes that code for the bacterial flagellar motor should be older than those that code for the proteins in the T3SS, and not the reverse. Alternatively, if the T3SS and the flagellar motor arose by design independently, the T3SS should have unique (nonhomologous) genes that are not present in the genome for the flagellar motor.

• The functional sequences of amino acids within amino acid–sequence space should be extremely rare rather than common.

Multiverse Cosmology and the Origin of Life

Scientists have increasingly recognized that the probabilistic resources of the observable universe are insufficient to explain—by chance alone—the origin of a minimally complex cell or even a self-replicating system of RNA molecules (or even, for that matter, a single protein of modest length). In response, some scientists have sought to explain the origin of life by invoking other materialistic mechanisms or self-organizational processes. But as noted in Chapters 11–14, theories of this kind have also fallen on hard times. As a result, a few scientists have looked beyond our universe for additional probabilistic resources by which to render a chance explanation for the origin of life more plausible.

In May 2007, Eugene Koonin, of the National Center for Biotechnology Information at the National Institutes of Health, published an article in *Biology Direct* entitled "The Cosmological Model of Eternal Inflation and the Transition from Chance to Biological Evolution in the History of Life." In it, Koonin acknowledges that neither the RNA world nor any other materialistic chemical evolutionary hypothesis can account for the origin of life, given the probabilistic resources of the entire universe. As he explains: "Despite considerable experimental and theoretical effort, no compelling scenarios currently exist for the origin of replication and translation, the key processes that together comprise the core of biological systems and the apparent prerequisite of biological evolution. The RNA World concept might offer the best chance for the resolution of this conundrum, but so far cannot account for the emergence of an efficient RNA replicase or the translation system."[1]

To address this problem, Koonin proposes an explanation for the origin of life based purely on chance. His particular chance explanation, however, does not refer to any process taking place on earth or even within the observable cosmos. Instead, he posits the existence of an infinite number of other life-compatible universes, since, he argues, the existence of such universes would render even fantastically improbable events (such as the origin of life) probable or even inevitable.

To justify his postulation of other universes, he invokes a model of cosmological origins based on inflationary cosmology dubbed the "many worlds in one" hypothesis by cosmologist Alexander Vilenkin. According to inflationary cosmology, in the first fraction of a second after the big bang our universe experienced an exponentially rapid rate of expansion, after which its expansion settled down to a more sedate pace. Inflationary cosmology was originally proposed in order to explain two features of the universe that were puzzling from the perspective of standard big-bang cosmology—its uniformity (homogeneity) and its flatness.

By homogeneity, cosmologists mean that the universe looks the same to all observers, no matter where they are located. One aspect of this homogeneity is the uniformity of cosmic background radiation, which has the same temperature throughout the observable cosmos. This is a problem in standard big-bang cosmology. According to the big-bang theory, up until about 300,000 years after the beginning of the universe, the photons in the background radiation would have been bouncing off the electrons in the hot plasma that filled the entire universe. At that point the universe would have cooled enough for electrically neutral atoms to form, releasing the background radiation. This radiation eventually reached us, giving us a picture of the universe at around 300,000 years of age.

The puzzling thing about this radiation is that it has the same temperature in every direction to about one part in a hundred thousand. This implies that the universe at 300,000 years old was incredibly uniform in temperature, which in turn would have required very precise initial conditions. It follows that the observed uniformity of the background radiation can be explained in the ordinary big-bang scenario only by postulating that the initial state of the universe was one of almost perfect uniformity in the temperature and distribution of the plasma, which in turn requires a very precisely fine-tuned initial explosion.[2]

A homogeneous universe is called "flat" if it is balanced between eventual gravitational collapse and eternal expansion; in such a case its geometry would be precisely Euclidean and space would not be curved. The universe achieves such flatness when the actual mass density in the universe is very close to the critical mass density (the density required to halt the expansion of the universe)—that is, if the ratio between the actual and critical mass densities is close to one. In our universe, the ratio of these two quantities is ever so slightly less than one. As a result, our universe will keep on expanding without a gravitational recollapse, and space has hardly any overall curvature. That these values were so precisely balanced is surprising from the standpoint of standard big-bang theory, because, again, for this balance to arise, the universe would have needed to have very finely tuned initial conditions.

Inflationary cosmology attempts to explain the horizon problem (homogeneity) not as the result of these finely tuned initial conditions (though it does invoke special conditions of its own; see below), but instead as a consequence of an early, exponentially rapid rate of cosmic expansion. According to the inflationary model, during the first fractions of a second after the big bang the temperature of the universe had a chance to homogenize. Then the rapid expansion of the universe distributed this homogeneous radiation throughout the observable universe. It also pushed any remaining inhomogeneity beyond the edge of the observable universe.[3]

In current models, inflation begins at around 10^{-37} seconds after the big bang and lasts until 10^{-35} seconds, during which space itself expands by a factor of 10^{60} or so. At the beginning of the inflationary epoch the observable universe was, say, about 10^{-60} meters in size and at the end of it about a meter across. At the start of inflation, however, the horizon size (the distance light traveled since the big bang) was 10^{-37} light-seconds, which is far larger than the tiny patch that was destined to grow into our observable universe. Thus, the inflationary process not only distributed the homogeneous background radiation throughout the observable universe, it also distributed any remaining inhomogeneity beyond the edge of the observable universe.

Inflation explains the near flatness of the universe as a consequence of the hyper-expansion as well. During the inflationary epoch, all the distances in the universe increased by a measure of 10^{60}, which means the radius of the observable universe increased by this factor as well. Suppose the four-dimensional space-time of the universe prior to inflation had positive curvature, like the surface of a balloon in three dimensions, and that its radius was a billionth of a meter (a nanometer). After inflation, its radius would be 10^{51} meters, or about 10 billion trillion trillion light-years. Just like inflating a balloon to larger and larger sizes makes a small patch of it look flatter, so inflating the whole universe makes the observable patch of space-time look flatter and flatter.

Inflationary cosmology is relevant to discussions of the origin of life, because some cosmologists think that it provides a mechanism for generating many universes other than our own, and also because one prominent molecular biologist has recently invoked those other universes in an attempt to explain the origin of life. According to the currently dominant "chaotic eternal inflationary model," the rapid expansion of the universe was driven by an "inflaton field"—a repulsive gravitational field. After an initial phase of expansion, the inflaton field decayed locally to produce our universe. However, it also continued to operate at full strength outside the local area to produce a wider expansion of space into which other universes were birthed as the inflaton field decayed at other locations. Thus, inflationary cosmologists postulate the decay of the inflaton field as a mechanism by which other "bubble universes" can be created. They also postulate that inflation can continue indefinitely into the future and, therefore, that the wider inflaton field will spawn an endless number of other universes as it decays in local pockets of the larger and larger expansions of space. Since the inflaton field continues to expand at a rate vastly greater than the bubble universes expanding within it, none of these bubble universes will ever interfere with each other. The one inflaton field therefore gives birth to endless bubble universes—"many worlds in one," as Vilenkin colorfully describes it.[4]

Koonin has appropriated this cosmology in order to explain the origin of life by chance. Following Vilenkin, he argues that since the inflaton field can produce an infinite number of other universes, every event that has occurred in our universe was bound to occur somewhere endlessly many times. Thus, events that appear to be extremely improbable when considering the probabilistic resources of our universe are actually highly probable—indeed, inevitable—given the plethora of other universes that do—and will—exist. As Koonin himself explains: "In an infinite multiverse with a finite number of distinct macroscopic histories (each repeated an infinite number of times), emergence of even highly complex systems by chance is not just possible but inevitable. . . . it becomes conceivable that the minimal requirement (the breakthrough stage) for the onset of biological evolution is a primitive coupled replication-translation system that emerged by chance. That this extremely rare event occurred on earth and gave rise to life as we know it is explained by anthropic selection alone."[5] By "anthropic selection," Koonin simply means that our perception that life is incredibly improbable is just an artifact of our particular vantage point. Since we observe only one bubble universe, we do not realize that the existence of other universes and the mechanism that produced them make life in a universe such as our own inevitable.

So has Koonin's use of inflationary cosmology solved the problem of the origin of life and the origin of the biological information necessary to it? Has he proposed a better explanation for the origin of biological information than intelligent design? There are several reasons to think not.

Do Inflaton Fields Exist?

First, there are good reasons to doubt that inflaton fields even exist. Inflaton fields were postulated mainly to explain the homogeneity and flatness problems, but they may not explain these features of the universe well at all, as several prominent physicists have long pointed out. In order to explain the homogeneity of the universe using inflaton fields, physicists have to make gratuitous assumptions about the singularity from which everything came. As Oxford physicist Roger Penrose points out, if the singularity were perfectly generic, expansion from it could yield many different kinds of irregular (inhomogeneous) universes, even if inflation had occurred.[6] Thus, inflation alone, without additional assumptions, does not solve the homogeneity problem. Getting workable results requires imposing the right metric (distance measure) on space-time.

Furthermore, as physicists Stephen Hawking and Don Page note, it has proven difficult to explain why inflaton fields and gravitational fields (as described by general relativity, which we have strong reasons to accept) should work together to produce the homogeneity of the background radiation and the flatness of space-time in our observable universe. Indeed, when the fields are linked, there is no guarantee that inflation will even take place.[7] Moreover, these inflaton fields, with their uncanny ability to decay at just the right time (between 10^{-37} to 10^{-35} seconds after the big bang) and in just the right measure, have properties associated with no other physical fields. (Instead, they have properties that were invented solely for the purpose of solving the horizon and flatness problems, which they can't solve without additional arbitrary assumptions and specifications of initial conditions.)

Causal Adequacy Considerations

There is another reason that inflationary cosmology does not provide a satisfying explanation or a better explanation than intelligent design for the origin of biological information. Inflationary cosmology relies for its explanatory power on the presumed causal powers of an entirely unknown entity—one posited solely to explain a mysterious class of effects—and one whose causal powers have not been demonstrated or observed. We do not know if inflaton fields exist. And we do not know, if they exist, what exactly they can actually do. Nevertheless, we do know (from direct first-person awareness, if nothing else) that conscious intelligent minds exist and what they can do.

Further, as philosopher of physics Robin Collins argues, all things being equal, we should prefer hypotheses "that are natural extrapolations from what we already know" about the causal powers of various kinds of entities.[8] In a slightly different context he argues that multiple-worlds hypotheses fail to meet this test in their explanations of the anthropic fine-tuning of the universe, whereas design hypotheses do not. To illustrate, Collins asks his readers to imagine a paleontologist who posits the existence of an electromagnetic "dinosaur bone–producing field," as opposed to actual dinosaurs, as the explanation for the origin of large fossilized bones. Although certainly such a field qualifies as a possible explanation for the origin of the fossil bones, we have no experience of such fields or of their producing fossilized bones. Yet we have observed animal remains in various phases of decay and preservation in sediments and sedimentary rock. Thus, most scientists rightly prefer the actual dinosaur hypothesis over the apparent dinosaur hypothesis (i.e., the "dinosaur bone–producing field" hypothesis) as an explanation for the origin of fossils.

In the same way, we have no experience of anything like an inflaton field generating infinitely many universes (or, for that matter, any experience of any machine or mechanism capable of producing something as finely tuned as our universe that is not itself designed). Yet we do have extensive experience of intelligent agents producing finely tuned machines or information-rich systems of alphabetic or digital code. Thus, Collins concludes, the postulation of mind to explain the fine-tuning of the universe constitutes a natural extrapolation from our experience-based knowledge of the causal powers of intelligent agency, whereas postulation of multiple universes (including those produced by inflaton fields) lacks a similar basis. It follows *a fortiori* that the design hypothesis is a better explanation than inflaton-field decay for the origin of the information necessary to produce the first life, because it depends upon the known causal powers of an entity familiar from repeated and direct experience. Inflationary cosmology depends upon an abstract entity whose causal powers have not been observed or demonstrated.

Return of the Displacement Problem

There is an additional problem with using inflaton fields to explain the origin of the information necessary to produce the first life. In order to explain the origin of certain features of our observable universe, and (as an unintended bonus) the origin of presumably innumerable life-friendly universes such as our own, inflationary cosmology must invoke a number of unexplained sources or infusions of information. For example, both inflaton fields, and the fields to which they are coupled, have to be finely tuned in order to produce new bubble universes of the right sort. The "shutoff" of the energy of the inflaton field (which occurs during its decay) alone has to be finely tuned to between one part in 10^{53} and one part in 10^{123} (depending on the model of inflation invoked) to produce a bubble universe compatible with life. Additionally, inflationary cosmology makes the already acute fine-tuning problem associated with the initial low-entropy state of our universe exponentially worse. According to calculations by Roger Penrose (who regards inflationary cosmology as a very dubious enterprise), the initial entropy of our universe is already finely tuned to an accuracy of one part in 10exp(10exp(123)).[9] Inflation not only does nothing to explain this fine tuning; it actually exacerbates it.

Some cosmologists argue, of course, that these improbabilities can be overcome by the number of bubble universes that the original inflaton field produces. But aside from the inelegance and lack of parsimony of this explanatory strategy, generating a larger inflaton field that produces the right results (i.e., a universe with the properties of our observable universe) itself depends on a number of gerrymandered assumptions and finely tuned initial conditions. As noted above, physicists make a number of gratuitous assumptions about the initial singularity in order to get the inflaton field to mesh with the theory of general relativity. For example, to get inflationary cosmology to harmonize with general relativity, cosmologists have to assume a specific way of measuring distance in space-time (a so-called metric) and reject others. In addition, there are a number of possible inflationary cosmological models, only some of which (when conjoined with general relativity) would actually cause universes to inflate. In order to ensure that inflaton fields will create bubble universes, physicists have to select some inflationary models and exclude others in their theoretical postulations. Each of these choices constitutes an informative intervention on the part of the modeler—one that reflects unexplained information that would have had to have been present in the initial conditions associated with the universal singularity.

Indeed, the need to make such assumptions and restrict theoretical postulations implies that the initial singularity itself would have had to have been finely tuned in order for any inflaton field to be capable of producing a universe such as our own. Yet we know that our universe exists. We also have good reasons for thinking that general relativity is true. Thus, if an inflaton field exists, it could operate the way that inflationary cosmologists envision only if the singularity from which the inflaton field emerged was itself finely tuned (and information-rich).

Thus, by relying on inflationary cosmology to explain the information necessary to produce the first life, Koonin has once again created an information problem in the act of purportedly solving one (see

Chapter 13). Even assuming that inflaton fields exist and that they can create an infinite number of universes (by no means a safe bet), Koonin solves the problem of the origin of biological information by creating a new problem of cosmological information—information that, in his model, is nevertheless entirely necessary to explain the origin of life. Additionally, all inflationary models assume that the inflaton field operates within and creates new universes with the same basic laws and constants of physics that exist within our universe. Yet the laws and constants of our universe are themselves extremely fine-tuned to allow for the possibility of life. This fine tuning represents another source of information that has to be accounted for in order to explain the origin of life in our cosmos. Yet inflationary theory presupposes, rather than explains, the existence of this fine tuning.

An Epistemological Cost

Inflationary cosmology has yet another liability: once permitted as a possible explanation for anything, it destroys practical and scientific reasoning about everything. Inflationary cosmology can explain the origin of all events, no matter how improbable, by reference to chance because of the infinite probabilistic resources it purports to generate. It follows that events we explain by reference to known causes based upon ordinary experience are just as readily explained in inflationary cosmology as chance occurrences without any causal antecedent. According to inflationary cosmology, all events consistent with our laws of nature will eventually arise as the result of random fluctuations in the quantum vacuum constituted by the inflaton field. This means that an exquisitely designed machine or an intricately crafted piece of poetry is just as likely to have been produced by chance fluctuations in the quantum vacuum as by a human being. It also means that events such as earthquakes or regular phenomena such as condensation are just as likely to have been the result of chance fluctuations in the quantum vacuum as they are to have been the result of an orderly progression of discernible material causes. In short, if inflationary cosmology is true, anything can happen for no reason at all, beyond the supposed random fluctuations in the quantum vacuum of the inflaton field.

To make matters worse, inflationary cosmology actually implies that certain explanations that we regard as extremely improbable are actually more likely to be true than explanations we ordinarily accept. Consider the "Boltzmann brain" phenomenon, for example, over which quantum cosmologists have been greatly exercised. Within inflationary cosmology, it is theoretically possible for a fully functioning human brain to pop spontaneously into existence, due to thermal fluctuations in the quantum vacuum, and then disappear again. Such an entity has been called a "Boltzmann brain." Under standard conditions for bubble-universe generation in inflationary cosmology, Boltzmann brains would be expected to arise as often, or more often, than normal occurrences in our universe. Indeed, calculations based upon some inflationary cosmological models lead to a situation in which these free-floating Boltzmann brains infinitely outnumber normal brains in people like us.[10]

The epistemological implications of this possibility have raised issues that cosmologists cannot ignore. If these inflationary cosmological models are accurate, it becomes infinitely more probable that we ourselves are free-floating Boltzmann brains than real persons with a history living in a universe 13.7 billion years old. In some models, it's even more probable that a whole universe like ours spontaneously fluctuated into existence than it is that our universe with its extraordinarily improbable initial conditions evolved in an orderly and lawlike way over billions of years. This means that the many-worlds-in-one hypothesis generates an absurdity. It implies that we are probably not the people we take ourselves to be and that our memories and perceptions are not reliable, but quite possibly chance fabrications of quantum fields. Neither is our universe itself what it appears to be according to the hypothesis of eternal inflation. In short, the proposal Koonin has adopted to solve the origin-of-life problem renders all scientific reasoning and explanation unreliable, thus undermining any basis for his own explanation of how life came to be. It would be hard to invent a more self-refuting hypothesis than that!

Notes

Prologue

1. Wilgoren, "Seattle Think Tank Behind 'Intelligent Design' Push."
2. Wilgoren, "Politicized Scholars Put Evolution on the Defensive."
3. Holden, "Random Samples," 1709; Giles, "Peer-Reviewed Paper Defends Theory of Intelligent Design"; Stokes, "Intelligent Design Study Appears"; Monastersky, "Society Disowns Paper Attacking Darwinism."
4. Klinghoffer, "The Branding of a Heretic"; Price, "Researcher Claims Bias by Smithsonian"; "Unintelligent Design Hostility."
5. "Intelligent Design and the Smithsonian," *New York Times,* August 20, 2005, Editorial/Letters, national edition; Powell, "Darwinian Debate Evolving."
6. Ostling, "Lifelong Atheist Changes Mind."
7. Powell, "Editor Explains Reason for 'Intelligent Design' Article"; "Intelligent Design and the Smithsonian," *New York Times,* August 20, 2005, Editorial/Letters, national edition; Wilgoren, "Politicized Scholars Put Evolution on the Defensive."
8. *The O'Reilly Factor* (Fox News), "Brutally Criticized," August 25, 2005.
9. Thaxton, Bradley, and Olsen, *The Mystery of Life's Origin.*
10. Meyer, "The Origin of Biological Information and the Higher Taxonomic Categories." See also Meyer, "The Cambrian Information Explosion"; "DNA and the Origin of Life"; "Evidence of Design in Physics and Biology," 53–111; "The Scientific Status of Intelligent Design"; "Teleological Evolution"; "DNA by Design"; "The Explanatory Power of Design"; Meyer, Nelson, and Chien, "The Cambrian Explosion."
11. Meyer, "The Origin of Biological Information and the Higher Taxonomic Categories."
12. Pera, *The Discourses of Science.*
13. Darwin, *On the Origin of Species,* 481–82.
14. Ayala, "Darwin's Greatest Discovery."

Chapter 1: DNA, Darwin, and the Appearance of Design

1. Newton, *Opticks*, 369–70.
2. Crick, *What Mad Pursuit*, 138.
3. Dawkins, *River out of Eden*, 17.
4. Gates, *The Road Ahead*, 188.
5. Küppers, *Information and the Origin of Life*, 170–72.
6. The Harvard Origins of Life Initiative, http://origins.harvard.edu.
7. Elizabeth Pennisi, "Finally, the Book of Life."
8. Interview with Williams, in Brockman, ed., *The Third Culture*, 42–43.
9. Interview with Williams, in Brockman, ed., *The Third Culture*, 42–43.
10. Klir and Wierman, *Uncertainty-Based Information*.
11. Gilder, *Telecosm*.
12. Watson and Crick, "A Structure for Deoxyribose Nucleic Acid."
13. Williams, *Natural Selection*, 11.
14. Yockey, "Origin of Life on Earth," 105.
15. Lewontin, "Adaptation."
16. Dawkins, *The Blind Watchmaker*, 1.
17. Mayr, "Darwin: Intellectual Revolutionary." The effort to explain biological organisms naturalistically was reinforced by a trend in science to provide fully naturalistic accounts for other phenomena such as the precise configuration of the planets in the solar system (Pierre Laplace) and the origin of geological features (Charles Lyell and James Hutton). It was also reinforced (and in large part made possible) by an emerging positivistic tradition in science that increasingly sought to exclude appeals to supernatural or intelligent causes from science *by definition* (see Gillespie, "Natural History, Natural Theology, and Social Order"). See also Darwin, *On the Origin of Species*, 481–82.
18. Darwin, *On the Origin of Species*, 188.
19. Darwin, *Life and Letters*, 1:278–79.
20. Ayala, "Darwin's Greatest Discovery," in Ruse and Dembski, eds., *Debating Design*, 58. As the late Harvard evolutionary biologist Ernst Mayr explained, "The real core of Darwinism . . . is the theory of natural selection. This theory is so important for the Darwinian because it permits the explanation of adaptation, the 'design' of the natural theologian, by natural means, instead of by divine intervention" (Foreword, in Ruse, ed., *Darwinism Defended*).
21. Ayala, "Darwin's Greatest Discovery," *Proceedings of the National Academy of Sciences*, 8573.
22. Bishop, "The Religious Worldview."
23. Crick, *What Mad Pursuit*, 138. Emphasis added.
24. Ruse, "Teleology in Biology."
25. Dawkins, *River Out of Eden*, 17.
26. Shapiro, review of *Darwin's Black Box*, by Michael Behe.
27. Lenoir, *The Strategy of Life*, ix.
28. Watson, et al., *Molecular Biology of the Gene*, 1:704.
29. I used the character converter at http://www.csgnetwork.com/asciiset.html for the binary conversion of the first few words of the Declaration of independence in ASCII code.

Chapter 2: The Evolution of a Mystery and Why It Matters

1. For a scholarly account of this discovery, along with a more detailed account of the series of discoveries that caused vitalism to fall from favor, see Aaron J. Ijde, *The Development of Modern Chemistry*, 164–65.

2. As quoted in Kauffman and Chooljian, "Friedrich Wöhler (1800–1882), on the Bicentennial of His Birth," 126.
3. Partington, *A History of Chemistry.*
4. Florkin, *A History of Biochemistry,* 251–52.
5. Hopkins, "BAAS Presidential Address," 382.
6. Darwin, *The Life and Letters,* 1:18.
7. For a detailed treatment of this perspective in ancient philosophy, see Sedley, *Creationism and Its Critics in Antiquity.*
8. Newton, General Scholium, *Mathematical Principles of Natural Philosophy.*
9. Kirk and Raven, *The Presocratic Philosophers.*
10. Hobbes, *Leviathan;* Hume, *Dialogues Concerning Natural Religion.*
11. Laplace, *(Vietnamese)Exposition du système du monde.*
12. Lyell, *Principles of Geology.*
13. Ernst Haeckel, for instance, in *The History of Creation,* stated: "We can, therefore, from these general outlines of the inorganic history of the earth's crust, deduce the important fact that at a certain definite time life had its beginning on earth, and that terrestrial organisms did not exist from eternity" (401).
14. Darwin, *More Letters of Charles Darwin,* 273.
15. Farley, *The Spontaneous Generation Controversy,* 103ff.; Lechevalier and Solotorovsky, *Three Centuries of Microbiology,* 35–37.
16. Farley, *Spontaneous Generation Controversy,* 103–7, 114, 172; Lanham, *Origins of Modern Biology,* 268.
17. Nevertheless, the doctrine of spontaneous generation did not die easily. Even after Pasteur's work, Henry Bastian continued to find microbial organisms in various substances that had been sealed and "sterilized" at 100 degrees C or higher. Not until the 1870s, when microbiologists like Cohn, Koch, and Tyndall perfected methods of killing heat-resistant spores, were Bastian's observations discredited. Despite an increasingly critical scientific response to his experimental methods and conclusions, Bastian continued to offer observational evidence for spontaneous generation from inorganic matter for another thirty years. Nevertheless, these and other attempts yielded to a familiar pattern of refutation. Experiments supposedly establishing the spontaneous occurrence of microorganisms remained tenable only as long as sterilization methods were inadequate to kill existing microorganisms or prevent bacterial contamination of experimental vessels from the surrounding environment. When sources of microorganisms were identified and methods of destroying them perfected, observational evidence for spontaneous generation was withdrawn or discredited. In the minds of some scientists, especially after the turn of the century, development seemed to confirm that living matter is too complex to organize itself spontaneously, whether beginning from organic or inorganic precursors. Although Huxley and Haeckel accepted Pasteur's results, both insisted that his work was not relevant to abiogenesis (life arising from nonliving matter), as his experiments discredited only theories of what Haeckel called "plasmogeny" or what Huxley called "heterogenesis," i.e., spontaneous generation from once living matter (Haeckel, *The Wonders of Life,* 115; Kamminga, "Studies in the History of Ideas," 55, 60).
18. Glas, *Chemistry and Physiology,* 118.
19. Coleman, *Biology in the Nineteenth Century,* 129.
20. Steffens, *James Prescott Joule and the Concept of Energy,* 139.
21. Steffens, *James Prescott Joule and the Concept of Energy,* 129–30; Glas, *Chemistry and Physiology,* 86.
22. Haeckel, *The Wonders of Life,* 27.
23. Haeckel, *The History of Creation,* 332.

24. Virchow, "On the Mechanistic Interpretation of Life," 114.

25. Virchow, "On the Mechanistic Interpretation of Life," 115.

26. "Haeckel's attitude, and that of other contemporary Darwinians, to the question of the origin of life was first and foremost an expression of their worldview. Abiogenesis was a necessary logical postulate within a consistent evolutionary conception that regarded matter and life as stages of a single historical continuum" (Fry, *The Emergence of Life on Earth*, 58).

27. Hull, "Darwin and the Nature of Science," 63–80. As Hull, a philosopher of biology, explains, Darwin posited "that species are not eternal but temporary, not immutable but quite changeable, not discrete but graduating imperceptibly through time one into another." As Darwin himself said in *On the Origin of Species*, "I was much struck how entirely vague and arbitrary is the distinction between species and varieties. . . . Certainly no clear line of demarcation has yet been drawn between species and subspecies . . . or, again, between subspecies and well-marked varieties" (104, 107).

28. For instance, John Tyndall argued, "There does not exist a barrier possessing the strength of a cobweb to oppose the hypothesis which ascribes the appearance of life to that 'potency of matter' which finds expression in natural evolution" (*Fragments of Science*, 434). As Kamminga explains, Tyndall held that "it was inconsistent to believe in evolution and at the same time reject abiogenesis" ("Studies in the History of Ideas," 57).

29. Darwin's work amplified a theme that evolutionists such as Lamarck and Matthew had articulated in various ways since early in the nineteenth century. The publication of *Origin* in 1859 reaffirmed the importance of environmental influences on the development of species. See Mckinney, ed., *Lamarck to Darwin*, 7, 9, 29, 41.

30. Darwin, "Letter to Hooker"; see also Darwin, *Life and Letters*, 18.

31. Oparin, *Genesis and Evolutionary Development of Life*, 7.

32. Haeckel, *The Wonders of Life*, 135.

33. This strengthened the conviction among many scientists that vital function was ultimately reducible to a "physical basis," as Thomas H. Huxley phrased it in 1868 ("On the Physical Basis of Life"). See also Geison, "The Protoplasmic Theory of Life"; and Hughes, *A History of Cytology*, 50.

34. Geison, "The Protoplasmic Theory of Life."

35. Geison, "The Protoplasmic Theory of Life," 274.

36. As cited in Geison, "The Protoplasmic Theory of Life," 274.

37. The descriptions matched those by Felix Dujardin in 1835 and Gabriel Gustav Valentin in 1836. See Geison, "The Protoplasmic Theory of Life"; Hughes, *A History of Cytology*, 40, 112–13.

38. Geison, "The Protoplasmic Theory of Life," 276. Shultze, in particular, emphasized the importance of protoplasm based on his realization that lower marine animals sometimes exist in a "primitive membraneless condition" and on his identification of protoplasm as the source of vital characteristics like contractility and irritability. See Suñer, *Classics of Biology*, 19–20. This conclusion was widely endorsed and amplified throughout the 1860s by influential thinkers like Ernst Wilhelm von Brucke, Wilhelm Kühne, Ernst Haeckel, T. H. Huxley, and others.

39. Hughes *A History of Cytology*, 50; see also Huxley, "On the Physical Basis of Life," 129; Wilson, *The Physical Basis of Life*, 1–2; Geison, "The Protoplasmic Theory of Life," 273, 276–91. Geison comments that during this period, "The conviction grew that the basic unit of life was essentially a protoplasmic unit" (278). It was during this period that the term "protoplasm" gained wide currency.

40. Huxley, "On the Physical Basis of Life," 138–39.

41. Haeckel, *Generelle Morphologie der Organismen*, 179–80; *The History of Creation*, 411–13; Kamminga, "Studies in the History of Ideas," 60–61.

42. Haeckel, *The History of Creation*, 411–13; Kamminga, "Studies in the History of Ideas," 62.

43. Haeckel, *Generelle Morphologie der Organismen*, 183; Kamminga, "Studies in the History of Ideas," 62.

44. Haeckel, *The History of Creation*, 421–22; Kamminga, "Studies in the History of Ideas," 60–62.

45. Haeckel, *The Wonders of Life*, 111. Even amid such bold claims, Haeckel and Huxley acknowledged that certain special but unspecified past conditions must have played an important role in the origin of protoplasm, since its spontaneous generation from nonliving matter had never been observed. However, neither Huxley nor Haeckel could demonstrate that the specific conditions enumerated in their prebiotic scenarios could be inferred from classes of evidence other than the existence of life itself. This deficiency ensured that, as Haeckel himself put it, "Any account of the primary generation of [life] must be considered premature" (as quoted in Oparin, *The Origin of Life* [1938], 105). Neither Huxley nor Haeckel attempted to identify the relevant past conditions. Haeckel acknowledged, "We have no satisfactory conception of the extremely peculiar state of our earth's surface at the time of the first appearance of organisms." Instead, both men tried to turn this absence of information to their advantage by intimating that the conditions specified in their scenarios were at least possible. They also relied on materialistic conceptions of life's nature to suggest that life's origin could be explained by reference to materialistic causes.

46. Kamminga, "Studies in the History of Ideas," 69–70, 73–74; Haeckel, *The Wonders of Life*, 111–14.

47. Kamminga, "Studies in the History of Ideas," 70–71; Haeckel, *The Wonders of Life*, 112.

48. Kamminga, "Studies in the History of Ideas," 70–71.

49. Kamminga, "Studies in the History of Ideas," 185, 188–92. As Robert E. Kohler argues, the rise of the new field of biochemistry after 1900 marked the emergence of a "common outlook on the physico-chemical nature of life" in which it was agreed "that enzymes were the key agents in life processes" ("The Enzyme Theory and the Origin of Biochemistry," 194).

50. Oparin, *The Origin of Life* (1924); Meyer, "Of Clues and Causes."

51. Kamminga, "Studies in the History of Ideas," 222.

52. Graham continues: "The Russian biologist with the securest claim to being an orthodox Darwinist was K. A. Timiriazev. If any Russian deserved the title of 'Darwin's Bulldog,' to match that of Huxley in England, it was Timiriazev. A plant physiologist, Timiriazev combined scientific knowledge with radical politics. Throughout his career he was engaged in a running battle with established authorities, even though, as a teacher in St. Petersburg University, and later, a professor at Moscow University, he became something of an authority himself, particularly among radical intellectuals. His political scraps included expulsion as a student from St. Petersburg University and, many years later, dismissal from the faculty of Moscow University as a result of his continuing radical sympathies. Timiriazev was the most popular defender of Darwinism in all of Russia. His books, *A Short Sketch of the Theory of Darwin* and *Charles Darwin and His Theory*, were published in fifteen editions between 1883 and 1941. His influence was so great that it could still be felt well into the middle of the twentieth century. In an interview in Moscow in 1970 academician A. I. Oparin, a well-known authority on origin of life, described the lectures on Darwinism Timiriazev gave at the Polytechnical Museum in Moscow when Oparin was a boy as the most important influence on his professional development" (*Science in Russia and the Soviet Union*, 67).

53. Graham, *Science and Philosophy in the Soviet Union*, 262–63.

54. In a letter to Ferdinand Lassalle, Marx wrote: "Darwin's work is most important and suits my purpose in that it provides a basis in natural science for the historical class struggle. One does, of course, have to put up with the clumsy English style of argument. Despite all shortcomings, it is here that, for the first time, 'teleology' in natural science is not only dealt a mortal blow but its rational meaning is empirically explained" (http://www.marxists.org/archive/marx/works/1861/letters/61_01_16.html).

55. Engels, *The Dialectics of Nature.*

56. Yockey, *Information Theory, Evolution, and the Origin of Life,* 151–52.

57. Graham, *Science and Philosophy in the Soviet Union,* 262–63. Oparin did not himself participate in the revolution or join the Communist Party. Moreover, in an interview near the end of his life, he stressed the importance of Darwinism over Marxism on his initial theorizing. When asked if he had been a Marxist in 1924, Oparin replied, "Politically, yes I was. But at that moment what exercised the major influence on my ideas about the origin of life was the postulates of Darwin's theory of evolution."

58. Although as Graham notes, "Oparin has not been one to attempt rigorous definitions of 'life.'" His theoretical work in 1924 did reflect a definite conception of life, which he expressed "in metaphors or in terms of varying combinations of characteristics necessary for life" (*Science and Philosophy in the Soviet Union,* 268).

59. Graham, *Science and Philosophy in the Soviet Union,* 214, 208.

60. Oparin not only rejected "idealistic" approaches to the origin of life that assumed an impassable divide between animate and inanimate. He also rejected the "theory of panspermia" advocated by various nineteenth-century scientists such as Helmholtz, Thomson, and Richter. Panspermia specified that life on earth had originated from spores or germs transmitted here from outer space. These germ species were then supposed either to have originated elsewhere through some unspecified process or to have existed eternally. Many nineteenth-century advocates of the theory opted for the latter of these two views on the grounds that the complicated structure of living cells, though only vaguely discernable with nineteenth-century microscopes, suggested an "impassable abyss" between animate and inanimate matter. In rejecting this assumption, Oparin regarded recourse to panspermia as motivated by an incorrect premise and that it was, as such, "philosophically useless." Even if life on earth had issued from outer space, he noted, the ultimate origin of life would remain unsolved, especially if one assumed that life had existed eternally. (See Graham, *Science and Philosophy in the Soviet Union,* 203–4, 206, 211.)

61. Graham, *Science and Philosophy in the Soviet Union,* 203.

62. Graham, *Science and Philosophy in the Soviet Union,* 199–202.

63. Oparin's hypothetical reconstruction actually began with an account of the earth's formation. Oparin accepted a contemporary version of the nebular hypothesis. He thought, therefore, that the planets formed from hot, incandescent clouds of gas. According to this view, as these nebular clouds condensed under the mutual gravitational attractions of their gas molecules they would have formed stars, with the vapors of heavier metals like iron, nickel, and cobalt settling to the center of the cloud. These vapors then cooled and concentrated in a liquid core of heavy metal. (See Graham, *Science and Philosophy in the Soviet Union,* 217–18, 221.)

64. Oparin supported this suggestion with spectroscopic studies confirming the presence of hydrogen in the sun's atmosphere and studies showing the presence of hydrocarbons in the atmospheres of dying red stars. He also cited the work of the Russian chemist D. I. Mendeleev, who earlier had proposed that methane and other naturally occurring hydrocarbons could have had a nonbiological origin. Mendeleev believed hydrocarbon production occurred naturally when metallic carbides combined with oxygen at high temperatures and pressures. Though Oparin did not accept Mendeleev's account of the origin of extant terrestrial hydrocarbons, he did

appropriate Mendeleev's suggested reaction pathway to account for the origin of high-energy biological precursors on the early earth. (See Oparin, *The Origin of Life* [1938], 98, 107–8; Graham, *Science and Philosophy in the Soviet Union*, 218–23.)

65. Oparin thought these hydrocarbon compounds would have included alcohols, ketones, aldehydes, and organic acids (*The Origin of Life* [1938], 108).
66. Oparin thought these nitrogen-rich compounds would have included amines, amides, and ammonium salts (*The Origin of Life* [1938], 108).
67. Oparin, *The Origin of Life* (1938), 229.
68. Oparin, *The Origin of Life* (1938), 229–31.
69. Miller, "A Production of Amino Acids."
70. Shapiro, *Origins*, 98.
71. William Day, *Genesis on Planet Earth*, 5.

Chapter 3: The Double Helix

1. As Crick noted: "Only gradually did I realize that this lack of qualification could be an advantage. By the time most scientists have reached age thirty they are trapped in their own expertise. They have invested so much effort in one particular field that it is often extremely difficult, at that time in their careers, to make a radical change. I, on the other hand, knew nothing, except for a basic training in somewhat old-fashioned physics and mathematics and an ability to turn my hand to new things" (*What Mad Pursuit*, 16).
2. Chamberlain, "The Method of Multiple Working Hypotheses."
3. Oparin, *Genesis and Evolutionary Development of Life*, 7.
4. Kamminga, "Protoplasm and the Gene," 1.
5. Haeckel, *The Wonders of Life*, 111.
6. Mendel called the discrete inheritance of the two traits for one characteristic the "law of segregation." See Jenkins, *Genetics*, 13–15.
7. Jenkins, *Genetics*, 238–39.
8. Jenkins, *Genetics*, 238–39.
9. Morgan, *The Physical Basis of Heredity*.
10. Olby, *The Path to the Double Helix*, 73–96, including chap. 6, "Kossel, Levene and the Tetranucleotide Hypothesis."
11. Avery, MacCleod, and McCarty, "Induction of Transformation by a Deoxyribonucleic Acid Fraction."
12. Griffith, "The Significance of Pneumococcal Types"; Dawson, "The Interconvertibility of 'R' and 'S' Forms of Pneumococcus."
13. Avery, MacLeod, and McCarty, "Induction of Transformation by a Deoxyribonucleic Acid Fraction"; for a detailed historical account of their work, see Judson, *The Eighth Day of Creation*, 13–23.
14. Chargaff, "Preface to a Grammar of Biology."
15. Chargaff, *Essays on Nucleic Acids*, 1–24, "Chemical Specificity of Nucleic Acids."
16. According to Chargaff: "The deoxypentose nucleic acids extracted from different species thus appear to be different substances or mixtures of closely related substances of a composition constant for different organs of the same species and characteristic of the species. The results serve to disprove the tetranucleotide hypothesis" (*Essays on Nucleic Acids*, 13).
17. Chargaff, *Essays on Nucleic Acids*, 21.
18. "The number of possible nucleic acids having the same analytical composition is truly enormous. For example, the number of combinations exhibiting the same molar proportions of individual purines and pyrimidines as the deoxyribonucleic acid of the ox is more than 10^{56}, if the nucleic acid is assumed to consist of only 100 nucleotides; if it consists of 2,500 nucleotides, which probably is much nearer the

truth, then the number of possible 'isomers' is not far from 10^{1500}" (Chargaff, *Essays on Nucleic Acids,* 21).

19. McElheny, *Watson and DNA,* 5–29; Watson, *The Double Helix,* 17.
20. Watson, *The Double Helix,* 17, 27.
21. Watson, *The Double Helix,* 9–12, 31.
22. Olby, *The Path to the Double Helix,* 310.
23. "Politeness, Francis Crick said over the BBC at the time he got the Nobel Prize, is the poison of all good collaboration in science. The soul of collaboration is candor, rudeness if need be. Its prerequisite, Crick said, is parity of standing in science, for if one figure is too much senior to the other, that's when the serpent politeness creeps in. A good scientist values criticism almost higher than friendship: no, in science criticism is the height and measure of friendship. The collaborator points out the obvious, with due impatience. He stops the nonsense, Crick said—speaking of James Watson" (Judson, *The Eighth Day of Creation,* 125).
24. "When Crick and Watson began, they knew very little about DNA for sure, and part of what they were most sure of was wrong. To consider DNA as a physical object, they wanted diameters, lengths, linkages and rotations, screw pitch, density, water content, bonds, and bonds and again bonds. The sport would be to see how little data they could make do with and still get it right: the less scaffolding visible, the more elegant and astonishing the structure. . . .'The point is [said Crick] that evidence can be unreliable, and therefore you should use as little of it as you can. . . . There isn't such a thing as a hard fact when you're trying to discover something. It's only afterwards that the facts become hard'" (Judson, *The Eighth Day of Creation,* 92–93).
25. Watson, *The Double Helix,* 57–59.
26. Watson, *The Double Helix,* 57–59.
27. Watson, *The Double Helix,* 81.
28. According to Judson: "On Friday of that week, January 30 [1953], Watson went to London to see Hayes. He took the copy of Pauling's manuscript with him, and about teatime that afternoon stopped at King's College" (*The Eighth Day of Creation* [1996], 135).
29. Watson, *The Double Helix,* 96.
30. "Afterwards, in the cold, almost unheated train compartment, I sketched on the blank edge of my newspaper what I remembered of the B pattern" (Watson, *The Double Helix,* 99).
31. Chargaff, *Essays on Nucleic Acids,* 1–24.
32. As Watson recalled in his memoirs, "The high point in Chargaff's scorn came when he led Francis into admitting that he did not remember the chemical differences among the four bases. . . . Francis's subsequent retort that he could always look them up got nowhere in persuading Chargaff that we knew where we going or how to get there" (Watson, *The Double Helix,* 78).
33. Chargaff, *Heraclitean Fire,* 101.
34. Judson, *The Eighth Day of Creation,* 120.
35. Judson, *The Eighth Day of Creation,* 135–38.
36. Judson, *The Eighth Day of Creation,* 141.
37. Olby, *The Path to the Double Helix,* 398.
38. Crick, quoted in Olby, *The Path to the Double Helix,* 400.
39. Olby, *The Path to the Double Helix,* 400.
40. "It must have been during the second week of February [1953] that Perutz passed on to Crick his copy of the report which Sir John Randall had circulated to all the members of the Biophysics Research Committee [of the Medical Research Council] who came to see round his Unit on December 15, 1952" (Olby, *The Path*

to the Double Helix, 402). This report contained a summary of Rosalind Franklin's crystallographic findings about the structure of DNA.

41. Watson, *The Double Helix*, 114.
42. Watson, *The Double Helix*, 110.
43. "I went back to my desk hoping that some gimmick might emerge to salvage the like-with-like idea. But it was obvious that the new assignments were its death blow. Shifting the hydrogen atoms to their keto locations made the size differences between the purines and pyrimidines even more important than would be the case if the enol forms existed. Only by the most special pleading could I imagine the polynucleotide backbone bending enough to accommodate irregular base sequences" (Watson, *The Double Helix*, 112).
44. Watson, *The Double Helix*, 114.
45. Watson, *The Double Helix*, 114.
46. "A few minutes later he [Crick] spotted the fact that two glycosidic bonds (joining base and sugar) of each base pair were systematically related by a diad axis perpendicular to the helical axis. Thus, both pairs could be flip-flopped over and still have their glycosidic bonds facing in the same direction. This had the important consequence that a given chain could contain both purines and pyrimidines. At the same time, it strongly suggested that the backbones of the two chains must run in opposite directions" (Watson, *The Double Helix*, 115).
47. Crick, *What Mad Pursuit*, 74. This aesthetic criterion—the beauty and, hence, rightness of the double helix—rings through many accounts of the encounters with the molecule. Watson himself, at the end of an early 1950s inebriated after-dinner lecture about DNA, was (in Crick's judgment) "at a loss for words. He gazed at the model, slightly bleary-eyed. All he could manage to say was 'It's so beautiful, you see, so beautiful!' But then, of course, it was" (Crick, *What Mad Pursuit*, 79). François Jacob, on first hearing from Watson about the structure of DNA, reflected that "this structure was of such simplicity, such perfection, such harmony, such beauty even, and biological advantages flowed from it with such rigor and clarity, that one could not believe it to be untrue" (*The Statue Within*, 270–71).
48. Watson, *The Double Helix*, 115.
49. Watson and Crick, "Genetical Implications," 965.

Chapter 4: Signature in the Cell

1. Watson and Crick, "A Structure for Deoxyribose Nucleic Acid"; "Genetical Implications," esp. 964; Schneider, "Information Content of Individual Genetic Sequences"; Loewenstein, *The Touchstone of Life*. Hood and Galas, "The Digital Code of DNA."
2. Lewbel, "A Personal Tribute to Claude Shannon."
3. Gardner, *The Mind's New Science*, 11.
4. Horgan, "Unicyclist, Juggler and Father of Information Theory."
5. Shannon, "A Mathematical Theory of Communication." Information theorists found it convenient to measure information additively rather than multiplicatively. Thus, the common mathematical expression ($I = -\log_2 p$) for calculating information converts probability values into informational measures through a negative logarithmic function, where the negative sign expresses an inverse relationship between information and probability.
6. Shannon and Weaver, *The Mathematical Theory of Communication*, 8.
7. Astbury and Street, "X-Ray Studies of the Structure of Hair, Wool and Related Fibres"; Judson, *The Eighth Day of Creation*, 61–62; Olby, *The Path to the Double Helix*, 63.
8. Bergmann and Niemann, "Newer Aspects of Protein Chemistry."

9. According to Judson: "The man who released the present-day understanding of molecular specificity in living processes was Frederick Sanger. His determination, beginning in the mid-forties, of the amino-acid sequences of bovine insulin proved that they have no general periodicities. His methods and this surprising result had many consequences, of course: the most general and profound was that proteins are entirely and uniquely specified" (*The Eighth Day of Creation*, 88–89, 585). See Sanger and Thompson, "The Amino Acid Sequence in the Glycyl Chain of Insulin."

10. Judson, *The Eighth Day of Creation*, 581–85.

11. Olby, *The Path to the Double Helix*, 265.

12. Rose, "No Assembly Required," 36, 28.

13. Judson, *The Eighth Day of Creation*, 538.

14. Judson, *The Eighth Day of Creation*, 539.

15. Kendrew, et al., "A Three-Dimensional Model of the Myoglobin Molecule," 662–66.

16. Lodish, et al., *Molecular Cell Biology*, 321–23.

17. Gitt, *In the Beginning Was Information*, 192–93.

18. Lodish, et al., *Molecular Cell Biology*, 322.

19. Alberts, et al., *Molecular Biology of the Cell*, 111–12, 127–31.

20. As Monod expressed the point in an interview with Judson: "The first determination of the exact amino-acid sequence of a protein by Sanger"—the sequencing of bovine insulin between 1949 and 1955—"was absolutely essential. One could not even have begun to think seriously about the genetic code until it had been revealed, to begin with, that a protein is beyond the shadow of doubt a polypeptide in which the amino acid residues really are arranged in a definite, constant, genetically determined sequence—and yet a sequence with no rule by which it determined itself. Therefore it had to have a code—that is, complete instructions expressed in some manner—to tell it how to exist, you see" (Judson, *The Eighth Day of Creation*, 188).

21. Watson and Crick, "Genetical Implications."

22. Judson, *The Eighth Day of Creation*, 332–33.

23. Judson, *The Eighth Day of Creation*, 332–35.

24. Beadle and Tatum, "Genetic Control of Biochemical Reactions in *Neurospora*." Though Beadle and Tatum's results helped to establish that genes encode proteins, we now know that in most cases one gene encodes multiple RNA and protein products. See Gerstein, et al. "What Is a Gene, Post-ENCODE?"; Carninci, Yasuda, and Hayashizaki, "Multifaceted Mammalian Transcriptome"; Kapranov, Willingham, and Gingeras, "Genome-wide Transcription"; Pesole, "What Is a Gene?"

25. De Chadarevian, "Protein Sequencing and the Making of Molecular Genetics."

26. Judson, *The Eighth Day of Creation*, 453–70; Matthei and Nirenberg, "Characteristics and Stabilization of DNAase-Sensitive Protein Synthesis in *E. coli* Extracts"; "The Dependence of Cell-Free Protein Synthesis"; Zamecnik, "From Protein Synthesis to Genetic Insertion"; Brenner, Jacob, and Meselson, "An Unstable Intermediate"; Portugal and Cohen, *A Century of DNA*, 272, 298–302; Hoagland, et al., "A Soluble Ribonucleic Acid Intermediate in Protein Synthesis."

27. Alberts, et al., *Molecular Biology of the Cell*, 106–8; Wolfe, *Molecular and Cellular Biology*, 48.

28. Alberts, et al., *Molecular Biology of the Cell*, 108.

29. Yockey, *Information Theory and Molecular Biology*, 110.

30. Küppers, "On the Prior Probability of the Existence of Life," 355–69.

31. See Schneider, "Information Content of Individual Genetic Sequences." See also Yockey, *Information Theory and Molecular Biology*, 246–58, for important refinements in the method of calculating the information-carrying capacity of proteins and DNA.

32. Crick, "On Protein Synthesis"; Sarkar, "Decoding 'Coding,'" esp. 857; "Biological Information."

33. Orgel, *The Origins of Life,* 189.
34. Dawkins, *River Out of Eden,* 17.
35. Gates, *The Road Ahead,* 188.
36. Yockey, "Origin of Life on Earth," esp. 105.

Chapter 5: The Molecular Labyrinth

1. Judson, *The Eighth Day of Creation,* 264.
2. Gamow, "Possible Relation Between Deoxyribonucleic Acid and Protein Structures"; "Possible Mathematical Relation Between Deoxyribonucleic Acid and Proteins."
3. Gamow, "Possible Relation Between Deoxyribonucleic Acid and Protein Structures." As Judson explains, Gamow thought that the "permutations of the bases formed holes of different shapes . . . into which various amino acids fit as specifically as keys into locks" (*The Eighth Day of Creation,* 256–57).
4. Crick, "On Degenerate Templates."
5. Judson, *The Eighth Day of Creation,* 291.
6. Judson, *The Eighth Day of Creation,* 258.
7. Judson, *The Eighth Day of Creation,* 291.
8. De Chadarevian, "Protein Sequencing and the Making of Molecular Genetics."
9. Judson, *The Eighth Day of Creation,* 585.
10. http://ascii-table.com/ (last accessed July 5, 2008).
11. Crick, "On Degenerate Templates," 8–9.
12. Crick, "On Degenerate Templates," 8–9.
13. Judson, *The Eighth Day of Creation,* 293.
14. Judson, *The Eighth Day of Creation,* 292.
15. Zamecnik, "From Protein Synthesis to Genetic Insertion."
16. Brenner, Jacob, and Meselson, "An Unstable Intermediate."
17. Lengyel, Speyer, and Ochoa, "Synthetic Polynucleotides and the Amino Acid Code"; Nirenberg and Matthaei, "The Dependence of Cell-Free Protein Synthesis."
18. Portugal and Cohen, *A Century of DNA,* 272, 298–302.
19. Hoagland, et al., "A Soluble Ribonucleic Acid Intermediate in Protein Synthesis."
20. Jones, *CAD/CAM.*
21. Beeby, "The Future of Integrated CAD/CAM Systems."
22. Watson, et al., *Molecular Biology of the Gene,* vol. 1:360–81.
23. Fraser, et al., "The Minimal Gene Complement."
24. Watson, et al., *Molecular Biology of the Gene,* vol. 1:368–80.
25. Wolfe, *Molecular and Cellular Biology,* 580–81, 639–48.
26. Wolfe, *Molecular and Cellular Biology,* 639, 731.
27. Lodish, et al., *Molecular Cell Biology,* 342–45; Wolfe, *Molecular and Cellular Biology,* 727–30.
28. Cramer, et al., "Structure of Eukaryotic RNA Polymerases."
29. Wolfe, *Molecular and Cellular Biology,* 580–81.
30. Introns are now known to perform many functions in the cell, including (a) encoding transcriptional (and mutational) control elements, (b) encoding nontranslated RNAs and their transcriptional regulatory sequences, and (c) encoding RNA splicing control elements. See Sowpati, et al., "An Intronic DNA Sequence within the Mouse Neuronatin Gene Exhibits Biochemical Characteristics of an ICR and Acts as a Transcriptional Activator Gene in Drosophila," 963–73; Ozsolak, et al. "Chromatin Structure Analyses Identify miRNA Promoters," 3172–83. Wang, et al. "Splicing Regulation: From a Parts List of Regulatory Elements to an Integrated Splicing Code," 802–13.
31. Lodish, et al., *Molecular Cell Biology,* 437–38.
32. Fraser, et al., "Minimal Gene Complement," 399.

33. Watson, et al., *Molecular Biology of the Gene,* vol. 1:393–430.
34. Watson, et al., *Molecular Biology of the Gene,* vol. 1:443.
35. Lodish, et al., *Molecular Cell Biology,* 116–34.
36. Lodish, et al., *Molecular Cell Biology,* 120–25.
37. Lodish, et al., *Molecular Cell Biology,* 120–25.
38. Fraser, et al., "Minimal Gene Complement," 399.
39. Fraser, et al., "Minimal Gene Complement," 618–22.
40. Lewontin, "The Dream of the Human Genome."
41. Goodsell, *The Machinery of Life,* 45.
42. Monod, *Chance and Necessity,* 143.
43. Bult, et al., "Complete Genome Sequence"; Glass, et al., "Essential Genes of a Minimal Bacterium."
44. Lewontin, "The Dream of the Human Genome," esp. 33.
45. Goodsell, *The Machinery of Life,* 45.
46. Popper, "Scientific Reduction."
47. Monod, *Chance and Necessity,* 143.

Chapter 6: The Origin of Science and the Possibility of Design

1. Dyson, *Origins of Life,* 18.
2. Biever, "The God Lab."
3. I'm not, of course, invoking the example of such genius to claim a similar stature, but rather to highlight the legitimacy of a particular form of scientific inquiry—one that I was inspired to pursue by my admiration for these examples of boldness and willingness to think "outside the box" and perceive possibilities that had not been considered before, even by others who had access to all the same facts.
4. Whitehead, *Science and the Modern World,* 2–4, 13.
5. Fuller, *Science vs. Religion?* 15. See also: Crombie, *Augustine to Galileo*; Jaki, "Science: Western or What?"; Butterfield, *The Origins of Modern Science*; Hooykaas, *Religion and the Rise of Modern Science.*
6. As cited in Holton, *Thematic Origins of Scientific Thought,* 86.
7. "God is a most free agent, and created the world not out of necessity but voluntarily, having framed it as he pleased and thought fit at the beginning of things, when there was no substance but himself and consequently no creature to which he could be obliged, or by which he could be limited" (Boyle, *A Free Enquiry,* 160).
8. Hodgson, "The Christian Origin of Science," 145.
9. Hodgson, "The Christian Origin of Science," 142.
10. Watson, *The Double Helix,* 133–34.
11. Agassiz, *Essay on Classification*; see also Rieppel, "Louis Agassiz (1807–1873) and the Reality of Natural Groups," 34.
12. "Rational inquiry must inevitably, in Linné's [Linnaeus] opinion, lead, not to skepticism or disbelief, but to the acknowledgment of and respect for an omniscient and omnipotent Creator" (Larson, *Reason and Experience,* 151).
13. The full passage reads: "For the characters and impressions of wisdom that are conspicuous in the curious fabric and orderly train of things can with no probability be referred to blind chance, but must be ascribed to a most intelligent and designing agent" (Boyle, *A Free Enquiry,* 101). See also Lennox, "Robert Boyle's Defense of Teleological Inference."
14. Newton, General Scholium, *Mathematical Principles of Natural Philosophy* (1969).
15. Newton, General Scholium, *Mathematical Principles of Natural Philosophy* (1969).
16. Scott, Testimony Before the U.S. Commission on Civil Rights.
17. Meyer, Testimony Before the U.S. Commission on Civil Rights.

18. I also remembered in my reading of the great chemist Robert Boyle that he thought that scientists and philosophers ask different kinds of questions about natural objects. Boyle noted that scientists usually ask what something is made of and how it works or operates, whereas philosophers or theologians ask what the purpose of something is. But Boyle noted that there was a third kind of question that scientists ask that stands at the boundary between science and philosophy. That question is, "How did this originate?" or "Where did this come from?" So I thought maybe there is a special class of scientific questions that required a special method of investigation.

Chapter 7: Of Clues to Causes

1. Gould, "Evolution and the Triumph of Homology." Following the nineteenth-century philosopher of science William Whewell, Gould describes the process of testing in the historical sciences as seeking "consilience." "Consilience" occurs when many facts can be explained well by a single proposition or theory. Gould also notes that despite the differences between historical and experimental sciences, both share "nomothetic undertones," meaning, in the case of the historical sciences, that they depend upon knowledge of the laws (Greek *nomos*) of nature to make inferences about the past (see esp. 64–65). Cleland, "Methodological and Epistemic Differences"; "Historical Science, Experimental Science, and the Scientific Method."
2. Meyer, "Of Clues and Causes."
3. Morris, *The Crucible of Creation*, 63–115.
4. Whewell, *The Philosophy of the Inductive Sciences; History of the Inductive Sciences*.
5. Whewell, *The Philosophy of the Inductive Sciences*, 1st ed., 95–123; 2nd ed., 637–65; Meyer, "Of Clues and Causes," 12–23.
6. Whewell, *The Philosophy of the Inductive Sciences*, 1st ed., 95–96; 2nd ed., 637–38; *History of the Inductive Sciences*, 3:397; Meyer, "Of Clues and Causes," 12–13.
7. Although most inductive sciences sought to *establish* general laws via induction, palaetiology sought to *use* knowledge of such generalizations *in order to* establish past causal conditions or events that could, in turn, be used to explain present events or "manifest effects."
8. Whewell, *The Philosophy of the Inductive Sciences*, 1st ed., 112–13, 121–22; 2nd ed., 654–55, 663–64; Meyer, "Of Clues and Causes," 18–22. For a contemporary study of how singular or past events can explain other events, see Alston, "The Place of the Explanation of Particular Facts in Science"; Martin, "Singular Causal Explanation"; Gallie, "Explanations in History and the Genetic Sciences."
9. Hodge, "The History of the Earth, Life, and Man," esp. 262.
10. Whewell, *The Philosophy of the Inductive Sciences*, 1st ed., 95, 101–4, 121; 2nd ed., 637, 643–46, 663. See also Meyer, "Of Clues and Causes," 14–18.
11. Whewell, *The Philosophy of the Inductive Sciences*, 1st ed., 113, 121; 2nd ed., 665, 663.
12. Gould, "Evolution and the Triumph of Homology," 62.
13. Morris, *The Crucible of Creation*, 63–115; Gould, *Wonderful Life*.
14. Meyer, "Of Clues and Causes," 36–76; Cleland, "Historical Science, Experimental Science, and the Scientific Method"; Gould, "Evolution and the Triumph of Homology."
15. Peirce, *Collected Papers*, 2: 372–88. Later Peirce adapted his mode of classifying forms of reasoning so it would not contradict his earlier classification scheme, but in so doing he made its implications for the methodology of historical science less apparent. Only his early classification scheme is appropriated in this discussion. See also Fann, *Peirce's Theory of Abduction*, 28–34.
16. Peirce, *Collected Papers*, 2:375; "Abduction and Induction."
17. Peirce, *Collected Papers*, 2:375, emphasis added.

18. Chamberlain, "The Method of Multiple Working Hypotheses." Chamberlain, though a contemporary of Peirce, does not address the limits of abductive reasoning by name. Nevertheless, he advances the method of multiple competing hypotheses to address the problem just the same. In his seminal article, he recommends that geologists (and presumably other historical scientists) use the method to protect against prematurely establishing a merely possible explanation as a single "ruling theory." Moreover, his concern arises specifically in the context of geological and historical sciences in which many possible causes may conceivably explain the origin of present features of the earth. Chamberlain uses as an example a geologist who faces this problem as he tries to explain the origin of the Great Lakes basin.

19. Lipton, *Inference to the Best Explanation*, 1.

20. Lipton, *Inference to the Best Explanation*, 1.

21. Lipton, *Inference to the Best Explanation*, 56–74.

22. Lipton, *Inference to the Best Explanation*, 133–57. A concrete example supportive of Lipton's claims can be found in a fascinating case study by University of Maryland historian of science Stephen Brush, who showed that, in the case of Einstein's theory of general relativity, the ability to explain many already known facts played a far greater role in general relativity's acceptance than did the few successful predictions it made. See Brush, "Prediction and Theory Evaluation."

23. Prior to Lipton's work, the most common conception of scientific reasoning had been advanced by Sir Karl Popper, a British-Hungarian philosopher of science. Popper taught that for a theory to be scientific it must be testable against the evidence. Furthermore, he insisted that to test a scientific theory, scientists must make predictions based upon the theory and then observe whether the predictions came true. Typically, scientists do this by establishing a set of highly controlled conditions in the laboratory and then conducting an experiment to see if the outcome conforms to their expectations based on the theory in question. If the predictions come true, the theory passes the experimental test and can be regarded as "confirmed," but not as "verified." If the prediction fails to materialize, however, the experiment is said to have falsified the theory—unless, of course, it can be determined that some other factor, such as faulty experimental apparatus, is responsible for the failed test. Another philosopher, Carl Hempel, had a similar idea. He argued that to explain an event scientifically, scientists need to be able to deduce an event from a set of initial conditions and scientific laws. Lipton's work showed that Popper's and Hempel's conceptions of scientific testing and explanation were far too limited (*Inference to the Best Explanation*, 29–31, 75–98). He showed that explanatory power played a key role in testing in many branches of science. I realized that Lipton's work provided an excellent description of how historical scientists formulate and test their theories—something that Popper's account of scientific testing could not do. Historical scientists usually do not make predictions or deduce outcomes in advance. Instead, historical scientific theories typically explain events after the fact and can be tested by comparing their explanatory power against that of their competitors.

24. Lipton summarized the problem this way: "One reason 'Inference to the Best Explanation' has been so little developed, in spite of its popularity, is clear. The model is an attempt to account for inference in terms of explanation, but our understanding of explanation is so patchy that the model seems to account for the obscure in terms of the equally obscure" (*Inference to the Best Explanation*, 2).

25. Lipton, *Inference to the Best Explanation*, 32–55, 75–98.

26. Lipton, *Inference to the Best Explanation*, 32–98. A particularly helpful example can be found in Lipton's extended discussion of Ignaz Semmelweis's well-known research on childbed fever, 75–98. See also Scriven, "Causes, Connections and Conditions in History," esp. 250.

27. For example, see Lyell's discussions in *Principles of Geology*, 1:75–91; 3:1–7.

28. William Whewell, *The Philosophy of the Inductive Sciences*, 2: 101.

29. For a discussion of the similarity between Lyell's principle of uniformitarianism and the *vera causa* principle adopted by Darwin, see Kavalovski, "The *Vera Causa* Principle," 78–103.

30. Darwin, *On the Origin of Species*, 80–130.

31. Darwin also argued that the process of descent with modification was a *vera causa* of certain kinds of patterns found among living organisms. He noted that diverse organisms share many common features. He called these "homologous" and noted that we know from experience that descendants, although they differ from their ancestors, also resemble them in many ways, usually more closely than others who are more distantly related. So he proposed descent with modification as a *vera causa* (a known cause) for homologous structures. That is, he argued that our uniform experience shows that the process of descent with modification from a common ancestor is "causally adequate" or capable of producing homologous features (*On the Origin of Species*, 131–70; esp. 159, where Darwin refers to the "community of descent" as a "*vera causa*" of homologies among plant species). See also Meyer, "Of Clues and Causes," 112–36; Kavalovski, "The *Vera Causa* Principle,"104–29; Hodge, "Darwin's Theory and Darwin's Argument"; Recker, "Causal Efficacy"; Kitcher, "Darwin's Achievement"; Hull, *The Philosophy of Biological Science*, 50–51.

32. Scriven, "Explanation and Prediction in Evolutionary Theory," 480–81; "Causes, Connections and Conditions in History," esp. 250.

33. Scriven, "Causes, Connections and Conditions in History," 250; "Explanation and Prediction in Evolutionary Theory," 481. As Hodge explains: "One should invoke only causes whose . . . competence to produce such an effect can be known independently of their putative responsibility for that phenomenon" ("The Structure and Strategy of Darwin's 'Long Argument,'" 239).

34. Scriven, "Explanation and Prediction in Evolutionary Theory," 481. See also Hodge, "The Structure and Strategy of Darwin's 'Long Argument,'" 239; Meyer, "Of Clues and Causes," 79–111; see esp. 108–10 for a discussion comparing Scriven's "retrospective causal analysis" and the *vera causa* principle.

35. See the discussion of Sober below. See also Scriven, "Explanation and Prediction in Evolutionary Theory," 480; "The Temporal Asymmetry of Explanations and Predictions."

36. Meyer, "Of Clues and Causes," 96–108.

37. Sober, *Reconstructing the Past*, 1–5.

38. Sober, *Reconstructing the Past*, 3–4. I have modified Sober's example in order to make its point clearer.

39. See also: Michael Scriven, "Causes, Connections and Conditions in History," *Philosophical Analysis and History*, 250.

40. Scriven, "Explanation and Prediction in Evolutionary Theory," 480; Sober, *Reconstructing the Past*, 1–5; Alston, "The Place of the Explanation of Particular Facts in Science," 23; Gallie, "Explanations in History and the Genetic Sciences," 392.

41. Gallie, "Explanations in History and the Genetic Sciences," 392.

42. Meyer, "Of Clues and Causes," 96–108. See also Whewell, *The Philosophy of the Inductive Sciences*, 2:240; Thagard, "The Best Explanation," esp. 85. For background on Whewell's ideas about consilience, see Butts, *William Whewell's Theory of Scientific Method;* Laudan, "William Whewell on the Consilience of Inductions."

43. Meyer, et al., *Explore Evolution*, 4–5.

44. Scriven, "Causes, Connections and Conditions in History," 250.

45. Scriven, "Explanation and Prediction in Evolutionary Theory," 481.

46. Scriven, "Causes, Connections and Conditions in History," 251.
47. Scriven, "Explanation and Prediction in Evolutionary Theory," 481.
48. Meyer, "Of Clues and Causes," 77–111.
49. Scriven, "Causes, Connections and Conditions in History," 249–50.
50. Scriven, "Explanation and Prediction in Evolutionary Theory," 481.
51. See the extended discussion of this claim in Meyer, "Of Clues and Causes," 112–36. For other relevant studies, see Kavalovski, "The *Vera Causa* Principle," 104–29; Hodge, "Darwin's Theory and Darwin's Argument," 167–74; Recker, "Causal Efficacy," 147–75; Kitcher, "Darwin's Achievement," 127–89; Ruse, *The Darwinian Revolution*, 160–201; Hull, *The Philosophy of Biological Science*, 50–51.
52. Darwin, *Life and Letters,* 1:437, emphasis added.
53. Quastler, *The Emergence of Biological Organization*, 16.
54. Küppers, *Information and the Origin of Life*, 170–72.

Chapter 8: Chance Elimination and Pattern Recognition

1. Monod, *Chance and Necessity*, 97–98.
2. See Hogg, Craig, and McKean, *Introduction to Mathematical Statistics,* esp. the sections on statistical testing.
3. For a quick synopsis of Fisher's views on hypothesis testing, see Fisher, *The Design of Experiments*, 13–17. For a fuller treatment, see Fisher, *Statistical Methods and Statistical Inference.* For Neyman and Pearson's extension of Fisher's ideas, which sets an explicit alternate hypothesis over and against a null hypothesis, see Ian Hacking, *The Logic of Statistical Inference*, 92–93. Hacking's text also has an excellent historical discussion of rejection regions ("rejection classes") going back to the eighteenth-century work of John Arbuthnot and Nicholas Bernoulli and relating it to the work of Fisher (78–83). For a Bayesian critique of Fisher's ideas about statistical significance and hypothesis testing, see Howson and Urbach, *Scientific Reasoning,* chap. 8. For Dembski's response to Bayesian criticisms of Fisher's theory and Dembski's approach more generally, see Dembski, *The Design Revolution*, chap. 33, "Design by Elimination Versus Design by Comparison."
4. The waiting time is computed in terms of a Poisson distribution. See Grimmett and Stirzaker, *Probability and Random Processes,* chap. 3.
5. Fisher, *The Design of Experiments*, 13–17.
6. Often scientists and statisticians have a wealth of data from their disciplines about how frequently certain events occur, data that helps them define these rejection regions. Other times they may define them based on their analysis of the relevant outcome-producing process, such as a roulette wheel, radioactive decay, or genetic recombination or mutation. Fisher argued that when events occur that fall within well-defined and prespecified rejection regions, the chance hypothesis could safely be eliminated from consideration. Given the chance hypothesis, it would be *more likely than not* that events would fall outside such a rejection region. In such cases, Fisher therefore concluded that something other than chance was operating—such as, in our previous illustration, Slick cheating at roulette or some physical defect in the roulette wheel (along with an apparently dim-witted croupier).
7. In this case, a gaming commissioner will be on the lookout for an improbable ratio of victories for the casino, such as a roulette wheel that typically has bets on about half its numbers, but consistently turns up winners only one-fourth of the time.
8. William A. Dembski, "The Logical Underpinnings of Intelligent Design," in William A. Dembski and Michael Ruse, eds., *Debating Design: From Darwin to DNA*, 314.
9. De Duve, *Singularities*, 233.

10. For the underlying mathematics, see Dembski, *The Design Inference,* 175–77. Dembski recycles this example for a more popular audience in his book *No Free Lunch,* 19.

11. Dembski puts the criminal's situation in perspective: "The probability of getting 100 heads in a row on a given trial is so small that the prisoner has no practical hope of getting out of prison, even if his life expectancy and coin-tossing ability were dramatically increased. If he could, for instance, make 10 billion attempts each year to obtain 100 heads in a row (this is coin-flipping at a rate of over 500 coin-flips per second every hour of every day for a full year), then he stands an even chance of getting out of prison in 10^{20} years (i.e., a hundred billion billion years). His probabilistic resources are so inadequate for obtaining the desired 100 heads that it is pointless for him to entertain hopes of freedom" (*No Free Lunch,* 19).

Chapter 9: Ends and Odds

1. Lehninger, *Biochemistry,* 782.
2. Wald, "The Origin of Life," 48; Shapiro, *Origins,* 121.
3. Crick, "The Origin of the Genetic Code," 369–70; Kamminga, "Studies in the History of Ideas," 303–4.
4. Cairns-Smith, *The Life Puzzle,* 95.
5. De Duve, "The Beginnings of Life on Earth," 437.
6. According to the advocates of chance, we are now observing in the DNA molecule or in proteins the outcome of a fortuitous sequence of random events. So, according to this hypothesis, if an observer had been on the early earth to watch this process, he or she would have witnessed an event very similar to a gambler turning up just the right card time after time after time. An astute observer would have witnessed a pattern, a pattern of events that produced a functionally significant outcome. De Duve's point is that just as we should be suspicious of chance in the case of the uncannily successful gambler, so should we be suspicious of chance in the game of life—and so should we be suspicious of chance as the explanation for the origin of life and the information it requires.
7. De Duve, "The Beginnings of Life on Earth," 437; Quastler, *The Emergence of Biological Organization,* 7; Morowitz, *Energy Flow in Biology,* 5–12; Hoyle and Wickramasinghe, *Evolution from Space,* 24–27; G. Cairns-Smith, *The Life Puzzle,* 91–96; Prigogine, Nicolis, and Babloyantz, "Thermodynamics of Evolution," 23–31; Yockey, *Information Theory and Molecular Biology,* 246–58; "Self-Organization Origin of Life Scenarios and Information Theory"; Shapiro, *Origins,* 117–31; Behe, "Experimental Support for Regarding Functional Classes of Proteins"; Crick, *Life Itself,* 88.
8. Bult, et al., "Complete Genome Sequence."
9. In prokaryotic cells, the information content of the genes and their protein products are nearly equivalent. In eukaryotic cells, posttranscriptional processing of the amino-acid chain requires information beyond the information that encodes amino-acid sequences. Nevertheless, since prokaryotes arose on the earth first and have a simpler structure, they are thought to more closely resemble the first living cells. Thus, those calculating the probability of the first minimally complex cell arising by chance correctly assume that the genes and the proteins they encode had closely comparable information content—just as they do in modern prokaryotes.
10. The redundancy in the genetic code makes it easier as a practical matter to calculate these odds using proteins.
11. Moorhead and Kaplan, *Mathematical Challenges to the Neo-Darwinian Interpretation of Evolution.*

12. Reidhaar-Olson and Sauer, "Functionally Acceptable Substitutions"; Bowie and Sauer, "Identifying the Determinants of Folding and Activity."

13. Oddly, Sauer's descriptions of his team's results often downplay the rarity of functional sequences within sequence space. Instead, he often emphasizes the tolerance for different amino acids that is allowable at each site. For example, the abstract of the paper reporting the figure of 1 in 10^{63} makes no mention of that figure or its potential significance, stating instead that their results "reveal the high level of degeneracy in the information that specifies a particular protein fold" (Reidhaar-Olson and Sauer, "Functionally Acceptable Substitutions").

14. Bowie and Sauer, "Identifying the Determinants of Folding and Activity"; Reidhaar-Olson and Sauer, "Functionally Acceptable Substitutions"; Chothia, Gelfand, and Kister, "Structural Determinants in the Sequences of Immunoglobulin Variable Domain"; Axe, "Extreme Functional Sensitivity"; Taylor, et al., " Searching Sequence Space for Protein Catalysts."

15. See, e.g., Perutz and Lehmann, "Molecular Pathology of Human Hemoglobin."

16. Axe, "Estimating the Prevalence of Protein Sequences."

17. Axe, "Estimating the Prevalence of Protein Sequences."

18. Cramer, Bushnell, and Kornberg, "Structural Basis of Transcription."

19. Hoyle and Wickramasinghe, *Evolution from Space,* 24–27.

20. Morowitz, *Energy Flow in Biology,* 5–12; Hoyle and Wickramasinghe, *Evolution from Space,* 24–27; Cairns-Smith, *The Life Puzzle,* 91–96; Prigogine, Nicolis, and Babloyantz, "Thermodynamics of Evolution," 23–31; Yockey, *Information Theory and Molecular Biology,* 246–58; "Self-Organization Origin of Life Scenarios and Information Theory"; De Duve, "The Beginnings of Life on Earth," 437; Shapiro, *Origins,* 117–31; Crick, *Life Itself,* 88.

Chapter 10: Beyond the Reach of Chance

1. The number of possible ways to combine elementary particles (and thus the number of combinatorial possible events) is actually much greater than the number of different events that could have taken place in the history of the universe. Why? Because the occurrence of each individual event precludes the occurrence of many other possible events within the larger combinatorial space. The number of combinatorial possible events represents the number of different events that might have occurred before the universe actually unfolded in the way that it did. Dembski correctly identifies the maximum number of events that could *actually* occur in any given history of the universe as the number that determines the probabilistic resources of the universe. This smaller number determines how many opportunities the universe has to produce a particular outcome by chance. As Dembski explains, it is not the total number of combinatorial possible events (or elementary particles) in the universe that determines the available probabilistic resources, but how many opportunities there are to "individuate" actual events. See Dembski, *The Design Inference,* chap. 6; see also 209, n. 15.

2. For Dembski's treatment of probabilistic resources at the scale of the known universe, see Dembski, *The Design Inference,* chap. 6.

3. The elementary particles enumerated in this calculation include only protons, neutrons, and electrons (fermions), because only these particles have what physicists call "half-integral spin," which allows them to form material structures. This calculation does not count bosons, which cannot form material structures, but instead only transmit energy. Nor does this calculation count the quarks out of which protons and neutrons are made, because quarks are necessarily bound together within these particles. Even if quarks were counted, however, the total number of elementary particles would change by less than one order of magnitude because there are only three quarks per proton or neutron.

4. To be safe, Dembski rounded the number he had calculated up a few orders of magnitude to 10^{150}, though without any physical or mathematical justification. Since he didn't need to do this, I decided to use his more accurate, if less round, number as the actual measure of the probabilistic resources of the universe in my evaluations of the chance hypothesis.

5. Cryptographers, for instance, have established 1 chance in 10^{94} as a universal limit. They interpret that improbability to mean that if it requires more than 10^{94} computational steps to decrypt a cryptosystem, then it is more likely than not that the system won't be cracked because of the limited probabilistic resources of the universe itself.

6. Borel, *Probabilities and Life*, 28–30.

7. Van de Sande, "Measuring Complexity in Dynamical Systems."

8. Dam and Lin, eds., *Cryptography's Role in Securing the Information Society*, 380, n. 17; Lloyd, "Computational Capacity of the Universe"; see also Kauffman, *Investigations*, 144.

9. Sometimes after I explain the odds against producing a protein of even modest length by chance, someone will ask why a shorter functional protein couldn't have arisen by chance and then gradually evolved into a larger one. Sometimes critics point out that some functional amino acid chains, such as peptide hormones, have fewer than 150 amino acids. Some peptide hormones, for example, are just a few tens of amino acids long, though some are much longer. And there are also some proteins that are shorter than 150 amino acids. Critics ask me, "Couldn't such molecules have arisen by chance and then evolved into longer functional molecules?"

   There are a number of problems with this scenario.

   First, functional proteins (including all enzymes) depend upon complex folds, or "tertiary structures." Attaining tertiary structure in proteins requires about 50 properly sequenced amino acids for the simplest structures and many more (typically hundreds) for more typical structures. Moreover, these thresholds of minimal function vary from protein to protein. Just because one protein fold or tertiary structure may need "only" 50 specifically sequenced amino acids does not mean that another can form with that few. Most can't. The protein equivalent of a ruler may form with only 50 amino acids, but the hammer and saw may need 150, the wrench 200, and the drill 300. Many of the functions that a minimally complex cell requires depend upon these longer proteins. Thus, the presence of some shorter proteins or peptide hormones in living systems does nothing to obviate the need for many larger proteins in the origin of life.

   Moreover, as protein chemist Doug Axe explains in more detail in a forthcoming article ("The Nature of Protein Folds: Quantifying the Difficulty of an Unguided Search Through Protein Sequence Space") there are physical reasons that short proteins with small tertiary structures can't be gradually transformed into larger tertiary structures. Short proteins typically exhibit a hydrophilic exterior. To build a larger structure around them, at least some of this hydrophilic exterior would have to become interior to the larger structure. But this requires, among other things, that a region of hydrophilic surface become hydrophobic, which in turn requires many simultaneous amino acid changes. Having a short protein to start with contributes little or nothing toward building a larger one. The same probabilistic hurdles have to be overcome in sequencing.

   In any case, it is important to distinguish between peptides that function without a folded structure at all and proteins that function only with a folded structure. The former (which includes the shorter peptide hormones) are functional only by virtue of binding to larger folded protein structures. But this implies that these shorter molecules have no function—and no selective advantage—apart from the prior existence of much larger protein molecules. Thus, citing functional peptide hormones as a starting point in evolution begs the question as to the origin of the larger protein molecules that give them functional significance. It only pushes the problem back to where it started—to the problem of explaining the origin of large functionally specified proteins by chance. Indeed, absent long functional protein molecules—and, realistically, a minimally complex self-reproducing cell—there would be no context to confer functional significance or advantage on either unfolded peptide hormones or shorter proteins (for that matter).

10. If Slick had replaced each marble before he searched again, the number of marbles he would have needed to sample to have a 50 percent chance of finding the red one would have been significantly more than 5,000. The probability of selecting a red marble in n random draws with replacement from the sack is given by the equation $1 - q^n$, where q equals the proportion of blue marbles. The variable q is given by the equation $q = 1 - p$ where p is the proportion of red marbles in the gunnysack. If $p = .0001$, then $q = 1 - .0001$, or .9999. It follows that Slick would have a 50 percent chance of finding a red marble only after he had sampled 6,931 marbles. Indeed, the probability of finding the red marble reaches $1/2$ only if $n = \log(.5) / \log(.9999) = 6{,}931$. For the relevant math see Feller, *An Introduction to Probability Theory*, 49.

11. For the record, most of the skeptics about the chance hypothesis to whom I refer here (Christian de Duve, Alexander G. Cairns-Smith, P .T. Mora, Hubert Yockey, Ilya Prigogine, Robert Shapiro, the later Francis Crick ca. 1980) were not advocates of intelligent design.

12. Mora, "Urge and Molecular Biology," 215.

13. Miller, "A Production of Amino Acids."

14. Abelson, "Chemical Events on the Primitive Earth"; Florkin, "Ideas and Experiments in the Field of Prebiological Chemical Evolution"; Fox and Dose, *Molecular Evolution and the Origin of Life*, 43, 74–76; Cohen, "Novel Center Seeks to Add Spark to Origins of Life"; Walker, *Evolution of the Atmosphere*, 210, 246; Kerr, "Origin of Life"; Thaxton, Bradley, and Olsen, *The Mystery of Life's Origin*, 73–94; Horgan, "In the Beginning"; Kasting, "Earth's Early Atmosphere."

15. Towe, "Environmental Oxygen Conditions"; Berkner and Marshall, "On the Origin and Rise of Oxygen Concentration," 225; Kasting, "Earth's Early Atmosphere"; Brinkman, "Dissociation of Water Vapor and Evolution of Oxygen," 5355; Dimroth and Kimberly, "Pre-Cambrian Atmospheric Oxygen," 1161; Carver, "Prebiotic Atmospheric Oxygen Levels," 136; Holland, Lazar, and McCaffrey, "Evolution of the Atmosphere and Oceans"; Kasting, Liu, and Donahue, "Oxygen Levels in the Prebiological Atmosphere"; Kerr, "Origin of Life"; Thaxton, Bradley, and Olsen, *The Mystery of Life's Origin*, 73–94.

16. Holland, *The Chemical Evolution of the Atmosphere and Oceans*, 99–100; Schlesinger and Miller, "Prebiotic Synthesis in Atmospheres Containing CH_4, CO, and CO_2," 376; Horgan, "In the Beginning."

17. Thaxton, Bradley, and Olsen, *The Mystery of Life's Origin*, 69–98; Dembski and Wells, *The Design of Life*, 222–24.

18. Thaxton and Bradley have shown that polymerizing amino acids under reducing conditions releases 200 kcal. of energy per mole, whereas polmerizing amino acids in neutral conditions requires an *input* of 50 kcal. of energy per mole ("Information and the Origin of Life," 184).

19. Brooks, *The Origins of Life*, 118.

20. Schwartz, "Intractable Mixtures and the Origin of Life," 656.

21. Dean Kenyon, in Thaxton, Bradley, and Olsen, *The Mystery of Life's Origin*, vi.

22. Thaxton, Bradley, and Olsen, *The Mystery of Life's Origin*, 102.

23. Thaxton and Bradley, "Information and the Origin of Life," 184; Thaxton, Bradley, and Olsen, *The Mystery of Life's Origin*, 100–101.

24. De Duve, "The Beginnings of Life on Earth," 437; emphasis added.
25. Koonin, "The Cosmological Model of Eternal Inflation."

Chapter 11: Self-Organization and Biochemical Predestination

1. Kenyon and Steinman, *Biochemical Predestination*, 31.
2. Kenyon and Steinman, *Biochemical Predestination*, 206.
3. Kenyon and Steinman, *Biochemical Predestination*, 207.
4. These differences in bonding affinity are due to steric (spatial) hindrances between the side chains of the different amino acids. Longer side chains block each other and impede the formation of peptide bonds.
5. Kenyon and Steinman, *Biochemical Predestination*, 207.
6. The most prominent advocate of the protein-first origin-of-life model during this time period was Sidney Fox. Fox not only thought that the first functional macromolecules would have been proteins, but he thought the first protocells would have been enclosed with membranelike enclosures made from polypeptides. He called these circular enclosures proteinoid microspheres. See Fox, "Simulated Natural Experiments."
7. *Biochemical Predestination* cemented Kenyon's reputation as a world-class origin-of-life researcher. For example, in 1974, Kenyon was invited to publish an article in a prestigious *Festschrift* in honor of Aleksandr I. Oparin. The volume included essays from most of the leading origin-of-life researchers at the time and included an introduction from Oparin himself. See Kenyon, "Prefigured Ordering and Proto-Selection in the Origin of Life," in Dose, et al., *The Origin of Life and the Evolutionary Biochemistry*, 207–20; see also table of contents, v–vii.
8. Solomon Darwin, http://www.haas.berkeley.edu/faculty/darwin.html (last accessed September 9, 2008). Professor Kenyon conveyed the details of this story to me in a personal interview.
9. Steinman and Cole, "Synthesis of Biologically Pertinent Peptides," 745–41. Though Kenyon and Steinman expected that the amino-acid sequences in the proteins in extant organisms would strongly resemble the original proteins from which they had evolved, they did not expect that they would exhibit an exact sequence identity. Instead, as Kenyon told me, "we thought that the primordial amino-acid sequences would have different mutation histories within the subsequently evolving lineages of cells, but thought that much of the primordial sequence specificity would still be intact today."
10. Kok, Taylor, and Bradley, "A Statistical Examination of Self-Ordering."
11. Kamminga, "Studies in the History of Ideas," 301–4, 308–12.
12. Crick, *Of Molecules and Men*, 10.
13. Dawkins, *River Out of Eden*, 17.
14. Polanyi, "Life Transcending Physics and Chemistry," 55.
15. Polanyi, "Life Transcending Physics and Chemistry," 55; "Life's Irreducible Structure," 1309.
16. Polanyi, "Life's Irreducible Structure," 1309.
17. Polanyi, "Life's Irreducible Structure," 1309.
18. Polanyi, "Life's Irreducible Structure," 1309.
19. The bases do participate in weak van der Waals and hydrophobic interactions, but these chemical affinities are so slight and nonspecific as to be incapable of determining the specific sequential arrangements of the bases in the DNA molecule.
20. Küppers, "On the Prior Probability of the Existence of Life," 364.
21. De Duve, "The Beginnings of Life on Earth," 428–37, esp. 437.
22. Note that the "RNA world" scenario was not devised to explain the origin of the sequence specificity of biomacromolecules. Rather, it was proposed as an explanation

for the origin of the interdependence of nucleic acids and proteins in the cellular information-processing system. By proposing an early earth environment in which RNA performed both the enzymatic functions of modern proteins and the information-storage function of modern DNA, "RNA first" advocates sought to formulate a scenario making the functional interdependence of DNA and proteins unnecessary to the first living cell. In so doing, they sought to make the origin of life a more tractable problem. Chapter 14 examines this hypothesis and the many problems it has encountered.

23. An article heralding a breakthrough for "RNA world" scenarios makes this clear. After telling how RNA researcher Jack Szostak had succeeded in engineering RNA molecules with a broader range of catalytic properties than previously known, science writer John Horgan makes a candid admission: "Szostak's work leaves a major question unanswered: How did RNA, self-catalyzing or not, arise in the first place?" ("The World According to RNA").

24. The National Center for Biotechnology Information (NCBI) maintains a Web archive where the currently known variant codes, both nuclear and mitochondrial, are listed; see http://www.ncbi.nlm.nih.gov/Taxonomy/Utils/wprintgc.cgi#SG4 (last accessed September 10, 2008).

25. Polanyi, "Life's Irreducible Structure," 1309.

26. This, in fact, happens where adenine and thymine interact chemically in the complementary base pairing *across* the information-bearing axis of the DNA molecule. *Along* the message-bearing axis, however, there are no chemical bonds or differential bonding affinities that determine sequencing.

27. Stalnaker, *Inquiry,* 85.

28. Dretske, *Knowledge and the Flow of Information* (1981), 12.

29. As noted in Chapter 4, the information-carrying capacity of any symbol in a sequence is inversely related to the probability of its occurrence. The informational capacity of a sequence as a whole is inversely proportional to the product of the individual probabilities of each member in the sequence. Since chemical affinities between constituents ("symbols") increase the probability of the occurrence of one, given another (i.e., necessity increases probability), such affinities decrease the information-carrying capacity of a system in proportion to the strength and relative frequency of such affinities within the system.

30. Yockey, "Self-Organization Origin of Life Scenarios."

31. Polanyi, "Life's Irreducible Structure," 1309.

Chapter 12: Thinking Outside the Bonds

1. Nicolis and Prigogine, *Self-Organization in Nonequilibrium Systems,* 339–53, 429–47.

2. Prigogine, Nicolis, and Babloyantz, "Thermodynamics of Evolution," 23–31. Prigogine's statement in full: "The probability that at ordinary temperatures a macroscopic number of molecules are assembled to give rise to the highly ordered structures and to the coordinated functions characterizing living organisms is vanishingly small. The idea of spontaneous genesis of life in its present form is therefore highly improbable, even on the scale of the billions of years during which prebiotic evolution occurred."

3. Yockey, "A Calculation of the Probability of Spontaneous Biogenesis," esp. 380.

4. Eigen, *Steps Toward Life,* 12.

5. Kauffman, *At Home in the Universe,* 274.

6. Fry, *The Emergence of Life on Earth,* 158.

7. Kauffman, *The Origins of Order,* 285–341.

8. Kauffman, *The Origins of Order*, 299. He also cited proteins in the blood-clotting cascade that "cleave essentially single-target polypeptides" to support his claim that some low-specificity proteins can perform biological functions.
9. Creighton, *Proteins*, 217–21.
10. See the entry for trypsin at http://www.expasy.org/cgi-bin/niceprot.pl?P07477 (last accessed September 15, 2008).
11. Kauffman, *The Origins of Order*, 298.
12. Thaxton, Bradley, and Olsen, *The Mystery of Life's Origin*, 127–43.
13. Shapiro, "A Simpler Origin of Life," 47–52.
14. Shapiro, "A Simpler Origin of Life," esp. 52.
15. Lifson, "On the Crucial Stages"; Joyce, "RNA Evolution and the Origin of Life"; Dover, "On the Edge."
16. Kauffman, *At Home in the Universe*, 47–92.
17. Kauffman, *At Home in the Universe*, 53, 89, 102.
18. Kauffman, *At Home in the Universe*, 86, 88.
19. De Duve, "The Beginnings of Life on Earth," 437, emphasis added. De Duve also notes: "The probability of any given card distribution is infinitesimally small. Yet no bridge player has ever exclaimed at being witness to a near-miracle. What would be a miracle, however, or, rather, unmistakable evidence of trickery, is if the same distribution should be dealt again, even once" (*Singularities*, 4).
20. De Duve, "The Beginnings of Life on Earth," 437.
21. De Duve, *Vital Dust*, 9–10, emphasis added. In April 2000, I presented my case for intelligent design in a talk titled "DNA by Design" in a session with Christian de Duve at the "Nature of Nature Conference" at Baylor University. In the discussion after my presentation, de Duve explained that he had agreed with everything I had presented *except* "the last slide," by which he meant my criticism of his statement on that slide. The slide to which he referred contained the quote cited in the main body of the text to which this note refers. De Duve's response to my presentation revealed that his differences with me were not scientific. Indeed, he has acknowledged the same problems with the main naturalistic explanations for the origin of biological information that I had described. Rather, our differences were mainly philosophical and methodological. Because of his understanding of the definition of science, he was unwilling to consider the design hypothesis. I was not.

Chapter 13: Chance and Necessity, or the Cat in the Hat Comes Back

1. Oparin, *The Origin of Life* (1938), 64–103, 107–8; Meyer, "Of Clues and Causes," 174–79, 194–98, 211–12.
2. Oparin, *The Origin of Life* (1938), 133–35.
3. Oparin, *The Origin of Life* (1938), 148–59.
4. Oparin, *The Origin of Life* (1938), 195.
5. Oparin, *Genesis and Evolutionary Development of Life*, 146–47.
6. Kamminga, "Studies in the History of Ideas," 326; Oparin, *Genesis and Evolutionary Development of Life*, 146–47.
7. Kamminga, "Studies in the History of Ideas," 300–301, 308–28.
8. Kamminga, "Studies in the History of Ideas," 301.
9. Kamminga, "Studies in the History of Ideas," 323–28; Oparin, *Origin of Life on Earth* (1957), 229–90; *Genesis and Evolutionary Development of Life*, 127–51.
10. Versions of the gene theory, such as Richard Dawkins's "selfish gene" hypothesis (see Dawkins, *The Selfish Gene*) and the theory that life began as self-replicating RNA, have persisted to the present day. The latter idea has been the subject of extensive investigation during the mid-1980s.

11. Pattee, "The Problem of Biological Hierarchy," esp. 123.

12. De Duve, *Blueprint for a Cell,* 187.

13. Dobzhansky, "Discussion of G. Schramm's Paper," 310.

14. Joyce and Orgel, "Prospects for Understanding," esp. 8–13.

15. Mora, "The Folly of Probability," 311–12; Bertalanffy, *Robots, Men and Minds,* 82.

16. Von Neumann, *Theory of Self-Reproducing Automata.*

17. Pennisi, "Seeking Life's Bare (Genetic) Necessities," 1098–99; Mushegian and Koonin, "A minimal gene set," 10268–73.

18. Wigner, "The Probability of the Existence of a Self-reproducing Unit," 231–35; Morowitz, "The Minimum Size of the Cell"; *Energy Flow in Biology,* 10–11.

19. Quastler, *The Emergence of Biological Organization,* ix.

20. Quastler, *The Emergence of Biological Organization,* 16.

21. Quastler, *The Emergence of Biological Organization,* 1.

22. Quastler, *The Emergence of Biological Organization,* 43.

23. Hubert Yockey, *Information Theory and Molecular Biology,* 247.

24. Though many regard hypercycles as a self-organizational model that produces information purely by deterministic chemical reactions, I prefer to address them as a model that relies on both chance and necessity, because the initial complement of information necessary to make them work (even hypothetically) clearly does not come from any self-organization process and thus presumably must come from chance.

25. Eigen and Schuster, "The Hypercycle."

26. Smith, "Hypercycles and the Origin of Life"; Dyson, *Origins of Life,* 9–11, 35–39, 65–66, 78; Shapiro, *Origins,* 161.

27. See the critical discussion of Eigen's hypercycle in Yockey, *Information Theory and Molecular Biology,* 275–77. Yockey cites Eigen as acknowledging that the "hypercyclic link would then become effective only after concentrations have risen to sufficiently high level. . . . There is only one solution to this problem: the hypercycle must have a precursor, present in high natural abundance, from which it originates gradually by a mechanism of mutation and selection" (275).

28. Dawkins, *The Blind Watchmaker,* 46–49; Küppers, "On the Prior Probability of the Existence of Life."

29. Küppers, "On the Prior Probability of the Existence of Life," 366.

30. Dawkins reports the following as "10th generation" phrases, arising as intermediate "progeny" of his Weasel program: MDLDMNLS ITJISWHRZREZ MECS P and Y YVMQKSPFTXWSHLIKEFV HQYSPY. This sequence contains a few small words. A few small English words (for example, IT and ME) are imbedded in these longer strings of gibberish. In the context of the surrounding gibberish, however, these short strings perform no communication function and thus provide nothing "selectable" by the logic of natural selection that Dawkins means to simulate. Even if they did, these short strings lack the complexity of genes and protein and thus do not simulate the amount of complexity that would need to arise by chance in biology before selection can play a role.

31. Berlinski, "On Assessing Genetic Algorithms."

32. Schneider, "Evolution of Biological Information."

33. See http://www-lmmb.ncifcrf.gov/~toms/paper/ev/ (last accessed September 17, 2008).

34. Schneider, "Evolution of Biological Information," 2794, emphasis added.

35. Schneider, "Evolution of Biological Information," 2796. In his simulation, Schneider treats 131 specified digital characters as equivalent to a protein-binding site on a gene. Schneider indicates that his computer organism has 265 digital characters, roughly 134 or so of which are dedicated to computing. That leaves 131 digital characters corresponding to the binding site, the generation of which *Ev*

attempts to simulate (see the caption to Figure 1, p. 2796). A successful simulation would, therefore, involve specifying 131 digital characters ("zeros" and "ones"), or 131 bits of information.

36. Schneider, "Evolution of Biological Information," 2795.

37. The amount of exogenous information is equal to the difference between the amount of endogenous information and the amount of active information. It represents the amount of information that a purely random unguided search must generate after all active information has been used to search for a target.

38. Links to Robert Marks's papers analyzing *Ev* and other evolutionary algorithms are available on the interactive Web site supporting this book, www.signatureinthecell. com. See also Dembski and Marks, "The Search for a Search." Dembski and Marks, "The Conservation of Information."

39. Adami and Brown, "Evolutionary Learning."

40. Lenski, et al., "The Evolutionary Origin of Complex Features."

41. Lenski, et al., "The Evolutionary Origin of Complex Features," 139.

42. Pennock, "Does Design Require a Designer?" See also Pennock, "Learning Evolution."

43. In these determinations, the character "1" typically represents true and the character "0" typically represents false. After reading a pair of binary digits, a computer applying an OR function will register a "1" if it is true that one "or" the other of the digits represented true, that is, that one "or" the other of the digits was a "1." In the same way, if the computer applies an AND to the pair of binary numbers it will register a "1" if both characters —first and second—represent true—i.e., if they are "1's." It will register a "0" otherwise.

44. In particular, the *Avida* organism generates an output indicating that the logic function EQU has been performed on the input. Computer scientists will recognize EQU as a compound logic function that is equivalent to AND, OR, and NOR being performed in sequence.

45. Lenski, et al., "The Evolutionary Origin of Complex Features," 139.

46. Even as a demonstration of how biological evolution—mutation and natural selection—might generate new information, *Avida* leaves much to be desired. To understand why requires knowing a bit more about the way *Avida* functions. After the evaluation algorithm has detected a logic function, *Avida* will replicate the loop of instructions that generated the logic function more frequently than it would have otherwise. *Avida* then randomly mutates the instructions in that loop before allowing them to operate again on the set of input strings. As noted, each of these instructions is, in fact, information- and content-rich. Yet *Avida* does not mutate the content of these instructions themselves. Instead, it changes only the instructions as discrete packets of specified information. Given the amount of information represented by each command, this presumably simulates the equivalent of shuffling whole genes. Therefore, this rearranging of instructions might simulate how a new structure and function made from several separate but preexisting gene products might have arisen (via mechanisms such as inversion, recombination, or lateral gene transfer). Nevertheless, *Avida* does not simulate how new genes and proteins might have arisen, even from preexisting genes and self-replicating organisms. Yet clearly, theories of biological evolution must account for the origin of such genetic information in order to account for the origin of new organisms.

Avida rewards small incremental steps as functional improvements, but it does so in a way that lacks biological realism. Although *Avida* rewards new functions that are qualitatively similar to Darwinian evolution, it lacks quantitative similarity to biology. *Avida* does not select for proximity to future function as *Ev* and earlier simulations did. Instead, it rewards incremental additions of function—simple logic functions—along the way to its desired end point, the compound logic function

EQU. In *Avida* these intermediate functional steps are relatively easy to generate by a random search. In biology, however, no comparably modest changes produce new protein structures. Yet such structures must arise before any truly novel function can arise during biological evolution. And any complete theory of biological evolution must explain the origin of such novel proteins and the genetic information required to produce them.

The authors of *Avida* hoped that their simulation would produce a compound logic function, namely, EQU (EQU represents "equal"). EQU is a compound logic function equivalent to a combination of other logic functions operating on bit pairs. For this reason, the authors of the *Avida* paper regard it as a reasonable representation of a functionally integrated or "irreducibly complex" biological system in their simulation. Nevertheless, the programmers of *Avida* knew that they could build EQU by concatenating a series of less complex logic functions together. So they designed their evaluation algorithm to reward each of these less complex logic functions (NAND, AND, OR, and NOR) as well. This, by itself, would not necessarily compromise the realism of the *Avida* simulation, were it not for the size of the steps that the *Avida* simulation rewards. In theory, at least, natural selection might select and preserve a series of small incremental changes provided each of the smaller increments confers a functional advantage along the way toward producing a more complex structure. To produce new biological information, however, even starting from a self-replicating organism, requires producing new genes that find expression as new proteins. Biological change is denominated in novel proteins. In other words, proteins represent the smallest functionally significant (and selectable) step in evolution. And building new proteins requires new genetic information.

Since natural selection acts upon or preserves new functional proteins only after the fact of a successful search, random mutations must do the work of generating these new genes and proteins. Yet finding new proteins (and the genetic information necessary to produce them) by a random search is a vastly more difficult proposition than *Avida* simulates. Instead, the logic functions that *Avida* rewards can be produced easily given the probabilistic resources available to it. Yet the same thing cannot be said of novel protein structures.

Bob Marks has written a similar simulation program to replicate the function of *Avida* for purposes of evaluation. He used loops with a fixed length of 50 without the instructions needed for organism duplication. He found the NAND logic function, out of which EQU can be made, arose roughly once every 2,297 times. The other logic functions out of which EQU can be built—AND, OR, and NOR—arose roughly once every 892,193; 3,605,769; and 157,894,737 (1.57×10^8 times), respectively. Though these frequencies are clearly small, *Avida* had more than enough trials (probabilistic resources) available to it to ensure that these functions would arise. *Avida* also had more than enough trials to ensure that these simple functions would combine in enough different ways to ensure that the compound EQU function would eventually arise from one of the combinations. (In Professor Marks's simulation of *Avida*, EQU arose 9 times in 6 billion queries.)

Yet, as noted in Chapter 9, site-directed mutagenesis experiments have shown that the odds of generating a functionally sequenced protein of a modest length (150 residues) via a random search stand at about 1 chance in 10^{74}. In other words, the ratio of functional proteins of modest length to nonfunctional amino-acid sequences of equivalent length is about 1 in 10^{74}. To sample a combinatorial space this size by a series of undirected trials would require probabilistic resources far in excess of those available to the entire history of the earth. (Recall that the probabilistic resources corresponding to the 4.6 billion year history of the earth stand at about 10^{41} possible events.) Yet many crucial evolutionary transitions, including the Cambrian explosion in which many novel body plans as well as proteins first arose, took place within 5 to 10 million years, in less than .2 percent of the history of the earth. I have argued elsewhere that these facts cast doubt on the mechanism of mutation and selection as an adequate explanation of the origin of new genes and proteins. They certainly also suggest that *Avida*, whatever its other merit, does not provide a realistic simulation of the origin of new genetic information in a biological context. That is, *Avida* does not realistically simulate the generation of new functional genetic information even in a context in which the information necessary to a self-replicating organism is already present. See Meyer, "The Origin of Biological Information."

47. Pennock, "Does Design Require a Designer?"
48. Wolpert and Macready, "No Free Lunch Theorems for Optimization."
49. Brillouin, *Science and Information Theory,* 269.
50. Personal interview with Robert Marks, June 2008.
51. William Dembski gives a more generalized expression and justification of this law in his paper "Intelligent Design as a Theory of Information."

Chapter 14: The RNA World

1. Gilbert, "Origin of Life."
2. Meyer, "DNA and Other Designs." See the write-up in the *Chronicle of Higher Education,* March 29, 2000. See also Correspondence, *First Things,* October (2000): 2–3.
3. Miller, "How Intelligent Is Intelligent Design?"
4. Szostak, Bartel, and Luisi, "Synthesizing Life," esp. 387–88.
5. Gilbert, "Origin of Life."
6. Gilbert, "Origin of Life."
7. Shapiro, "Prebiotic Cytosine Synthesis"; Waldrop, "Did Life Really Start Out in an RNA World?"
8. Levy and Miller, "The Stability of the RNA Bases; Shapiro, "A Simpler Origin of Life," 49.
9. Levy and Miller, "The Stability of the RNA Bases," 7933.
10. Shapiro, "Prebiotic Cytosine Synthesis," 4396.
11. Shapiro, "Prebiotic Ribose Synthesis," esp. 71.
12. Prebiotic chemists have proposed that the nucleotide bases, particularly adenine, could have been produced on the early earth from hydrogen cyanide, HCN, reacting in solutions of ammonium hydroxide, $NH_4(OH)$. This reaction does produce small yields of adenine and other nitrogenous bases, at least if the reaction starts with extremely high (though probably unrealistic) concentrations of hydrogen cyanide. But this reaction also produces a variety of nitrogenous substances that quash the formose reaction. This fact creates a formidable difficulty for the RNA-world hypothesis: the main reaction proposed for producing nucleotide bases in a prebiotic environment also prevents synthesis of ribose. *Yet both substances are needed to make RNA.* See Shapiro, "Prebiotic Ribose Synthesis," 81–82.
13. Kenyon and Mills, "The RNA World."
14. Shapiro, "Prebiotic Ribose Synthesis," 71.
15. Noller, Hoffarth, and Zimniak, "Unusual Resistance of Peptidyl Transferase."
16. Zhang and Cech, "Peptide Bond Formation."
17. Illangasekare, et al., "Aminoacyl-RNA Synthesis Catalyzed by an RNA."
18. Joyce, "RNA Evolution and the Origins of Life."
19. Gil, et al., "Determination of the Core of a Minimal Bacterial Gene Set," Table 1, 521–22.
20. Wolf and Koonin, "On the Origin of the Translation System."
21. Wolf and Koonin, "On the Origin of the Translation System," 6.

22. Kumar and Yarus, "RNA-Catalyzed Amino Acid Activation."

23. There is another difficulty associated with generating the enzymatic capacities of synthetases in an RNA world. The probability of ribozymes arising with even the limited capacity to catalyze aminoacyl bonds is very small. The first researchers who found an RNA molecule capable of self-aminoacylation with phenylalanine had to sift through a preengineered pool of 170 trillion (or 1.7×10^{14}) RNA molecules (see Illangasekare, et al., "Aminoacyl-RNA Synthesis Catalyzed by an RNA"). This suggests that the probability of finding a single RNA molecule that could catalyze the formation of this bond is roughly one chance in 10^{14}. But to generate an RNA-based genetic code equivalent to that in the modern translation system would require not just one such ribozyme, but nineteen others (corresponding to each aminoacyl-tRNA synthetase enzyme) working together as a system, each with its own specific role. And *that* would require sequestering all the components of the system in a compartment that prevents interference from useless RNAs. If the other necessary ribozymes were roughly as rare as the first, then the probability of sequestering one additional ribozyme that performs the same function with a different amino acid would be the square of the original probability, or less than chance 1 in 10^{28}. The probability of sequestering three such ribozymes in close quarters would be the cube of that initial probability, or less than one chance in 10^{42}. The probability of sequestering twenty such ribozymes in close enough proximity to function as a system—as a part of a genetic code—would be prohibitively small, no better than 1 chance in 10^{280}. Overcoming these odds would require a huge infusion of information (930 bits). And still these ribozymes would not be capable of coordinating the complex two-stage reaction that actual synthetase enzymes perform in extant cells.

24. Some researchers cite the ability of certain RNA molecules to catalyze peptide bonds outside the ribosome (albeit in the presence of another catalyst) as support for the RNA-world hypothesis (see, e.g., Zhang and Cech, "Peptide Bond Formation"). But the mere presence of an RNA molecule capable of catalyzing peptide bonding would not obviate the need for a system of proteins to translate mRNA into proteins. This scenario has two obvious problems, however. (1) Ribozymes facilitating peptide-bond formation in the absence of an mRNA transcript produce random, not sequence-specific, arrangements of amino acids. Yet as shown in Chapters 8–10, formidable probabilistic hurdles face any chance-based scenarios for the origin of sequence-specific functional proteins. (2) RNA-catalyzed peptide bonding outside of a ribosome will lead to the production of many improperly formed and folded peptide chains. Because the side chains of several amino acids feature amino groups and/or carboxyl groups, many branching peptide linkages can form. This would essentially preclude structures like natural protein folds. Indeed, papers that have acknowledged the possibility of rRNA catalyzing peptide bonds in the absence of ribosomal proteins have also acknowledged that the ribosome as a whole provides critical "substrate positioning" during protein synthesis. (See Rodnina, Beringer, and Bieling, "Ten Remarks on Peptide Bond Formation.")

25. The twenty protein-forming amino acids are capable of a much wider range of chemical interactions than are the four nucleotide bases. The four bases are exclusively hydrophilic (water-attracting) molecules, and they form mainly hydrogen bonds with each other. On the other hand, some amino acids are hydrophilic; some are hydrophobic; some are acidic; and some are basic. This diversity of properties among its constitutive chemical groups allows proteins to attain more complex three-dimensional geometries than do the simple pairing interactions between bases in RNA molecules. As a result, proteins play functional roles that RNA does not and, in many cases, probably cannot.

26. By coupling and enabling energetically favorable and unfavorable reactions, synthetase enzymes catalyze the production of a molecule—a charged tRNA ready

for translation—that would not form in any appreciable quantities otherwise—even with the help of two separate RNA catalysts. An RNA catalyst might drive forward the energetically favorable first reaction that activates amino acids with AMP. But even a separate RNA catalyst will not drive forward the energetically unfavorable second aminoacylation reaction (unless massive amounts of the reactants are provided to overcome the unfavorable energetics).

27. Quastler, *The Emergence of Biological Organization*, 16.

28. Recently, Lincoln and Joyce claim to have produced a fully self-replicating RNA molecule. Nevertheless, their claim trades upon an ambiguity in the meaning of self-replication and constitutes little more than a gimmick. True template-directed polymerases can copy any template using free-floating bases from their surroundings.

  Polymerases do the work of copying a template by sequestering, aligning, and linking bases on a template strand. For a polymerase to function as a true replicase, it would likewise have to do the work of replicating a template, in this case the template provided by itself.

  The RNA molecules that Lincoln and Joyce devise do not do this work. Instead, they simply joined together via a single bond two *presynthesized, specifically sequenced* RNA chains to form a longer chain. After the formation of a single phosphate bond, these linked chains resulted in a copy of the original RNA molecule, but *only because Lincoln and Joyce first designed the original RNA molecule and then directed the synthesis of two specifically sequenced, complementary partial strands* to match it. Thus, Lincoln and Joyce provided the information (did the sequencing work) required to make even this limited form of replication possible. Further, instead of demonstrating RNA-directed *self*-replication, they demonstrated investigator-directed replication. Lincoln and Joyce, "Self-Sustained Replication of an RNA Enzyme," 1–6. Hayden, "A Never-Ending Dance of RNA: The Recreation of Life's Origins Comes a Self-Catalysing Step Closer."

29. Johnston, et al., "RNA-Catalyzed RNA Polymerization."

30. Joyce and Orgel, "Progress Toward Understanding the Origin of the RNA World," esp. 33.

31. Szostak, Bartel, and Luisi, "Synthesizing Life," 389.

32. Johnston, et al., "RNA-Catalyzed RNA Polymerization," 1321.

33. Because of the complex chemical interactions of their amino-acid side chains, proteins form much more complex three-dimensional shapes than RNA molecules do. The inability of RNA to form the same kind of subtle three-dimensional geometries that proteins form may limit the capacity of ribozymes to perform many of the functions that proteins perform. It also appears to limit the efficiency and usefulness of known ribozymes in comparison to their protein counterparts.

  Gerald Joyce has noted, for example, that ribozyme polymerases are not nearly as stable and do not perform nearly as well as wild-type polymerases. He notes that the RNA-only polymerases have several limitations that naturally occurring polymerases do not. First, they copy templates of genetic information more slowly. Second, they can copy only certain templates of information without stalling. Third, they copy with less fidelity than wild-type RNA polymerases. Fourth, ribozymes require a higher concentration of RNA monomers to facilitate copying than do their wild-type counterparts.

  Joyce summarizes the functional problems with the Bartel lab-engineered 189-nucleotide class 1 ligase: "Closer inspection of the polymerase ribozyme reveals some of its key limitations. First, although the k_{cat} for NTP [nucleoside triphosphate, a unit of RNA that includes the sugar-phosphate backbone and the nucleotide base] addition is greater than 1 min^{-1}, the K_m for the separate template-primer complex is immeasurably high and in excess of 1 mM. In practical terms this means that if one employs typical concentrations of 1 μM template-primer, about 2 hours are required for a productive substrate-binding event. For most templates the ribozyme has little processivity, that is, little ability to add multiple NTPs before dissociating from the template-primer complex. Thus another 2 hours are required for the next productive binding event. The ribozyme is susceptible to hydrolysis of its component phosphodiester linkages, and under the preferred reaction conditions of 200 mM of MgCl$_2$ at pH 8.5 and 22°C, suffers nonspecific cleavage of one of its phosphodiesters at a rate of about 10^{-2} min^{-1}. Thus in the race between NTP addition and degradation of the ribozyme, it is possible to achieve about 12 NTP additions in 24 hours, but not many more, because by then the ribozyme is largely degraded.

"A second limitation of the class-I-derived polymerase ribozyme is that, although for one special template it can add up to 14 successive NTPs, for more typical templates it adds only a few NTPs. Even a very subtle change in the sequence of the preferred template dramatically reduces the extent of NTP addition. A third limitation is that, although the fidelity of template copying is high when measured for the full-length products, the overall fidelity is considerably lower because incorporation of the wrong NTP reduces the rate of subsequent extension. A fourth limitation is that the affinity of NTP binding to the template is determined largely by the strength of Watson-Crick pairing, thus requiring high concentrations of NTPs and providing an inherent advantage for GC pairs" ("Forty Years of *In Vitro* Evolution," esp. 6430–31).

34. De Duve, *Vital Dust*, 23.
35. Joyce and Orgel, "Progress Toward Understanding the Origin of the RNA World," 35–36.
36. Joyce and Orgel, "Progress Toward Understanding the Origin of the RNA World," 33.
37. In theory, at least, Joyce and Orgel may have assumed an overly restrictive condition on the origin of a self-replicating RNA system. They are correct to note that self-replication requires two RNA molecules, one to act as replicator and one to act as template. Moreover, for a ribozyme to replicate itself in one step, the template RNA would have to be the exact reverse complement of the replicase RNA.

Nevertheless, it might be possible to copy the original replicator in two steps if the original replicator had come into contact with another RNA molecule capable of copying any RNA. This molecule would not need to be the exact reverse complement of the original, but it would have to have a sequence and structure making it capable of general replicase function. That way, after this second replicase had produced a transcript of the original replicase, the original one could transcribe this transcript. The result would be an exact copy of the original RNA replicase.

Extending the RNA world this extra benefit of the doubt, however, does not improve the odds of it occurring or eliminate the displacement problem. It only illustrates the displacement problem in a slightly different way. In this scenario, RNA self-replication would arise after two long functional RNA replicators had arisen in close spatial and temporal proximity to one another. These two replicators would not need to be identically sequenced. But both would need to be sequenced specifically to ensure replicase function. Thus, instead of displacing a *sequence-identity* problem from one molecule to another, this revised RNA-first hypothesis just displaces a *sequence-specificity* problem from one molecule to another.

Because these two sequences would likely possess a roughly equal degree of functional specificity, the probability of achieving a system of two molecules capable of RNA self-replication would be roughly the square of the probability of achieving one such molecule. Though scientists don't know exactly how specified such sequences would have to be in order to achieve RNA-catalyzed self-replication, the experimental evidence cited above suggests that RNA molecules with this capacity are exceedingly rare within the space of possible RNA base sequences.

Since, additionally, Orgel and Joyce made an overly optimistic assumption about the number of bases necessary (only 50) to make an RNA capable of true self-replication, they probably vastly underestimated the probability of two sequence-specific replicases of any kind (whether identical or not) arising in close proximity to one another.

Beyond that, ribozyme engineers are finding it more difficult to design an RNA molecule that can copy *any* RNA sequence whatsoever. As Joyce notes, all known RNA polymerases are template dependent, meaning that they can copy only complements of themselves with appreciable fidelity. This means that the scenario that Joyce and Orgel envision—in which an RNA replicase and its exact complement arise in close proximity to each other—is still the only realistic one from a chemical—though perhaps not a theoretical—point of view.

38. Joyce and Orgel, "Prospects for Understanding the Origin of the RNA World," 35.
39. Shapiro, as quoted in Brockman, ed., *Life: What a Concept!,* 90.
40. Szostak, Bartel, and Luisi, "Synthesizing Life."
41. Szostak, Bartel, and Luisi, "Synthesizing Life," 388; Wolf and Koonin, "On the Origin of the Translation System."
42. Ekland, Szostak, and Bartel have claimed that functional ribozymes are more common in pools of RNA sequences than previously thought ("Structurally Complex and Highly Active RNA Ligases," esp. 364). The sequence space (the total number of possible combinations of nucleotides) corresponding to the 220-base RNA molecules that they investigated is vast, 4^{220}, or roughly 1×10^{132} possibilities. Since they were able to isolate a ribozyme ligase by searching a tiny fraction of the whole space (about 1.4×10^{15} distinct sequences), this does imply that the whole space would contain a vast number of comparable ligases. But, for such a simple function to require a one-in-a-trillion sequence raises serious questions about the rarity of more complex functions. As noted, for example, ribozyme engineering has not yet produced a single ribozyme capable of true polymerase activity. Nor have researchers discovered ribozymes capable of most other essential enzymatic functions.
43. As Wolf and Koonin note, "Because evolution has no foresight, no system can evolve in anticipation of becoming useful once the requisite level of complexity is attained" ("On the Origin of the Translation System").
44. Crick, *Life Itself,* 88.
45. Dose, "The Origin of Life."
46. Joyce and Orgel, "Prospects for Understanding the Origin of the RNA World," 19.
47. Because of the improbability of producing an RNA molecule with the specificity necessary to perform self-replication, Christian de Duve and Robert Shapiro have proposed "pre-RNA world" scenarios. These scenarios envision self-replication arising first in systems of smaller molecules, thus reducing the probabilistic hurdles facing the RNA-first approach. These models face a different problem than that of the RNA world, however. RNA-first models fail to explain the origin of biological information because of the improbability of specific sequencing arising in molecules so large (that is, in molecules large enough to perform replicase function). Small molecule–first hypotheses avoid this probability problem by envisioning smaller molecules arising first. Nevertheless, pre-RNA-world scenarios encounter the difficulty of getting sufficient three-dimensional complexity (and therefore biologically relevant specificity) to arise in molecules so small. Roughly speaking, an RNA world can produce complex molecules that aren't sufficiently specific (to perform biologically relevant functions). A pre-RNA world can produce specific molecules that aren't sufficiently complex (to perform relevant functions). Thus, neither actually solves the problem of the origin of biological information. These theories are relatively new, however. I will track their progress as necessary at:

http://www.signatureinthecell.com. See de Duve, *Vital Dust,* 20–45; Shapiro, "Small Molecule Interactions Were Central to the Origin of Life"; "A Simpler Origin of Life."

Chapter 15: The Best Explanation

1. Gian Capretti and others explore the use of abductive reasoning by Sherlock Holmes in the detective fiction of Sir Arthur Conan Doyle. Capretti attributes the success of Holmesian abductive "reconstructions" to a willingness to employ a method of "progressively eliminating hypotheses" ("Peirce, Holmes, Popper").
2. Peirce, *Collected Papers,* 2:375, emphasis added; "Abduction and Induction."
3. Capretti, "Peirce, Holmes, Popper," 143.
4. Whewell, *The Philosophy of the Inductive Sciences,* 2nd ed., 1:24–25, 78–80; 2:96, 101–4, 120–22, 397; Chamberlain, "The Method of Multiple Working Hypotheses"; Scriven, "Explanation and Prediction in Evolutionary Theory," 480; "Causes, Connections and Conditions in History," 238–46; Meyer, "Of Clues and Causes," chaps. 1–3; Lipton, *Inference to the Best Explanation;* Miller, "The Similarity of Theory Testing in the Historical and 'Hard' Sciences"; Cleland, "Historical Science, Experimental Science, and the Scientific Method."
5. See Meyer, "Of Clues and Causes," chaps. 4–5.
6. Kamminga, "Protoplasm and the Gene," 1.
7. Shannon, "A Mathematical Theory of Communication"; Shannon and Weaver, *The Mathematical Theory of Communication.*
8. Crick, "On Protein Synthesis," esp. 144, 153.
9. Lipton, *Inference to the Best Explanation,* 32–98.
10. See Chapter 7. See, e.g., Scriven, "Explanation and Prediction in Evolutionary Theory"; "Causes, Connections and Conditions in History," 238–64; Sober, *Reconstructing the Past,* 1–5, 36–69; Lipton, *Inference to the Best Explanation,* 32–88; Meyer, "Of Clues and Causes," chaps. 1–3.
11. Meyer, "The Scientific Status of Intelligent Design," 151–212; "The Demarcation of Science and Religion," 17–23; "Of Clues and Causes," 77–140; Sober, *The Philosophy of Biology.*
12. See, e.g., Charles Lyell's discussions in *Principles of Geology,* 1:75–91; 3:1–7.
13. Monod, *Chance and Necessity.* Chance-based theories invoke processes that produce particular outcomes with a low probability. Theories of necessity invoke processes that produce specific outcomes with a high probability, typically with a probability of one. For this reason, these two general categories of explanation plus explanations combining them are generally considered to represent a logically exhaustive set of possible explanatory approaches, at least from within a materialistic framework.
14. See Ben Stein's final interview of Richard Dawkins in the documentary film *Expelled: No Intelligence Allowed.*
15. Thaxton, Bradley, and Olsen, *The Mystery of Life's Origin,* vi.
16. Thaxton, Bradley, and Olsen, *The Mystery of Life's Origin,* 102.
17. Shapiro, "Prebiotic Ribose Synthesis."
18. Thaxton, Bradley, and Olsen, *The Mystery of Life's Origin,* 184.
19. Polanyi, "Life Transcending Physics and Chemistry," esp. 64.
20. Dawkins, *The Blind Watchmaker,* 47–49; Küppers, "On the Prior Probability of the Existence of Life"; Lenski, et al., "The Evolutionary Origin of Complex Features"; Schneider, "Evolution of Biological Information."
21. Denton, *Evolution,* 309–11.
22. Lawrence and Bartel, "New Ligase-Derived RNA Polymerase Ribozymes."
23. Polanyi, "Life Transcending Physics and Chemistry," esp. 64.
24. Paul and Joyce, "A Self-Replicating Ligase Ribozyme."

25. "The RNA molecules A, B, and T (Fig. 1) were prepared either by automated solid-phase synthesis or *in vitro* transcription [i.e., experimenter intervention]. B and T were obtained by transcription of the corresponding DNA template to yield 5'–pppGAGACCGCAAUCC–3' and 5'–pppGGAUUGUGCUCGAUUGUUCGUAA GAACAGUUUGAAUGGGUUGAAUAUAGAGACCGCAAUCC–3', respectively. Due to poor yields obtained for the transcription of A, this oligonucleotide, having the sequence 5'–GGAUUGUGCUCGAUUGUUCGUAAGAACAGUUUGAAU GGGUUGAAUAUA–3', was prepared synthetically" (Paul and Joyce, "A Self-Replicating Ligase Ribozyme," 12734).

26. Dose, "The Origin of Life"; Yockey, *Information Theory and Molecular Biology*, 259–93; Thaxton, Bradley and Olsen, *The Mystery of Life's Origin*, 42–172; Thaxton and Bradley, "Information and the Origin of Life"; Shapiro, *Origins*.

27. Of course, the phrase "large amounts of specified information" again begs a quantitative question, namely, "How much specified information or complexity would the minimally complex cell have to have before it implied design?" Recall that Dembski has calculated a universal probability bound of 1 out of 10^{139} (which he now rounds to 1 chance in 10^{150}) corresponding to the probabilistic/specificational resources of the known universe. Recall further that probability is inversely related to information by a logarithmic function. Thus, the universal small probability bound of 1 out of 10^{150} translates into roughly 500 bits of information. Thus, chance alone does not constitute a sufficient explanation for the de novo origin of any specified sequence or system containing more than 500 bits of (specified) information. Further, since systems characterized by complexity (a lack of redundant order) defy explanation by self-organizational laws, and since appeals to prebiotic natural selection presuppose, but do not explain, the origin of the specified information necessary to a minimally complex self-replicating system, intelligent design best explains the origin of the more than 500 bits of specified information required to produce the first minimally complex living system. Thus, assuming a nonbiological starting point (see Chapter 13, section headed "The Conservation of Information"), the de novo emergence of 500 or more bits of specified information will reliably indicate design.

28. See the relevant discussion in Peirce, *Collected Papers*, 2:372–88; "Abduction and Induction"; Fann, *Peirce's Theory of Abduction*, 28–34; Scriven, "Explanation and Prediction in Evolutionary Theory," 480; Sober, *Reconstructing the Past*, 1–5; Alston, "The Place of the Explanation," 23; Gallie, "Explanations in History and the Genetic Sciences," 392.

29. Sober, *Reconstructing the Past*, 1–5; Scriven, "Causes, Connections, and Conditions in History," esp. 249–50.

30. *McNeil-Lehrer News Hour*, Transcript 19 (May 1992).

31. Moreover, this generalization holds not only for the semantically specified information present in natural languages, but also for other forms of information or specified complexity, whether present in machine codes, machines, or works of art. Like the letters in a section of meaningful text, the parts in a working engine represent a highly improbable yet functionally specified configuration. Similarly, the highly improbable shapes in the rock on Mt. Rushmore conform to an independently given pattern: the faces of American presidents known from books and paintings. Thus, both systems have a large amount of *specified* complexity or information so defined. Not coincidentally, they also originated by intelligent design, not by chance and/or physical-chemical necessity.

32. Again, this claim applies at least in cases where the competing causal entities or conditions are nonbiological—or where the mechanism of natural selection can be safely eliminated as an inadequate means of producing requisite specified information.

33. Meyer, "Of Clues and Causes," 77–140.

34. Less exotic (and more successful) design detection occurs routinely in both science and industry. Fraud detection, forensic science, and cryptography all depend on the application of probabilistic or information-theoretic criteria of intelligent design. See Dembski, *The Design Inference*, 1–35.

35. Many would admit that we may justifiably infer a past human intelligence operating (within the scope of human history) from an information-rich artifact or event, but only because we already know that human minds exist. But, they argue, since we do not know whether an intelligent agent(s) existed prior to humans, inferring the action of a designing agent that antedates humans cannot be justified, even if we observe an information-rich effect. Note, however, that SETI scientists do not already know whether an extraterrestrial intelligence exists. Yet they assume that the presence of a large amount of specified information (or even just an unnaturally modulated radio signal) would establish the existence of one. Indeed, SETI seeks precisely to establish the existence of other intelligences in an unknown domain. Similarly, anthropologists have often revised their estimates for the beginning of human history or civilization, because they discovered information-rich artifacts dating from times that antedate their previous estimates. Most inferences to design establish the existence or activity of a mental agent operating in a time or place where the presence of such an agency was previously unknown. Thus, to infer the activity of a designing intelligence from a time prior to the advent of humans on earth does not have a qualitatively different epistemological status than other design inferences that critics already accept as legitimate. See McDonough, *The Search for Extraterrestrial Intelligence*.

36. Meyer, "The Explanatory Power of Design"; "DNA by Design"; "DNA and Other Designs"; "DNA and the Origin of Life."

Chapter 16: Another Road to Rome

1. Dembski's work on elucidating the forms of reasoning by which we infer design goes back to a seminal paper he wrote on the nature of randomness, "Randomness by Design." It was this and another paper that first brought Dembski to my attention and led to our initial collaboration. In it he shows that randomness cannot properly be understood except with reference to patterns that random objects systematically violate—once a pattern is matched, randomness dissolves. From studying the nature of the patterns used to defeat randomness, Dembski came upon the design inference. He presented an outline of the design inference in 1993 in Seattle in a paper, "Theoretical Basis for the Design Inference." He developed these ideas into a doctoral dissertation about the foundations of probability theory (1996) and then went on to publish that dissertation in 1998: *The Design Inference*. He expanded on this work in his 2002 sequel, *No Free Lunch*. His idea of specification, which sits at the heart of the design inference, is subtle and requires some technical sophistication to elucidate in full, though the essence of the idea can be illustrated clearly with examples of the kind used in this chapter. Dembski's most current formulation of specification appears in his article "Specification," available at http://www.designinference.com/documents/2005.06.Specification.pdf. The most user-friendly treatment of his work on design inferences, with an application to biology, appears in Dembski and Wells, *The Design of Life*, chap. 7.

2. Dembski, *The Design Inference*, 1–35.

3. Dembski, *The Design Inference*, 1–35, 136–223.

4. Dembski, *The Design Inference*, 37.

5. Dembski's initial work on specification focused on two requirements, a conditional independence condition of patterns from background knowledge and a tractability condition for the descriptive complexity of patterns. The former

could be characterized using conventional probability theory. The latter could be characterized using Andrei Kolmogorov's work on algorithmic information theory; see Kolmogorov, "Three Approaches to the Quantitative Definition of Information." Dembski held to both these requirements throughout his books *The Design Inference* and *No Free Lunch*. In 2005, however, he wrote a paper on specification that significantly simplified its characterization, "Specification." In that paper, he was able to characterize specification simply in terms of tractability by showing that conditional independence comes as a consequence of tractability (by tractability he means a pattern of low-descriptive complexity). Specified complexity as Dembski now defines it signifies high-probabilistic complexity of an event (i.e., improbability) combined with low-descriptive complexity of a pattern to which the event conforms.

6. Dembski illustrates this distinction with an example of an archer shooting an arrow. If an observer witnesses an archer shooting an arrow and then the arrow hitting a target, the observer can correctly infer something about the skill and intent (design) of the archer. If, however, the archer shoots an arrow and then the observer goes and draws a circle around the arrow and calls it the target, the observer can infer nothing about the skill of the archer. In that case, the pattern was fabricated to match the event; it did not exist independently of the event.

7. Of course, intelligent agents can also produce repetitive patterns. Nevertheless, so can natural causes and processes. For example, electrostatic forces of attraction between ions form highly repetitive crystal structures. Wave action on the beach produces patterns of interlocking arcs of sand. Indeed, what scientists call laws of nature typically describe natural processes that produce—of necessity—highly regular and repetitive outcomes. For this reason, repetitive patterns do not necessarily indicate the activity of intelligent causes, though they may and commonly do indicate undirected lawlike processes. Knowing this, the croupier rightly suspected a physical cause for the regularity he observed in the outcomes at the roulette wheel. When he found a physical cause that explained the regularity that he observed—the ball falling in the same hole every time—he rejected chance as an explanation and elected physical necessity as, at least, part of the explanation. A smart croupier would realize that design might still have played a role in the observed regularity, for example, in the case that someone had intentionally scarred the wheel to produce the defect responsible for the ball landing in the same hole every time. This illustration suggests that explanations involving lawlike necessity and design do not necessarily mutually exclude each other, even though the presence of a regularity by itself does not point decisively or exclusively to an intelligent cause, but instead typically points to lawlike necessity.

8. Dembski, *The Design Inference,* 136–74.

9. Dembski, *The Design Inference,* 1–35, 136–74.

10. Axe, "Estimating the Prevalence of Protein Sequences." See also calculations after section head "Symposium at the Wistar Institute" in Chapter 9.

11. Gamma, et al., *Design Patterns.*

12. Yockey, "Origin of Life on Earth," esp. 105.

13. Crick, "On Protein Synthesis," 144, 153.

Chapter 17: But Does It Explain?

1. In this regard, see, e.g., Elsberry and Wilkins, "The Advantages of Theft over Honest Toil"; Pennock, *Tower of Babel,* 171–72; Wilson, "Intelligent Evolution"; Scott and Branch, "'Intelligent Design' Not Accepted by Most Scientists."

2. Quastler, *The Emergence of Biological Organization,* 16.

3. Lipton, *Inference to the Best Explanation,* 32–88.

4. Lipton, *Inference to the Best Explanation*, 32–88; Meyer, "The Scientific Status of Intelligent Design"; "The Demarcation of Science and Religion"; "Of Clues and Causes," 77–140; Sober, *The Philosophy of Biology*.

5. Shermer, Michael, "ID Works in Mysterious Ways."

6. See, e.g., Miller "In Defense of Evolution"; Jones, Decision in *Kitzmiller et al. v. Dover Area School Board*, No.04cv2688; Ruse, "Keep Intelligent Design Out of Science Classes."

7. Rothman, *The Science Gap*, 65–92; Popper, *Conjectures and Refutations*, 35–37. See also Oparin's discussion of the impossibility of spontaneous generation. Oparin, *Origin of Life* (1938), 27–28.

8. Meyer, "Of Clues and Causes," 77–140; Sober, *Reconstructing the Past*, 4–5; De Duve, "The Beginnings of Life on Earth," 249–50.

9. Harre and Madden, *Causal Powers*.

10. Barrow and Tipler, *The Anthropic Cosmological Principle*, 69.

11. See, e.g., G. S. Hurd, "The Explanatory Filter, Archaeology, and Forensics"; Pennock, "The Wizards of ID."

12. Arguably, the unnaturally modulated electromagnetic signals that SETI scientists are looking for represent an improbable pattern, but not necessarily evidence of digitally encoded, functionally specified information. Some have noted this difference in the criterion that SETI uses to detect intelligence in order to discredit ID proponents who have cited SETI to legitimate design reasoning in biology. But, if anything, the SETI standard for detecting intelligence constitutes a less demanding threshold for detecting intelligence than that used by ID advocates. Whereas SETI requires only evidence of a channel of communication (i.e., an unnaturally modulated signal), I have argued for design based upon the presence of functionally specified digital code within the communication channel represented by DNA and the gene-expression system. If the inference to design is legitimate based upon the former criterion, it is a fortiori certainly legitimate based upon the latter. Thus, SETI validates, rather than undermines, the reasoning used to infer intelligent design as the best explanation for the origin of biological information. See Sagan, *Contact*; McDonough, *The Search for Extraterrestrial Intelligence*.

13. See Hume, *Dialogues Concerning Natural Religion* (1980), 61–66.

14. See, e.g., Forrest and Gross, *Creationism's Trojan Horse*, 120ff.; Pigliucci, "Intelligent Design—The Classical Argument."

15. Kay, "Cybernetics, Information, Life"; *Who Wrote the Book of Life?*

16. Kay, *Who Wrote the Book of Life?*

17. Kay, *Who Wrote the Book of Life?* 611–12, 629.

18. Sarkar, "Biological Information," 199–202.

19. Schrödinger, *What Is Life?* 82.

20. Alberts, et al., *Molecular Biology of the Cell*, 21.

21. Watson and Crick, "A Structure for Deoxyribose Nucleic Acid"; "Genetical Implications"; Crick, "On Protein Synthesis."

22. Judson, *The Eighth Day of Creation*, 611.

23. Orgel, *The Origins of Life*, 189.

24. Davies, *The Fifth Miracle*, 120.

25. Orgel, *The Origins of Life*, 189.

26. Dawkins, *The God Delusion*, 147, 185–86, 188.

27. Dawkins, *The God Delusion*, 147, 185–186, 188. See also Gleiser, "Who Designed the Designer?"; Rosenhouse, "Who Designed the Designer?"

28. Moreover, explanations of particular events not only explain particular events; they also provide information about conditions in the past, in particular, the conditions that gave rise to the event in question. That is also part of what causal explanations do.

29. A second problem with the objection is that the theory of intelligent design makes no claims about the nature of the designer. Moreover, some philosophers question

whether agents qua agents must necessarily be complex. For the sake of argument, however, I grant the assumption that an agent sophisticated enough to design the first life is both complex and highly specified.

30. Tommyrot, "The Dawkins Delusion."

31. Nor has evolutionary theory, or any other theory, given a causal explanation of the origin of consciousness, the origin of mind itself. Contemporary Darwinism may provide an explanation for the origin of the physical features of higher primates including ourselves, but it does not give anything like an adequate account of how consciousness first arose in *Homo sapiens*. Indeed, even explaining what generates consciousness in individual human agents now has proven utterly intractable from a materialistic point of view. How does the flow of electricity in our neurons or the reaction of chemicals in the brain produce my subjective experience of "red" or my comprehension of mathematical truth or my awareness of my own thoughts? How does chemistry produce consciousness? Leading brain physiologists and philosophers of mind have long acknowledged that they do not know the answer to this question. Many doubt there is an answer to it. See, e.g., Chalmers, *The Conscious Mind;* Eccles, *Facing Reality;* Koestler, *The Ghost in the Machine;* Penfield, *The Mystery of the Mind,* xiii; Popper and Eccles, *The Self and Its Brain,* viii; Searle, *Mind;* Sherrington, *Man on His Nature;* Smythies, "Some Aspects of Consciousness," 235.

32. Additionally, ID advocates do not necessarily conceive of minds as complex (and functionally specified) collections of parts (or even symbols). Some would argue that complexity (whether specified complexity or irreducible complexity) is a feature that characterizes material effects, not self-conscious minds. Indeed, a long tradition in Western philosophy conceives of mind as an immaterial entity consisting of no material parts, in which case it cannot be characterized as complex at all. This is, at least, a logically coherent conception of mind. Thus, positing such a mind as the cause of the origin of biological information neither entails a contradiction nor commits the ID advocate to any inconsistency in reasoning. The rule "specified complexity always originates from intelligent design" is not violated if advocates of intelligent design posit a designing mind that is not complex (or specified) in the way that material artifacts or communication systems are.

33. This claim of inconsistency subtly misrepresents the basis of the argument to intelligent design. Uniform and repeated experience does not show that specified complexity always points to an intelligent cause. Instead, uniform and repeated experience shows that whenever specified information arises—*originates*—a mind always plays a causal role. Thus, we typically infer that specified complexity points to an antecedent intelligence because of this principle. In those many cases where we have good reason to think that specified complexity not only exists, but also first originated in time, we always infer an antecedent intelligence. Since an uncaused mind did not originate in time, such a mind would not point to a prior designing intelligence, even conceding for the sake of argument that minds are complex (and specified; see n. 32).

Experience says nothing about whether there might be a self-existent uncaused mind capable of initiating a novel chain of cause and effect. Even so, that is a logically possible hypothesis for the origin of biological information. Nor would the existence of such an uncaused mind violate what we know from experience about information-rich systems always *originating* from intelligence. Since we have good reason to think that life and biological information originated at a time in the finite past, we have good reason to think that an intelligent cause played a role in that event. Nevertheless, inferring an intelligent designer for the origin of that information does not entail an infinite regress of other designing intelligences, since the designing intelligence in question may not itself have originated, that is, begun to exist. The designing mind may have existed eternally, as theists think the mind of

God has done. Dawkins may not like this possibility, but an uncaused self-existent designer is, at least, a logically possible candidate cause. Thus, inferring intelligent design from the information that *originates* with the first life does not entail an infinite regress of similar designing minds.

Chapter 18: But Is It Science?

1. Jones, *Kitzmiller et al. v. Dover Area School District*, 400 F.Supp.2d 707, 720–21, 735 (M. D. Pa. 2005).
2. Jones, *Kitzmiller et al. v. Dover Area School District*, 400 F.Supp.2d 707, 764 (M. D. Pa. 2005).
3. Prominent news reports include Wallis, "The Evolution Wars"; Ratliff, "The Crusade Against Evolution"; Wilgoren, "Politicized Scholars Put Evolution on the Defensive."
4. Judge Jones falsely claimed that: (1) the theory of intelligent design affirms a "supernatural creation"—a position that ID proponents who testified in court denied during the trial; (2) proponents of intelligent design make their case solely by arguing against Darwinian evolution; (3) no peer-reviewed scientific publications in support of the theory of intelligent design have been published in the scientific literature; and (4) the theory of intelligent design has been refuted. See DeWolf, et al., *Traipsing into Evolution;* DeWolf, West, and Luskin, "Intelligent Design Will Survive *Kitzmiller v. Dover*"; Luskin, "Will Americans United Retract Their Demonstrably False Claims?".
5. See Worden, "Bad Frog Beer to 'Intelligent Design.'"
6. See West and DeWolf, "A Comparison of Judge Jones's Opinion in *Kitzmiller v. Dover* with Plaintiffs' Proposed 'Findings of Fact and Conclusions of Law.'"
7. The stories of many such scientists are told in the documentary film *Expelled: No Intelligence Allowed*. See also Wells, *The Politically Incorrect Guide to Darwinism and Intelligent Design.*
8. There are numerous examples of this claim. Some include: "The claim that life is the result of a design created by an intelligent cause cannot be tested and is not within the realm of science" (Skoog, "A View from the Past," 1–2); "ID has never produced an empirically testable hypothesis" (Forrest and Gross, *Creationism's Trojan Horse*, 235); "The hypothesis of design is compatible with any conceivable data, makes no testable predictions, and suggests no new avenues for research" (Miller, *Only a Theory*, 87).
9. "There is no demarcation line between science and nonscience, or between science and pseudo-science, which would win assent from a majority of philosophers" (Laudan, *Beyond Positivism and Relativism*, 210).
10. Laudan, "The Demise of the Demarcation Problem."
11. Gould, "Evolution and the Triumph of Homology, or Why History Matters," 60–69.
12. For example, Pennock testified, "Intelligent design needs to have for it to be a science a way of offering a specific hypothesis that one could then test in an ordinary way. They failed to do that, and so they really don't get off the ground with regard to science" (*Kitzmiller v. Dover* testimony, September 28, 2005, 39).
13. Behe, *Darwin's Black Box.*
14. Meyer, et al., "The Cambrian Explosion."
15. See, generally, Craig, "God, Creation and Mr. Davies," 163; "Barrow and Tipler on the Anthropic Principle vs. Divine Design," 389.
16. Gonzalez and Richards, *The Privileged Planet.*
17. See Nelson and Wells, "Homology in Biology."
18. Dembski, *The Design Inference*, 36.

19. Dembski, *The Design Inference*, 36–66.
20. As noted in Chapters 7 and 15, considerations of causal existence also play a role in the evaluation—and testing—of historical scientific theories. Indeed, historical scientific theories can fail by being unable to meet this critical test as well.
21. Shermer, *Why Darwin Matters*, 75.
22. Miller, *Only a Theory*, 37, 96–97.
23. See also Kitcher, *Living with Darwin*, 57.
24. Some advocates of intelligent design think that an intelligent cause is directly responsible for only the information present in the first living organisms; other ID advocates think intelligent design is responsible for the information necessary to produce subsequent forms of life as well. Those who hold the latter view predict that the nonprotein-coding DNA in both eukaryotes and prokaryotes should perform functional roles. Those who hold the former view predict that only the noncoding DNA in prokaryotes should perform functional roles. The discovery that noncoding DNA plays an important functional role in both prokaryotic and eukaryotic organisms confirms the prediction of the more expansive ID hypothesis. I hold this latter view. See Meyer, "The Origin of Biological Information." In this book, however, I have argued only for intelligent design as the best explanation of the origin of the information necessary to build the first living cell.
25. Dembski, "Intelligent Science and Design." Here's what Dembski writes about junk DNA: "[Intelligent] design is not a science stopper. Indeed, design can foster inquiry where traditional evolutionary approaches obstruct it. Consider the term 'junk DNA.' Implicit in this term is the view that because the genome of an organism has been cobbled together through a long, undirected evolutionary process, the genome is a patchwork of which only limited portions are essential to the organism. Thus on an evolutionary view we expect a lot of useless DNA. If, on the other hand, organisms are designed, we expect DNA, as much as possible, to exhibit function."
26. ENCODE Project Consortium, "Identification and Analysis of Functional Elements."
27. Von Sternberg and Shapiro, "How Repeated Retroelements Format Genome Function."
28. Han, Szak, and Boeke, "Transcriptional Disruption by the L1 Retrotransposon"; Bethany Janowski, et al., "Inhibiting Gene Expression at Transcription Start Sites"; Goodrich and Kugel, "Non-coding-RNA Regulators of RNA Polymerase II Transcription"; Li, et al., "Small dsRNAs Induce Transcriptional Activation in Human Cells"; Pagano, et al., "New Small Nuclear RNA Gene-like Transcriptional Units"; Van de Lagemaat, et al., "Transposable Elements in Mammals"; Donnelly, Hawkins, and Moss, "A Conserved Nuclear Element"; Dunn, Medstrand, and Mager, "An Endogenous Retroviral Long Terminal Repeat"; Burgess-Beusse, et al., "The Insulation of Genes"; Medstrand, Landry, and Mager, "Long Terminal Repeats Are Used as Alternative Promoters"; Mariño-Ramírez, et al., "Transposable Elements Donate Lineage-Specific Regulatory Sequences to Host Genomes."
29. Green, "The Role of Translocation and Selection"; Figueiredo, et al., "A Central Role for *Plasmodium Falciparum* Subtelomeric Regions."
30. Henikoff, Ahmad, and Malik, "The Centromere Paradox"; Bell, West, and Felsenfeld, "Insulators and Boundaries"; Pardue and DeBaryshe, "Drosophila Telomeres"; Henikoff, "Heterochromatin Function in Complex Genomes"; Figueiredo, et al., "A Central Role for *Plasmodium Falciparum*"; Schueler, et al., "Genomic and Genetic Definition of a Functional Human Centromere."
31. Jordan, et al., "Origin of a Substantial Fraction"; Henikoff, Ahmad, and Malik, "The Centromere Paradox"; Schueler, et al., "Genomic and Genetic Definition of a Functional Human Centromere."

32. Chen, DeCerbo, and Carmichael, "Alu Element-Mediated Gene Silencing"; Jurka, "Evolutionary Impact of Human Alu Repetitive Elements."; Lev-Maor, et al., "The Birth of an Alternatively Spliced Exon"; Kondo-Iida, et al., "Novel Mutations and Genotype–Phenotype Relationships"; Mattick and Makunin, "Non-coding RNA."

33. McKenzie and Brennan, "The Two Small Introns of the Drosophila Affinidisjuncta Adh Gene"; Arnaud, et al., "SINE Retroposons Can Be Used In Vivo"; Rubin, Kimura, and Schmid, "Selective Stimulation of Translational Expression"; Bartel, "MicroRNAs"; Mattick and Makunin, "Small Regulatory RNAs in Mammals."

34. Dunlap, et al., "Endogenous Retroviruses"; Hyslop, et al., "Downregulation of NANOG Induces Differentiation"; Peaston, et al., "Retrotransposons Regulate Host Genes."

35. Morrish, et al., "DNA Repair Mediated"; Tremblay, Jasin, and Chartrand, "A Double-Strand Break in a Chromosomal LINE Element"; Grawunder, et al., "Activity of DNA Ligase IV"; Wilson, Grawunder, and Liebe, "Yeast DNA Ligase IV."

36. Mura, et al., "Late Viral Interference Induced"; Kandouz, et al., "Connexin43 Pseudogene Is Expressed."

37. Goh, et al., "A Newly Discovered Human Alpha Globin Gene"; Kandouz, et al., "Connexin43 Pseudogene Is Expressed"; Tam, et al., "Pseudogene-Derived Small Interfering RNAs"; Watanabe, et al., "Endogenous siRNAs from Naturally Formed dsRNAs"; Piehler, et al., "The Human ABC Transporter Pseudogene Family."

38. Mattick and Gagen, "The Evolution of Controlled Multitasked Gene Networks"; Von Sternberg and Shapiro, "How Repeated Retroelements Format Genome Function."

39. In 1994, pro-ID scientist and writer Forrest M. Mims III submitted a letter to the journal *Science* (which was rejected) predicting function for junk DNA: "DNA that molecular biologists refer to as 'junk' don't necessarily appear so useless to those of us who have designed and written code for digital controllers. They have always reminded me of strings of NOP (No OPeration) instructions. A do-nothing string of NOPs might appear as 'junk code' to the uninitiated, but, when inserted in a program loop, a string of NOPs can be used to achieve a precise time delay. Perhaps the 'junk DNA' puzzle would be solved more rapidly if a few more computer scientists would make the switch to molecular biology" ("Rejected Publications"). See also Dembski's prediction already cited.

    In 2004, Jonathan Wells argued that the theory of intelligent design provides a fruitful heuristic (guide to discovery) for genomic research precisely because it predicts that noncoding DNA should have latent function. As he explained: "The fact that 'junk DNA' is not junk has emerged not because of evolutionary theory but in spite of it. On the other hand, people asking research questions in an ID framework would presumably have been looking for the functions of non-coding regions of DNA all along, and we might now know considerably more about them" ("Using Intelligent Design Theory to Guide Scientific Research").

    Other scientists have noted how materialistic evolutionary theories have impeded scientific progress in the study of the genome. In 2002, Richard von Sternberg reported extensive evidence for functional junk-DNA, noting that "neo-Darwinian 'narratives' have been the primary obstacle to elucidating the effects of these enigmatic components of chromosomes" and concluding that "the selfish DNA narrative and allied frameworks must join the other 'icons' of neo-Darwinian evolutionary theory that, despite their variance with empirical evidence, nevertheless persist in the literature" ("On the Roles of Repetitive DNA Elements").

40. Though historical scientists focus *primarily* on questions about the past, they clearly also have a secondary interest in questions about the present operation and cause-and-effect structure of the world. Indeed, the uniformitarian method requires that

historical scientists use knowledge of the present cause-and-effect structure of the world to reconstruct what happened in the past. Nevertheless, that the historical sciences address questions about the past at all distinguishes them from many sciences that focus wholly on questions about how nature *generally* operates.

41. Gould, "Evolution and the Triumph of Homology," 61.

42. Meyer, "DNA and the Origin of Life"; Meyer, et al., "The Cambrian Explosion"; Behe, *Darwin's Black Box;* Gonzalez and Richards, *The Privileged Planet;* Craig, "Barrow and Tipler on the Anthropic Principle vs. Divine Design," 389.

43. This is not to deny that laws or process theories may play roles in support of causal explanation, as even opponents of the covering-law model, such as Scriven, admit. Scriven notes that laws and other types of general-process theories may play an important role in justifying the causal status of an explanatory antecedent and may provide the means of inferring plausible causal antecedents from observed consequents. Nevertheless, as both Scriven and I have argued elsewhere, laws are not necessary to the explanation of particular events or facts; and even when laws are present, antecedent events function as the primary causal or explanatory entity in historical explanations. See Scriven, "Truisms as the Grounds," 448–50; "Explanation and Prediction," 480; "Causes, Connections and Conditions," 249–50; Meyer, "Of Clues and Causes," 18–24, 36–72, 84–92.

44. Dawkins, *The Blind Watchmaker,* 1.

45. Ayala, "Darwin's Revolution."

46. Nagel, "Public Education and Intelligent Design."

47. For example, Judge Jones asserted in his decision: "We find that ID is not science and cannot be adjudged a valid, accepted scientific theory as it has failed to publish in peer-reviewed journals" (*Kitzmiller et al. v. Dover Area School District,* 400 F.Supp.2d).

48. Forrest, quoted in Vergano and Toppo, "'Call to Arms' on Evolution."

49. Forrest, Expert Witness Report.

50. For example, Judge Jones asserted: (1) "We find that ID is not science and cannot be adjudged a valid, accepted scientific theory as it has failed to publish in peer-reviewed journals, engage in research and testing, and gain acceptance in the scientific community"; (2) "A final indicator of how ID has failed to demonstrate scientific warrant is the complete absence of peer-reviewed publications supporting the theory"; and (3) "The evidence presented in this case demonstrates that ID is not supported by any peer-reviewed research, data or publications" (*Kitzmiller et al. v. Dover Area School District,* 400 F.Supp.2d).

51. Holden, "Random Samples"; Giles, "Peer-Reviewed Paper Defends Theory of Intelligent Design"; Klinghoffer, "The Branding of a Heretic"; Powell, "Editor Explains Reasons for 'Intelligent Design' Article."

52. See Discovery Institute, Brief of Amicus Curiae (Revised).

53. E.g., Behe, *Darwin's Black Box;* Gonzalez and Richards, *The Privileged Planet;* Thaxton, Bradley, and Olsen, *The Mystery of Life's Origin.*

54. E.g., Dembski, *The Design Inference;* Campbell and Meyer, eds., *Darwinism, Design and Public Education;* Dembski and Ruse, eds., *Debating Design.*

55. See the five articles advancing the case for the theory of intelligent design published in Campbell and Meyer, eds., *Darwinism, Design and Public Education,* and the four articles published in Dembski and Ruse, eds., *Debating Design.*

56. Minnich and Meyer, "Genetic Analysis of Coordinate Flagellar and Type III Regulatory Circuits"; Dembski, ed., *Mere Creation.*

57. Craig, "God, Creation and Mr. Davies," 163; "Barrow and Tipler on the Anthropic Principle vs. Divine Design"; Behe, "Self-Organization and Irreducibly Complex Systems."

58. Wells, "Do Centrioles Generate a Polar Ejection Force?"; Behe and Snoke, "Simulating Evolution by Gene Duplication"; Dembski and Marks, "The Conservation of Information: Measuring the Information Cost of a Successful Search"; Voie, "Biological Function and the Genetic Code Are Interdependent"; Davison, "A Prescribed Evolutionary Hypothesis." See also Lönnig and Saedler, "Chromosomal Rearrangements and Transposable Elements."

59. In 2005 my colleagues and I at the Discovery Institute actually opposed the policy of the Dover school board and urged the school board to withdraw it. Though we think there is nothing unconstitutional about teaching students about the theory of intelligent design, we feared that politicizing the issue would result in reprisals against ID proponents in university science departments.

    We also objected to the way some members of the Dover board attempted to justify their policy by invoking religious authority as a reason to vote for the policy. This justification guaranteed the policy would run afoul the establishment clause in the courts. From our point of view, this justification was entirely gratuitous—and incongruous—since the theory of intelligent design is not based upon a religious authority, but upon empirical evidence and standard scientific methods of reasoning. Unfortunately, the school board did not heed our advice. It lost the case, just as we predicted, ultimately causing trouble for ID proponents at universities around the country, just as we, alas, also predicted.

60. "The Council [of the Biological Society of Washington, which oversees the publication of the *Proceedings of the Biological Society of Washington*] endorses a resolution on ID published by the American Association for the Advancement of Science (www.aaas.org/news/releases/2002/1106id2.shtml), which observes that there is no credible scientific evidence supporting ID as a testable hypothesis to explain the origin of organic diversity. Accordingly, the Meyer paper does not meet the scientific standards of the Proceedings." See http://www.ncseweb.org/resources/news/2004/US/294_bsw_strengthens_statement_repu_10_4_2004.asp.

61. These charges were clearly misplaced. The president of the council that oversaw the publication of the *Proceedings* admitted to the editor in an e-mail: "I have seen the review file and comments from 3 reviewers on the Meyer paper. All three with some differences among the comments recommended or suggested publication. I was surprised but concluded that there was not inappropriate behavior vs a vis [*sic*] the review process" (Roy McDiarmid, "Re: Request for information," January 28, 2005, 2:25 p.m., to Hans Sues, Congressional Staff Report, "Intolerance and the Politicization of Science at the Smithsonian" [December 2006]: 26, at http://www.discovery.org/scripts/viewDB/filesDB-download.php?command=download&id=1489). See also Rick Sternberg, "Statement of Facts / Response to Misinformation," at http://www.richardsternberg.org/smithsonian.php?page=statement.

62. As Stephen Jay Gould wrote with other scientists and historians of science in a brief to the U.S. Supreme Court in 1993: "Judgments based on scientific evidence, whether made in a laboratory or a courtroom, are undermined by a categorical refusal even to consider research or views that contradict someone's notion of the prevailing 'consensus' of scientific opinion. . . . Automatically rejecting dissenting views that challenge the conventional wisdom is a dangerous fallacy, for almost every generally accepted view was once deemed eccentric or heretical. Perpetuating the reign of a supposed scientific orthodoxy in this way, whether in a research laboratory or in a courtroom, is profoundly inimical to the search for truth. . . . The quality of a scientific approach or opinion depends on the strength of its factual premises and on the depth and consistency of its reasoning, not on its appearance in a particular journal or on its popularity among other scientists." Brief Amici Curiae of Physicians, Scientists, and Historians of Science in Support of Petitioners, *Daubert v. Merrell Dow Pharmaceuticals, Inc.*, 509 U.S. 579 (1993).

63. For a comprehensive annotated bibliography of ID publications, see http://
discovery.org/a/2640. Some critics of design have attempted to discredit the theory
because its advocates have published their case primarily in books rather than in
scientific articles. But as I noted in Chapter 6 this argument ignores the important
role that books have played in the history of science in establishing new scientific
ideas. Anyone who understands the role that technical journals play in science will
understand why this is so. Science journals are a highly specialized and conservative
genre. They publish research designed to fill out an established scientific research
program. They are part of what philosopher of science Thomas Kuhn calls "normal
science" (*The Structure of Scientific Revolutions*). New and revolutionary ideas in
science are unlikely to appear first in their pages. If the history of science is any
indication, then we should expect most of the initial work in any fundamentally
new scientific perspective to appear first in books. And this is precisely the pattern
of publication we see in the case of intelligent design. In the last decade or so, new
evidence-based arguments for the theory have made their initial appearance in books.
More recently, scientific articles have begun to appear, elucidating the theory in
more detail.

Chapter 19: Sauce for the Goose

1. *McLean v. Arkansas Board of Education,* 529 F.Supp. 1255 (E.D. Ark. 1982).
2. For the purposes of litigation the court defined creation science as a belief system
that affirms: "(1) [The] sudden creation of the universe, energy, and life from
nothing; (2) the insufficiency of mutation and natural selection in bringing about
development of all living kinds from a single organism; (3) changes only within fixed
limits of originally created kinds of plants and animals; (4) separate ancestry for man
and apes; (5) explanation of the earth's geology by catastrophism, including the
occurrence of a worldwide flood; and (6) a relatively recent inception of the earth
and living kinds" (*McLean v. Arkansas Board of Education,* 529 F.Supp. 1255, 1264
[E.D.Ark.1982]).
3. *McLean v. Arkansas Board of Education,* 529 F.Supp. 1255, 1267 (E.D.Ark. 1982).
4. Ruse, "Darwinism," 21.
5. Ruse, "Darwinism," 23.
6. Ruse, "A Philosopher's Day in Court"; "Witness Testimony Sheet," 301;
"Darwinism," 21–26.
7. Meyer, "Laws, Causes and Facts."
8. A law of nature typically describes a general relationship between two or more
different types of events, entities, or properties. Many have the form, "If A occurs,
then B will always follow under conditions C," such as the laws, "Pure water at
sea level heated to 100 degrees C will boil" and "All unsuspended bodies will
fall." Other laws describe mathematical relationships between different entities
or properties that apply universally, such as the law, "Force equals mass times
acceleration," or the law, "The pressure of a gas is proportional to its volume times
its temperature." In any case, laws are not events; they describe *relationships* (causal,
logical, or mathematical) between different types of events, entities, or properties.
9. Scriven, "Causation as Explanation," 14; Lipton, *Inference to the Best Explanation,*
47–81.
10. In the *On the Origin of Species,* Darwin proposed both a mechanism (natural
selection) and a historical theory—the theory of universal common descent.
Evolutionary biologists debate whether natural selection can be formulated as a
law. But within Darwin's argument, the theory of common descent had its own
explanatory power. Yet it did not explain by reference to a law of nature. Instead,
the theory of common descent explains by postulating a hypothetical pattern of

events (as depicted in Darwin's famous tree of life). The theory of common descent makes a claim about what happened in the past—namely, that a series of unobserved transitional organisms existed, forming a genealogical bridge between presently existing life-forms—to account for a variety of presently observed evidence (such as the similarity of anatomical structures in different organisms or the pattern of fossil progression). Darwin himself referred to common descent as the *vera causa* (i.e., the actual cause) of a diverse set of biological observations (*On the Origin of Species*, 195, 481–82). And in the theory of common descent, a pattern of events, not a law, does what I call the "primary explanatory work."

11. Ruse, *Darwinism Defended*, 59.
12. Grinnell, "Radical Intersubjectivity"; Scott, "Keep Science Free from Creationism."
13. Miller, *Only a Theory*, 87; Skoog, "View from the Past"; Sober, "What Is Wrong with Intelligent Design?"
14. Ruse, "Darwinism."
15. Miller, *Only a Theory*, 87.
16. *Kitzmiller v. Dover School District* 04 cv 2688 (December 20, 2005), 22, 77; Riordan, "Stringing Physics Along," 38.
17. *Kitzmiller v. Dover School District* 04 cv 2688 (December 20, 2005), 81; Jack Krebs, "A Summary of Objections to 'Intelligent Design,'" *Kansas Citizens for Science*, June 30, 2001 (http://www.sunflower.com/~jkrebs/JCCC/04%20Summary_Objections.html).
18. Michigan Science Teachers Association, "Evolution Education and the Nature of Science," February 3, 2007 (http://www.msta-mich.org/downloads/about/2007–02–03.doc).
19. See www.signatureinthecell.com and, specifically, "The Scientific Status of Intelligent Design: The Methodological Equivalence of Naturalistic and Non-Naturalistic Origins Theories" and "The Demarcation of Science and Religion." See also Philip Kitcher's new book, *Living with Darwin*, 9–14. Kitcher is a leading philosopher of science who rejects intelligent design, but thinks that the attempts to refute it using demarcation arguments fail. Kitcher argues that the problem with intelligent design is not that it isn't scientific, but that it is "discarded" or "dead science." He argues this largely because he thinks that intelligent design cannot explain the accumulation of "junk DNA." He does not, however, address the empirical argument for intelligent design based upon the information-bearing properties of DNA or the evidence showing that nonprotein-coding DNA plays a crucial functional role in the cell.
20. Meyer, "A Scopes Trial for the '90s"; "Open Debate on Life's Origin."
21. Scott, "Keep Science Free from Creationism."
22. Grinnell, "Radical Intersubjectivity."
23. Meyer, "Of Clues and Causes," 120; Darwin, *On the Origin of Species*, 398; Hull, *Darwin and His Critics*, 45.
24. For example, in 1999 the U.S. National Academy of Sciences (NAS) issued a statement against intelligent design that claimed that ID is "not science because [it is] not testable by the methods of science." Yet the NAS simultaneously stated, "Molecular data counter a recent proposition called 'intelligent design theory'" and asserted that "scientists have considered the hypotheses" of intelligent design and "rejected them because of a lack of evidence." National Academy of Sciences, *Science and Creationism*, 21, ix.

Similarly, Gerald Skoog argues that "the claim that life is the result of a design created by an intelligent cause cannot be tested and is not within the realm of science." Then in the next paragraph he states, "Observations of the natural world also make these dicta [concerning the theory of intelligent design] suspect" ("View from the Past"). Yet clearly something cannot be both untestable in principle and made suspect by empirical observations.

25. Pennock, Expert Witness Report. Emphasis added.
26. Scott, "Keep Science Free from Creationism."
27. Judson, *The Eighth Day of Creation,* 157–90.
28. Francis Darwin, ed., *Life and Letters,* 1:437, emphasis added.
29. Miller, *Only a Theory,* 87.
30. Meyer, "The Return of the God Hypothesis."
31. Miller, *Finding Darwin's God,* 238.
32. Similarly, neither type of theory describes events that will necessarily occur *repeatedly,* but both use uniform and *repeated* experience of cause and effect to make inferences about the most likely cause of various singular happenings.
33. Laudan, "The Demise of the Demarcation Problem"; "Science at the Bar—Causes for Concern"; Quinn, "The Philosopher of Science as Expert Witness."
34. For example, in the Arkansas trial Ruse had asserted that to qualify as scientific a theory must be falsifiable. He argued that young-earth creationism (creation science) wasn't scientific because it wasn't falsifiable. But other philosophers have pointed out that young-earth creationism made many testable claims and that these claims were nothing if not falsifiable. As one of Ruse's critics, Larry Laudan, points out, to claim that "Creationism is neither falsifiable nor testable is to assert that Creationism makes no empirical assertions whatever. That is surely false. Creationists make a wide range of testable assertions. . . . Creationists say that the earth is of very recent origin. . . . They argue that most of the geological features of the earth's surface are diluvial in character. . . . They assert the limited variability of species. They are committed to the view that, since animals and man were created at the same time, the human fossil record must be paleontologically coextensive with the record of lower animals." But as Laudan argues, "no one has shown how to reconcile such claims with the available evidence—evidence which speaks persuasively to a long earth history, among other things. Thus, these claims are testable, they have been tested, and they have failed those tests." It follows from Ruse's definition that creation science should qualify as scientific theory. Since it had been falsified, it was falsifiable. Since it was falsifiable, it qualified—paradoxically—as a scientific theory. (Laudan, "Science at the Bar—Causes for Concern.")

    Conversely, Laudan shows that many of Ruse's criteria would also, depending on how strictly they were applied, have the opposite effect of disqualifying mainstream evolutionary theories. Advancing his own criticism of Ruse's use of the "must explain by natural law" criterion, Laudan notes that scientists often make "existence claims" about past events or present processes without knowing the natural laws on which they depend. As he says, "Darwin took himself to have established the existence of [the mechanism of] natural selection almost a half century before geneticists were able to lay out the laws of heredity on which natural selection depended." Thus, Ruse's second demarcation criterion would require, if applied consistently, classifying classical Darwinism (as well as much of neo-Darwinism) as unscientific. As Laudan notes, if we took Ruse's second criterion seriously "we should have to say that . . . Darwin [was] unscientific; and, to take an example from our own time, it would follow that plate tectonics is unscientific because we have not yet identified the laws of physics and chemistry which account for the dynamics of crustal motion."
35. Laudan, "The Demise of the Demarcation Problem," 349; "Science at the Bar— Causes for Concern"; Quinn, "The Philosopher of Science as Expert Witness"; Kitcher, *Abusing Science,* 125–27. Kitcher allows for the possibility of a testable theory of divine creation. He argues that the presence of unobservable elements in theories, even ones involving an unobservable Divine Creator, do not mean that such theories cannot be evaluated empirically. He writes: "Even postulating an unobserved Creator need be no more unscientific than postulating unobserved particles. What matters is the character of the proposals and the ways in which they

are articulated and defended." Nevertheless, Kitcher argues creationism was tested and found wanting in the nineteenth century.

36. Speech at Annual Meeting of the American Association for the Advancement of Science.

37. Ruse, *Monad to Man*, 511–17.

38. Eger, quoted in Buell, "Broaden Science Curriculum," A21. Larry Laudan also categorically rejects the efficacy of demarcation arguments. As he summarizes, "If we could stand up on the side of reason, we ought to drop terms like 'pseudoscience' . . . They do only emotive work for us" ("The Demise of the Demarcation Problem").

39. Jones, *Kitzmiller et al. v. Dover Area School Board*. As Judge Jones explained, "This self-imposed convention of science, which limits inquiry to testable, natural explanations about the natural world, is referred to by philosophers as 'methodological naturalism.'"

40. Murphy, "Phillip Johnson on Trial," 33. Nancey Murphy is a philosopher and seminary professor who strongly affirms methodological naturalism. Here's what she says in full: "Science *qua* science seeks naturalistic explanations for all natural processes. Christians and atheists alike must pursue scientific questions in our era without invoking a Creator. . . . Anyone who attributes the characteristics of living things to creative intelligence has by definition stepped into the arena of either metaphysics or theology."

41. Some might object to my description of methodological naturalism as a principle that excludes all intelligent causes from science. They could point out, correctly, that some scholars construe the principle of methodological naturalism (MN) as forbidding, invoking only *supernatural* intelligent causes, not intelligent causes in general, within science. Nevertheless, nothing follows from this objection. Interpreting the principle of MN in this more limited way doesn't justify disqualifying intelligent design from consideration as a scientific theory. If methodological naturalism merely forbids reference to supernatural causes in science, then the theory of intelligent design should qualify as a scientific theory. Why? The theory itself claims to do no more than establish an intelligent cause, not a supernatural intelligent cause, as the best explanation for the origin of biological information.

Clearly, some advocates of the theory of intelligent design think (as I do) that the designing intelligence responsible for life is most likely to be a supernatural deity. But again, not much follows from that about the scientific status of the theory itself. Some advocates of the theory of intelligent design do indeed think that the theory has theistic *implications*. So what? Even if one concedes a definition of science that forbids reference to supernatural entities, all that follows is that the putative *implications* of intelligent design, not the claims of the theory itself, are unscientific.

Some scientists may even offer arguments in support of their contention that the designing intelligence responsible for life is most likely to be a supernatural deity. A definition of science that forbids reference to supernatural entities would, in that case, require classifying those arguments as unscientific. Nevertheless, classifying such arguments as unscientific or metaphysical or religious would not necessarily refute them.

42. Some scientists attempt to justify methodological naturalism by using another demarcation argument against intelligent design: ID does not cite a mechanism as the cause of the origin of biological form or information. But this demarcation argument assumes without justification that all scientifically acceptable causes are *mechanistic* causes. To insist that all causal explanations in science must be mechanistic is to insist that all causal theories must refer only to material entities. A "mechanism" is, after all, just another word for a material cause. Yet this requirement

is merely another expression of the principle of methodological naturalism, for which there is no noncircular justification.

As I have argued throughout this chapter, scientists have tried to justify a categorical exclusion of intelligent causes from science (i.e., methodological naturalism) by reference to specific demarcation criteria such as "testability," "observability," or "must explain by natural law." But, as I have shown, these arguments have failed to justify the exclusion of intelligent causes from science and, thus, the principle of methodological naturalism as a rule for science.

The failure of demarcation arguments against intelligent design (or in favor of methodological naturalism) has left scientists without any justification for treating methodological naturalism as a normative rule of scientific practice. Attempts to justify this convention simply restate it in another form: "Scientific theories must cite mechanisms." Thus, it provides no grounds for excluding consideration of intelligent causes in scientific theories, even if the intelligence in question is ultimately immaterial.

This demarcation argument against intelligent design clearly assumes the point at issue, which is whether there are independent and metaphysically neutral reasons for preferring exclusively materialistic causal explanations of origins over explanations that invoke entities such as creative intelligence, conscious agency, mental action, intelligent design or mind—entities that may not ultimately be reducible to matter alone. Since demarcation arguments have failed to provide such reasons, and since we know from first-person experience that our choices and actions as conscious intelligent agents do cause certain kinds of events, structures, and systems to arise—that minds have real causal powers—there does not seem to be any reason for prohibiting scientists from considering this type of cause as a possible explanation for certain kinds of effects. Instead, there is every reason to consider intelligent causes as explanations for effects that are known to arise only from intelligent activity.

43. Some have argued that the theory of intelligent design is unscientific because it doesn't cite a mechanism to explain how the designing intelligence responsible for life arranged the constituent parts of life. But this is also true in our own experience. We do not know how our minds influence the material substrate of our brains, the actions of our bodies, or through them the material world around us. Nevertheless, we have good reason to think *that* our conscious thoughts and decisions do influence the material world. Moreover, we can often know or infer *that* intelligent thought played a role in the arrangement of matter or outcome of events without knowing exactly *how* mind influences matter.

It's hard to see how this limitation in our understanding makes the theory of intelligent design unscientific. Many scientific theories do not explain events, evidence, or phenomena by reference to any cause whatsoever, let alone a mechanistic one. Newton's universal law of gravitation was no less a scientific theory because Newton failed—indeed, refused—to postulate a mechanism or cause for the regular pattern of attraction his law describes.

In addition, many historical theories about *what* happened in the past stand on their own without any mechanistic or causal theory about *how* the events to which such theories attest could have occurred. The theory of universal common descent is generally regarded as a scientific theory even though some scientists do not think there is currently an adequate mechanism to explain how transmutation between lines of descent was achieved. In the same way, there seems little justification for asserting that the geological theory of continental drift became scientific only after the advent of plate tectonics. Although the mechanism provided by plate tectonics certainly helped render continental drift a more persuasive theory, it was nevertheless not strictly necessary to know the mechanism by which continental drift *occurs* (1) to know or theorize that drift *had occurred* or (2) to regard a theory of continental drift as scientific.

In a similar way, advocates of design can affirm and detect *that* intelligence played a causal role in the design of life without knowing exactly *how* mind exerts its influence over matter. All that follows from this admission is that intelligent design does not provide an answer to every question, not that it is an unscientific (or unjustified) answer to the question that it does answer.

44. Jones, *Kitzmiller et al. v. Dover Area School Board.*
45. There are many types of scientific inquiry in which the convention of methodological naturalism does no damage to the intellectual freedom of scientists. Consider a scientist investigating the question, "How does atmospheric pressure affect crystal growth?" The answer, "Crystals were designed by a creative intelligence" (or, for that matter, "Crystals evolved via undirected natural processes") entirely misses the point of the question. The question demands an answer expressed as a general relationship describing the interaction of two physical entities: gases and crystals. A scientific law expressing such a relationship necessarily describes entirely materialistic entities.

Methodological naturalism does not limit the freedom of scientists to theorize in this case. Instead, the question motivating their inquiry ensures that scientists will only consider certain kinds of answers and that these answers will necessarily describe materialistic entities. In this case, methodological naturalism does no harm, though neither is it necessary to guide scientists to appropriate theories. Instead, the implicit question motivating the inquiry will do this on its own.

Methodological naturalism and its prohibition against inferring creative intelligence does inhibit inquiry in historical sciences such as archaeology, forensics, paleobiology, cosmology, anthropology, and origin-of-life studies, however, because in the historical sciences researchers are addressing different kinds of questions. They are asking about the causal histories of particular events, events in which the purposeful design of intelligent agents might have played a role. Scientists investigating the origin of life, for example, are motivated by the question, "What happened to cause life to arise on earth?" Since, conceivably, an intelligent designer could have played a causal role in the origin of life, any rule that prevents scientists from considering that possibility prevents scientists from considering a possibly true hypothesis.

46. Bridgman, *Reflections of a Physicist*, 535.

Chapter 20: Why It Matters

1. This quotation is from a court case in which the 3rd Circuit Court of Appeals defined religion for legal purposes. See *Africa v. Pennsylvania,* 662 F.2d 1025, 1035–36 (3d Cir. 1981), cert. denied, 456 U.S. 908 (1982). See also *Alvarado v. City of San Jose,* 94 F.3d 1223, 1227 (9th Cir. 1996). For a more extensive analysis of whether intelligent design qualifies as an establishment of religion, see DeWolf, Meyer, and DeForrest, "Teaching the Origins Controversy."
2. See, e.g., Center for Science and Culture, Discovery Institute, http://www.discovery. org/csc. See also the Biologic Institute at http://www.biologicinstitute.org.
3. Skoog, "A View from the Past"; Pennock, Expert Witness Report.
4. Michael Denton, an agnostic, argues for intelligent design in *Evolution: A Theory in Crisis,* 326–43. Antony Flew, a longtime champion of atheism, recently announced his abandonment of atheism based on the evidence of intelligent design, but emphasized that his religion was far from conventional (much less sectarian): "I'm thinking of a God very different from the God of the Christian and far and away from the God of Islam, because both are depicted as omnipotent Oriental despots, cosmic Saddam Husseins" ("Famous Atheist Now Believes in God," http:// abcnews.go.com/US/wireStory?id=315976). See also Flew, *There Is a God.*

5. Logically, a statement or proposition (call it A) *implies* another proposition (call it B), if B is more plausible or *more likely* to be true if A is true. Logically, a proposition (call it A) *entails* another proposition (call it B), if B is *necessarily* true if A is true. Implication refers to any kind of epistemic support short of deductive proof, whether that support is provided by weak forms of abductive inference, induction, confirmation of hypothesis, Bayesian inference, or a strong abductive inference to the best explanation.

6. Similarly, archaeology often yields data or artifacts that may prove supportive or subversive of particular religious traditions. Yet the implications of archaeological theories do not make archaeology a religion or its theories necessarily false (or true).

7. Hawking and Penrose, "The Singularities of Gravitational Collapse and Cosmology."

8. Eddington, "The End of the World."

9. Kenyon and Steinman, *Biochemical Predestination*, 6.

10. Dawkins, *The Blind Watchmaker*, 6.

11. Dawkins, *The Blind Watchmaker*, 6.

12. Simpson, *The Meaning of Evolution*, 344, emphasis added.

13. Futuyma, *Evolutionary Biology*, 5.

14. "The implications of modern evolutionary biology are inescapable. . . . Evolutionary biology undermines the fundamental assumptions underlying ethical systems in almost all cultures, Western civilization in particular" (Provine, "Evolution and the Foundation of Ethics").

15. Gould, *Ever Since Darwin*, 33.

16. Dawkins, *The Blind Watchmaker*, 6.

17. Dawkins, *The Blind Watchmaker*, 6.

18. Gould, *Ever Since Darwin*, 33, 147, 267.

19. Miller and Levine, *Biology*, 658. A later edition of the textbook deleted this language.

20. Purvis, Orians, and Heller, *Life*, 14.

21. Futuyma, *Evolutionary Biology*, 5.

22. "In many respects, evolution is the key to understanding our relationship with God" (Miller, *Finding Darwin's God*, 291).

23. Dawkins, *The Blind Watchmaker*, 6.

24. Miller, *Finding Darwin's God*, 291.

25. Russell, *Mysticism and Logic, Including a Free Man's Worship*, 10–11.

Epilogue: A Living Science

1. Kitcher, *Living with Darwin*, 11.

2. Newton, *Newton's Principia*.

3. Carninci, Yasuda, and Hayashizaki, "Multifaceted Mammalian Transcriptome"; Gerstein, Bruce, Rozowsky, et al. "What Is a Gene, Post-ENCODE?"

4. Amaral and Mattick, "Noncoding RNA in Development."

5. Squires and Berry, "Eukaryotic Selenoprotein Synthesis."

6. Pesole, "What Is a Gene?"; Gerstein, Bruce, Rozowsky, et al. "What Is a Gene, Post-ENCODE?"

7. Savarese and Grosschedl, "Blurring Cis and Trans in Gene Regulation."

8. Kapranov, Willingham, and Gingeras, "Genome-wide Transcription and the Implications for Genomic Organization."

9. Sowden, Ballatori, Jensen, Reed, and Smith, "The Editosome for Cytidine to Uridine mRNA Editing has a Native Complexity of 27S"; Sperling, Azubel, and Sperling, "Structure and Function of the Pre-mRNA Splicing Machine."

10. Fischer, Butler, Pan, and Ruvkun, "*Trans*-splicing in *C. elegans* Generates the Negative RNAi Regulator ERI–6/7."

11. Hudson and Goldstein, "The Gene Structure of the *Drosophila melanogaster* Proto-

Oncogene, Kayak, and Its Nested Gene, fos-Intronic Gene."

12. Huang, Zhou, He, Chen, Liang, and Qu, "Genome-wide Analyses of Two Families of snoRNA Genes from *Drosophila melanogaster*"; Saini, Enright, and Griffiths-Jones, "Annotation of Mammalian Primary microRNAs."

13. Chung, et al., "A First Look at the ARFome."

14. Ohno, "So Much 'Junk' DNA in Our Genome"; Dawkins, *The Selfish Gene*; Orgel, Crick, Sapienza, "Selfish DNA"; Orgel and Crick, "Selfish DNA."

15. Shapiro and von Sternberg, "Why Repetitive DNA Is Essential to Genome Function."

16. Denoeud, et al., "Prominent Use of Distal 5^Transcription Start Sites."

17. Michalak, "Coexpression, Coregulation, and Cofunctionality of Neighboring Genes in Eukaryotic Genomes."

18. Miles, et al., "Intergenic Transcription, Cell-Cycle and the Developmentally Regulated Epigenetic Profile of the Human Beta-Globin Locus."

19. Gierman, et al., "Domain-wide Regulation of Gene Expression in the Human Genome"; Goetze, et al., "The Three-Dimensional Structure of Human Interphase Chromosomes."

20. Miller, *Only a Theory,* 37, 56.

21. Kapranov, Willingham, and Gingeras, "Genome-wide Transcription and the Implications for Genomic Organization"; Wu, et al., "Systematic Analysis of Transcribed Loci in ENCODE Regions."

22. Costantini, et al., "Human Chromosomal Bands."

23. Carvalho, et al., "Chromosomal G-dark Bands Determine the Spatial Organization of Centromeric Heterochromatin in the Nucleus."

24. Lynch writes that he is striving for "a general theory of the evolution of the gene that incorporates the universal properties of random genetic drift and mutation pressure" and that "at this point it is difficult to reject the hypothesis that the basic embellishments of the eukaryotic gene originated largely as a consequence of nonadaptive processes operating contrary to the expected direction of natural selection" ("Origins of Eukaryotic Gene Structure," 464).

25. According to Gehring, " . . . the mouse gene [*Pax*-6] was also shown to be capable of inducing eyes [in *Drosophila*]" (*Master Control Genes in Development and Evolution,* 204).

26. Panganiban and Rubenstein, "Developmental Functions of the *Distal-less*/Dlx Homeobox Genes."

27. Newman, "The Developmental Genetic Toolkit and the Molecular Homology-Analogy Paradox."

28. Carroll, *Endless Forms Most Beautiful: The New Science of Evo Devo.* "The discovery that the same sets of genes control the formation and pattern of body regions and body parts with similar functions (but very different designs) in insects, vertebrates, and other animals has forced a complete rethinking of animal history, the origins of structures, and the nature of diversity. Comparative and evolutionary biologists had long assumed that different groups of animals, separated by vast amounts of evolutionary time, were constructed and had evolved by entirely different means . . . . So prevalent was this view of great evolutionary distance that in the 1960s the evolutionary biologist (and an architect of the Modern Synthesis) Ernst Mayr remarked: 'Much that has been learned about gene physiology makes it evident that the search for homologous genes is quite futile except in very close relatives . . . .' This view was entirely incorrect. The late Stephen Jay Gould, in his monumental work *The Structure of Evolutionary Theory,* saw the discovery of Hox clusters and common body-building genes as overturning a major view of the Modern Synthesis" (71–72). See also Ron Amundson, "Development and Evolution," in Sarkar and Plutynski, eds., *A*

Companion to the Philosophy of Biology, 248–68; esp. 264: "By the early 1990s molecular geneticists were beginning to identify genes on the basis of their molecular composition. The first shock was the discovery that certain genes were widely shared among nearly all animal groups, from mammals to insects to flatworms. The mere fact of widely shared genes was inconsistent with the expectations of major adaptationists. But the nature of those genes was even more surprising. They acted at the deepest and earliest stages of embryonic development . . . Adaptationists like Dobzhansky and Mayr had predicted just the opposite—the absence of any important homologous genes."

29. "Insect and vertebrate legs," note Erwin and Davidson, "are constructed entirely differently and it is difficult to imagine a morphogenetic process that would produce a version ancestral to both, i.e., beyond the initial patterning stage." Erwin and Davidson, "The Last Common Bilaterian Ancestor."

30. Erwin and Davidson, "The Last Common Bilaterian Ancestor."

31. Nelson and Wells, "Homology in Biology," esp. 317.

32. Goodwin, "What Are the Causes of Morphogenesis?"; Nijhout, "Metaphors and the Role of Genes in Development"; Sapp, *Beyond the Gene;* Müller and Newman, "Origination of Organismal Form."

33. Harold, "From Morphogenes to Morphogenesis"; Moss, *What Genes Can't Do.*

34. Harold, *The Way of the Cell,* 125.

35. Harold, *The Way of the Cell,* 125.

36. McNiven and Porter, "The Centrosome."

37. Lange, Faragher, March, and Gull, "Centriole Duplication and Maturation in Animal Cells"; Marshall and Rosenbaum, "Are There Nucleic Acids in the Centrosome?"

38. Lange, Faragher, March, and Gull, "Centriole Duplication and Maturation in Animal Cells."

39. Sonneborn, "Determination, Development, and Inheritance of the Structure of the Cell Cortex"; Frankel, "Propagation of Cortical Differences in *Tetrahymena*"; Nanney, "The Ciliates and the Cytoplasm."

40. Moss, *What Genes Can't Do.*

41. Harold, "From Morphogenes to Morphogenesis."

42. On his Michigan State University Web page, philosopher of biology Robert Pennock notes that his major 2003 *Nature* paper using the program *Avida* was intended to rebut Michael Behe's arguments about the origin of irreducibly complex systems. "Though not discussed explicitly in the paper," he writes, "the experiment refutes Behe's claim that evolution cannot produce 'irreducible complexity.'" This claim occurs under Pennock's Web-page heading "Rebutting Michael Behe and Irreducible Complexity." See https://www.msu.edu (last accessed February 9, 2009). See also Thornton, "Implications for Intelligent Design," and Minnich and Meyer, "Genetic Analysis of Coordinate Flagellar and Type III Regulatory Circuits in Pathogenetic Bacteria."

Appendix A: Some Predictions of Intelligent Design

1. Dembski and Marks, "The Conservation of Information"; "The Search for a Search."

2. Shapiro, "A 21st Century View of Evolution," 124.

3. Wells, "Do Centrioles Generate a Polar Ejection Force?"; "Using Intelligent Design Theory to Guide Scientific Research"; "A Possible Link Between Centrioles, Calcium Deficiency and Cancer."

4. Meyer, Ross, Nelson, and Chien, "The Cambrian Explosion."

5. Ayoub, "On the Design of the Vertebrate Retina"; Denton, "Selected Excerpts: The Inverted Retina."

6. It might be objected that a beneficent designer skilled enough to produce biological systems would not allow natural processes to degrade them. But this objection inevitably raises a theological question about what a theistic designer would or would not have done, one moreover addressed by traditional theological formulations. For more on the subject see Hunter, *Darwin's God;* Wiker and Witt, *A Meaningful World.*

7. Minnich and Rohde, "A Rationale for Repression and/or Loss of Motility by Pathogenic Yersinia in the Mammalian Host."

8. Chain, et al., "Insights into the Evolution of Yersinia pestis Through Whole-Genome Comparison with Yersinia pseudotuberculosis."

9. Miller, "The Flagellum Unspun."

10. As microbiologist Milton H. Saier Jr. has noted: "In considering the logistics of the conclusions made by Gophna et al. [10], it should be noted that Fla[gellar] systems have been able to diverge in structure so as to span either one membrane or two in the envelopes of Gram-positive and Gram-negative bacteria, respectively. So why haven't T3SSs? The most plausible explanation considers that T3SSs arose late from preexisting Gram-negative bacterial Fla[gellar] systems" ("Evolution of Bacterial Type III Protein Secretion Systems"). Additionally, as Saier observes, mutation-density studies have shown that the T3SS and flagellar motor genes have roughly equivalent ages. Saier regards these studies as less definitive than phylogenetic studies because of their assumption of a constant rate of evolutionary change, i.e., that mutation rates function as a consistent "molecular clock." But even if these studies yield accurate conclusions, they are consistent with the possibility of independent design or with an extremely rapid devolution of the T3SS from the flagellar motor, both of which are possibilities that would challenge the co-option hypothesis. This is the case, because it would presumably take longer to build a new system than to offload parts of an old one. Saier himself has recently changed his mind about his 2004 claim that the flagellar motor is older than the T3SS. He now favors the view that both the T3SS and the flagellar motor share a common ancestor. Nevertheless, he offers little empirical justification for this new view.

11. Axe, "Estimating the Prevalence of Protein Sequences Adopting Functional Enzyme Folds."

Appendix B: Multiverse Cosmology and the Origin of Life

1. Koonin, "The Cosmological Model."

2. Calculations predicated on the standard big-bang cosmology tell us that radiation arriving from opposite directions in the sky would have originally been separated, when the universe was 300,000 years old, by about 100 horizon distances. (A horizon distance is the distance light could have traveled since the beginning of the universe.) Thus, these calculations imply that there would have been no opportunity for the background radiation in far flung corners of the universe to have "thermalized" or come into a thermal equilibrium by mixing. Instead, the only way to explain the homogeneity in the temperature of the background radiation given the standard big-bang model is to posit that the initial conditions of the universe were extremely finely tuned.

3. In current models, inflation begins at around 10^{-37} seconds after the big bang and lasts until 10^{-35} seconds, during which space itself expands by a factor of 10^{60} or so. At the beginning of the inflationary epoch the observable universe was, say, about 10^{-60} meters in size and at the end of it about a meter across. At the start of inflation, however, the horizon distance (the distance light could have traveled since the big bang) was 10^{-37} light-seconds, which is far larger than the tiny patch that was destined to grow into our observable universe. According to the inflationary model,

some residual inhomogeneity in the background radiation might have remained even after the initial thermalization occurred. Nevertheless, if it did it would have existed only in the parts of the early universe that lay beyond the patch that would become our observable universe. Thus, the inflationary process not only distributed the homogeneous background radiation throughout the observable universe, it also would have distributed all remaining inhomogeneity beyond the edge of the observable universe as well.

4. See Garriga and Vilenkin, "Many Worlds in One"; Vilenkin, *Many Worlds in One.* The many-worlds-in-one model has the consequence that all macroscopic sequences of events not forbidden by physical conservation laws not only occur somewhere in an eternally inflating universe, but occur over and over again without limit as inflation endlessly spawns new expanding regions of space-time. For instance, the model suggests there are an unlimited number of macroscopically exact copies of the earth and everything that exists on it, even though the probability of any given observable region of the universe containing such a copy is vanishingly small.

5. Koonin, "The Cosmological Model of Eternal Inflation and the Transition from Chance to Biological Evolution in the History of Life."

6. Penrose, "Difficulties with Inflationary Cosmology," 249–64. Penrose, *The Road to Reality: A Complete Guide to the Laws of the Universe,* 746–57.

7. Hawking and Page, "How Probable Is Inflation?"

8. Collins, "The Fine-tuning Design Argument," esp. 61.

9. Penrose, "Difficulties with Inflationary Cosmology," 249–64. Penrose, *The Road to Reality: A Complete Guide to the Laws of the Universe,* 746–57, esp. 730, 755.

10. Dyson, Kleban, and Susskind, "Disturbing Implications of a Cosmological Constant."

Bibliography

Abelson, Philip H. "Chemical Events on the Primitive Earth." *Proceedings of the National Academy of Sciences USA* 55 (1966): 1365–72.

Adami, Christopher, and C. Titus Brown. "Evolutionary Learning in the 2D Artificial Life System 'Avida.'" In *Proceedings of "Artificial Life IV,"* edited by Rodney Brooks and Pattie Maes, 377–81. Cambridge, MA: MIT Press, 1994.

Adams, Melissa D., et al. "The Genome Sequence of *Drosophila Melanogaster.*" *Science* 287 (2000): 2185–95.

Agassiz, Louis. *Essay on Classification.* Edited by Edward Lurie. Cambridge, MA: Harvard University Press, Belknap Press, 1962.

Alberts, Bruce D., Dennis Bray, Julian Lewis, Martin Raff, Keith Roberts, and James D. Watson. *Molecular Biology of the Cell.* New York: Garland, 1983.

Allen, Timothy E., Nathan D. Price, Andrew R. Joyce, and Berhard O. Palsson. "Long-Range Periodic Patterns in Microbial Genomes Indicate Significant Multi-Scale Chromosomal Organization." *PLoS Computational Biology* 2 (2006): 0013–21.

Alston, William P. "The Place of the Explanation of Particular Facts in Science." *Philosophy of Science* 38 (1971): 13–34.

Amaral, P. P., and J. S. Mattick. "Noncoding RNA in Development." *Mammalian Genome* 19 (2008): 454–92.

Amundson, Ron. "Development and Evolution." In *A Companion to the Philosophy of Biology,* edited by S. Sarkar and A. Plutynski, 248–68, esp. 264. London: Blackwell, 2008.

Arnaud, Phillipe, Chantal Goubely, Thierry Pe'Lissier, and Jean-Marc Deragon. "SINE Retroposons Can Be Used In Vivo as Nucleation Centers for De Novo Methylation." *Molecular and Cellular Biology* 20 (2000): 3434–41.

Arthur, Wallace. *The Origin of Animal Body Plans.* Cambridge: Cambridge University Press, 1997.

Astbury, William T., and A. Street. "X-Ray Studies of the Structure of Hair, Wool and Related Fibres." *Philosophical Transactions of the Royal Society of London* 230 (1932): 75–101.

Avery, Oswald T., Colin M. MacCleod, and Maclyn McCarty. "Induction of Transformation by a Deoxyribonucleic Acid Fraction Isolated from Pneumococcus Type III." *Journal of Experimental Medicine* 79 (1944): 137–58.

Axe, Douglas. "Extreme Functional Sensitivity to Conservative Amino Acid Changes on Enzyme Exteriors." *Journal of Molecular Biology* 301 (2000): 585–95.

_____. "Estimating the Prevalence of Protein Sequences Adopting Functional Enzyme Folds." *Journal of Molecular Biology* 341 (2004): 1295–315.

Axe, Douglas D., Nicholas Foster, and Alan Fersht. "Active Barnase Variants with Completely Random Hydrophobic Cores." *Proceedings of the National Academy of Sciences USA* 93 (1996): 5590–94.

Ayala, Francisco. "Darwin's Revolution." In *Creative Evolution?!* edited by J. Campbell and J. Schopf, 4–5. Boston: Jones and Bartlett, 1994.

_____. "Design Without Designer: Darwin's Greatest Discovery," In *Debating Design: From Darwin to DNA,* edited by W. Dembski and M. Ruse, 55–80. Cambridge: Cambridge University Press, 2004.

_____. "Darwin's Greatest Discovery: Design Without Designer." *Proceedings of the National Academy of Sciences* 104 (2007): 8567–73.

Ayoub, George. "On the Design of the Vertebrate Retina." *Origins & Design* 17:1 (1996). http://www.arn.org/docs/odesign/od171/retina171.htm.

Barrow, John D. *Theories of Everything: The Quest for Ultimate Explanation.* New York: Fawcett, 1991.

Barrow, John D., and Frank J. Tipler. *The Anthropic Cosmological Principle.* Oxford: Oxford University Press, 1986.

Bartel, David. "MicroRNAs: Genomics, Biogenesis, Mechanism, and Function." *Cell* 116 (2004): 281–97.

Bartel, David P., and Jack W. Szostak. "Isolation of New Ribozymes from a Large Pool of Random Sequences." *Science* 261 (1993): 1411–18.

Bastian, H. Charlton. *The Evolution of Life.* London: Dutton, 1907.

_____. *Origin of Life.* London: Dutton, 1911.

Beadle, George, and Edward Tatum. "Genetic Control of Biochemical Reactions in *Neurospora.*" *Proceedings of the National Academy of Sciences USA* 27 (1941): 499–506.

Beeby, William. "The Future of Integrated CAD/CAM Systems: The Boeing Perspective." *IEEE Computer Graphics and Applications* 2 (1982): 51–56.

Behe, Michael. "Experimental Support for Regarding Functional Classes of Proteins to Be Highly Isolated from Each Other." In *Darwinism: Science or Philosophy?* edited by Jon Buell and Virginia Hearn, 60–71. Richardson, TX: Foundation for Thought and Ethics, 1994.

_____. *Darwin's Black Box: The Biochemical Challenge to Evolution.* New York: Free Press, 1996.

_____. "Self-Organization and Irreducibly Complex Systems: A Reply to Shanks and Joplin." *Philosophy of Science* 67 (2000): 155–62.

_____. "Irreducible Complexity: Obstacle to Darwinian Evolution." In *Debating Design: From Darwin to DNA,* edited by W. Dembski and M. Ruse, 352–70. Cambridge: Cambridge University Press, 2004.

Behe, Michael J., and D. W. Snoke. "Simulating Evolution by Gene Duplication of Protein Features That Require Multiple Amino Acid Residues." *Protein Science* 13 (2004): 2651–64.

Bell, C., A. G. West, and G. Felsenfeld. "Insulators and Boundaries: Versatile Regulatory Elements in the Eukaryotic Genome." *Science* 291 (2001): 447–50.

Bergmann, Max, and Carl Niemann. "Newer Aspects of Protein Chemistry." *Science* 86 (1937): 187–90.

Berkner, L. V., and L. C. Marshall. "On the Origin and Rise of Oxygen Concentration in the Earth's Atmosphere." *Journal of Atmospheric Science* 22 (1965): 225–61.

Berlinski, David. "The Deniable Darwin." *Commentary* 101.6 (1996): 19–29.

———. "On Assessing Genetic Algorithms." Public lecture. "Science and Evidence of Design in the Universe" Conference. Yale University, November 4, 2000.

Bernal, John D. *The Physical Basis of Life*. London: Routledge and Kegan Paul, 1951.

———. *The Origin of Life*. London: World Publishing Company, 1967.

Bertalanffy, Ludwig von. *Robots, Men and Minds*. New York: Braziller, 1967.

Biever, Celeste. "The God Lab." *New Scientist* (2006): 8–11.

Bishop, George F. "The Religious Worldview and American Beliefs About Human Origins." *Public Perspective: A Roper Center Review of Public Opinion and Polling* 9 (1998): 39–44.

Blanco, F. J., I. Angrand, and L. Serrano. "Exploring the Confirmational Properties of the Sequence Space Between Two Proteins with Different Folds: An Experimental Study." *Journal of Molecular Biology* 285 (1999): 741–53.

Blum, Harold F. *Time's Arrow and Evolution*. Princeton, NJ: Princeton University Press, 1951.

———. "Introductory Remarks." *Annals of the New York Academy of Science* 66 (1957): 257–59.

Borel, Emile. *Probabilities and Life*. Translated by M. Baudin. New York: Dover, 1962.

———. *Probability and Certainty*. Translated by D. Scott. New York: Walker, 1963.

Bowie, James, and Robert Sauer. "Identifying Determinants of Folding and Activity for a Protein of Unknown Structure." *Proceedings of the National Academy of Sciences USA* 86 (1989): 2152–56

Bowie, James, John Reidhaar-Olson, W. Lim, and Robert Sauer. "Deciphering the Message in Protein Sequences: Tolerance to Amino Acid Substitution." *Science* 247 (1990): 1306–10.

Bowring, S. A., J. P. Grotzinger, C. E. Isachsen, A. H. Knoll, S. M. Pelechaty, and P. Kolosov. "Calibrating Rates of Early Cambrian Evolution." *Science* 261 (1993): 1293–98.

———. *A Free Enquiry into the Vulgarly Received Notion of Nature*. Edited by Edward B. Davis and Michael Hunter. Cambridge Texts in the History of Philosophy. Cambridge: Cambridge University Press, 1996.

Bradley, Walter L. "Information, Entropy and the Origin of Life." In *Debating Design: From Darwin to DNA*, edited by W. Dembski and M. Ruse, 331–51. Cambridge: Cambridge University Press, 2004.

Brenner, Sydney, François Jacob, and Matthew Meselson. "An Unstable Intermediate for Carrying Information from Genes to Ribosomes for Protein Synthesis." *Nature* 190 (1961): 576–81.

Bridgewater, J. H., et al. "Microfossil-like Objects from the Archaean of Greenland: A Cautionary Note." *Nature* 289 (1981): 51–56.

Bridgman, Percy W. *Reflections of a Physicist*. 2nd ed. New York: Philosophical Library, 1955.

Brillouin, Leon. *Science and Information Theory*. New York: Courier Dover, 2004.

Brinkman, R. T. "Dissociation of Water Vapor and Evolution of Oxygen in the Terrestrial Atmosphere." *Journal of Geophysical Research* 74 (1969): 5354–68.

British Association for the Advancement of Science. "Meeting Report on the Origin of Life." *Nature* 90 (1912): 261–62.

———. "Address to Physiology Section of 16 October, 1913." *Nature* 92 (1913): 213–14.

———. "Presidential Address of 6 September, 1933." *Nature* 132 (1933): 381–94.

Brockman, John, ed. *The Third Culture: Beyond the Scientific Revolution*. New York: Simon & Schuster, 1995.

———. *Life: What a Concept!* 2007. http://www.edge.org/documents/life/Life.pdf (last accessed October 1, 2008).

Brodbeck, M. "Explanation, Prediction and Imperfect Knowledge." In *Minnesota Studies in the Philosophy of Science III*, edited by Herbert Feigl and Grover Maxwell, 252–72. Minneapolis: University of Minnesota Press, 1962.

Brooks, Daniel R., and E. O. Wiley. *Evolution as Entropy.* 2nd ed. Chicago: University of Chicago Press, 1988.

Brooks, James, and Gordon Shaw. *Origin and Development of Living Systems.* London: Academic, 1973.

Brooks, Jim. *Origins of Life.* Belleville, MI: Lion, 1985.

Brush, Steven G. "Prediction and Theory Evaluation: The Case of Light Bending." *Science* 246 (1989): 1124–29.

Buell Jon. "Broaden Science Curriculum." *Dallas Morning News,* March 10, 1989.

Bult, Carol, et al. "Complete Genome Sequence of the Methanogenic Archaeon *Methanococcus jannaschii.*" *Science* 273 (1996): 1058–73.

Bungenberg de Jong, H. G. "Die Koazeruation und ihre Bedeutung für die Biologie." *Protoplasma* 15 (1932): 110–73.

Burgess-Beusse, B., C. Farrell, M. Gaszner, M. Litt, V. Mutskov, F. Recillas-Targa, M. Simpson, A. West, and G. Felsenfeld. "The Insulation of Genes from External Enhancers and Silencing Chromatin." *Proceedings of the National Academy of Sciences USA* 99 (2002): 16433–37.

Butterfield, Herbert. *The Origins of Modern Science.* New York: Free Press, 1965.

Butts, Robert E. *William Whewell's Theory of Scientific Method.* Pittsburgh, PA: University of Pittsburgh Press, 1968.

———. "Consilience of Inductions and the Problem of Conceptual Change in Science." In *Logic, Laws and Life,* edited by R. Colodny, 71–88. Pittsburgh, PA: University of Pittsburgh Press, 1977.

Cairns-Smith, Alexander G. *The Life Puzzle.* Edinburgh: Oliver and Boyd, 1971.

———. *Genetic Takeover and the Mineral Origins of Life.* Cambridge: Cambridge University Press, 1982.

———. "The First Organisms." *Scientific American* 253 (1985): 90–100.

———. *Seven Clues to the Origin of Life.* Cambridge: Cambridge University Press, 1985.

Calvin, M. *Chemical Evolution.* New York: Oxford University Press, 1970.

Campbell, John Angus, and Stephen C. Meyer, eds. *Darwinism, Design and Public Education.* East Lansing: Michigan State University Press, 2003.

Capretti, Gian. "Peirce, Holmes, Popper." In *The Sign of Three,* edited by U. Eco and T. Sebeok, 135–53. Bloomington: Indiana University Press, 1983.

Carninci, Piero, Jun Yasuda, and Yoshihide Hayashizaki. "Multifaceted Mammalian Transcriptome." *Current Opinion in Cell Biology* 20 (2008): 274–80.

Carpentier, Anne-Sophie, Bruno Torrésani, Alex Grossmann, and Alain Hénaut. "Decoding the Nucleoid Organization of *Bacillus subtilis* and *Escherichia coli* Through Gene Expression Data." *BMC Genomics* 6 (2005):1–11. http://www.biomedcentral.com/1471-2164/6/84.

Carr, B. J., and M. J. Rees. "The Anthropic Principle and the Structure of the Physical World." *Nature* 278 (1979): 605–12.

Carroll, Sean. *Endless Forms Most Beautiful: The New Science of Evo Devo and the Making of the Animal Kingdom.* New York: Norton, 2005.

Carvalho, C. H., M. Pereira, J. Ferreira, C. Pina, D. Mendonça, A. C. Rosa, and M. Carmo-Fonseca. "Chromosomal G-dark Bands Determine the Spatial Organization of Centromeric Heterochromatin in the Nucleus." *Molecular Biology of the Cell* 12 (2001): 3563–72.

Carver, J. H. "Prebiotic Atmospheric Oxygen Levels." *Nature* 292 (1981): 136–38.

Cech, Thomas R. "Ribozyme Self-Replication?" *Nature* 339 (1989): 507–8.

Chain, P. S. G., E. Carniel, F. W. Larimer, J. Lamerdin, P. O. Stoutland, W. M. Regala, A. M. Georgescu, L. M. Vergez, M. L. Land, V. L. Motin, et al. "Insights into the Evolution of *Yersinia pestis* Through Whole-Genome Comparison with *Yersinia pseudotuberculosis.*" *Proceedings of the National Academy of Sciences USA* 101 (2004): 13826–31.

Chaitin, G. J. "On the Length of Programs for Computing Finite Binary Sequences." *Journal of the Association for Computing Machinery* 13 (1966): 547–69.

Chalmers, David J. *The Conscious Mind: In Search of a Fundamental Theory.* New York: Oxford University Press, 1996.

Chamberlain, Thomas C. "The Method of Multiple Working Hypotheses." *Science* (old series) 15 (1890): 92–96. Reprinted in *Science* 148 (1965): 754–59. Also reprinted in *Journal of Geology* (1931): 155–65.

Chargaff, Erwin. *Essays on Nucleic Acids.* Amsterdam, New York: Elsevier, 1963.

———. "Preface to a Grammar of Biology." *Science* 172 (1971): 637–42.

———. *Heraclitean Fire: Sketches from a Life Before Nature.* New York: Rockefeller University Press, 1978.

Chen, Ling-Ling, Joshua N. DeCerbo, and Gordon G. Carmichael. "Alu Element-Mediated Gene Silencing." *EMBO Journal* (2008): 1–12.

Chothia, Cyrus, Israel Gelfand, and Alexander Kister. "Structural Determinants in the Sequences of Immunoglobulin Variable Domain." *Journal of Molecular Biology* 278 (1998): 457–79.

Chung, W.-Y., et al. "A First Look at the ARFome: Dual-coding Genes in Mammalian Genomes." *PLoS Computational Biology* 3 (2007): e91.

Cicero, M. T. *De Natura Deorum.* Translated by Harris Rackham. Cambridge, MA: Harvard University Press, 1933.

Cleland, Carol E. "Historical Science, Experimental Science, and the Scientific Method." *Geology* 29 (2001): 987–90.

———. "Methodological and Epistemic Differences Between Historical Science and Experimental Science." *Philosophy of Science* 69 (2002): 474–96.

Cohen, Jon. "Novel Center Seeks to Add Spark to Origins of Life." *Science* 270 (1995): 1925–26.

Coleman, William. *Biology in the Nineteenth Century.* New York: Wiley, 1971.

Collingridge, D., and M. Earthy. "Science Under Stress: Crisis in Neo-Darwinism." *History and Philosophy of the Life Sciences* 12 (1990): 3–26.

Collins, Robin. "The Fine-tuning Design Argument: A Scientific Argument for the Existence of God." In *Reason for the Hope Within,* edited by Michael Murray, 47–75. Grand Rapids, MI: Eerdmans, 1999.

Costantini, M., O. Clay, C. Federico, S. Saccone, F. Auletta, and G. Bernardi. "Human Chromosomal Bands: Nested Structure, High-Definition Map and Molecular Basis." *Chromosoma* 116 (2007): 29–40.

Courtenay, W. "The Dialectic of Omnipotence in the High and Late Middle Ages." In *Divine Omniscience and Omnipotence in Medieval Philosophy.* Edited by T. Ruduvsky. Dordrecht: Reidel, 1985: 243–69.

Craig, William Lane. "God, Creation and Mr. Davies." *British Journal for the Philosophy of Science* 37 (1986): 163–75.

———. "Barrow and Tipler on the Anthropic Principle vs. Divine Design." *British Journal for the Philosophy of Science* 38 (1988): 389–95.

Cramer, Patrick, K.-J. Armache, S. Baumli, S. Benkert, F. Brueckner, C. Buchen, G. E. Damsma, et al. "Structure of Eukaryotic RNA Polymerases." *Annual Review of Biophysics* 37 (2008): 337–52.

Cramer, Patrick, David A. Bushnell, and Roger D. Kornberg. "Structural Basis of Transcription: RNA Polymerase II at 2.8 Angstrom Resolution." *Science* 292 (2001): 1863–76.

Creighton, Thomas E. *Proteins: Structures and Molecular Properties.* 2nd ed. New York: Freeman, 1993.

Crick, Francis. "On Degenerate Templates and the Adaptor Hypothesis: A Note for the RNA Tie Club." 1955. http://profiles.nlm.nih.gov/SC/B/B/G/F/_/scbbgf.pdf (last accessed July 5, 2008).

———. "On Protein Synthesis." *Symposium for the Society of Experimental Biology* 12 (1958): 138–63.

———. *Of Molecules and Men*. Seattle: University of Washington Press, 1966.

———. "The Origin of the Genetic Code." *Journal of Molecular Biology* 38 (1968): 367–79.

———. *Life Itself*. New York: Simon & Schuster, 1981.

———. *What Mad Pursuit: A Personal View of Scientific Discovery*. New York: Basic Books, 1988.

Crick, Francis, and Leslie Orgel. "Directed Panspermia." *Icarus* 19 (1973): 341–46.

Crombie, A. C. *Augustine to Galileo*. London: Heinemann, 1952.

Dam, Kenneth W., and Herbert S. Lin, eds. *Cryptography's Role in Securing the Information Society*. Washington, DC: National Academy Press, 1996.

Darwin, Charles. *On the Origin of Species by Means of Natural Selection*. A facsimile of the first edition, published by John Murray, London, 1859. Reprint, Cambridge, MA: Harvard University Press, 1964.

———. Letter to Hooker. 1871. Courtesy of Mr. Peter Gautrey. Cambridge University Library, Darwin Archives, Manuscripts Room.

———. *Life and Letters of Charles Darwin, Including an Autobiographical Chapter*. Edited by Francis Darwin. 2 vols. New York: Appleton, 1898.

———. *More Letters of Charles Darwin*. Edited by Francis Darwin. 2 vols. London: John Murray, 1903.

Darwin, Solomon. http://www.haas.berkeley.edu/faculty/darwin.html (last accessed September 9, 2008).

Davidson, E. *Genomic Regulatory Systems: Development and Evolution*. New York: Academic, 2001.

Davies, Paul. *The Fifth Miracle*. New York: Simon & Schuster, 1999.

Davison, John A. "A Prescribed Evolutionary Hypothesis," *Rivista di Biologia/Biology Forum* 98 (2005): 155–166.

Dawkins, Richard. *The Selfish Gene*. New York: Oxford University Press, 1976.

———. *The Blind Watchmaker: Why the Evidence Reveals a Universe Without Design*. New York: Norton, 1987.

———. *River Out of Eden: A Darwinian View of Life*. New York: Basic Books, 1995.

———. "The 'Information Challenge': How Evolution Increases Information in the Genome." *Skeptic* (1999): 64–69.

———. *The God Delusion*. New York: Houghton Mifflin, 2008.

Dawson, Martin H. "The Interconvertibility of 'R' and 'S' Forms of Pneumococcus." *Journal of Experimental Medicine* 47 (1928): 577–91.

Day, William. *Genesis on Planet Earth*. New Haven, CT: Yale University Press, 1984.

De Beer, G. *Homology: An Unsolved Problem*. London: Oxford University Press, 1971.

De Chadarevian, Soraya. "Protein Sequencing and the Making of Molecular Genetics." *Trends in Biochemical Sciences* 24 (1999): 203–6.

De Duve, Christian. *Blueprint for a Cell: The Nature and Origin of Life*. Burlington, NC: Neil Patterson, 1991.

———. "The Beginnings of Life on Earth." *American Scientist* 83 (1995): 249–50, 428–37.

———. *Vital Dust: Life as a Cosmic Imperative*. New York: Basic Books, 1995.

———. "The Constraints of Chance," *Scientific American* 271(1996): 112.

———. *Singularities: Landmarks on the Pathways of Life*. Cambridge: Cambridge University Press, 2005.

Dembski, William A. "Randomness by Design." *Noûs* (1991): 75–106.

———. "Theoretical Basis for the Design Inference." The 48th Annual Meeting of the American Scientific Affiliation, Seattle Pacific University, August 9, 1993.

————. "Intelligent Design as a Theory of Information." Paper presented to the Naturalism, Theism and the Scientific Enterprise Conference, University of Texas, Austin, February 22, 1997. http://www.dla.utexas.edu/depts/philosophy/faculty/ koons/ntse/papers/Dembski.txt.

————. *The Design Inference: Eliminating Chance Through Small Probabilities.* Cambridge: Cambridge University Press, 1998.

————. "Intelligent Science and Design." *First Things* 86 (1998): 21–27.

————. *No Free Lunch: Why Specified Complexity Cannot be Purchased Without Intelligence.* Boston: Rowman & Littlefield, 2002.

————. "The Logical Underpinnings of Intelligent Design." In *Debating Design: From Darwin to DNA,* edited by W. Dembski and M. Ruse, 311–30. Cambridge: Cambridge University Press, 2004.

————. *The Design Revolution: Answering the Toughest Questions About Intelligent Design.* Downers Grove, IL: InterVarsity, 2004.

————. "Specification: The Pattern That Signifies Intelligence." *Philosophia Christi* (2005): 299–343. http://www.designinference.com/documents/2005.06. Specification.pdf.

————, ed. *Mere Creation: Science, Faith and Intelligent Design.* Downers Grove, IL: InterVarsity, 1998.

Dembski, William A., and Robert J. Marks II. "The Conservation of Information: Measuring the Information Cost of Successful Search." *Transactions on Systems, Man and Cybernetics, Part A. http://ieeexplore.ieee.org/xpl/RecentIssue. jsp?puNumber=3468/.* Forthcoming.

————. "The Search for a Search: Measuring the Information Cost of Higher Level Search." *The International Journal of Information Technology and Intelligent Computing* (2008). http://itic.wshe.lodz.pl/.

Dembski, William A., and Michael Ruse, eds. *Debating Design: From Darwin to DNA.* Cambridge: Cambridge University Press, 2004.

Dembski, William A., and Jonathan Wells. *The Design of Life: Discovering Signs of Intelligence in Biological Systems.* Dallas: Foundation for Thought and Ethics, 2008.

Denoeud, F., P. Kapranov, C. Ucla, A. Frankish, R. Castelo, J. Drenkow, J. Lagarde, et al. "Prominent Use of Distal 5' Transcription Start Sites and Discovery of a Large Number of Additional Exons in ENCODE Regions." *Genome Research* 17 (2007): 746–59.

Denton, Michael. *Evolution: A Theory in Crisis.* London: Adler and Adler, 1985.

————. *Nature's Destiny: How the Laws of Biology Reveal Purpose in the Universe.* New York: Free Press, 1986.

————. "Selected Excerpts: The Inverted Retina: Maladaptation or Pre-adaptation?" *Origins & Design* 19:2 (1999). http://www.arn.org/docs/odesign/od192/ invertedretina192.htm.

DeWolf, David, Stephen C. Meyer, and Mark DeForrest. "Teaching the Origins Controversy: Science, Religion or Speech?" *University of Utah Law Review* 39 (2000): 39–110.

DeWolf, David K., John West, and Casey Luskin. "Intelligent Design Will Survive *Kitzmiller v. Dover.*" *Montana Law Review* 68 (2007): 7–57.

DeWolf, David K., John G. West, Casey Luskin, and Jonathan Witt. *Traipsing into Evolution: Intelligent Design and the* Kitzmiller vs. Dover *Decision.* Seattle, WA: Discovery Institute Press, 2006.

Dickerson, R. E. "Chemical Evolution and the Origin of Life." *Scientific American* 239 (1978): 70–85.

Dimroth, Erich, and M. M. Kimberly. "Pre-Cambrian Atmospheric Oxygen: Evidence in Sedimentary Distribution of Carbon, Sulfur, Uranium and Iron." *Canadian Journal of Earth Sciences* 13 (1976): 1161–85.

Discovery Institute. Brief of Amicus Curiae (Revised). In *Kitzmiller v. Dover*, 400 F.Supp. 707 (M.D.Pa. 2005), 17. http://www.discovery.org/scripts/viewDB/ filesDB-download.php?command=download&id=646. Appendix A, http://www. discovery.org/scripts/viewDB/filesDB-download.php?command=download&id=647.

Dobzhansky, Theodosius. "Discussion of G. Schramm's Paper." In *The Origins of Prebiological Systems and of Their Molecular Matrices*, edited by Sidney W. Fox, 309–15. New York: Academic, 1965.

Donnelly, S. R., T. E. Hawkins, and S. E. Moss. "A Conserved Nuclear Element with a Role in Mammalian Gene Regulation." *Human Molecular Genetics* 8 (1999): 1723–28.

Dose, Klaus. "The Origin of Life: More Questions Than Answers." *Interdisciplinary Science Review* 13 (1988): 348–56.

———. *The Origin of Life and Evolutionary Biochemistry*. Edited by K. Dose, S. W. Fox, G. A. Deborin, and T. E. Pavlosvskaya. New York: Plenum, 1974.

Doudna, Jennifer A., and Jack W. Szostak. "RNA-Catalyzed Synthesis of Complementary-Strand RNA." *Nature* 339 (1989): 519–22.

Dover, Gabriel. "On the Edge." 365 *Nature* (1993): 704–6.

Doyle, Sir A. C. "The Boscome Valley Mystery." In *The Sign of Three: Peirce, Holmes, Popper*, edited by T. Sebeok, 145. Bloomington: Indiana University Press, 1983.

Dray, W. *Laws and Explanation in History*. London: Oxford University Press, 1957.

———. "'Explaining What' in History." In *Theories of History*, edited by P. Gardiner, 402–8. Glencoe, IL: Free Press, 1959.

Dretske, Fred I. *Knowledge and the Flow of Information*. Cambridge, MA: MIT Press, 1981.

Dunlap, K. A., M. Palmarini, M. Varela, R. C. Burghardt, K. Hayashi, J. L. Farmer, and T. E. Spencer. "Endogenous Retroviruses Regulate Periimplantation Placental Growth and Differentiation." *Proceedings of the National Academy of Sciences USA* 103 (2006): 14390–95.

Dunn, C. A., P. Medstrand, and D. L. Mager. "An Endogenous Retroviral Long Terminal Repeat Is the Dominant Promoter for Human B1, 3-galactosyltransferase 5 in the Colon." *Proceedings of the National Academy of Sciences USA* 100 (2003):12841–46.

Dyson, Freeman F. *Origins of Life*. Cambridge: Cambridge University Press, 1985.

Dyson, Lisa, Matthew Kleban, and Leonard Susskind. "Disturbing Implications of a Cosmological Constant." *Journal of High Energy Physics* 10 (2002): 11.

Eccles, John C. *Facing Reality*. New York: Springer International, 1970.

Eckland, Eric H., Jack W. Szostak, and David P. Bartel. "Structurally Complex and Highly Active RNA Ligases Derived from Random RNA Sequences." *Science* 269 (1995): 364–70.

Eddington, Arthur S. "The End of the World: From the Standpoint of Mathematical Physics." *Nature* 127 (1956): 450.

Eden, M. "Inadequacies of Neo-Darwinian Evolution as a Scientific Theory." In *Mathematical Challenges to the Neo-Darwinian Interpretation of Evolution*, edited by P. S. Moorhead and M. M. Kaplan, 11. Wistar Institute Symposium Monograph. New York: Liss, 1967.

Eigen, Manfred. *Steps Toward Life: A Perspective on Evolution*. Translated by P. Woolley. Oxford: Oxford University Press, 1992.

Eigen, Manfred, W. Gardner, Peter Schuster, and R. Winkler-Oswaititich. "The Origin of Genetic Information." *Scientific American* 244 (1981): 88–118.

Eigen, Manfred, and Peter Schuster. "The Hypercycle: A Principle of Natural Self-Organization." *Naturwissenschaften* 65 (1978): 7–41.

Ekland, Erik H., Jack W. Szostak, and David P. Bartel. "Structurally Complex and Highly Active RNA Ligases Derived from Random RNA Sequences." *Science* 269 (1995): 364–70.

Eldredge, N., and S. J. Gould. "Punctuated Equilibria: An Alternative to Phyletic Gradualism." In *Models in Paleobiology,* edited by T. Schopf, 82–115. San Francisco: Freeman, 1973.

Elsberry, Wesley, and John Wilkins. "The Advantages of Theft over Honest Toil: The Design Inference and Arguing from Ignorance." *Biology and Philosophy* 16 (2001): 709–22.

ENCODE Project Consortium. "Identification and Analysis of Functional Elements in 1% of the Human Genome by the ENCODE Pilot Project." *Nature* 447 (2007): 799–816.

Engels, Friedrich. *The Dialectics of Nature.* New York: International, 1940.

Erwin, D. H. "Early Introduction of Major Morphological Innovations." *Acta Palaeontologica Polonica* 38 (1994): 281–94.

_____. "Macroevolution Is More Than Repeated Rounds of Microevolution." *Evolution and Development* 2 (2000): 78–84.

Erwin, Douglas, and Eric Davidson. "The Last Common Bilaterian Ancestor." *Development* 129 (2002): 3021–32.

Expelled: No Intelligence Allowed. Documentary film. Premise Media, 2008.

Fann, K. T. *Peirce's Theory of Abduction.* The Hague: Martinus Nijhoff, 1970.

Farley, James. *The Spontaneous Generation Controversy from Descartes to Oparin.* Baltimore, MD: Johns Hopkins University Press, 1974.

Feller, William. *An Introduction to Probability Theory and Its Applications.* 3rd ed. Hoboken, NJ: John Wiley and Sons, 1968.

Feyerabend, Paul. *Against Method.* London: Verso, 1978.

Figueiredo, L. M., L. H. Freitas-Junior, E. Bottius, J.-C. Olivo-Marin, and A. Scherf. "A Central Role for *Plasmodium Falciparum* Subtelomeric Regions in Spatial Positioning and Telomere Length Regulation." *EMBO Journal* 21 (2002): L815–24.

Fischer, S. E., M. D. Butler, Q. Pan, and G. Ruvkun. "*Trans*-splicing in *C. elegans* Generates the Negative RNAi Regulator ERI–6/7." *Nature* 455 (2008): 491–96.

Fisher, Ronald A. *The Design of Experiments.* New York: Hafner, 1935.

_____. *Statistical Methods and Statistical Inference.* Edinburgh: Oliver and Boyd, 1956.

Flew, Antony. *There Is a God: How the World's Most Notorious Atheist Changed His Mind.* San Francisco: HarperOne, 2007.

Florkin, Marcel. *A History of Biochemistry.* New York: American Elsevier, 1972.

_____. "Ideas and Experiments in the Field of Prebiological Chemical Evolution." *Comprehensive Biochemistry* (1975): 241–42.

Folsome, C. E. *The Origin of Life.* San Francisco: W. H. Freeman, 1979.

Forrest, Barbara. Expert Witness Report. *Kitzmiller v. Dover,* 45. http://www2.ncseweb.org/kvd/experts/2005_04_01_Forrest_expert_report_P.pdf.

Forrest, Barbara, and Paul R. Gross. *Creationism's Trojan Horse: The Wedge of Intelligent Design.* New York: Oxford University Press, 2004.

Fox, Sidney W. "Simulated Natural Experiments in Spontaneous Organization of Morphological Units from Proteinoid." In *The Origins of Prebiological Systems and of Their Molecular Matrices,* edited by Sidney W. Fox, 361–82. New York: Academic, 1965.

_____. "Proteinoid Experiments and Evolutionary Theory." In *Beyond Neo-Darwinism,* edited by M. W. Ho and P. T. Saunders, 15–60. New York: Academic, 1984.

_____, ed. *The Origins of Prebiological Systems and of Their Molecular Matrices.* New York: Academic, 1965.

Fox, Sidney W., and Klaus Dose. *Molecular Evolution and the Origin of Life.* Rev. ed. San Francisco: Freeman, 1972.

Fox, Sidney W., K. Harada, G. Krampitz, and G. Mueller. "Chemical Origins of Cells." *Chemical Engineering News,* June 22, 1970, 80.

Frankel, J. "Propagation of Cortical Differences in *Tetrahymena.*" *Genetics* 94 (1980): 607–23.

Fraser, Claire M., Jeannine D. Gocayne, Owen White, Mark D. Adams, Rebecca A. Clayton, Robert D. Fleischmann, Carol J. Bult, et al. "The Minimal Gene Complement of *Mycoplasma Genitalium.*" *Science* 270 (1995): 397–403.

Fry, Iris. *The Emergence of Life on Earth: A Historical and Scientific Overview.* New Brunswick, NJ: Rutgers University Press, 2000.

Fuller, Steve. *Science vs. Religion? Intelligent Design and the Problem of Evolution.* Oxford: Polity, 2007.

Futuyma, Douglas J. *Evolutionary Biology.* 3rd ed. Sunderland, MA: Sinauer, 1998.

Gallie, Walter Bryce. "Explanations in History and the Genetic Sciences." In *Theories of History,* edited by P. Gardiner, 386–402. Glencoe, IL: Free Press, 1959.

Gamma, Erich, Richard Helm, Ralph Johnson, and John Vlissides. *Design Patterns: Elements of Reusable Object-Oriented Software.* Reading, MA: Addison-Wesley, 1995.

Gamow, George. "Possible Relation Between Deoxyribonucleic Acid and Protein Structures." *Nature* 173 (1954): 318.

———. "Possible Mathematical Relation Between Deoxyribonucleic Acid and Proteins." *Det Kongelige Danske Videnskabernes Selskab: Biologiske Meddelelser* 22 (1954): 1–13.

Gardner, Howard. *The Mind's New Science: A History of the Cognitive Revolution.* New York: Basic Books, 1995.

Garriga, Jaume, and Alexander Vilenkin. "Many Worlds in One." *Physical Review* D64 (2001): 43511. http://xxx.tau.ac.il/PS_cache/gr-qc/pdf/0102/0102010v2.pdf.

Gates, Bill. *The Road Ahead.* Rev. ed. New York: Viking, Penguin Group, 1996.

Gautier, A. *Chemistry of the Living Cell.* Translated by C. M. Stern. Paris, 1894.

Gehring, Walter. *Master Control Genes in Development and Evolution: The Homeobox Story.* New Haven, CT: Yale University Press, 1998.

Geison, Gerald. "The Protoplasmic Theory of Life and the Vitalist-Mechanist Debate." *Isis* 60 (1969): 273–92.

Gerhart, J., and M. Kirschner. "Cells, Embryos, and Evolution." London: Blackwell Science, 1997.

Gerstein, Mark B., et al. "What Is a Gene, Post-ENCODE? History and Updated Definition." *Genome Research* 17 (2007): 669–81.

Gesteland, Raymond F., and John F. Atkins. *The RNA World.* Cold Spring Harbor, NY: Cold Spring Harbor Laboratory Press, 1993.

Gesteland, Raymond F., Thomas R. Cech, and John F. Atkins, eds. *The RNA World.* 2nd ed. Cold Spring Harbor, NY: Cold Spring Harbor Laboratory Press, 1999.

Gibbs, W. W. "The Unseen Genome: Gems Among the Junk." *Scientific American* 289 (2003): 46–53.

Gierman, H. J., M. H. Indemans, J. Koster, S. Goetze, J. Seppen, D. Geerts, R. van Driel, and R. Versteeg. "Domain-wide Regulation of Gene Expression in the Human Genome." *Genome Research* 17 (2007): 1286–95.

Gil, Rosario, Francisco J. Silva, Juli Peretó, and Andrés Moyal. "Determination of the Core of a Minimal Bacterial Gene Set." *Microbiology and Molecular Biology Reviews* 68 (2004): 518–37.

Gilbert, S. F., J. M. Opitz, and R. A. Raff. "Resynthesizing Evolutionary and Developmental Biology." *Developmental Biology* 173 (1996): 357–72.

Gilbert, Walter. "Origin of Life: The RNA World." *Nature* 319 (1986): 618.

Gilder, George. *Telecosm: How Infinite Bandwidth Will Revolutionize Our World.* New York: Free Press, 2000.

Giles, Jim. "Peer-Reviewed Paper Defends Theory of Intelligent Design." *Nature* 431 (2004): 114.

Gillespie, N. C. *Charles Darwin and the Problem with Creation.* Chicago: University of Chicago Press, 1979.

————. "Natural History, Natural Theology, and Social Order: John Ray and the 'Newtonian Ideology.'" *Journal of the History of Biology* 20 (1987): 1–49.

Gitt, Werner. *In the Beginning Was Information*. Bielefeld, Germany: Christliche-Literatur Verbreitung, 1997.

Glas, Eduard. *Chemistry and Physiology in Their Historical and Philosophical Relations*. Delft: Delft University Press, 1979.

Glass, John I., et al. "Essential Genes of a Minimal Bacterium." *Proceedings of the National Academy of Sciences USA* 13 (2006): 425–30.

Gleiser, Marcello. "Who Designed the Designer?" *Boston Globe,* op. ed., August 29, 2005.

Goetze, S., J. Mateos-Langerak, H. J. Gierman, W. de Leeuw, O. Giromus, M. H. Indemans, J. Koster, V. Ondrej, R. Versteeg, and R. van Driel. "The Three-Dimensional Structure of Human Interphase Chromosomes Is Related to the Transcriptome Map." *Molecular Biology of the Cell* 27 (2007): 4475–87.

Goh, S.-H., Y. T. Lee, N. V. Bhanu, M. C. Cam, R. Desper, B. M. Martin, R. Moharram, R. B. Gherman, and J. L. Miller. "A Newly Discovered Human Alpha Globin Gene." *Blood* DOI 10.1182/blood-2005-03-0948.

Gonzalez, Guillermo, and Jay W. Richards. *The Privileged Planet: How Our Place in the Cosmos Was Designed for Discovery*. Washington, DC: Regnery, 2004.

Goodrich, J. A., and J. F. Kugel. "Non-coding-RNA Regulators of RNA Polymerase II Transcription." *Nature Reviews Molecular and Cell Biology* 7 (2006): 612–16.

Goodsell, David. *The Machinery of Life*. New York: Springer, 1998.

Goodwin, Brian C. "What Are the Causes of Morphogenesis?" *BioEssays* 3 (1985): 32–36.

————. *How the Leopard Changed Its Spots: The Evolution of Complexity*. New York: Scribner, 1995.

Goudge, T. A. *The Ascent of Life*. Toronto: University of Toronto Press, 1961.

Gould, Stephen Jay. "Is Uniformitarianism Necessary?" *American Journal of Science* 263 (1965): 223–28.

————. *Ever Since Darwin*. New York: Norton, 1977.

————. "Is a New Theory of Evolution Emerging?" *Paleobiology* 6 (1980): 119–30.

————. "The Senseless Signs of History." In *The Panda's Thumb*. New York: Norton, 1984.

————. "Evolution as Fact and Theory." In *Science and Creationism,* edited by Ashley Montagu, 118–24. New York: Oxford University Press, 1984.

————. "Genesis and Geology." In *Science and Creationism,* edited by Ashley Montagu, 130–31. New York: Oxford University Press, 1984.

————. "Evolution and the Triumph of Homology: Or, Why History Matters." *American Scientist* 74 (1986): 60–69.

————. "Darwinism Defined: The Difference Between Theory and Fact." *Discovery* (January 1987): 64–70.

————. "Response by Stephen Jay Gould." *Geological Society of America* 101 (1989): 998–1000.

————. *Wonderful Life: The Burgess Shale and the Nature of History*. New York: Norton, 1990.

————. *The Structure of Evolutionary Theory*. Cambridge, MA: Harvard University Press, 2002.

Graham, G. *Historical Explanation Reconsidered*. Aberdeen: Aberdeen University Press, 1983.

Graham, Loren R. *Science and Philosophy in the Soviet Union*. London: Allen Lane, 1973.

————. *Science in Russia and the Soviet Union: A Short History*. Cambridge: Cambridge University Press, 1993.

Grawunder, U., M. Wilm, X. Wu, P. Kulesza, T. E. Wilson, M. Mann, and M. R. Lieber. "Activity of DNA Ligase IV Stimulated by Complex Formation with XRCC4 Protein in Mammalian Cells." *Nature* 388 (1997): 492–95.

Green, David G. "The Role of Translocation and Selection in the Emergence of Genetic Clusters and Modules." *Artificial Life* 13 (2007): 249–58.

Green, Michael R. "Mobile RNA Catalysts." *Nature* 336 (1988): 716–18.

Gregory, F. *Scientific Materialism in Nineteenth Century Germany.* Boston: Springer, 1977.

Griffith, Frederick. "The Significance of Pneumococcal Types." *Journal of Hygiene* 27 (1928): 113–59.

Grimes, G. W., and K. J. Aufderheide. *Cellular Aspects of Pattern Formation: The Problem of Assembly.* Monographs in Developmental Biology. Vol. 22. Basel: Karger, 1991.

Grimmett, Geoffrey R., and D. R. Stirzaker. *Probability and Random Processes.* 3rd ed. Oxford: Oxford University Press, 2001.

Grinnell, Frederick. "Radical Intersubjectivity: Why Naturalism Is an Assumption Necessary for Doing Science." In *Darwinism: Science or Philosophy?* edited by Jon Buell and Virginia Hearn, 99–105. Richardson, TX: Foundation for Thought and Ethics, 1994.

Hacking, Ian. *The Logic of Statistical Inference.* Cambridge: Cambridge University Press, 1965.

Haeckel, Ernst. *Generelle Morphologie der Organismen.* Vol. 1. Berlin: Reimer, 1866.

———. *The History of Creation.* Translated by E. Ray Lankester. New York: Appleton, 1876.

———. *The Riddle of the Universe.* Translated by J. McCabe. London: Harper & Brothers, 1904.

———. *The Wonders of Life.* Translated by J. McCabe. London: Harper, 1905.

———. *Last Words on Evolution.* Translated by J. McCabe. London: Kissinger Publishing, LLC, 1910.

Hager, Alicia J., Jack D. Polland Jr., and Jack W. Szostak. "Ribozymes: Aiming at RNA Replication and Protein Synthesis." *Chemistry and Biology* 3 (1996): 717–25.

Han, Jeffrey S., Suzanne T. Szak, and Jef D. Boeke. "Transcriptional Disruption by the L1 Retrotransposon and Implications for Mammalian Transcriptomes." *Nature* 429 (2004): 268–74.

Harada, K., and S. Fox. "Thermal Synthesis of Amino Acids from a Postulated Primitive Terrestrial Atmosphere." *Nature* 201 (1964): 335–37.

Harold, Franklin M. "From Morphogenes to Morphogenesis." *Microbiology* 141 (1995): 2765–78.

———. *The Way of the Cell: Molecules, Organisms, and the Order of Life.* New York: Oxford University Press, 2001.

Harre, Rom, and Edward H. Madden. *Causal Powers.* London: Blackwell, 1975.

Harvard Origins of Life Initiative, The. http://origins.harvard.edu.

Hawking, Stephen W. *A Brief History of Time: From the Big Bang to Black Holes.* New York: Bantam, 1988.

Hawking, Stephen, and Donald Page. "How Probable Is Inflation?" *Nuclear Physics* B 298 (1988): 789–809.

Hayden, Erika Check. "A Never-Ending Dance of RNA: The Recreation of Life's Origins Comes a Self-Catalyzing Step Closer." *Nature.* January 8, 2009. doi:10.1038/news.2009.5.

Hempel, Carl G. "The Function of General Laws in History." *Journal of Philosophy* 39 (1942): 35–48.

Henikoff, Steven. "Heterochromatin Function in Complex Genomes." *Biochimica et Biophysica Acta* 1470 (2000): O1–O8.

Henikoff, Steven, Kami Ahmad, and Harmit S. Malik. "The Centromere Paradox: Stable Inheritance with Rapidly Evolving DNA." *Science* 293 (2001): 1098–1102.

Hick, J. *Arguments for the Existence of God.* London: Palgrave Macmillan, 1970.

Hoagland, Mahlon B., Mary Louise Stephenson, Jesse F. Scott, Liselotte I. Hecht, and Paul C. Zamecnik. "A Soluble Ribonucleic Acid Intermediate in Protein Synthesis." *Journal of Biological Chemistry* 231 (1958): 241–57.

Hobbes, Thomas. *Leviathan*. Baltimore, MD: Penguin, 1968.

Hodge, M. Jonathan S. "The Structure and Strategy of Darwin's 'Long Argument.'" *British Journal for the History of Science* 10 (1977): 237–45.

———. "Darwin's Theory and Darwin's Argument." In *What the Philosophy of Biology Is*, edited by M. Ruse, 167–74. Dordrecht: Kluwer, 1989.

———. "The History of the Earth, Life, and Man: Whewell and Palaetiological Science." In *William Whewell: A Composite Portrait*, edited by Menachem Fisch and Simon Schaffer, 255–88. Oxford: Clarendon, 1990.

Hodgson, Peter. "The Christian Origin of Science." *Logos* 4 (2001):138–59.

Hogg, Robert V., Allen Craig, and Joseph W. McKean. *Introduction to Mathematical Statistics*. 6th ed. Upper Saddle River, NJ: Prentice Hall, 2004.

Holden, Constance. "Random Samples." *Science* 305 (2004): 1709.

Holland, Heinrich D. *The Chemical Evolution of the Atmosphere and Oceans*. Princeton, NJ: Princeton University Press, 1984.

Holland, Heinrich D., B. Lazar, and Mark McCaffrey. "Evolution of the Atmosphere and Oceans." *Nature* 320 (1986): 27–33.

Holton, Gerald. *Thematic Origins of Scientific Thought: Kepler to Einstein*. Cambridge, MA: Harvard University Press, 1973.

Hood, Leroy, and David Galas. "The Digital Code of DNA." *Nature* 421 (2003): 444–48.

Hooykaas, R. *Religion and the Rise of Modern Science*. Grand Rapids, MI: Eerdmans, 1972.

———. "Catastrophism in Geology: Its Scientific Character in Relation to Actualism and Uniformitarianism." In *Philosophy of Geohistory (1785–1970)*, edited by C. Albritton, 270–316. Stroudsburg, PA: John Wiley & Sons, Inc., 1975.

Hopkins, Frederick G. "BAAS Presidential Address: Some Chemical Aspects of Life." *Nature* 132 (1933): 381–94.

Horgan, John. "Unicyclist, Juggler and Father of Information Theory." *Scientific American* 262 (1990): 22–22B.

———. "'In the Beginning . . .': Report on Attempts to Solve the Mystery of Life's Origin." *Scientific American* 264 (1991): 116–25.

———. "The World According to RNA." *Scientific American* 274 (1996): 29–30.

Howson, Colin, and Peter Urbach. *Scientific Reasoning: The Bayesian Approach*. 2nd ed. La Salle, IL: Open Court, 1996.

Hoyle, Fred. *The Intelligent Universe*. London: Michael Joseph Limited, 1983.

Hoyle, Fred, and Chandra Wickramasinghe. *Evolution from Space*. London: Dent, 1981.

Huang, Z. P., H. Zhou, H. L. He, C. L. Chen, D. Liang, and L. H. Qu. "Genome-wide Analyses of Two Families of snoRNA Genes from *Drosophila melanogaster*, Demonstrating the Extensive Utilization of Introns for Coding of snoRNAs." *RNA* 11, 8 (2005): 1303–16.

Hubbert, M. K. "Critique of the Principle of Uniformity." In *Philosophy of Geohistory (1785–1970)*, edited by C. Albritton, 225–55. Stroudsburg, PA: John Wiley & Sons, Inc., 1975.

Hudson, S. G., and E. S. Goldstein. "The Gene Structure of the *Drosophila melanogaster* Proto-Oncogene, Kayak, and Its Nested Gene, fos-Intronic Gene." *Gene* 420 (2008): 76–81.

Hughes, Arthur. *A History of Cytology*. London: Abelard-Schuman, 1959.

Hull, David L. *Darwin and His Critics*. Chicago: University of Chicago Press, 1973.

———. *Philosophy of Biological Science*. Englewood Cliffs, NJ: Prentice-Hall, 1974. Portuguese translation, 1975; Japanese, 1994.

———. "Darwin and the Nature of Science." In *Evolution from Molecules to Men*, edited by David Bendall, 63–80. Cambridge: Cambridge University Press, 1983.

Hume, David. *Dialogues Concerning Natural Religion*. Edited, with an introduction, by Richard H. Popkin. Indianapolis, IN: Hackett, 1980.

Hunter, Cornelius. *Darwin's God: Evolution and the Problem of Evil.* Grand Rapids, MI: Brazos, 2001.

Hurd, Gary S. "The Explanatory Filter, Archaeology, and Forensics." In *Why Intelligent Design Fails: A Scientific Critique of the New Creationism,* edited by Matt Young and Taner Edis, 107–20. Piscataway, NJ: Rutgers University Press, 2004.

Huxley, Thomas H. "On the Physical Basis of Life." *Fortnightly Review* 5 (1869): 129–45.

———. "Biogenesis and Abiogenesis." Presidential address to the British Association of the Advancement of Science for 1870. *Discourses: Biological and Geological* 8 (1896): 229–71.

Hyslop, L., M. Stojkovic, L. Armstrong, T. Walter, P. Stojkovic, S. Przyborski, M. Herbert, A. Murdoch, T. Strachan, and M. Lakoa. "Downregulation of NANOG Induces Differentiation of Human Embryonic Stem Cells to Extraembryonic Lineages." *Stem Cells* 23 (2005): 1035–43.

Ijde, Aaron J. *The Development of Modern Chemistry.* New York: Harper & Row, 1964.

Illangasekare, Mali, Oleh Kovalchuke, and Michael Yarus. "Essential Structures of a Self-aminoacylating RNA." *Journal of Molecular Biology* 274 (1997): 519–29.

Illangasekare, Mali, Giselle Sanchez, Tim Nickles, and Michael Yarus. "Aminoacyl-RNA Synthesis Catalyzed by an RNA." *Science* 267 (1995): 643–47.

Illangasekare, Mali, and Michael Yarus. "Specific, Rapid Synthesis of Phe-RNA by RNA." *Proceedings of the National Academy of Sciences USA* 96 (1999): 5470–75.

Jacob, François. *The Statue Within: An Autobiography.* Translated by Franklin Phillip. Cold Spring Harbor, NY: Cold Spring Harbor Laboratory Press, 1995.

Jaki, Stanley. "Science: Western or What?" *Intercollegiate Review* 26 (1990): 3–12.

Janowski, B. A., K. E. Huffman, J. C. Schwartz, R. Ram, D. Hardy, D. S. Shames, J. D. Minna, and D. R. Corey. "Inhibiting Gene Expression at Transcription Start Sites in Chromosomal DNA with Antigene RNAs." *Nature Chemical Biology* 1 (2005): 216–22.

Jenkins, John B. *Genetics.* 2nd ed. Boston: Houghton Mifflin, 1979.

Jenkins, John, B., and G. Miklos. *The Eukaryote Genome in Development and Evolution.* London: Allen & Unwin, 1988.

Johnson, Phillip E. *Darwin on Trial.* 2nd ed. Downers Grove, IL: InterVarsity, 1993.

Johnston, Wendy K., Peter J. Unrau, Michael S. Lawrence, Margaret E. Glasner, and David P. Bartel. "RNA-Catalyzed RNA Polymerization: Accurate and General RNA-Templated Primer Extension." *Science* 292 (2001): 1319–25.

Jones, Judge John E., III. Decision in *Kitzmiller et al. v. Dover Area School Board.* No.04cv2688, 2005 WL 2465563, *66 (M.D.Pa. Dec. 20, 2005). http://www.pamd.uscourts.gov/kitzmiller/kitzmiller_342.pdf.

———. *Kitzmiller et al. v. Dover Area School District.* 400 F.Supp.2d 707, 745 (M.D.Pa. 2005).

Jones, Peter F. *CAD/CAM: Features, Applications and Management.* New York: Macmillan, 1992.

Jordan, I. K., I. B. Rogozin, G. V. Glazko, and E. V. Koonin. "Origin of a Substantial Fraction of Human Regulatory Sequences from Transposable Elements." *Trends in Genetics* 19 (2003): 68–72.

Joyce, Gerald F. "RNA Evolution and the Origin of Life." *Nature* 338 (1989): 217–24.

———. "Forty Years of *In Vitro* Evolution." *Angewandte Chemie,* International Edition 46 (2007): 6420–36.

Joyce, Gerald F., and Leslie Orgel. "Prospects for Understanding the Origin of the RNA World." In *The RNA World,* edited by Raymond F. Gesteland and John J. Atkins, 1–25. Cold Spring Harbor, NY: Cold Spring Harbor Laboratory Press, 1993.

———. "Progress Toward Understanding the Origin of the RNA World." In *The RNA World: The Nature of Modern RNA Suggests a Prebiotic RNA World,* edited by

Raymond F. Gesteland, Thomas Cech, and John F. Atkins, 23–56. Woodbury, NY: Cold Spring Harbor Laboratory Press, 2006.

Joynt, C. B., and N. Rescher. "The Problem of Uniqueness in History." *History and Theory* 1 (1961): 151–54.

Judson, Horace Freeland. *The Eighth Day of Creation: Makers of the Revolution in Biology.* Exp. ed. Plainview, NY: Cold Spring Harbor Laboratory Press, 1996.

Jurka, Jerzy. "Evolutionary Impact of Human Alu Repetitive Elements." *Current Opinion in Genetics and Development* 14 (2004): 603–8.

Kamminga, Harmke. "Studies in the History of Ideas on the Origin of Life." Ph.D. dissertation. University of London, 1980.

———. "Protoplasm and the Gene." In *Clay Minerals and the Origin of Life,* edited by A. G. Cairns-Smith and H. Hartman, 1–10. Cambridge: Cambridge University Press, 1986.

Kandouz, M., A. Bier, G. D. Carystinos, M. A. Alaoui-Jamali, and G. Batist. "Connexin43 Pseudogene Is Expressed in Tumor Cells and Inhibits Growth." *Oncogene* 23 (2004): 4763–70.

Kapranov, Philipp, Aaron T. Willingham, and Thomas R Gingeras. "Genome-wide Transcription and the Implications for Genomic Organization." *Nature Reviews Genetics* 8 (2007): 413–23.

Kasting, James F. "Earth's Early Atmosphere." *Science* 295 (1993): 920–26.

Kasting, James F., and T. M. Donahue. "The Evolution of Atmospheric Ozone." *Journal of Geophysical Research* 85 (1980): 3255–63.

Kasting, James F., S. C. Liu, and T. M. Donahue. "Oxygen Levels in the Prebiological Atmosphere." *Journal of Geophysical Research* 84 (1979): 3097–3102.

Kasting, James F., and C. G. Walker. "Limits on Oxygen Concentration in the Prebiological Atmosphere and the Rate of Abiotic Fixation of Nitrogen." *Journal of Geophysical Research* 86 (1981): 1147–56.

Kauffman, George B., and Steven H. Chooljian. "Friedrich Wöhler (1800–1882), on the Bicentennial of His Birth." *Chemical Educator* 6 (2001):121–33.

Kauffman, Stuart A. *The Origins of Order: Self-Organization and Selection in Evolution.* Oxford: Oxford University Press, 1993.

———. *At Home in the Universe: The Search for the Laws of Self-Organization and Complexity.* Oxford: Oxford University Press, 1995.

———. *Investigations.* New York: Oxford University Press, 2000.

Kavalovski, V. "The *Vera Causa* Principle: A Historico-Philosophical Study of a Meta-theoretical Concept from Newton Through Darwin." Ph.D. dissertation. University of Chicago, 1974.

Kay, Lily E. "Who Wrote the Book of Life? Information and the Transformation of Molecular Biology." *Science in Context* 8 (1994): 601–34.

———. "Cybernetics, Information, Life: The Emergence of Scriptural Representations of Heredity." *Configurations* 5 (1999): 23–91.

———. *Who Wrote the Book of Life?* Stanford, CA: Stanford University Press, 2000.

Kehoe, A. "Modern Anti-evolutionism: The Scientific Creationists." In *What Darwin Began,* edited by Laurie R. Godfrey, 165–85. Boston: Allyn and Bacon, 1985.

Kendrew, John C., G. Bodo, Howard M. Dintzis, R. G. Parrish, and H. Wyckoff. "A Three-Dimensional Model of the Myoglobin Molecule Obtained by X-Ray Analysis." *Nature* 181 (1958): 662–66.

Kenyon, Dean. "Prefigured Ordering and Proto-Selection in the Origin of Life." In *The Origin of Life and Evolutionary Biochemistry,* edited by K. Dose, S. W. Fox, G. A. Deborin, and T. E. Pavlosvskaya, 207–20. New York: Plenum, 1974.

———. "A Comparison of Proteinoid and Aldocyanoin Microsystems as Models of the Primordial Cell." In *Molecular Evolution and Protobiology,* edited by K. Matsuno, K. Dose, K. Harada, and D. L. Rohlfing, 163–88. New York: Plenum, 1984.

————. Foreword. In *The Mystery of Life's Origin,* by Charles Thaxton, Walter Bradley, and Roger Olsen, v–viii. New York: Philosophical Library, 1984.

————. "Going Beyond the Naturalistic Mindset in Origin-of-Life Research." Paper presented to the conference on Christianity and the University, Dallas, February 9–10, 1985.

Kenyon, Dean, and Gordon Mills. "The RNA World: A Critique." *Origins and Design* 17 (1996): 9–16. http://www.arn.org/docs/odesign/od171/rnaworld171.htm (last accessed September 30, 2008).

Kenyon, Dean, and Gary Steinman. *Biochemical Predestination.* New York: McGraw-Hill, 1969.

Kepler, Johannes. *Mysterium Cosmographicum: The Secret of the Universe.* Translated by A. M. Duncan. New York: Abaris Books, 1981.

Kerr, Richard A. "Origin of Life: New Ingredients Suggested." *Science* 210 (1980): 42–43.

Kimura, M. *The Neutral Theory of Molecular Evolution.* Cambridge: Cambridge University Press, 1983.

Kirk, G. S., and J. E. Raven. *The Presocratic Philosophers.* Cambridge: Cambridge University Press, 1982.

Kitcher, Philip. *Abusing Science: The Case Against Creationism.* Cambridge, MA: MIT Press, 1982.

————. "Darwin's Achievement." In *Reason and Rationality in Science,* edited by Nicholas Rescher, 127–89. Washington, DC: University Press of America, 1985.

————. *Living with Darwin: Evolution, Design, and the Future of Faith.* New York: Oxford University Press, 2007.

Klinghoffer, David. "The Branding of a Heretic." *Wall Street Journal,* January 28, 2005, national edition, W11.

Klir, George J., and Mark J. Wierman. *Uncertainty-Based Information: Elements of Generalized Information Theory.* 2nd ed. New York: Physica-Verlag, 1999.

Koestler, Arthur. *The Ghost in the Machine.* London: Hutchinson, 1967; New York: Macmillan, 1968.

Koestler, Arthur, and J. R. Smythies, eds. *Beyond Reductionism.* London: Hutchinson, 1967; New York: Macmillan, 1970.

Kohler, Robert E. "The Enzyme Theory and the Origin of Biochemistry." *Isis* 64 (1973): 181–96.

Kok, Randall A., John A. Taylor, and Walter L. Bradley. "A Statistical Examination of Self-Ordering of Amino Acids in Proteins." *Origins of Life and Evolution of the Biosphere* 18 (1988): 135–42.

Kolmogorov, Andrei. "Three Approaches to the Quantitative Definition of Information. *Problemy Peredachi Informatsii* (in translation) 1 (1965): 3–11.

Kondo-Iida, E., K. Kobayashi, M. Watanabe, J. Sasaki, T. Kumagai, H. Koide, K. Saito, M. Osawa, Y. Nakamura, and T. Toda. "Novel Mutations and Genotype–Phenotype Relationships in 107 Families with Fukuyama-Type Congenital Muscular Dystrophy (FCMD)." *Human Molecular Genetics* 8 (1999): 2303–9.

Koonin, Eugene V. "How Many Genes Can Make a Cell? The Minimal Genome Concept." *Annual Review of Genomics and Human Genetics* 1 (2002): 99–116.

————. "The Cosmological Model of Eternal Inflation and the Transition from Chance to Biological Evolution in the History of Life." *Biology Direct* 2 (2007): 15. http://www.biology-direct.com/content/2/1/15.

Kuhn, Thomas. *The Structure of Scientific Revolutions.* 3rd ed. Chicago: University of Chicago Press, 1996.

Kumar, Raju K., and Michael Yarus. "RNA-Catalyzed Amino Acid Activation." *Biochemistry* 40 (2001): 6998–7004.

Küppers, Bernd-Olaf. "On the Prior Probability of the Existence of Life." In *The Probabilistic Revolution,* vol. 2, edited by Lorenz Krüger, Gerg Gigerenzer, and Mary S. Morgan, 355–69. Cambridge, MA: MIT Press, 1987.

———. *Information and the Origin of Life.* Cambridge, MA: MIT Press, 1990.

Lakatos, Imre. "Falsification and the Methodology of Scientific Research Programmes." In *Criticism and the Growth of Knowledge,* edited by Imre Lakatos and Alan Musgrave, 91–95. Cambridge: Cambridge University Press, 1970.

Landsberg, Peter T. "Does Quantum Mechanics Exclude Life?" *Nature* 203 (1964): 928–30.

Lange, B. M. H., A. J. Faragher, P. March, and K. Gull. "Centriole Duplication and Maturation in Animal Cells." In *The Centrosome in Cell Replication and Early Development,* edited by R. E. Palazzo and G. P. Schatten, 235–49. Current Topics in Developmental Biology, vol. 49. San Diego, CA: Academic, 2000.

Lanham, Url. *Origins of Modern Biology.* New York: Columbia University Press, 1968.

Laplace, Pierre Simon de. *Exposition du systéme du monde.* Paris: Courcier, 1808.

Larralde, Rosa, Michael P. Robertson, and Stanley L. Miller. "Rates of Decomposition of Ribose and Other Sugars: Implications for Chemical Evolution." *Proceedings of the National Academy of Sciences USA* 92 (1995): 8158–60.

Larson, Edward J. *Summer for the Gods: The Scopes Trial and America's Continuing Debate over Science and Religion.* New York: Basic Books, 1997.

Larson, James L. *Reason and Experience: The Representation of Natural Order in the Work of Carl von Linné.* Berkeley: University of California Press, 1971.

Laudan, Larry. "William Whewell on the Consilience of Inductions." *The Monist* 55 (1971): 368–91.

———. "The Demise of the Demarcation Problem." In *But Is It Science?* edited by Michael Ruse, 337–50. Buffalo, NY: Prometheus, 1988.

———. "More on Creationism." In *But Is It Science?* edited by Michael Ruse, 363–66. Buffalo, NY: Prometheus, 1988.

———. "Science at the Bar—Causes for Concern." In *But Is It Science?* edited by Michael Ruse, 351–55. Buffalo, NY: Prometheus, 1988.

———. *Beyond Positivism and Relativism: Theory, Method, and Evidence.* Boulder, CO: Westview, 1996.

Lawrence, Michael S., and David P. Bartel. "Processivity of Ribozyme-Catalyzed RNA Polymerization." *Biochemistry* 42 (2003): 8748–55.

———. "New Ligase-Derived RNA Polymerase Ribozymes." *RNA* 11 (2005): 1173–80.

Lawrence, P. A., and G. Struhl. "Morphogens, Compartments and Pattern: Lessons from Drosophila?" *Cell* 85 (1996): 951–61.

Lechevalier, Hubert A., and Morris Solotorovsky. *Three Centuries of Microbiology.* New York: Dover, 1974.

Lehninger, Albert L. *Biochemistry: The Molecular Basis of Cell Structure and Function.* New York: Worth Publishers, 1970.

Lemmon, R. "Chemical Evolution." *Chemical Reviews* 70 (1970): 95–96.

Lengyel, Peter, Joseph F. Speyer, and Severo Ochoa. "Synthetic Polynucleotides and the Amino Acid Code." *Proceedings of the National Academy of Sciences USA* 47 (1961): 1936–42.

Lennox, James. "Robert Boyle's Defense of Teleological Inference in Experimental Science." *Isis* 74 (1983): 38–52.

Lenoir, Timothy. *The Strategy of Life: Teleology and Mechanics in Nineteenth-Century German Biology.* Dordrecht: Reidel, 1982.

Lenski, Richard, Charles Ofria, Robert T. Pennock, and Christopher Adami. "The Evolutionary Origin of Complex Features." *Nature* 423 (2003): 139–44.

Leslie, John. "Modern Cosmology and the Creation of Life." In *Evolution and Creation,* edited by Ernan McMullin, 112. Notre Dame, IN: University of Notre Dame Press, 1985.

Levinton, J. *Genetics, Paleontology, and Macroevolution.* Cambridge: Cambridge University Press, 1988.

Lev-Maor, G., et al. "The Birth of an Alternatively Spliced Exon: 3' Splice-Site Selection in Alu Exons." *Science* 300 (2003): 1288–91.

Levy, Matthew, and Stanley L. Miller. "The Stability of the RNA Bases: Implications for the Origin of Life." *Proceedings of the National Academy of Sciences USA* 95 (1998): 7933–38.

Lewbel, Arthur. "A Personal Tribute to Claude Shannon." http://www2.bc.edu/~lewbel/Shannon.html (last accessed July 2, 2008).

Lewis, D. "Causal Explanation." *Philosophical Papers* 2 (1986): 214–40.

Lewontin, Richard. "Adaptation." In *Evolution: A Scientific American Book*, 114–25. San Francisco: Freeman, 1978.

———. "The Dream of the Human Genome." *New York Review of Books,* May 28, 1992, 31–40.

Li, Long-Cheng, S. T. Okino, H. Zhao, H., D. Pookot, R. F. Place, S. Urakami, H. Enokida, and R. Dahiya. "Small dsRNAs Induce Transcriptional Activation in Human Cells." *Proceedings of the National Academy of Sciences USA* 103 (2006): 17337–42.

Lifson, Shneior. "On the Crucial Stages in the Origin of Animate Matter." *Journal of Molecular Evolution* 44 (1997): 1–8.

Lincoln, Tracey A., and Gerald R. Joyce. "Self-Sustained Replication of an RNA Enzyme." *Science,* January 8, 2009. http://www.scienceexpress.org/8%20January%202009/Page%201/10.1126/science.1167856.

Lipton, Peter. *Inference to the Best Explanation.* London and New York: Routledge, 1991.

Lloyd, E. "The Nature of Darwin's Support for the Theory of Natural Selection." *Philosophy of Science* 51 (1983): 242–64.

Lloyd, Seth. "Computational Capacity of the Universe." *Physical Review Letters* 88 (2002): 7901–4.

Lodish, Harvey, Arnold Berk, S. Lawrence Zipursky, Paul Matsudaira, David Baltimore, and James Darnell. *Molecular Cell Biology.* New York: Freeman, 2000.

Loew, O. *The Energy of Living Protoplasm.* London, 1896.

Loewenstein, Werner R. *The Touchstone of Life: Molecular Information, Cell Communication, and the Foundations of Life.* New York: Oxford University Press, 1999.

Lönnig, Wolf-Ekkehard. "Natural Selection." In *The Corsini Encyclopedia of Psychology and Behavioral Sciences,* 3rd ed., edited by W. E. Craighead and C. B. Nemeroff, 3: 1008–16. New York: John Wiley and Sons, Inc., 2001.

Lönnig, Wolf-Ekkehard, and Heinz Saedler. "Chromosome Rearrangements and Transposable Elements." *Annual Review of Genetics* 36 (2002): 389–410.

Lowe, D. R. "Stromatolites 3,400-Myr Old from the Archean of Western Australia." *Nature* 284 (1980): 441–43.

Luskin, Casey. "Will Americans United Retract Their Demonstrably False Claims?" http://www.opposingviews.com/counters/will-americans-united-retract-their-demonstrably-false-claims.

Lyell, Charles. *Principles of Geology: Being an Attempt to Explain the Former Changes of the Earth's Surface, by Reference to Causes Now in Operation.* 3 vols. London: Murray, 1830–33.

Lynch, Michael. "The Origins of Eukaryotic Gene Structure." *Molecular Biology and Evolution* 23 (2006): 450–68.

Macnab, R. "Bacterial Mobility and Chemotaxis: The Molecular Biology of a Behavioral System." *CRC Critical Reviews in Biochemistry* 5 (1978): 291–341.

Maher, K., and D. Stevenson. "Impact Frustration of the Origin of Life." *Nature* 331 (1988): 612–14.

Mandelbaum, M. "Historical Explanation: The Problem of Covering Laws." *History Theory* 1 (1961): 229–42.

Margulis, Lynn, J. C. Walker, and M. Rambler. "Reassessment of Roles of Oxygen and Ultraviolet Light in Precambrian Evolution." *Nature* 264 (1976): 620–24.

Mariño-Ramírez, L., K. C. Lewis, D. Landsman, and I. K. Jordan. "Transposable Elements Donate Lineage-Specific Regulatory Sequences to Host Genomes." *Cytogenetic and Genome Research* 110 (2005): 333–41.

Marshall, W. F., and J. L. Rosenbaum. "Are There Nucleic Acids in the Centrosome?" In *The Centrosome in Cell Replication and Early Development,* edited by R. E. Palazzo and G. P. Schatten, 187–205. Current Topics in Developmental Biology, vol. 49. San Diego, CA: Academic, 2000.

Martin, Raymond. "Singular Causal Explanation." *Theory and Decision* 2 (1972): 221–37.

Marx, Karl. Letter to Ferdinand Lassalle, 1861. http://www.marxists.org/archive/marx/works/1861/letters/61_01_16.html.

Matthaei, J. Heinrich, and Marshall W. Nirenberg. "Characteristics and Stabilization of DNAase-Sensitive Protein Synthesis in *E. coli* Extracts." *Proceedings of the National Academy of Sciences USA* 47 (1961): 1580–88.

Matthews, C. N. "Chemical Evolution: Protons to Proteins." *Proceedings of the Royal Institution* 55 (1982): 199–206.

Mattick, John S., and Michael J. Gagen. "The Evolution of Controlled Multitasked Gene Networks: The Role of Introns and Other Noncoding RNAs in the Development of Complex Organisms." *Molecular Biology and Evolution* 18 (2001): 1611–30.

Mattick, J. S., and I. V. Makunin. "Small Regulatory RNAs in Mammals." *Human Molecular Genetics* 14 (2005): R121–32.

———. "Non-coding RNA." *Human Molecular Genetics* 15 (2006): R17–29.

Maulitz, R. C. "Schwann's Way: Cells and Crystals." *Journal of the History of Medicine* (October 1971): 422–37.

Mayr, Ernst. Foreword. In *Darwinism Defended,* edited by M. Ruse, xi–xii. Reading, MA: Addison-Wesley, 1982.

———. "Darwin: Intellectual Revolutionary." In *Evolution from Molecules to Men,* edited by D. S. Bendall, 23–41. Cambridge: Cambridge University Press, 1983.

McDonald, J. F. "The Molecular Basis of Adaptation: A Critical Review of Relevant Ideas and Observations." *Annual Review of Ecology and Systematics* 14 (1983): 77–102.

McDonough, Thomas R. *The Search for Extraterrestrial Intelligence: Listening for Life in the Cosmos.* New York: Wiley, 1988.

McElheny, Victor K. *Watson and DNA: Making a Scientific Revolution.* Cambridge, MA: Perseus, 2003.

McKenzie, Richard W., and Mark D. Brennan. "The Two Small Introns of the Drosophila Affinidisjuncta Adh Gene Are Required for Normal Transcription." *Nucleic Acids Research* 24 (1996): 3635–42.

Mckinney, H. Lewis, ed. *Lamarck to Darwin: Contributions to Evolutionary Biology 1809–1859.* Lawrence: University of Kansas Press, 1971.

McNeil-Lehrer News Hour, Transcript 19 (May 1992).

McNiven, M. A., and K. R. Porter. "The Centrosome: Contributions to Cell Form." In *The Centrosome,* edited by V. I. Kalnins, 313–29. San Diego, CA: Academic, 1992.

Medstrand, P., J.-R. Landry, and D. L. Mager. "Long Terminal Repeats Are Used as Alternative Promoters for the Endothelin B Receptor and Apolipoprotein C-I Genes in Humans." *Journal of Biological Chemistry* 276 (2001): 1896–1903.

Meyer, Stephen C. "An Interdisciplinary History of Scientific Ideas About the Origin of Life." M.Phil. thesis. University of Cambridge, 1986.

———. "Of Clues and Causes: A Methodological Interpretation of Origin of Life Studies." Ph.D. dissertation. Cambridge University, 1990.

———. "A Scopes Trial for the '90s." *Wall Street Journal,* December 6, 1993, A14.

———. "Open Debate on Life's Origin." *Insight,* February 21, 1994, 27–29.

———. "Laws, Causes and Facts: A Response to Michael Ruse." In *Darwinism: Science or Philosophy?* edited by Jon Buell and Virginia Hearn, 29–40. Richardson, TX: Foundation for Thought and Ethics, 1994.

———. "The Nature of Historical Science and the Demarcation of Design and Descent." *Facets of Faith and Science: Interpreting God's Action in the World.* Vol. 4. Lanham, MD: University Press of America, 1996.

———. "The Origin of Life and the Death of Materialism." *Intercollegiate Review.* 31 (1996): 24–43.

———. Testimony Before the U.S. Commission on Civil Rights, Hearing on Curriculum Controversies in Biology, August 21, 1998. Reprinted in *Darwinism, Design and Public Education,* edited by John Angus Campbell and Stephen C. Meyer, 567. East Lansing: Michigan State University Press, 2003.

———. "The Explanatory Power of Design: DNA and the Origin of Information." In *Mere Creation: Science, Faith and Intelligent Design,* edited by William A. Dembski, 114–47. Downers Grove, IL: InterVarsity, 1998.

———. "DNA by Design: An Inference to the Best Explanation for the Origin of Biological Information." In *Rhetoric and Public Affairs,* edited by J. A. Campbell and S. C. Meyer, 519–55. East Lansing: Michigan State University Press, 1999.

———. "Teleological Evolution: The Difference It Doesn't Make." In *Darwinism Defeated? The Johnson-Lamoureux Debate on Biological Origins,* by Phillip E. Johnson and Denis O. Lamoureux, edited by Robert Clements, 89–100. Vancouver, BC: Regent, 1999.

———. "The Return of the God Hypothesis." *Journal of Interdisciplinary Studies* 11 (1999): 1–38.

———. "DNA and Other Designs." *First Things* 102 (2000): 30–38.

———. "The Scientific Status of Intelligent Design: The Methodological Equivalence of Naturalistic and Non-Naturalistic Origins Theories." In *Science and Evidence for Design in the Universe,* by Michael Behe, William Dembski, and Stephen C. Meyer. Proceedings of the Wethersfield Institute, vol. 9, 151–211. San Francisco: Ignatius, 2000.

———. "The Demarcation of Science and Religion." In *The History of Science and Religion in the Western Tradition: An Encyclopedia,* edited by G. Ferngren, 17–23. New York: Garland, 2000.

———. "Evidence for Design in Physics and Biology: From the Origin of the Universe to the Origin of Life." In *Science and Evidence for Design in the Universe,* by Michael Behe, William Dembski, and Stephen C. Meyer. Proceedings of the Wethersfield Institute, vol. 9, 53–111. San Francisco: Ignatius, 2000.

———. "Word Games: DNA, Design and Intelligence." In *Signs of Intelligence,* edited by William A. Dembski and James Kushiner, 102–11. Grand Rapids, MI: Brazos, 2001.

———. "DNA and the Origin of Life: Information, Specification and Explanation." In *Darwinism, Design and Public Education,* edited by John Angus Campbell and Stephen C. Meyer, 223–85. East Lansing: Michigan State University Press, 2003.

———. "The Origin of Biological Information and the Higher Taxonomic Categories." *Proceedings of the Biological Society of Washington* 117 (2004): 213–39.

———. "The Cambrian Information Explosion: Evidence for Intelligent Design." In *Debating Design: From Darwin to DNA,* edited by W. Dembski and M. Ruse, 371–91. Cambridge: Cambridge University Press, 2004.

———. "A Scientific History—and Philosophical Defense—of the Theory of Intelligent Design." *Religion-Staat-Gesellschaft* 7, iss. 2 (2006): 203–47.

————. "The Origin of Biological Information and the Higher Taxonomic Categories." *Darwin's Nemesis: Phillip Johnson and the Intelligent Design Movement*, edited by William A. Dembski, 174–213. Downers Grove, IL: InterVarsity, 2006.

Meyer, Stephen C., and Michael N. Keas. "The Meanings of Evolution." In *Darwinism, Design and Public Education*, edited by John Angus Campbell and Stephen C. Meyer, 135–56. East Lansing: Michigan State University Press, 2003.

Meyer, Stephen C., Scott Minnich, Jonathan Moneymaker, Paul A. Nelson, and Ralph Seelke. *Explore Evolution: The Arguments for and Against Neo-Darwinism*. Melbourne and London: Hill House, 2007.

Meyer, Stephen C., Marcus Ross, Paul Nelson, and Paul Chien. "The Cambrian Explosion: Biology's Big Bang." In *Darwinism, Design and Public Education*, edited by John Angus Campbell and Stephen C. Meyer, 323–402. East Lansing: Michigan State University Press, 2003.

————. "Stratigraphic First Appearance of Phyla Body Plans." In *Darwinism, Design and Public Education*, edited by John Angus Campbell and Stephen C. Meyer, Appendix C. East Lansing: Michigan State University Press, 2003.

Michalak, Pawel. "Coexpression, Coregulation, and Cofunctionality of Neighboring Genes in Eukaryotic Genomes." *Genomics* 91 (2008): 243–48.

Miklos, G. L. G. "Emergence of Organizational Complexities During Metazoan Evolution: Perspectives from Molecular Biology, Palaeontology and Neo-Darwinism." *Memoirs of the Association of Australasian Palaeontologists* 15 (1993): 7–41.

Miles, J., J. A. Mitchell, L. Chakalova, B. Goyenechea, C. S. Osborne, L. O'Neill, K. Tanimoto, J. D. Engel, and P. Fraser. "Intergenic Transcription, Cell-Cycle and the Developmentally Regulated Epigenetic Profile of the Human Beta-Globin Locus." *PLoS ONE* 2 (2007): e630.

Miller, Keith. "The Similarity of Theory Testing in the Historical and 'Hard' Sciences." *Perspectives on Science and Christian Faith* 54 (2002): 119–22.

————. "The Flagellum Unspun," In *Debating Design: From Darwin to DNA*, edited by W. Dembski and M. Ruse, 81–97. Cambridge: Cambridge University Press, 2004.

Miller, Kenneth R. *Finding Darwin's God: A Scientist's Search for Common Ground Between God and Evolution*. New York: HarperCollins, 1999.

————. "How Intelligent Is Intelligent Design?" *First Things* 106 (October 2000): 2–3. http://www.firstthings.com/article.php3?id_article=2663 (last accessed September 30, 2008).

————. "In Defense of Evolution." *Judgment Day: Intelligent Design on Trial*, April 19, 2007. http://www.pbs.org/wgbh/nova/id/defense-ev.html.

————. *Only a Theory: Evolution and the Battle for America's Soul*. New York: Viking, 2008.

Miller, Kenneth R., and Joseph Levine. *Biology*. 4th ed. Upper Saddle River, NJ: Prentice-Hall, 1998.

Miller, Stanley L. "A Production of Amino Acids Under Possible Primitive Earth Conditions." *Science* 117 (1953): 528–29.

Miller, Stanley, and J. Bada. "Submarine Hotsprings and the Origin of Life." *Nature* 334 (1988): 609–10.

Miller, Stanley, and Leslie Orgel. *The Origins of Life on the Earth*. Englewood Cliffs, NJ: Prentice Hall, 1974.

Mills, Gordon C., and Dean Kenyon. "What Do Ribozyme Engineering Experiments Really Tell Us About the Origin of Life?" *Origins and Design* 17 (1996): 1.

Mims, Forrest, III. "Rejected Publications." http://www.forrestmims.org/publications.html.

Minnich, Scott A., and Harold N. Rohde. "A Rationale for Repression and/or Loss of Motility by Pathogenic *Yersinia* in the Mammalian Host." *Advances in Experimental Medicine and Biology* 603 (2007): 298–310.

Minnich, Scott A., and Stephen C. Meyer. "Genetic Analysis of Coordinate Flagellar and Type III Regulatory Circuits in Pathogenic Bacteria." In *Design and Nature II: Comparing Design in Nature with Science and Engineering*, edited by M. W. Collins and C. A. Brebbia, 295–304. Southampton: Wessex Institute of Technology, 2004.

Monastersky, Richard. "Society Disowns Paper Attacking Darwinism." *Chronicle of Higher Education* 51, no. 5 (2004): A16.

Monod, Jacques. *Chance and Necessity: An Essay on the Natural Philosophy of Modern Biology.* New York: Vintage, 1972.

Moore, Peter B., and Thomas A. Steitz. "After the Ribosome Structures: How Does Peptidyl Transferase Work?" *RNA Society* 9 (2003): 155–59.

Moorhead, P. S., and M. M. Kaplan. *Mathematical Challenges to the Neo-Darwinian Interpretation of Evolution.* Philadelphia: Wistar Institute Press, 1967.

Mora, P. T. "Urge and Molecular Biology." *Nature* 199 (1963): 212–19.

———. "The Folly of Probability." In *The Origins of Prebiological Systems and of Their Molecular Matrices,* edited by Sidney W. Fox, 39–52. New York: Academic, 1965.

Moreland, J. P. *The Creation Hypothesis: Scientific Evidence for an Intelligent Designer.* Downers Grove, IL: InterVarsity, 1994.

Morgan, Thomas Hunt. *The Physical Basis of Heredity.* Philadelphia: Lippincott, 1919.

Morowitz, Harold J. "The Minimum Size of the Cell." In *Principles of Biomolecular Organization,* edited by Gordon E. W. Wolstenholme and Maeve O'Connor, 446–59. London: Lehmann, 1966.

———. *Energy Flow in Biology: Biological Organization as a Problem in Thermal Physics.* New York: Academic, 1968.

Morris, Simon Conway. "The Question of Metazoan Monophyly and the Fossil Record." *Progress in Molecular and Subcellular Biology* 1 (1998): 21:1–9.

———. *The Crucible of Creation: The Burgess Shale and the Rise of Animals.* Oxford: Oxford University Press, 2000.

———. "Evolution: Bringing Molecules into the Fold." *Cell* 100 (2000): 1–11.

———. "The Cambrian 'Explosion' of Metazoans." In *Origination of Organismal Form: Beyond the Gene in Developmental and Evolutionary Biology,* edited by G. B. Müller and S. A. Newman, 13–32. Cambridge, MA: MIT Press, 2003.

———. *Life's Solution: Inevitable Humans in a Lonely Universe.* Cambridge: Cambridge University Press, 2003.

Morrish, Tammy A., Nicolas Gilbert, Jeremy S. Myers, Bethaney J. Vincent, Thomas D. Stamato, Guillermo E. Taccioli, Mark A. Batzer, and John V. Moran. "DNA Repair Mediated by Endonuclease-Independent LINE-1 Retrotransposition." *Nature Genetics* 31 (2002): 159–65.

Moss, Lenny. *What Genes Can't Do.* Cambridge, MA: MIT Press, 2004.

Müller, Gerd B., and Stuart A. Newman. "Origination of Organismal Form: The Forgotten Cause in Evolutionary Theory." In *Origination of Organismal Form: Beyond the Gene in Developmental and Evolutionary Biology,* edited by G. B. Müller and S. A. Newman, 3–10. Cambridge, MA: MIT Press, 2003.

Mura, M., P. Murcia, M. Caporale, T. E. Spencer, K. Nagashima, A. Rein, and M. Palmarini. "Late Viral Interference Induced by Transdominant Gag of an Endogenous Retrovirus." *Proceedings of the National Academy of Sciences USA* 101 (2004): 11117–22.

Murphy, Nancey. "Phillip Johnson on Trial: A Critique of His Critique of Darwin." *Perspectives on Science and Christian Faith* 45 (1993): 26–36.

Mushegian, Arcady, and Eugene Koonin. "A Minimal Gene Set for Cellular Life Derived by Comparison of Complete Bacterial Genomes." *Proceedings of the National Academy of Sciences USA* 93 (1996): 10268–73.

Nagel, Thomas. "Public Education and Intelligent Design." *Philosophy and Public Affairs* 36 (2008): 187–205.

Nagy, B., et al. "Amino Acids and Hydrocarbons ~3,800-Myr Old in the Isua Rocks, Southwestern Greenland." *Nature* 289 (1981): 53–56.

Nanney, D. L. "The Ciliates and the Cytoplasm." *Journal of Heredity* 74 (1983): 163–70.

National Academy of Sciences. *Science and Creationism: A View from the National Academy of Sciences.* 2nd ed. Washington, DC: National Academy Press, 1999.

National Center for Biotechnology Information (NCBI). Listing of currently known variant codes, both nuclear and mitochondrial. http://www.ncbi.nlm.nih.gov/Taxonomy/Utils/wprintgc.cgi#SG4 (last accessed September 10, 2008).

Nelson, Paul. "Anatomy of a Still-Born Analogy." *Origins and Design* 17 (1996): 12.

Nelson, Paul, and Jonathan Wells. "Homology in Biology: Problem for Naturalistic Science and Prospect for Intelligent Design." In *Darwinism, Design and Public Education,* edited by John Angus Campbell and Stephen C. Meyer, 303–22. East Lansing: Michigan State University Press, 2003.

Newman, Stuart. "The Developmental Genetic Toolkit and the Molecular Homology-Analogy Paradox." *Biological Theory* 1 (2006): 12–16, esp. 13.

Newton, Isaac. *Newton's Principia.* Translated by Andrew Motte (1686). Translation revised by Florian Cajori. Berkeley, CA: University of California Press, 1934.

———. *Opticks; or, A Treatise of the Reflections, Refractions, Inflections and Colours of Light.* Based on the 4th ed., London, 1730. With a foreword by Albert Einstein, an introduction by Sir Edmund Whittaker, a preface by I. Bernard Cohen, and an analytical table of contents by Duane H. D. Roller. Mineaola, NY: Dover Publications, Inc., 1952.

———. General Scholium, *Mathematical Principles of Natural Philosophy.* Translated by Andrew Motte (1729). Revised by Florian Cajori. New York: Greenwood, 1969.

———. *Mathematical Principles of Natural Philosophy.* Translated by Andrew Motte. Edited by Florian Cajori. Berkeley: University of California Press, 1978.

Nicolis, Grégoire, and Ilya Prigogine. *Self-Organization in Nonequilibrium Systems.* New York: Wiley, 1977.

Nijhout, H. Frederik. "Metaphors and the Role of Genes in Development." *BioEssays* 12 (1990): 441–46.

Nirenberg, Marshall W., and J. Heinrich Matthaei. "The Dependence of Cell-Free Protein Synthesis in *E. coli* upon Naturally Occurring or Synthetic Polyribonucleotides." *Proceedings of the National Academy of Sciences of the USA* 47 (1961): 1588–602.

Nissenbaum, Arie, Dean H. Kenyon, and Joan Oró. "On the Possible Role of Organic Melanoidin Polymers As Matrices for Prebiotic Activity." *Journal of Molecular Evolution* 6 (1975): 253–70.

Noller, H. F., V. Hoffarth, and L. Zimniak. "Unusual Resistance of Peptidyl Transferase to Protein Extraction Procedures." *Science* 256 (1992): 1416–19.

Numbers, Ron L. *Creation by Natural Law: Laplace's Nebular Hypothesis in American Thought.* Seattle: University of Washington Press, 1977.

Nusslein-Volhard, C., and E. Wieschaus. "Mutations Affecting Segment Number and Polarity in Drosophila." *Nature* 287 (1980): 795–801.

Ohno, S. "So Much 'Junk' DNA in Our Genome." *Brookhaven Symposia in Biology* 23 (1972): 366–70.

———. "The Notion of the Cambrian Pananimalia Genome." *Proceedings of the National Academy of Sciences, USA* 93 (1996): 8475–78.

Olby, Robert. *The Path to the Double Helix.* Seattle: University of Washington Press, 1974.

Oparin, Aleksandr I. "The Origin of Life." 1924. Translated by Ann Synge. In *The Origin of Life,* edited by John D. Bernal. London: Weidenfeld and Nicolson Natural History, 1967.

——. *The Origin of Life*. Translated by S. Morgulis. New York: Macmillan, 1938.

——. *Genesis and Evolutionary Development of Life*. New York: Academic, 1968.

——. Interview. *Uno Mas Uno* (Mexico City), May 7, 1981.

O'Reilly Factor, The. "Brutally Criticized." Fox News. August 25, 2005.

Orgel, Leslie E. *The Origins of Life*. New York: Wiley, 1973.

——. "The Origin of Life on the Earth." *Scientific American* 271 (1994): 76–83.

——. "The Origin of Life—A Review of Facts and Speculations." *Trends in Biochemical Sciences* 12 (1998): 491–95.

Orgel, Leslie E., and Francis H. Crick. "Selfish DNA: The Ultimate Parasite." *Nature* 284 (1980): 604–7.

Orgel, Leslie E., Francis H. Crick, and C. Sapienza. "Selfish DNA." *Nature* 288 (1980): 645–46.

Ostling, Richard N. "Lifelong Atheist Changes Mind About Divine Creator." *Washington Times*, December 10, 2004, national edition.

Oswald, Avery T., Colin M. MacCleod, and Maclyn McCarty. "Induction of Transformation by a Deoxyribonucleic Acid Fraction Isolated from Pneumococcus Type III." *Journal of Experimental Medicine* 79 (1944): 137–58.

Ozsolak, Fatih, Laura L. Poling, Zheng Wang, Hui Liu, X. Shirley Liu, Robert G. Roeder, Xinmin Zhang, Jun S. Song, and David E. Fisher. "Chromatin Structure Analyses Identify miRNA Promoters." *Genes and Development* 22 (2008): 3172–83.

Pagano, A., M. Castelnuovo, F. Tortelli, R. Ferrari, G. Dieci, and R. Cancedda. "New Small Nuclear RNA Gene-like Transcriptional Units as Sources of Regulatory Transcripts." *PLoS Genetics* 3 (2007): e1.

Paley, William. *Natural Theology: Or Evidences of the Existence and Attributes of the Deity Collected from the Appearances of Nature*. 1802. Reprint, Boston: Gould and Lincoln, 1852.

Panganiban, Grace, and John L. R. Rubenstein. "Developmental Functions of the *Distal-less*/Dlx Homeobox Genes." *Development* 129 (2002): 4371–86.

Pardue, M.-L., and P. G. DeBaryshe. "Drosophila Telomeres: Two Transposable Elements with Important Roles in Chromosomes." *Genetica* 107 (1999): 189–96.

Partington, James Riddick. *A History of Chemistry*. Vol. 4. 1964. Reprint, London: Macmillan, 1972.

Pattee, Howard H. "The Problem of Biological Hierarchy." In *Towards a Theoretical Biology*, vol. 3, edited by Conrad H. Waddington, 117–36. Edinburgh: Edinburgh University Press, 1970.

Paul, Natasha, and Gerald F. Joyce. "A Self-Replicating Ligase Ribozyme." *Proceedings of the National Academy of Sciences USA* 99 (2002): 12733–40.

Peaston, E., A. V. Evsikov, J. H. Graber, W. N. de Vries, A. E. Holbrook, D. Solter, and B. B. Knowles. "Retrotransposons Regulate Host Genes in Mouse Oocytes and Preimplantation Embryos." *Developmental Cell* 7 (2004): 597–606.

Peirce, Charles S. *Collected Papers*. Edited by Charles Hartshorne and P. Weiss. 6 vols. Cambridge, MA: Harvard University Press, 1931–35.

——. "Abduction and Induction." In *The Philosophy of Peirce*, edited by J. Buchler, 150–54. London: Routledge, 1956.

Penfield, Wilder. *The Mystery of the Mind*. Princeton, NJ: Princeton University Press, 1975.

Pennisi, Elizabeth. "Seeking Life's Bare (Genetic) Necessities." *Science* 272 (1996): 1098–99.

——. "Finally, the Book of Life and Instructions for Navigating It." *Science* 288 (2000): 2304–7.

Pennock, Robert. "The Wizards of ID." In *Intelligent Design Creationism and Its Critics: Philosophical, Theological and Scientific Perspectives,* edited by Robert Pennock, 645–68. Cambridge, MA: MIT Press, 2001.

———. *Tower of Babel: The Evidence Against the New Creationism.* Cambridge, MA: MIT Press, 2005.

———. Expert Witness Report. *Kitzmiller v. Dover.* 400 F.Supp.2d 707 (M.D.Pa. 2005), 20. https://www.msu.edu/~pennock5/research/papers/Pennock_DoverExptRpt.pdf.

———. "Learning Evolution and the Nature of Science Using Evolutionary Computing and Artificial Life." *McGill Journal of Education* 42 (2007): 211–23.

———. "Does Design Require a Designer? What Digital Experiments Can Tell Us." Lecture presented at the "God, Nature and Design Conference," St. Anne's College, Oxford University, July 10–13, 2008.

———. List of Projects "Defending the Integrity of Science." https://www.msu. edu/~pennock5/research/DISE_PennockVsIDC.html#Projects (last accessed February 9, 2009).

Penrose, Roger. "Difficulties with Inflationary Cosmology," in E. Fergus, ed., Proceedings of the 14th Texas Symposium on Relativistic Astrophysics, Annals of the New York Academy of Sciences 571 (1989): 249–264.

———. *The Road to Reality: A Complete Guide to the Laws of the Universe.* New York: Alfred A. Knopf, 2005.

Pera, Marcello. *The Discourses of Science.* Translated by Clarissa Botsford. Chicago: University of Chicago Press, 1994.

Perutz, Max F., and Hermann Lehmann. "Molecular Pathology of Human Hemoglobin." *Nature* 219 (1968): 902–9.

Pesole, Graziano. "What Is a Gene? An Updated Operational Definition." *Gene* 417 (2008): 1–4.

Piehler, A. P., M. Hellum, J. J. Wenzel, E. Kaminski, K. B. Haug, P. Kierulf, and W. E. Kaminski. "The Human ABC Transporter Pseudogene Family: Evidence for Transcription and Gene-Pseudogene Interference." *BMC Genomics* 9 (2008): 165.

Pigliucci, Massimo. "Intelligent Design—The Classical Argument." *Rationally Speaking,* n. 4, November 2000. http://nyhumanist.org/RationallySpeaking/RS2000-11.htm.

Plantinga, Alvin. "Methodological Naturalism?" Part 1. *Origins and Design* 18.1 (1997): 18–26.

———. "Methodological Naturalism?" Part 2. *Origins and Design* 18.2 (1997): 22–34.

Plato. *The Laws.* Translated by A. E. Taylor. London: J. M. Dent & Sons, Ltd., 1960.

Polanyi, Michael. "Life Transcending Physics and Chemistry." *Chemical and Engineering News* 45 (1967): 54–66.

———. "Life's Irreducible Structure" *Science* 160 (1968): 1308–12.

Ponnamperuma, C. "Chemical Evolution and the Origin of Life." *Nature* 201 (1964): 337–40.

Popper, Karl. *Poverty of Historicism.* London: Routledge, 1957.

———. *Conjectures and Refutations: The Growth of Scientific Knowledge.* London: Routledge and Kegan Paul, 1962.

———. "Darwinism as a Metaphysical Research Program." In *But Is It Science?* edited by Michael Ruse, 144–55. Buffalo, NY: Prometheus, 1988.

———. "Scientific Reduction and the Essential Incompleteness of All Science." In *Studies in the Philosophy of Biology: Reduction and Related Problems,* edited by F. J. Ayala and T. Dobzhansky, 259–84. Berkeley: University of California Press, 1974.

Popper, Karl, and John C. Eccles. *The Self and Its Brain.* New York: Springer International, 1977.

Portugal, Franklin H., and Jack S. Cohen. *A Century of DNA*. Cambridge, MA: MIT Press, 1977.

Pourquie, O. "Vertebrate Somitogenesis: A Novel Paradigm for Animal Segmentation?" *International Journal of Developmental Biology* 47 (2003): 597–603.

Powell, Michael. "Editor Explains Reasons for 'Intelligent Design' Article." *Washington Post*, August 19, 2005, district edition. http://www.washingtonpost.com/wp-dyn/content/article/2005/08/18/AR2005081801680_pf.html.

———. "Darwinian Debate Evolving, Scientists Argue over Intelligent-Design Idea." *Seattle Times*, August 20, 2005, local edition.

Price, Joyce Howard. "Researcher Claims Bias by Smithsonian." *Washington Times*, February 14, 2005, national edition.

———. "Unintelligent Design Hostility Toward Religious Believers at the Nation's Museum." *National Review Online*, August 16, 2005.

Prigogine, Ilya, Grégoire Nicolis, and Agnessa Babloyantz. "Thermodynamics of Evolution." *Physics Today* 25 (1972): 23–31.

Prohaska, Sonja J., and Peter F. Stadler. "Genes." *Theory in Biosciences* DOI 10.1007/s12064-008-0025-0.

Provine, William. "Evolution and the Foundation of Ethics." *MBL Science* 3 (1988): 25–26.

Purvis, W. K., G. H. Orians, and H. C. Heller. *Life: The Science of Biology*. 4th ed. Sunderland, MA: Sinauer, 1995.

Quastler, Henry. *The Emergence of Biological Organization*. New Haven, CT: Yale University Press, 1964.

Quinn, Philip L. "Creationism, Methodology and Politics." In *But Is It Science?* edited by Michael Ruse, 395–400. Buffalo, NY: Prometheus, 1988.

———. "The Philosopher of Science as Expert Witness." In *But Is It Science?* edited by Michael Ruse, 367–85. Buffalo, NY: Prometheus, 1988.

Raff, R. "Larval Homologies and Radical Evolutionary Changes in Early Development." *Homology. Novartis Symposium*, vol. 222. Chichester, UK: Wiley, 1999.

Ratliff, Evan. "The Crusade Against Evolution." *Wired*, October 2004. http://www.wired.com/wired/archive/12.10/evolution.html.

Raup, David. "Conflicts Between Darwin and Paleontology." *Field Museum of Natural History Bulletin* 50 (1979): 24–25.

Ray, J. *The Wisdom of God Manifested in the Works of the Creation*. 3rd ed. London, 1701.

Recker, Doren A. "Causal Efficacy: The Structure of Darwin's Argument Strategy in the *Origin of Species*." *Philosophy of Science* 54 (1987): 147–75.

Reid, Thomas. *Lectures on Natural Theology*. 1780. Edited by E. Duncan and W. R. Eakin. Reprint, Washington, DC: University Press of America, 1981.

Reidhaar-Olson, John, and Robert Sauer. "Functionally Acceptable Solutions in Two Alpha-Helical Regions of Lambda Repressor." *Proteins: Structure, Function, and Genetics* 7 (1990): 306–16.

Rieppel, Olivier. "Louis Agassiz (1807–1873) and the Reality of Natural Groups." *Biology and Philosophy* 3 (1988): 29–47.

Riordan, Michael. "Stringing Physics Along." *Physics World* 20 (2007): 38–39.

Rodnina, Marina V., Malte Beringer, and P. Bieling. "Ten Remarks on Peptide Bond Formation on the Ribosome." *Biochemical Society Transactions* 33 (2005): 493–98.

Rose, George D. "No Assembly Required." *The Sciences* 36 (1996): 26–31.

Rosenfield, Israel, Edward B. Ziff, and Boris von Loon. *DNA for Beginners*. London: Writers and Readers Publishing, 1983.

Rosenhouse, Jason. "Who Designed the Designer?" http://www.csicop.org/intelligentdesignwatch/designer.html.

Rothman, Milton A. *The Science Gap*. Buffalo, NY: Prometheus, 1992.

Rubin, C. M., R. H. Kimura, and C. W. Schmid. "Selective Stimulation of Translational Expression by Alu RNA." *Nucleic Acids Research* 30 (2002): 3253–61.

Ruse, Michael. *The Philosophy of Biology.* London: Hutchinson's University Library, 1973.

———. "Charles Darwin's Theory of Evolution: An Analysis." *Journal of the History of Biology* 8 (1975): 219–41.

———. "Darwin's Debt to Philosophy." *Studies in History and Philosophy of Science* 6 (1975): 159–81.

———. *The Darwinian Revolution: Science Red in Tooth and Claw.* Chicago: University of Chicago Press, 1979.

———. *Darwinism Defended: A Guide to the Evolution Controversies.* London: Addison-Wesley, 1982.

———. "Creation Science Is Not Science." *Science, Technology and Human Values* 7 (1982): 72–78.

———. "Commentary: The Academic as Expert Witness." *Science, Technology and Human Values* 11 (1986): 66–73.

———. "Teleology in Biology: Cause for Concern?" *Trends in Ecology and Evolution* 2 (1987): 51–54.

———. "Karl Popper's Philosophy of Biology." In *But Is It Science?* edited by Michael Ruse, 159–76. Buffalo, NY: Prometheus, 1988.

———. "Charles Darwin and the 'Origin of Species.'" In *But Is It Science?* edited by Michael Ruse, 96–98. Buffalo, NY: Prometheus, 1988.

———. "A Philosopher's Day in Court." In *But Is It Science?* edited by Michael Ruse, 13–38. Buffalo, NY: Prometheus, 1988.

———. "The Relationship Between Science and Religion in Britain, 1830–1870." In *But Is It Science?* edited by Michael Ruse, 50–70. Buffalo, NY: Prometheus, 1988.

———. "Scientific Creationism." In *But Is It Science?* edited by Michael Ruse, 257–65. Buffalo, NY: Prometheus, 1988.

———. "Witness Testimony Sheet: *McLean* v. *Arkansas.*" In *But Is It Science?* edited by Michael Ruse, 287–306. Buffalo, NY: Prometheus, 1988.

———. "They're Here!" *Bookwatch Reviews* 2 (1989): 4.

———. Speech at the Annual Meeting of the American Association for the Advancement of Science. February 13, 1993. http://www.leaderv-.com/orgs/am/orpages/or151/mr93tran.html (accessed February 17, 2000).

———. "Darwinism: Philosophical Preference, Scientific Inference and Good Research Strategy." In *Darwinism: Science or Philosophy?* edited by Jon Buell and Virginia Hearn, 21–28. Richardson, TX: Foundation for Thought and Ethics, 1994.

———. *Monad to Man: The Concept of Progress in Evolutionary Biology.* Cambridge, MA: Harvard University Press, 1996.

———. "Keep Intelligent Design Out of Science Classes." Beliefnet 2008. http://www.beliefnet.com/Story/172/story_17244.html.

Russell, Bertrand. *Mysticism and Logic, Including a Free Man's Worship.* London: Unwin Paperbacks, 1986.

Sagan, Carl. *Contact.* New York: Simon & Schuster, 1985.

Saier, M. H. "Evolution of Bacterial Type III Protein Secretion Systems." *Trends in Microbiology* 12 (2004): 113–15.

Saini, H. K., A. J. Enright, and S. Griffiths-Jones. "Annotation of Mammalian Primary microRNAs." *BMC Genomics* 9 (2008): 564.

Sanger, Frederick, and E. O. P. Thompson. "The Amino Acid Sequence in the Glycyl Chain of Insulin." *Biochemical Journal* 53 (1953): 353–74.

Sapp, J. *Beyond the Gene.* New York: Oxford University Press, 1987.

Sarkar, Sahotra. "Decoding 'Coding': Information and DNA." *BioScience* 46 (1996): 857–64.

———. "Biological Information: A Skeptical Look at Some Central Dogmas of Molecular Biology." In *The Philosophy and History of Molecular Biology: New Perspectives,* edited by S. Sarkar, 187–233. Dordrecht: Kluwer Academic, 1996.

Savarese, F., and R. Grosschedl. "Blurring Cis and Trans in Gene Regulation." *Cell* 126 (2006): 248–50.

Schlesinger, Gordon, and Stanley L. Miller, "Prebiotic Synthesis in Atmospheres Containing CH_4, CO, and CO_2: I. Amino Acids." *Journal of Molecular Evolution* 19 (1983): 376–82.

Schneider, Thomas D. "Information Content of Individual Genetic Sequences." *Journal of Theoretical Biology* 189 (1997): 427–41.

———. "Evolution of Biological Information." *Nucleic Acids Research* 28 (2000): 2794–99.

Schopf, J. W., and E. S. Barghoorn. "Alga-Like Fossils from the Early Precambrian of South Africa." *Science* 156 (1967): 508–11.

Schrödinger, Erwin. *What Is Life? Mind and Matter.* Cambridge: Cambridge University Press, 1967.

Schueler, Mary G., Anne W. Higgins, M. Katharine Rudd, Karen Gustashaw, Huntington F. Willard, "Genomic and Genetic Definition of a Functional Human Centromere." *Science,* 294 (2001): 109–15.

Schuster, P. "Prebiotic Evolution." In *Biochemical Evolution,* edited by H. Gutfreund, 15–87. Cambridge: Cambridge University Press, 1981.

Schützenberger, M. "Algorithms and the Neo-Darwinian Theory of Evolution." In *Mathematical Challenges to the Darwinian Interpretation of Evolution,* edited by P. S. Morehead and M. M. Kaplan, 73–80. Wistar Institute Symposium Monograph. Philadelphia: Wistar Institute Press, 1967.

Schwartz, Alan W. "Intractable Mixtures and the Origin of Life." *Chemistry and Biodiversity* 4 (2007): 656–64.

Scott, A. *The Creation of Life.* Oxford: Oxford University Press, 1986.

Scott, Eugenie C. "Keep Science Free from Creationism." *Insight,* February 21, 1994, 30.

———. Testimony Before the U.S. Commission on Civil Rights, Hearing on Curriculum Controversies in Biology, August 21, 1998. Reprinted in *Darwinism, Design and Public Education,* edited by John Angus Campbell and Stephen C. Meyer, 555–86. East Lansing: Michigan State University Press, 2003.

Scott, Eugenie, and Glenn Branch. "'Intelligent Design' Not Accepted by Most Scientists." *School Board News,* August 13, 2002. http://www.ncseweb.org/resources/articles/996_intelligent_design_not_accep_9_10_2002.asp.

Scriven, Michael. "Truisms as the Grounds for Historical Explanations." In *Theories of History,* edited by P. Gardiner. Glencoe, IL: Free Press, 1959.

———. "Explanation and Prediction in Evolutionary Theory." *Science* 130 (1959): 477–82.

———. "New Issues in the Logic of Explanation." In *Philosophy and History,* edited by S. Hook, 339–69. New York: New York University Press, 1963.

———. "The Temporal Asymmetry of Explanations and Predictions." In *Delaware Studies in the Philosophy of Science,* vol. 1, edited by S. Baumrin, 99. Newark: University of Delaware Press, 1963.

———. "Causes, Connections, and Conditions in History." In *Philosophical Analysis and History,* edited by W. Dray, 238–64. New York: Harper & Row, 1966.

———. "The Logic of Cause." *Theory and Decision* 2 (1971): 49–66.

———. "Causation as Explanation." *Nous* 9 (1975): 3–15.

Searle, John R. *Mind: A Brief Introduction.* New York: Oxford University Press, 2004.

Sedley, David. *Creationism and Its Critics in Antiquity.* Berkeley: University of California Press, 2008.

Semmelweis, Ignaz. *Etiology, Concept and Prophylaxis of Childbed Fever.* Translated by K. Codell Carter. Madison: University of Wisconsin Press, 1983.

Shannon, Claude E. "A Mathematical Theory of Communication." *Bell System Technical Journal* 27 (1948): 379–423, 623–56.

Shannon, Claude E., and Warren Weaver. *The Mathematical Theory of Communication.* Urbana: University of Illinois Press, 1949.

Shapiro, James A. Review of *Darwin's Black Box,* by Michael Behe. *National Review,* September 16, 1996.

———. "A 21st Century View of Evolution: Genome System Architecture, Repetitive DNA, and Natural Genetic Engineering." *Gene* 345 (2005): 91–100.

Shapiro, James, and Richard von Sternberg. "Why Repetitive DNA Is Essential to Genome Function." *Biological Reviews of the Cambridge Philosophical Society* 80 (2005): 227–50.

Shapiro, Robert. *Origins: A Skeptic's Guide to the Creation of Life on Earth.* New York: Summit, 1986.

———. "Prebiotic Ribose Synthesis: A Critical Analysis." *Origins of Life and Evolution of the Biosphere* 18 (1988): 71–85.

———. "Prebiotic Role of Adenine: A Critical Analysis." *Origins of Life and Evolution of the Biosphere* 25 (1995): 83–98.

———. "Prebiotic Cytosine Synthesis: A Critical Analysis and Implications for the Origin of Life." *Proceedings of the National Academy of Sciences, USA* 96 (1999): 4396–401.

———. "Small Molecule Interactions Were Central to the Origin of Life." *Quarterly Review of Biology* 81 (2006): 105–25.

———. "A Simpler Origin of Life." *Scientific American* 296 (2007): 46–53.

Shermer, Michael. "ID Works in Mysterious Ways," canada.com, July 9, 2008.

———. *Why Darwin Matters: The Case Against Intelligent Design.* New York: Times/ Holt, 2006.

Sherrington, Charles. *Man on His Nature.* 2nd ed. Cambridge: Cambridge University Press, 1951.

Shubin, N. H., and C. R. Marshall. "Fossils, Genes, and the Origin of Novelty." *Paleobiology* 26(4) (2000): 324–40.

Simpson, George Gaylord. *The Meaning of Evolution.* New Haven, CT: Yale University Press, 1967.

———. "Uniformitarianism: An Inquiry into Principle, Theory, and Method in Geohistory and Biohistory." In *Philosophy of Geohistory (1785–1970),* edited by C. Albritton, 256–307. Stroudsburg, PA: John Wiley & Sons, Inc., 1975.

Skoog, Gerald. "A View from the Past." *Bookwatch Reviews* 2 (1989): 1–2.

Smith, John Maynard. "Hypercycles and the Origin of Life." *Nature* 280 (1979): 445–46.

Sober, Elliott. *Reconstructing the Past.* Cambridge, MA: MIT Press, 1988.

———. *The Philosophy of Biology.* San Francisco: Westview, 1993.

———. "What Is Wrong with Intelligent Design?" *Quarterly Review of Biology* 82 (2007): 3–8.

———. *Evidence and Evolution: the Logic Behind the Science.* Cambridge: Cambridge University Press, 2008.

Sonneborn, Tracy M. "Determination, Development, and Inheritance of the Structure of the Cell Cortex." *Symposia of the International Society for Cell Biology* 9 (1970): 1–13.

Sowden, M. P., N. Ballatori, K. L. Jensen, L. H. Reed, and H. C. Smith. "The Editosome for Cytidine to Uridine mRNA Editing Has a Native Complexity of 27S: Identification of Intracellular Domains Containing Active and Inactive Editing Factors." *Journal of Cell Science* 115 (2002): 1027–39.

Sowpati, Divya T., Devi Thiagarajan, Sudhish Sharma, Hina Sultana, Rosalind John, Azim Surani, Rakesh K. Mishra, and Sanjeev Khosla. "An Intronic DNA Sequence Within the Mouse Neuronatin Gene Exhibits Biochemical Characteristics of an ICR and Acts as a Transcriptional Activator in *Drosophila.*" *Mechanisms of Development* 125 (2008): 963–73.

Sperling. J., M. Azubel, and R. Sperling. "Structure and Function of the Pre-mRNA Splicing Machine." *Structure* 16 (2008): 1605–15.

Squires, J. E., and M. J. Berry. "Eukaryotic Selenoprotein Synthesis: Mechanistic Insight Incorporating New Factors and New Functions for Old Factors." *IUBMB Life* 60 (2008): 232–35.

Stadler, B. M. R., P. F. Stadler, G. P. Wagner, and W. Fontana. "The Topology of the Possible: Formal Spaces Underlying Patterns of Evolutionary Change." *Journal of Theoretical Biology* 213 (2001): 241–74.

Stalnaker, Robert C. *Inquiry*. Cambridge, MA: MIT Press, 1984.

Stark, Rodney. *For the Glory of God: How Monotheism Led to Reformations, Science, Witch-Hunts, and the End of Slavery*. Princeton and Oxford: Princeton University Press, 2003.

Steffens, Henry John. *James Prescott Joule and the Concept of Energy*. New York: Science History Publications, 1979.

Steinman, Gary. "Sequence Generation in Prebiological Peptide Synthesis." *Archives of Biochemistry and Biophysics* 121 (1967): 533–39.

Steinman, Gary, and Marian N. Cole. "Synthesis of Biologically Pertinent Peptides Under Possible Primordial Conditions." *Proceedings of the National Academy of Sciences USA* 58 (1967): 735–41.

Stokes, Trevor. "Intelligent Design Study Appears." *The Scientist,* September 3, 2004. www.the-scientist.com, in association with BioMed Central.

Suñer, August Pi. *Classics of Biology*. Translated by C. M. Stern. New York: Philosophical Library, 1955.

Szostak, Jack W., David P. Bartel, and P. Luigi Luisi. "Synthesizing Life." *Nature* 209 (2001): 387–90.

Tam, O. H., A. A. Aravin, P. Stein, A. Girard, E. P. Murchison, S. Chelofi, E. Hodges, M. Anger, R. Sachidanandam, R. M. Schultz, and G. J. Hannon. "Pseudogene-Derived Small Interfering RNAs Regulate Gene Expression in Mouse Oocytes." *Nature* 453 (2008): 534–38.

Tamura, Koji, and Paul Schimmel. "Peptide Synthesis with a Template-Like RNA Guide and Aminoacyl Phosphate Adaptors." *Proceedings of the National Academy of Sciences USA* 100 (2003): 8666–69. 10.1073/pnas.1432909100.

Taylor, Sean V., Kai U. Walter, Peter Kast, and Donald Hilvert. "Searching Sequence Space for Protein Catalysts." *Proceedings of the National Academy of Sciences USA* 98 (2001): 10596–601.

Thagard, Paul. "The Best Explanation: Criteria for Theory Choice." *Journal of Philosophy* 75 (1978): 77–92.

Thaxton, Charles B., and Walter L. Bradley. "Information and the Origin of Life." In *The Creation Hypothesis: Scientific Evidence for an Intelligent Designer,* edited by J. P. Moreland, 193–97. Downers Grove, IL: InterVarsity, 1994.

Thaxton, Charles, Walter L. Bradley, and Roger L. Olsen. *The Mystery of Life's Origin: Reassessing Current Theories*. New York: Philosophical Library, 1984.

Thompson, D. W. *On Growth and Form*. 2nd ed. Cambridge: Cambridge University Press, 1942.

Thomson, K. S. "The Meanings of Evolution." *American Scientist* 70 (1982): 529–31.

———. "Macroevolution: The Morphological Problem." *American Zoologist* 32 (1992): 106–12.

Thornton, Joe. "Implications for Intelligent Design." http://www.uoregon.edu/~joet/news/news.htm (last accessed February 9, 2009).

Timiriazev, Kliment Arkadievich. *A Short Sketch of the Theory of Darwin* and *Charles Darwin and His Theory*. Fifteen editions, 1883–1941.

Tommyrot, Terry [pseud.]. "The Dawkins Delusion." YouTube. Added February 2, 2007. http://www.youtube.com/watch?v=QERyh9YYEis.

Towe, Kenneth M. "Environmental Oxygen Conditions During the Origin and Early Evolution of Life." *Advances in Space Research* 18 (1996): 7–15.

Tremblay, A., M. Jasin, and P. Chartrand. "A Double-Strand Break in a Chromosomal LINE Element Can Be Repaired by Gene Conversion with Various Endogenous LINE Elements in Mouse Cells." *Molecular and Cellular Biology* 20 (2000): 54–60.

Tyndall, John. *Fragments of Science: A Series.* 6th ed. 2 vols. London. New York: Collier, 1871.

Valentine, J. W. *On the Origin of Phyla.* Chicago: University of Chicago Press, 2004.

Van de Lagemaat, L. N., J. R. Landry, D. L. Mager, and P. Medstrand. "Transposable Elements in Mammals Promote Regulatory Variation and Diversification of Genes with Specialized Functions." *Trends in Genetics* 19 (2003): 530–36.

Vandervliet, G. *Microbiology and the Spontaneous Generation Debate During the 1870s.* Lawrence, KS: Coronado, 1971.

Van de Sande, Bret. "Measuring Complexity in Dynamical Systems." Paper presented to RAPID II (Research and Progress in Intelligent Design) Conference, Biola University, May 2006.

Van Fraassen, B. *The Scientific Image.* Oxford: Oxford University Press, 1980.

Vergano, Dan, and Greg Toppo. "'Call to Arms' on Evolution." *USA Today,* March 23, 2005. http://www.usatoday.com/news/education/2005-03-23-evolution_x.htm.

Vilenkin, Alexander. *Many Worlds in One: The Search for Other Universes.* New York: Hill and Wang, 2006.

Virchow, Rudolf Ludwig Karl. "On the Mechanistic Interpretation of Life." In *Disease, Life, and Man: Selected Essays,* translated and edited by Lelland J. Rather, 102–19. Stanford Studies in the Medical Sciences. Palo Alto, CA: Stanford University Press, 1958.

Voie, Albert. "Biological Function and the Genetic Code Are Interdependent," *Chaos, Solutions and Fractals* 28 (2006): 1000–1004.

Von Neumann, John. *Theory of Self-Reproducing Automata.* Completed and edited by A. Burks. Urbana: University of Illinois Press, 1966.

Von Sternberg, Richard. "On the Roles of Repetitive DNA Elements in the Context of a Unified Genomic-Epigenetic System." *Annals of the New York Academy of Sciences* 981 (2002): 154–88.

Von Sternberg, Richard, and James A. Shapiro. "How Repeated Retroelements Format Genome Function." *Cytogenetic and Genome Research* 110 (2005): 108–16.

Wagner, G. P. "What Is the Promise of Developmental Evolution? Part II: A Causal Explanation of Evolutionary Innovations May Be Impossible." *Journal of Experimental Zoology (Molecular and Developmental Evolution)* 291 (2001): 305–9.

Wagner, G. P., and P. F. Stadler. "Quasi-Independence, Homology and the Unity-C of Type: A Topological Theory of Characters." *Journal of Theoretical Biology* 220 (2003): 505–27.

Wald, George. "The Origin of Life." *Scientific American* 191 (1954): 44–53.

Waldrop, M. Mitchell. "Did Life Really Start Out in an RNA World?" *Science* 246 (1989): 1248–49.

Walker, James C. G. *Evolution of the Atmosphere.* New York: Macmillan, 1977.

Wallis, Claudia. "The Evolution Wars." *Time,* August 7, 2005. http://www.time.com/time/magazine/article/0,9171,1090909,00.html.

Walter, M. R., R. Buick, and J. S. R. Dunlop. "Stromatolites 3,400-3,500 Myr Old from the North Pole Area, Western Australia." *Nature* 284 (1980): 443–45.

Walton, J. C. "Organization and the Origin of Life." *Origins* 4 (1977): 16–35.

Wang, Jue D., Melanie B. Berkmen, and Alan D. Grossman. "Genome-Wide Coorientation of Replication and Transcription Reduces Adverse Effects on Replication in *Bacillus subtilis.*" *Proceedings of the National Academy of Sciences USA* 104 (2007): 5608–13.

Wang, Zefeng, and Christopher B. Burge. "Splicing Regulation: From a Parts List of Regulatory Elements to an Integrated Splicing Code." *RNA* 14 (2008): 802–13.

Watanabe T., Y. Totoki, A. Toyoda, M. Kaneda, S. Kuramochi-Miyagawa, Y. Obata, H. Chiba, Y. Kohara, T. Kono, T. Nakano, M.A. Surani, Y. Sakaki, and H. Sasaki. "Endogenous siRNAs from Naturally Formed dsRNAs Regulate Transcripts in Mouse Oocytes." *Nature* 453 (2008): 539–43.

Watson, James D. *Molecular Biology of the Gene.* New York: Benjamin, 1965.

——. *The Double Helix.* Edited by Gunther Stent. New York: Norton, 1980.

Watson, James D., and Francis H. C. Crick, "A Structure for Deoxyribose Nucleic Acid." *Nature* 171 (1953): 737–38.

——. "Genetical Implications of the Structure of Deoxyribonucleic Acid." *Nature* 171 (1953): 964–67.

Watson, James D., Nancy H. Hopkins, Jeffrey W. Roberts, Joan Steitz, and Alan M. Weiner. *Molecular Biology of the Gene.* 4th ed. 2 vols. Menlo Park, CA: Benjamin/Cummings, 1987.

Watson, Richard. *Compositional Evolution.* Cambridge, MA: MIT Press, 2006.

Webster, G., and B. Goodwin. "A Structuralist Approach to Morphology." *Rivista di Biologia* 77 (1984): 503–10.

——. *Form and Transformation: Generative and Relational Principles in Biology.* Cambridge: Cambridge University Press, 1996.

Wells, Jonathan. "Using Intelligent Design Theory to Guide Scientific Research." *Progress in Complexity, Information, and Design* (November 2004): 3.1.2.

——. "Do Centrioles Generate a Polar Ejection Force?" *Rivista di Biologia/Biology Forum* 98 (2005): 71–96.

——. *The Politically Incorrect Guide to Darwinism and Intelligent Design.* Washington, DC: Regnery, 2006.

——. "A Possible Link Between Centrioles, Calcium Deficiency and Cancer." Abstract. *American Society for Cell Biology, Annual Meeting,* December 2005. http://www.abstractsonline.com/.

West, John G., and David K. DeWolf. "A Comparison of Judge Jones's Opinion in *Kitzmiller v. Dover* with Plaintiffs' Proposed 'Findings of Fact and Conclusions of Law.'" http://www.discovery.org/scripts/viewDB/filesDB-download.php?command=download&id=1186 (December 12, 2006).

Whewell, William. "Lyell's Principles of Geology." *British Critic* 9 (1830): 180–206.

——. *The Philosophy of the Inductive Sciences, Founded upon Their History.* 1840. 2nd ed. 2 vols. London: Parker, 1847.

Whitehead, Alfred North. *Science and the Modern World.* 1925. Reprint, New York: Free Press, 1967.

Wicken, J. *Evolution, Thermodynamics and Information.* Oxford: Oxford University Press, 1987.

Wigner, Eugene. "The Probability of the Existence of a Self-Reproducing Unit." In *The Logic of Personal Knowledge: Essays Presented to Michael Polanyi,* edited by Edward Shils, 231–35. London: Routledge and Kegan Paul, 1961.

Wiker, Benjamin, and Jonathan Witt. *A Meaningful World: How the Arts and Sciences Reveal the Genius of Nature.* Downers Grove, IL: InterVarsity, 2007.

Wilgoren, Jodi. "Politicized Scholars Put Evolution on the Defensive." *New York Times,* August 21, 2005, national edition. http://www.nytimes.com/2005/08/21/national/21evolve.html.

——. "Seattle Think Tank Behind 'Intelligent Design' Push." *Seattle Post-Intelligence,* August 22, 2005, local edition.

Williams, George. *Natural Selection: Domains, Levels and Challenges.* New York: Oxford University Press, 1992.

Willmer, P. *Invertebrate Relationships: Patterns in Animal Evolution.* Cambridge: Cambridge University Press, 1990.

———. "Convergence and Homoplasy in the Evolution of Organismal Form." In *Origination of Organismal Form: Beyond the Gene in Developmental and Evolutionary Biology,* edited by G. B. Muller and S. A. Newman, 33–49. Cambridge, MA: MIT Press, 2003.

Wilson, David. "Butts on Whewell's View of True Causes." *Philosophy of Science* 40 (1973): 121–24.

Wilson, Edmund B. *The Physical Basis of Life.* New Haven, CT: Yale University Press, 1923.

Wilson, Edward O. "Intelligent Evolution: The Consequences of Charles Darwin's 'One Long Argument.'" *Harvard Magazine* (November–December 2005): 29–33.

Wilson, T. E., U. Grawunder, and M. R. Liebe. "Yeast DNA Ligase IV Mediates Non-Homologous DNA End Joining." *Nature* 388 (1997): 495–98.

Wolf, Yuri I., and Eugene V. Koonin. "On the Origin of the Translation System and the Genetic Code in the RNA World by Means of Natural Selection, Exaptation, and Subfunctionalization." *Biology Direct* 2 (2007): 1–25.

Wolfe, Stephen L. *Molecular and Cellular Biology.* Belmont, CA: Wadsworth, 1993.

Wolpert, David H., and William G. Macready. "No Free Lunch Theorems for Optimization." *IEEE Transactions on Evolutionary Computation* 1 (1997): 67–82.

Wood, T. B. V., and M. H. Thiemens. "The Fate of the Hydroxyl Radical in the Earth's Primitive Atmosphere and Implications for the Production of Molecular Oxygen." *Journal of Geophysical Research* 85 (1980): 1605–10.

Woodward, T. *Doubts About Darwin: A History of Intelligent Design.* Grand Rapids, MI: Baker Books, 2003.

Worden, Amy. "Bad Frog Beer to 'Intelligent Design': The Controversial Ex-Pa. Liquor Board Chief Is Now U.S. Judge in the Closely Watched Trial." *Philadelphia Inquirer,* October 16, 2005.

Wright, Matthew A., Peter Kharchenko, George M. Church, and Daniel Segre. "Chromosomal Periodicity of Evolutionarily Conserved Gene Pairs." *Proceedings of the National Academy of Sciences USA* 104 (2007) 10559–64.

Wu, J. Q., J. Du, J. Rozowsky, Z. Zhang, A. E. Urban, G. Euskirchen, S. Weissman, M. Gerstein, and M. Snyder. "Systematic Analysis of Transcribed Loci in ENCODE Regions Using RACE Sequencing Reveals Extensive Transcription in the Human Genome." *Genome Biology* 9 (2008): R3.

Yockey, Hubert P. "A Calculation of the Probability of Spontaneous Biogenesis by Information Theory." *Journal of Theoretical Biology* 67 (1977): 377–98.

———. "Self-Organization Origin of Life Scenarios and Information Theory." *Journal of Theoretical Biology* 91 (1981): 13–31.

———. *Information Theory and Molecular Biology.* Cambridge: Cambridge University Press, 1992.

———. "Origin of Life on Earth and Shannon's Theory of Communication." *Computers and Chemistry* 24 (2000): 105–23.

———. *Information Theory, Evolution, and the Origin of Life.* Cambridge: Cambridge University Press, 2005.

Zamecnik, Paul. "From Protein Synthesis to Genetic Insertion." *Annual Reviews of Biochemistry* 74 (2005): 1–28.

Zaug, A. J., and T. R. Cech. "The Intervening Sequence RNA of Tetrahymena Is an Enzyme." *Science* 231 (1986): 470–75.

Zhang, Biliang, and Thomas R. Cech. "Peptide Bond Formation by *In Vitro* Selected Ribozymes." *Nature* 390 (1997): 96–100.

Zilsel, Edgar. "Genesis of the Concept of Physical Law." *Philosophical Review* 51 (1942): 245–79.

Index

Page references followed by *fig* indicate an illustrated figure.

Acknowledgments

I would like to thank my editor, Roger Freet, at HarperOne for his expert guidance and patience and for permitting this manuscript to bear a big burden of proof. I'd also like to acknowledge the Harper production staff, particularly Lisa Zuniga and Ann Moru, for their professionalism and exquisite attention to detail. I would also like to thank my agent, Giles Anderson, for believing in this project and for connecting me to the good people at Harper.

This book was extensively reviewed for scientific and technical accuracy. For their work reviewing chapters I'd like to acknowledge: Doug Axe, Bruce Gordon, Ann Gauger, William Dembski, Tony Mega, Dean Kenyon, Robert Marks, Richard Sternberg, Jonathan Wells, Paul Nelson, and Alistair Noble.

I'd also like to acknowledge Jonathan Witt and David Klinghoffer for their coaching, editing, and help in framing the narrative structure of the book, and Bruce Chapman and Logan Gage for their careful editing of late chapter drafts. Thanks to John West for his willingness to assume additional burdens of management and for creating space for me to write, and to Janet Oberembt for her invaluable assistance in proofreading, entering sources, and managing chapter drafts, the bibliography, and the overall flow of communication. I'd also like to express my gratitude to Joseph Condeelis for his inspired animation work.

I'd like to express my deep gratitude to my parents, Chuck and Pat Meyer, for their encouragement and support of my scientific education and philosophical interests from my earliest years.

And finally, I'd like to thank Elaine—who has held so much together during the last two years—for her *courage* as well as her love, and for her partnership in this adventure from the beginning.